# Polarons and Excitons in Polar Semiconductors and Ionic Crystals

# NATO ASI Series

## Advanced Science Institutes Series

*A series presenting the results of activities sponsored by the NATO Science Committee, which aims at the dissemination of advanced scientific and technological knowledge, with a view to strengthening links between scientific communities.*

The series is published by an international board of publishers in conjunction with the NATO Scientific Affairs Division

| | | |
|---|---|---|
| A | **Life Sciences** | Plenum Publishing Corporation |
| B | **Physics** | New York and London |
| C | **Mathematical and Physical Sciences** | D. Reidel Publishing Company<br>Fordrecht Boston, and Lancaster |
| D | **Behavioral and Social Sciences** | Martinus Nijhoff Publishers |
| E | **Engineering and Materials Sciences** | The Hague, Boston, and Lancaster |
| F | **Computer and Systems Sciences** | Springer-Verlag |
| G | **Ecological Sciences** | Berlin, Heidelberg, New York, and Tokyo |


***Recent Volumes in this Series***

*Volume 102*—Magnetic Monopoles
edited by Richard A. Carrigan, Jr., and W. Peter Trower

*Volume 103*—Fundamental Processes in Energetic Atomic Collisions
edited by H. O. Lutz, J. S. Briggs, and H. Kleinpoppen

*Volume 104*—Short-Distance Phenomena in Nuclear Physics
edited by David H. Boal and Richard M. Woloshyn

*Volume 105*—Laser Applications in Chemistry
edited by K. L. Kompa and J. Wanner

*Volume 106*—Multicritical Phenomena
edited by R. Pynn and A. Skjeltorp

*Volume 107*—Positron Scattering in Gases
edited by John W. Humberston and M. R. C. McDowell

*Volume 108*—Polarons and Excitons in Polar Semiconductors and Ionic Crystals
edited by J. T. Devreese and F. Peeters

***Series B: Physics***

# Polarons and Excitons in Polar Semiconductors and Ionic Crystals

Edited by

J. T. Devreese

and

F. Peeters

University of Antwerp
Antwerp, Belgium

Plenum Press
New York and London
Published in cooperation with NATO Scientific Affairs Division

Proceedings of the Antwerp Advanced Study Institute on
Physics of Polarons and Excitons in Polar Semiconductors and Ionic Crystals,
held July 26–August 5, 1982,
at the Conference Center, Priorij Corsendonk, Belgium

Library of Congress Cataloging in Publication Data

Antwerp Advanced Study Institute on Physics of Polarons and Excitons in Polar Semiconductors and Ionic Crystals (1982)
Polarons and excitons in polar semiconductors and ionic crystals.

(NATO ASI series. Series B: Physics; v. 108)
Includes bibliographical references and index.
1. Polarons—Congresses. 2. Exciton theory—Congresses. 3. Ionic crystals—Congresses. 4. Semiconductors—Congresses. I. Devreese, J. T. (Jozef T.) II. Peeters, F. III. Title. IV. Series.
QC176.8.E4A58 1982 530.4′1 84-2144

ISBN-13: 978-1-4612-9674-4 e-ISBN-13: 978-1-4613-2693-9
DOI: 10.1007/978-1-4613-2693-9


Softcover reprint of the hardcover 1st edition 1984

A Division of Plenum Publishing Corporation
233 Spring Street, New York, N.Y. 10013

## PREFACE

The 1982 Antwerp Advanced Study Institute on "Physics of Polarons and Excitons in Polar Semiconductors and Ionic Crystals" took place from July 26 till August 5 at the Conference Center Priorij Corsendonk, a restored monastery, close to the city of Antwerp. It was the seventh Institute in our series which started in 1971.

This Advanced Study Institute, which was held fifty years after Landau introduced the polaron concept, can be considered as the third major international symposium devoted to the physics of polarons.

The first such symposium took place in St. Andrews in 1962 under the title "Polarons and Excitons" [1]. The early theoretical developments related to polarons were reviewed in depth at this meeting; the derivation of the polaron hamiltonian by Fröhlich, the Fröhlich weak coupling theory (and the equivalent weak coupling canonical transformations), the Landau-Pekar and Bogolubov strong coupling theory and the Feynman polaron model formulated with his path integrals. The main emphasis was on the polaron self-energy, effective mass and mobility. From the experimental side the first evidence for polaron effects was provided by the pioneering cyclotron and mobility measurements on the silver halides by F.C. Brown and his group. Also the significance of polaron effects for the understanding of excitons in ionic crystals was a central topic in St. Andrews.

The second Advanced Study Institute concerning polaron physics was organized at the University of Antwerp (R.U.C.A.) in 1971 [2]. The magneto-optical properties of polarons had been explored in the sixties both theoretically and experimentally. The cyclotron resonance studies of F.C. Brown's group were extended to the alkali halides. At Lincoln Laboratory and M.I.T. weak coupling magneto-optical polaron phenomena were observed in InSb, CdS, CdTe, ... Non-Ohmic polaron transport ("hot polarons") was measured by Brown, J. Hodby and others in alkali halides and the high field mobility of Feynman polarons (under the assumption of a Maxwellian distribution for the electron velocities) was calculated. Further studies concerned screening effects in the many polaron system (G. Mahan) and our work on relaxed excited states of polarons (including their role in magneto-optics and how they can be calculated from the Feynman model. All these developments were reviewed at the 1971 Antwerp Advanced Study Institute.

The recent developments include the following:

- Peculiar polaron effects have been observed in the silver halides under crossed electrical and magnetic fields (streaming motion, population inversion). These phenomena are well described by the transport theory for Fröhlich polarons.
- Collective polaron effects have been revealed experimentally in GaAs; plasmon - L.O. phonon modes, coupled via the Fröhlich interaction and their oscillator strength, were observed as a function of wavenumber, by using Resonant Raman Scattering. Earlier theories on the dielectric function of a (weak-coupling) polaron gas were applied and generalized to analyze these collective polaron modes.
- Generalized quadratic trial actions were used in the path integral formulation of the Fröhlich polaron. The Feynman variational result for the energy could not be improved by more than one percent. These generalized quadratic trial actions are a sound basis for a model of excitons.
- The Feynman model of the polaron was generalized to study the effect of a static magnetic field. The calculations suggest that at high magnetic fields (42 T for AgBr at 0 K) the polaron becomes unstable and makes a transition to a band-like electron state ("stripping of the polaron"). This "transition" occurs over a relatively narrow region of magnetic field strengths.
- "Polaron techniques" have been applied to describe two-dimensional electrons coupled to ripplons in liquid helium and to study spin polarized hydrogen on liquid He.
- Magneto-optical studies of polarons (e.g. in InSb) have been refined and the possibility that electrons couple to transverse phonons in InSb was analyzed in some beautiful experiments.
- Another challenging problem concerns the magneto-optical spectra of polarons in two dimensions, especially in the pinning region. Fröhlich interaction, screening and the two-dimensional character of the system are of importance here.
- Furthermore we mention several studies of the Boltzmann equations for polarons which is now well understood at weak coupling. It has also been established that the Thornber-Feynman transport theory corresponds to the choice of a "drifted Maxwellian" momentum distribution function for the electron.

The developments mentioned above, except the two-dimensional polaron pinning, are covered in the present volume. Also special attention is given in this volume to recent progress on excitons and the electron-hole liquid. The polaron concept, of course, is useful for the study of excitons in polar materials.

The term "polaron" has been used also in the context of electrons in polyacetylenes and to describe the so-called "magnetic polaron". In fact in some such cases the concept is quite removed from the original Fröhlich problem.

---

It is a pleasure to express my thanks to several instances and people who have made the 1982 Antwerp Advanced Study Institute possible.

First of all we thank the NATO Science Committee for sponsoring this Institute. Special thanks are due to the co-sponsors: Agfa-Gevaert, Bell Telephone Mfg. Co. N.V., CDC Belgium, European Research Office (U.S. Army), Ferstenberg, GBM Antwerpen, IBM Belgium and IBM Europe, Janssens Pharmaceutica, Kredietbank, Leunen en Partners, Tabacofina N.V.

Over several months Dr. F. Peeters, Mr. M. Mariën and Mrs. R.M. Vandekerkhof have greatly contributed to the practical organization of this Institute. I would like to extend special thanks to them.

Special thanks are due to Mrs. H. Evans for the excellent typing of one of the manuscripts and of the author and subject index. I also thank Mrs. M. Cuyvers for assistance with the organisation of the subject index.

Last but not least I express my gratitude to Mrs. F. Nedée for putting the magnificent Corsendonk Conference Center at our disposal and to Mr. D. Van Der Brempt, Director of the Corsendonk Conference Center for the unique way in which he and his staff made us feel at home in Corsendonk.

Brasschaat - Antwerpen
July 21, 1983

Jozef T. Devreese
Professor of Theoretical Solid State Physics
University of Antwerp (R.U.C.A. and U.I.A.)
and Eindhoven University of Technology

[1] "Polarons and Excitons", eds. C.G. Kuper and G.D. Whitfield, Oliver and Boyd, Edinburgh and London (1963).

[2] "Polarons in Ionic Crystals and Polar Semiconductors", ed. J.T. Devreese, North-Holland (1972).

## ADDENDUM

For two series of lectures no manuscripts are included in this volume:

- M. Lax gave a series of lectures on "Phonon optics in semiconductors: Phonon generation and electron-phonon scattering in n-GaAs epilayers". The interested reader is referred to the following publications:
  - M. Lax and V. Narayanamurti, Phys. Rev. B24, 4692 (1981).
  - M. Lax, V. Narayanamurti, P. Hu and W. Weber, Journ. de Phys. C6, 161 (1981).
  - M. Lax, P. Hu and V. Narayanamurti, Phys. Rev. B23, 3095 (1981).

- A. Baldereschi gave a series of lectures on "Exciton states in atoms and solids". The following references are useful for his lectures:
  - W.J. Hunt and W.A. Goddard III, Chem. Phys. Lett. 3, 414 (1969).
  - S. Huzinga and C. Arnau, J. Chem. Phys. 54, 1948 (1971).
  - K. Jürao and S. Huzinga, Chem. Phys. Lett. 45, 55 (1977).
  - S. Baroni, A. Quattropani and A. Baldereschi, Chem. Phys. Lett. 79, 509 (1981); Phys. Rev. A25, 2869 (1982).
  - A. Baldereschi and N.C. Lipari, Phys. Rev. B3, 439 (1971).

CONTENTS

# FROHLICH POLARONS

STUDIES OF THE FREE AND BOUND MAGNETO-POLARON AND ASSOCIATED TRANSPORT EXPERIMENTS IN n-InSb and OTHER SEMICONDUCTORS

R.A. Stradling

University of St Andrews
Physics Department
North Haugh
St Andrews KY16 9SS, Scotland

## INTRODUCTION

Although only of limited technological application InSb is of great interest to semiconductor physicists on account of its small forbidden energy gap (0.2eV) and its metallurgical properties which allow samples to be prepared with background concentrations of inadvertently-introduced electrically-active impurities of the order of $10^{14}$ $cm^{-3}$. The small energy gap gives rise to a very low effective mass at the conduction band edge, large band non-parabolicity and a very large effective g-factor on account of the strong spin-orbit interaction. The small effective mass gives rise to a very high electron mobility which can approach $10^6$ $cm^2/Vs$ at liquid nitrogen temperatures and to a very large cyclotron energy ($\hbar\omega_c$ where $\omega_c = eB/m^*$), such that this energy becomes equal to that of the longitudinal optical phonons at the relatively modest magnetic field of 3 T. As a consequence, even though the Frohlich coupling constant $\alpha$ is quite small on account of the low mass, much of pioneering studies of magneto-polaron and associated effects such as the magnetophonon effect observed in the electrical resistance of semiconductors were performed with this material. (see for example reviews by Harper et al (1) and by Levinson and Rashba (2)).

Other properties of InSb which make it a particularly interesting test material for transport experiments are that hot electron effects can be observed particularly readily (at 4 K electric fields of the order of 100mV/cm are sufficient to make the current-voltage relationship non-linear) and that neutron transmutation doping can be used to obtain extremely homogeneous samples. (3) Additional artefacts of the low effective mass are (i) that electron-electron scattering is strong which means that a reasonable approximation to a Maxwell-Boltzmann distribution can be maintained to quite high electric fields even at quite low electron concentrations, and (ii) that screening effects are generally substantial. The low effective mass means that the binding energy for the shallow donors is small and that the associated Bohr radius is large. Hence even at a concentration of donors of $10^{14}cm^{-3}$, bound states are not found in the absence of a field. However a field of the order of 0.3T is sufficient to shrink the orbits sufficiently for a metal-insulator transition to be observed at low temperatures and freeze out onto the donor states is found. This metal-insulator transition has been the subject of much study(4) and it has been suggested that the electron-electron interaction can give rise to a form of Wigner ordering(5). Because of its intrinsic purity, InSb is the lowest-mass material where bound states have been unambiguously identified. A particularly interesting implication of impurity states of small binding energy is that it is possible to obtain Zeeman splittings in the laboratory which are substantially larger than the ground state energy involved. This situation is normally characterised by a parameter $\gamma$ which is the ratio of the zero-point cyclotron energy to the effective Rydberg $R^*(\gamma = \frac{1}{2}\hbar\omega_c/R^*$ where $R^* = 13.6(m^*/m/\varepsilon^2))$. With InSb $\gamma \sim 70$ at 10T whereas $\gamma \sim 10^{-4}$ for a gas atom in the same field. In fact comparable splittings can only be obtained in two situations with real atoms i.e. near to neutron stars where fields of $10^8$ T are found so that $\gamma$ can exceed 100 and for highly excited states of atoms where the values of approaching unity occur at 10 T for principal quantum numbers n $\gtrsim$ 50 if R* is replaced by the binding energy of the state concerned ($\propto 1/n^2$).

This paper will review magneto-polaron effects in general with an emphasis on impurity related effects and on the results obtained with InSb, finishing up with very recent magneto-optical measurements of the shallow donor states which have revealed a new-type of polaronic interaction involving single TA phonons.

## 2. TRANSPORT EXPERIMENTS SHOWING STRUCTURE DUE TO THE ELECTRON-LO PHONON INTERACTION

### 2.1 Oscillatory Photoconductivity.

One of the first optical experiments to show distinct structure due to the electron-optical phonon interaction was 'oscillatory photoconductivity (6). As can be seen from figure (1), in the case of extrinsic excitation, dips in the photoresponse can be observed due to the very rapid emission of optical phonons which occurs subsequent to excitation provided that the carriers are injected at energies of more than an optical phonon into the band. These dips are expected at excitation frequencies ($\omega$) given by

$$\hbar\omega = N\hbar\omega_{LO} + E_I \qquad (1)$$

where $\omega_{LO}$ is the frequency of the LO phonons and $E_I$ is the binding

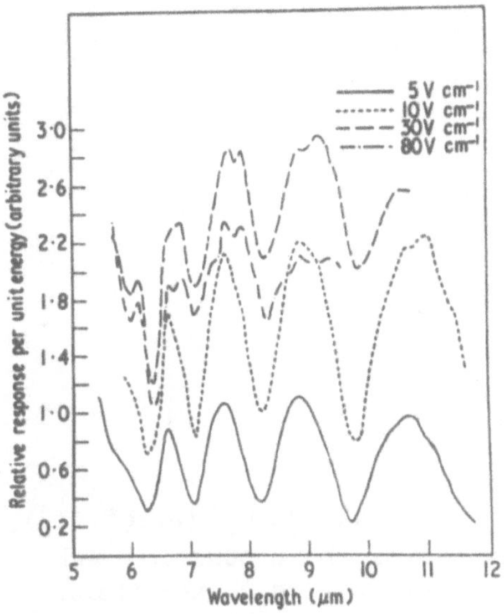


Fig.1 The extrinsic oscillatory photoconductivity of p-InSb at 7K from ref.(6)

energy of the impurity from which the carrier is initially photo-excited. The dips are interpreted (7) as arising from a strong distortion of the distribution function for carriers which find themselves just below the threshold energy for LO phonon emission on injection or after the emission of a cascade of phonons. The distortion is caused by the bias electric field which accelerates the carriers moving in the forward direction to states where phonon emission becomes energetically possible. These carriers are then returned to the band edge leaving a group of fast-moving carriers moving in the reverse direction. If momentum relaxation is not sufficiently rapid for the latter group of carriers or electron-electron scattering does not smooth out the distribution function, negative photoconductivity may result.(8) The ideas involved in this mechanism which give rise to an accumulation of carriers in limited regions of phase space below the threshold states for phonon emission are similar to those involved in "streaming motion" discussed elsewhere in the Advanced Study Institute.(9) Other contributory mechanisms which have also been used to interpret the phenomenon of oscillatory photoconductivity are that carriers, finding themselves near to the band edge, can be captured rapidly by the shallow impurities or that states near to the band edge have low mobility because of impurity scattering or band-tailing effects. With intrinsic excitation the situation is more complicated than in the extrinsic case because two bands of different curvature are involved. The photon wavevector is quite negligible in comparison with those of the photoexcited carriers and the transitions between bands centred at the same point in the Brillouin zone can be considered to be vertical. In this case, provided that the electrons have substantially higher mobility than the holes, the period of the oscillatory photoconducvity becomes $(1 + m_c^*/m_v^*)\hbar\omega_{LO}$

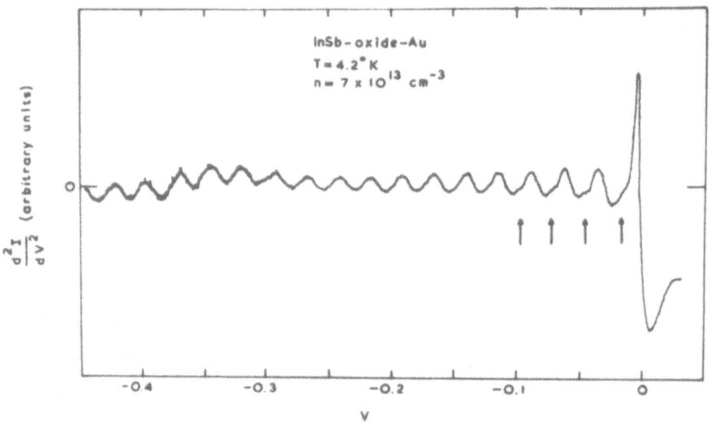


Fig.2.Tunnel current into InSb showing LO phonon Structure.ref.(12)

where $m_c$* and $m_v$* are the effective masses of the conduction and valence bands respectively. In the case of intrinsic excitation the processes are further complicated by the possibility of exciton formation during the excitation or subsequent decay and the periodicity in some circumstances can revert to $\hbar\omega_{LO}$. For materials where the bands are not centred at the $\Gamma$ point, intervalley relaxation involving large changes in wavevector become possible with the consequent replacement of $\omega_{LO}$ in equation (1) by the frequency of the appropriate phonon (10). As is the case for the magnetophonon effect discussed in the next section, the spectrum can become quite complex because of the multiplicity of phonons which may be involved. It has also been suggested that the coupling to two-TA phonons may be sufficiently strong to generate similar dips in the photoresponse with a periodicity of $2\hbar\omega_{TA}$(11). This type of coupling is strong enough to be an effective process for energy relaxation as demonstrated in the hot-electron magnetophonon spectra discussed also in the next section.
Finally it should be noted that, if carriers are injected electrically deep in to a band by means of a tunnel-junction or a Schottky barrier, similar relaxation phenomena and associated oscillatory structure can be observed if the tunnel current is measured as a function of bias voltage (12) as can be seen from figure two.

The first realisation that impurity states could be involved directly in the relaxation processes responsible for oscillatory photoconductivity came in the case of diamond.(13) It was noted that the photocurrent was quenched if the carriers found themselves at the correct energy to be captured at impurity states with the emission of a final LO phonon in the cascade. In this case $E_I$ in equation (1) has to be replaced by the difference in energy between the ground state energy and the excited state energy involved($\Delta E_I$) The limiting case of this is when the capture process involves the ground state itself and equation (1) then simplifies to

$$\hbar\omega = N\hbar\omega_{LO} \qquad (2)$$

Capture processes with LO phonon emission were found to both the ground state and n=2 excited state in the case of CdTe(14.15).

The application of a magnetic field introduces the possibility of additional resonances as the band states become quantised into Landau levels. In this case further singularities in the photoresponse are expected when the carriers after photoexcitation from the impurity states and subsequent photoemission of N optical phonons find themselves in the high density of states region near to the bottom of the mth Landau state. When this happens a minimum in the photoresponse occurs when

$$\hbar\omega = N\hbar\omega_{LO} + m\hbar\omega_c + E_I(B) \qquad (3)$$

where $\omega_c$ is the cyclotron frequency and $E_I$ is now dependent on the magnetic field B. A better known resonance effect which occurs without photoexcitation is the magnetophonon or Gurevich-Firshov effect (14) which occurs when the carrier is scattered between th m and the m + nth Landau level with the emission or absorption of LO phonons. In this case extrema are observed in the magnetoresistance at magnetic fields given by

$$\hbar\omega_{LO} = n\hbar\omega_c = n\hbar eB_{n}/m^* \qquad (4)$$

where $B_n$ are the resonance fields and m* is the effective mass of the band states concerned.

## 2.2 Magnetophonon effect

The magnetophonon effect is an oscillatory variation of electrical resistance with magnetic field generally observed with high purity samples when high-energy phonons (usually optical phonons) are limiting the mobility. The effect was first observed in InSb as reported by Puri and Geballe (16) and by Firsov et al (17), and can be readily understood in terms of the following simple model which, however, neglects any polaronic nature of the phenomenon.

As with other quantum magnetoresistance phenomena the effect arises from the redistribution of electronic states by the applied magnetic field into Landau levels whose energy is given by

$$E = \hbar\omega_c(n+\tfrac{1}{2}) + \frac{\hbar^2 k_B^2}{2m^*} \qquad (5)$$

and where the density of electronic states has the form

$$\frac{dg}{dE} = \frac{eB}{h^2}(2m^*)^{\frac{1}{2}} \sum_n \frac{1}{\{E-\hbar\omega_c(n+\frac{1}{2})\}^{\frac{1}{2}}} \qquad (6)$$

The predominant feature of the density of states in a magnetic field which arises from the one-dimensional nature of the longitudinal waves defined by the wave-vector $k_B$ is its characteristically singular behaviour near the regions where $E \sim \hbar\omega_c(n+\frac{1}{2})$. This is in sharp contrast to the density of states in the absence of a magnetic field where the dependence on energy is of the form $E^{\frac{1}{2}}$. The contribution to the electrical conductivity of electrons with energy E in a level with quantum number n is proportional to the number of such electrons, the probability of their transition to all other states and to the number of unoccupied final states. The total conductivity for a group of electrons interacting with optical phonons or other high energy phonons can therefore be written as

$$j_x \sim \int \sum_{nn'} \frac{C_{nn'} f(E)(1-f(E+\hbar\omega_L))dE}{\{E-\hbar\omega_c(n+\frac{1}{2})\}^{\frac{1}{2}}\{E+\hbar\omega_{LO}-\hbar\omega_c(n'+\frac{1}{2})\}^{\frac{1}{2}}} \qquad (7)$$

where the distribution function f and the transition probability $C_{nn'}$ are both smoothly varying functions of energy. Each term in the denominator of equation (7) contains an intergral singularity with the exception of magnetic field values such that $\omega_c(n-n')=M\omega_c$ is close to $\omega_{LO}$. At these fields both terms are simultaneously close to zero and the transverse conductivity for classical statistics contains a logarithmic divergence of the form

$$\Delta\sigma_{xx} \propto - \ln \frac{\hbar\omega_c}{kT}\delta \qquad (8)$$

where

$$\delta = \left|M - \frac{\omega_{LO}}{\omega_c}\right| \qquad (9)$$

At sufficiently low values of $\delta$ some mechanism will act to limit the height of the peak in the magnetoconductivity. In practice the amplitude of the magnetophonon terms is most likely to be determined by the finite lifetime of the electron which smears out the singularity in the density of states. The lifetime of the electrons is usually limited by the impurities present within the sample or by the phonons themselves, and the changes in the magnetoconductivity remain small and of the order of only a few per cent of the zero-field conductivity.

Hence $\rho_{xx} \simeq \sigma_{xx}/\sigma_{xy}^2$ so that peaks are observed in the transverse resistivity at fields given by

$$\omega_{LO} = M\omega_c. \qquad (10)$$

Physically,maxima are observed in the transverse resistance whenever the magnetic field is such that the optical phonons can cause transitions between the very high density of states in two Landau levels close to $k_B = 0$ with the enhanced scattering causing the orbit centres to diffuse more rapidly through the sample with the result that the transverse magnetoconductivity $\sigma_{xx}$ shows peaks. However the longitudinal conductivity $\sigma_{zz}$ shows no first-order effect as the inter-Landau level transitions responsible for the appearance of peaks only change the transverse momentum on re-

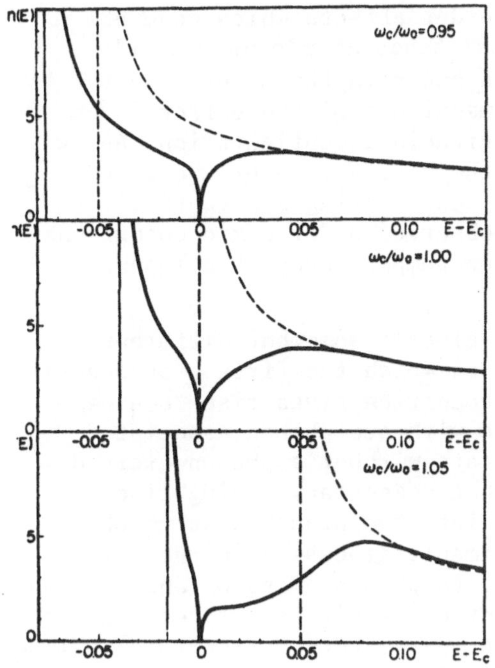


Fig.3.

Combined density of states near to fundamental magnetophonon peak on inclusion of polaronic interaction. ref.18.

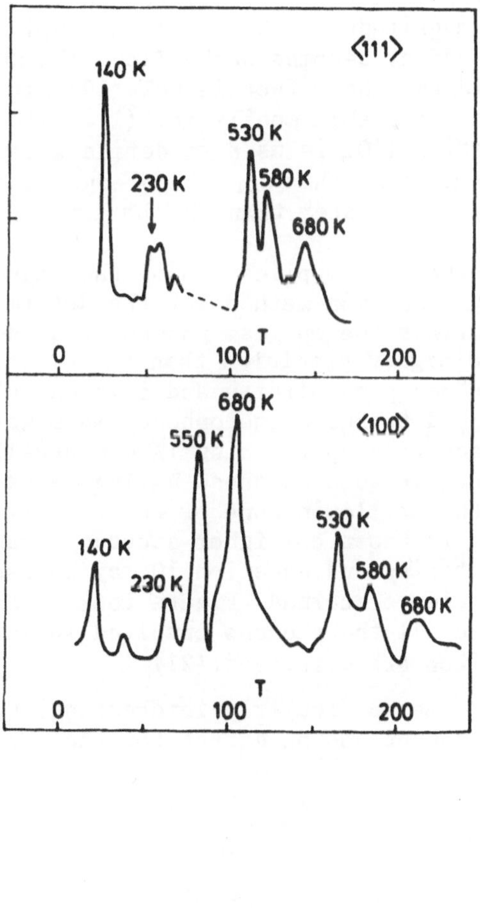


Fig.4.

Fourier analysis of magnetophonon spectrum in silicon showing fundamental fields of series present (ref. 21). The equivalent temperatures of the phonons involved are given next to the peaks concerned. Note the similarity of the amplitudes for different field directions.

sonance. There are, however, second-order effects which generate minima in the longitudinal magnetoresistance at fields slightly different to those at which the peaks occur in the transverse re-resistance. In addition the overall magnitude of the effect is an order of magnitude weaker in the longitudinal configuration. As is clear from the ideas outlined here, the observation of the effect implies that high energy phonons are contributing substantially to the momentum relaxation. The effect observed under ohmic conditions therefore invariably disappears at low temperatures when impurity scattering limits the mobility.

The inclusion of polaronic ideas gives a somewhat different picture to the phenomenon. The field at which the first peak occurs is given by $\omega_{LO} = \omega_c$. However this condition gives rise to a degeneracy between the n=0 Landau state with one phonon associated with that state and the n=1 Landau state without a phonon excited. The electron-phonon coupling lifts that degeneracy giving rise to a splitting of the degenerate states into two branches as is observed directly in the optical experiments discussed in the next section. The upper branch is unstable to phonon emission and is thus broader. Figure 3 shows the one-electron density of states function close to the fundamental resonance for parameters appropriate to InSb. It is clear that the asymmetrical lifetime will shift the magnetophonon peak away from the resonance field defined by $\omega_{LO}=eB/m^*$ if $m^*$ is assumed to be the band edge polaromic mass. The broadening of the levels and associated shift of the peak will increase with the magnitude of the Frohlich coupling constant and the magnitude of the effect becomes weaker(18). The most polar material which has exhibited the effect is CdTe(19) although the silver halides can have comparable mobilities (1). With CdTe, it is found that when equation (10) is used to define a magnetophonon mass, this mass is ∿20% greater than bare band edge mass although the band mass is only ∿6% greater than the bare mass.

The main application of the magnetophonon effect has been as a high precision method for the determination of effective mass. In most materials the optical phonon frequency is known by Raman scattering to a higher precision than the effective mass. After corrections for band non-parabolicity and polaron-self energy shift have been applied to the magnetophonon mass derived from equation 10 an accuracy approaching 1% can usually be obtained. For certain materials such as those having bands located away from the centre of the Brillouin zone or for ternary or quaternary alloys (20) the mass is known to higher accuracy than the phonons involved in scattering and equation 10 may be used to find the phonon frequencies concerned. Figure four is an example of this procedure where all the phonons involved in intervalley scattering in silicon are displayed.(21)

Some particularly interesting questions are raised concerning the magnetophonon effect for the quasi-two dimensional electron systems which are formed by size quantisation in space-charge layers formed in MOS-type of structures or in quantum-wells formed by semiconductor heterojunction structures. If a magnetic field is applied perpendicular to the thin conducting layer, the normal degree of freedom for motion along the magnetic field no longer exists because of the size quantisation, $k_B$ is no longer a good quantum number and a simple-harmonic oscillator ladder of cyclotron levels is formed. Without scattering the density of states function is simply a series of $\delta$-functions. It would therefore seem at first sight that the magnetophonon effect should be larger than in the three-dimensional case as 'non-vertical' transitions are not possible and scattering cannot occur when equation 10 is not obeyed. However, the existence of an interface or surface implies that the phonon wavevector perpendicular to the interface is also not a good quantum number. Consequently phonons which were originally forbidden by k-conservation become allowed with the result that the magnetophonon oscillations are likely to be blurred out. In the event the observed effect seems to be no stronger than in the three dimensional case.(22) A further intriguing possibility arises because the phonon spectrum should be modified by the presence of the surface or interface itself. Consequently it may prove possible to detect and measure zone-folding effects (23) or interfacial phonons themselves with samples where the conducting layer is only a few atoms thick.

An additional feature of the quasi-two dimensional system is that higher sub-bands are formed when the quantum number describing the size quantisation is greater than one or, in the case of a multivalley or degenerate band structures, from different combinations of band extrema. The separation of the sub-bands can be tuned by changing the electric field perpendicular to the interface or, if the masses are different in the various sub-bands, by varying the magnetic field. In these cases an 'electro-phonon resonance' (24) becomes possible where the separation of the quantum states is changed by an electric field rather than by a magnetic field. However the presence of a magnetic field still plays an important role in enhancing the electro-phonon effect because, without such a magnetic field, the density of states at the bottom of a sub-band does not diverge and, provided the sub-band separation is less than the phonon frequency, intersub-band scattering is possible with interchange of electron and phonon wavevectors ($k_{11}$). Nevertheless, even without a magnetic field, an abrupt change in phonon scattering rate should occur if the sub-band separation is tuned through the phonon energy.

## 2.2 Hot-electron magnetophonon effect

The normal magnetophonon effect discussed in the previous section involves momentum relaxation by means of high energy phonons and invariably disappears at low temperatures when elastic scattering becomes dominant. However energy relaxation usually requires the involvement of high energy phonons even when acoustic phonons or impurities limit the mobility. Thus, even when the lattice is maintained at very low temperatures magnetophonon structure can be caused to reappear by the application of an electric field sufficient to raise the electron temperature a few degrees above that of the lattice. When an equilibrium is established the power gained by the electrons from the applied electric field is transferred to the lattice by the emission of phonons. If a magnetic field is applied of sufficient magnitude to form well defined Landau levels, the strength of the link between the electrons and the lattice will vary in an oscillatory manner with the value of the applied magnetic field and the mean temperature of the electron system will show a corresponding variation if electron-electron interactions are sufficiently numerous to maintain a Boltzmann-like distribution. The electrical resistance, which depends directly on the momentum relaxation time, will then follow the variation in electron temperature as ionized impurity scattering, which is still the dominant scattering process for the vast majority of the electrons, is strongly energy dependent. The reappearance of the magnetophonon effect at low temperatures under nonohmic conditions therefore usually reflects the magnetic field dependence of the energy relaxation time rather than any direct variation in the momentum relaxation time.

In general the magnetophonon structure observed under hot-electron conditions at low temperatures is much more complex and richer in peaks than that observed at high temperatures under ohmic conditions and a number of new mechanisms for energy relaxation have been identified from these experiments. Magnetophonon extrema have been reported (see list of experimental references in (1)) in all three major components of the resistivity tensor ($\rho_{xx}, \rho_{xy}, \rho_{zz}$). Under hot-electron conditions the results obtained in the longitudinal magnetoresistance resemble quite closely those for the transverse configuration in contrast to the magnitude and sign differences obtained at high temperatures, confirming the view that hot-electron magnetophonon extrema reflect the energy relaxation processes within the material concerned. Experimentally, therefore, it is rather easier to observe hot-electron magnetophonon peaks in $\rho_{zz}$ as the monotonic magnetoresistance is much greater in the transverse configuration. The near equivalence of the magnetophonon series observed in the two configurations can be understood if momentum randomization by elastic scattering takes place much more rapidly than energy relaxation of the electron system as a whole (eg with InSb at low temperatures $\tau_p \sim 10^{-12}$s whereas $\tau_E \sim 10^{-8}$s). In this case the electron distribution heats up almost uniformly in all directions and energy relaxation can proceed by means of the very high density of states close to $k_B=0$ irrespective of the orientation of the current with respect to the magnetic field. At first sight, if ionized impurity scattering is dominant in relaxing the momentum of the electrons, the resistance should be a maximum at the resonant fields because the mean electron temperature should have minimum values at these points. However, Yamada and Kurosawa (25) show that a pronounced distortion of the electron distribution function away from the simple Boltzmann function could occur under hot-electron conditions at energies near to the bottom of each Landau state and above the optical phonon energy. This type of distortion is particularly strong under 'streaming' conditions. This distortion could act to cause minima to occur at the resonance fields, as is found experimentally with a number of materials. Furthermore, when the distribution is not in thermal equilibrium with the lattice, magnetophonon extrema could arise from a periodic variation of the number of conduction electrons with magnetic field, either because the decrease in energy of the electron system at the resonant fields decreases the probability of impact ionization of any electrons trapped on the donor sites, or because of an increase in the recombination rate of the electrons at the donors associated with the emission of optical phonons.

The early magnetophonon results obtained under hot-electron conditions have been reviewed by Harper et al (1). It should be noted however that, presumably because of the considerable distortion of the distribution function at low temperatures, extrema are rarely observed at the fields expected for the normal magnetophonon effect as given by equation (10). The first magnetophonon experiments involving hot electrons (26) showed that capture at impurity states was an alternative process which was favoured at low temperatures. In this case peaks in the magnetoresistance could be observed at fields given by

$$\hbar\omega_{LO} = N\hbar\omega_c + E_I(B) \tag{11}$$

where $E_I(B)$ is the magnetic-field dependent binding energy of the state concerned. In most cases the series observed corresponded to the 1s ground state. However, the more lightly doped of a number of n-CdTe samples showed a series best fitted by relaxation to the 2s state instead (15). With all the materials investigated to date extrema could be seen at fields predicted by equation (11) at lattice temperatures below 20K and in no case could the normal magnetophonon series given by equation (10) be observed at low temperatures for electric fields up to at least an order of magnitude greater than those required to generate oscillatory structure. With InP and GaAs the change-over in amplitude of the hot electron series from peak positions given by equation (11) to those given by equation (10) was studied in detail by third harmonic techniques (27,28). In both cases the coexistence of the two series was restricted to a very narrow range of temperatures and electric fields and the temperatures concerned coincided roughly with those for the appearance of the normal magnetophonon effect.

A further common process for energy relaxation which can be detected by means of hot-electron magnetophonon measurements involves the simultaneous emission of pairs of band-edge acoustic phonons according to the relation

$$2\hbar\omega_{TA} = N\hbar\omega_c \qquad (12)$$

Although weakly coupled compared with single phonon processes, two-phonon processes can dominate the energy relaxation over a substantial range of electron temperatures because of the much lower threshold energy required for emission. In addition the very high density of two-phonon states resulting from the extremely flat dispersion relation for the TA branches near to the zone boundary along the <100> and to a lesser extent along the <111> directions in III-V compounds in particular and for most diamond and zinc-blend semiconductors in general compensates to a considerable extent for the weak coupling.

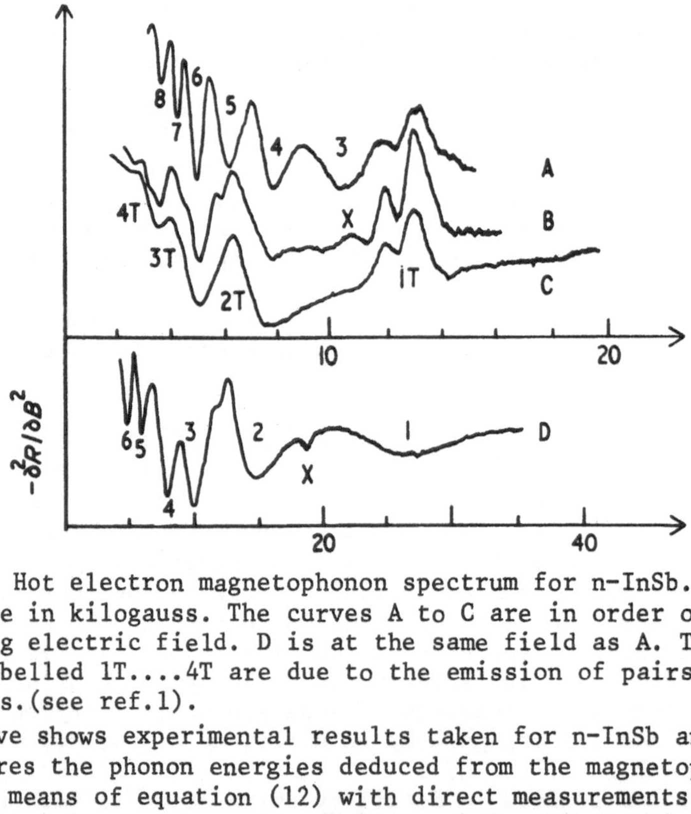


Fig.5. Hot electron magnetophonon spectrum for n-InSb. The fields are in kilogauss. The curves A to C are in order of decreasing electric field. D is at the same field as A. The series labelled 1T....4T are due to the emission of pairs of TA phonons.(see ref.1).

Figure five shows experimental results taken for n-InSb and table one compares the phonon energies deduced from the magnetophonon series by means of equation (12) with direct measurements of the phonon energies by neutron scattering or infrared techniques. In the case of InSb a second peak is observed at slightly lower energy than the strongest two-phonon peak and this is thought to arise from TA-phonons near to the L-point. The TA phonon energy required to fit this series is 9.5meV compared with 8.4 meV from infrared measurements.[1]

Table 1

| Material | Two-Phonon energy derived from magneto-phonon peaks + eqn.12 | 2TA(X) energy from neutron or i.r. techniques. |
|---|---|---|
| InSb | 10.3 meV | 10.3 meV |
| GaAs | 19.2 meV | 19.4 meV |
| InP | 16.0 meV | 16.5 meV |
| CdTe | 17.7 meV | 17.7 meV |

## 2.3 The magneto-impurity effect

A final resonance phenomenon is the magneto-impurity effect which frequently can be observed in transport experiments in conjunction with hot-electron magnetophonon peaks. The magneto-impurity effect occurs when an impurity is resonantly excited or de-excited on scattering a carrier between Landau states. In this case a phonon is not involved and the resonance is given by

$$\Delta E_I(B) = n\hbar\omega_c \qquad (13)$$

where $\Delta E_I(B)$ can either be the energy between the ground and excited states of the impurity or the separation between the ground state and the bottom of the conduction band. The magneto-impurity effect was first observed with n-InP(29) and subsequent measurements with this material showed splitting of the fundamental (n=1) peak due to central cell splitting of the ground state of the shallow donors. As can be seen from figure six a remarkable situation exists where a d.c. measurement can detect a difference of 0.1 meV (or 1%) in the binding energy of two impurity species with a resolution equivalent to that obtained with a high-performance infrared spectrometer. Similar resolution has been achieved with magneto-impurity measurements on the acceptors in tellurium (30), Various authors (31,32) have used photoexcitation rather than d.c. electric field heating of the carriers to observe the magneto-impurity effect in germanium. On changing the experimental conditions such as illumination intensity and temperature it is possible to cause a change in sign of the resistance change observed at a magneto-impurity resonance. From this result it was deduced that it was possible to produce a resonance heating of cold carriers in the band by de-excitation of the impurities(effectively an Auger type of process)rather than a cooling of carriers by impact ionisation or excitation of the impurities. A similar conclusion arose from studies of the third-harmonic in the current-voltage relationship (28,29). The magneto impurity effect has been reviewed by Eaves and Portal(33) and a detailed discussion of hot-carrier transport in quantising magnetic fields has been given by Nicholas and Portal (34).

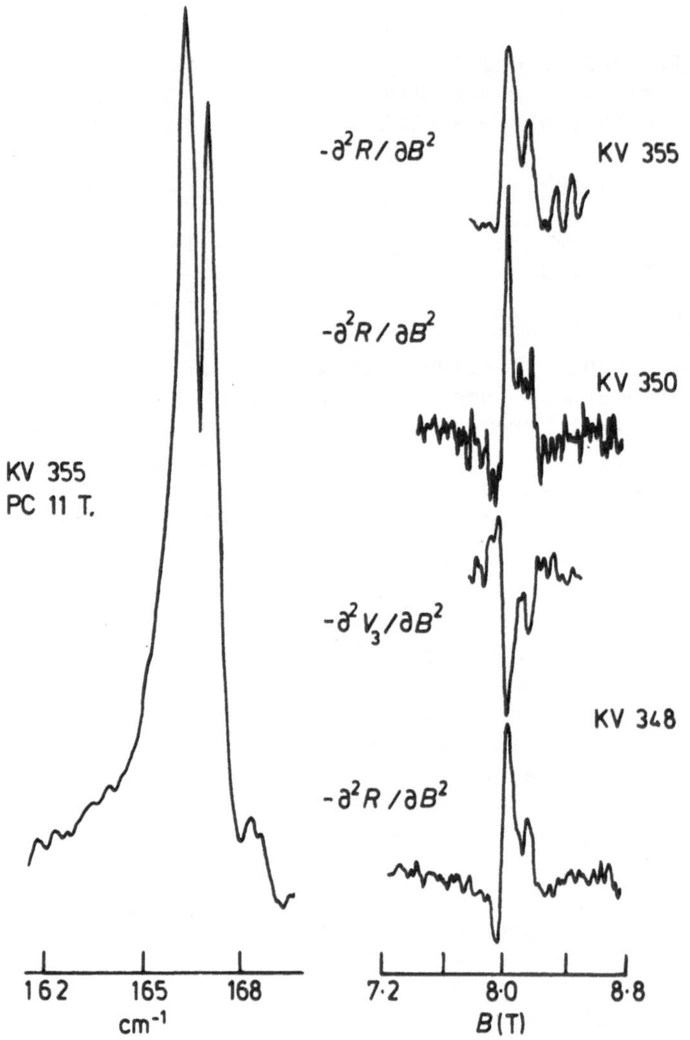


Fig.6. A comparison of the fundamental magneto-impurity peak(n=1) observed in the magnetoresistance and in the third harmonic component for three samples of n-InP with the 1s-$2p_+$ transition of the shallow donors for one of the samples obtained with a FIR Fourier Spectrometer. Similar central cell splitting of the shallow donor ground state is observed in all recordings.(ref.29).

## 3. MAGNETO-OPTICAL STUDIES OF RESONANT POLARON EFFECTS

All electronic energy levels are shifted by the electron-phonon interaction with the shifts being much stronger for polar rather than deformation potential coupling. Self energy effects will be magnified when a pair of electronic levels are close to resonance with the energy of the phonons concerned and the upper of the two electronic levels will be split by the electron-phonon coupling. The magnetophonon effect discussed in the previous section is one example of this type of resonance. However, although magnetotransport experiments have the advantage of simplicity, optical experiments have the feature of being able to probe different branches of the coupled electron-phonon system separately. The magnitude of the energy shift and line shape changes can therefore be measured for each branch. Resonance effects of this nature

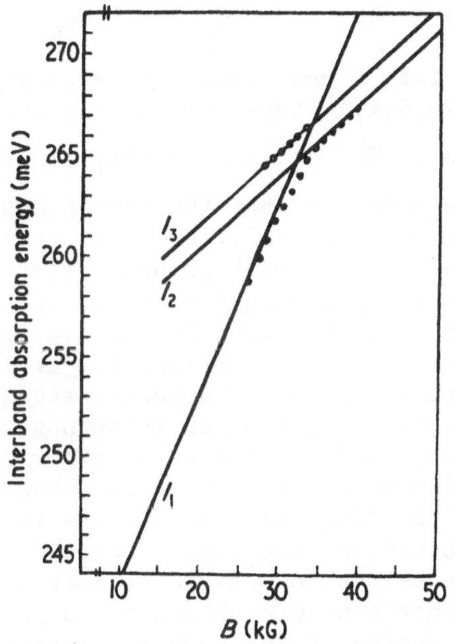


Fig.7. First observation of the splitting of electronic modes for a resonant polaron. The observation concerns the interband magneto-absorption of n-InSb (ref.35).

were first detected optically in the interband magneto-absorption experiments of Johnson and Larsen (35). Their results shown in figure seven demonstrated clearly the repulsion between the phonon associated level and the solely electronic state as they attempt to cross. As pointed out in the discussion of the magnetophonon effect in the previous section, the upper of the two branches is unstable against phonon emission as can be understood from the following argument. The energy of an electron in the nth Landau level in the absence of phonon coupling is given by

$$E\,(n,0,k_B) = (n+\tfrac{1}{2})\hbar\omega_c + \frac{\hbar^2 k_B^{\,2}}{2m^*} \qquad (14)$$

whereas the energy for the ground Landau state associated with one LO phonon is

$$E(0,1,k'_B) = \tfrac{1}{2}\hbar\omega_c + \frac{\hbar^2 k'^{\,2}_B}{2m^*} + \hbar\omega_\omega \qquad (15)$$

with density of states given by $(E-(n+\frac{1}{2})\hbar\omega_c)^{\frac{1}{2}}$ and $(E-\frac{1}{2}\hbar\omega_c-\hbar\omega_{LO})^{-\frac{1}{2}}$ respectively. States $k_B$ and $k'_B$ may be linked by a phonon whose component of wavevector along the magnetic field is $q_B$ if $q_B=k_B-k'_B$. In this case (15) can be rewritten as

$$E(0,1,k'_B) = \tfrac{1}{2}\hbar\omega_c + \frac{\hbar^2(k_B-q_B)^2}{2m^*} + \hbar\omega_{LO} \qquad (16)$$

The $E(0,1,k'_B)$ states form a continuum with a minimum energy of $\frac{1}{2}\hbar\omega_c + \hbar\omega_{LO}$ and the $E(n,0,k_B)$ states cross this threshold at $n\hbar\omega_c = \hbar\omega_{LO} - \hbar^2 k_B^{\,2}/2m^*$. The unmixed density of states functions for the levels concerned ensure that the levels couple strongly in the region of resonance $\hbar\omega_{LO} \sim n\hbar\omega_c$. The upper branch of the mixed states is broadened because of the high probability of phonon emission and this broadening diverges close to resonance. In contrast the lower branch is unbroadened by the L.O. phonon interaction and any width arises only from impurity or low energy phonon scattering. Experimentally Johnson and Larsen studied the resonance between the 0 and 1 Landau levels which is stronger than the n>1 resonances and also benefited by chosing a rather simple region of the interband absorption spectrum. Furthermore the photon energies were far removed from the spectral region close to the Restrahl ($\omega\to\omega_{LO}$). The strong absorption and rapid variation of the optical constants complicate attempts to measure cyclotron resonance in this region. However intraband measurements can be performed well clear of the Restrahl region if the combined resonance is studied instead of the stronger cyclotron resonance.(35). An alternative is to study the phonon-assisted cyclotron resonance (36) or harmonics of the cyclotron resonance.(37) where

$$\omega = r\omega_c \pm \omega_{LO} \qquad (17)$$

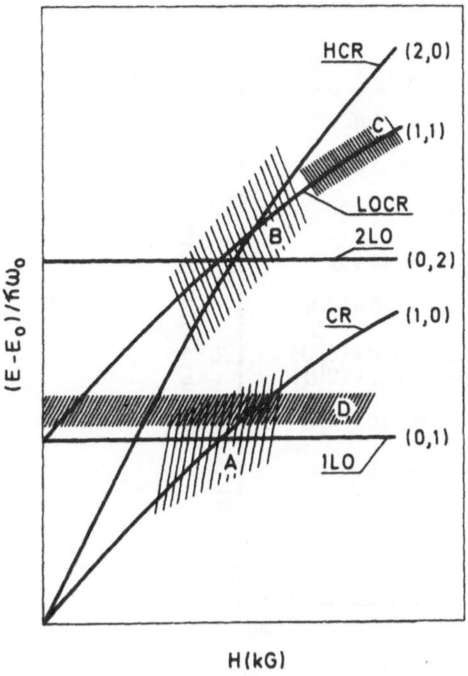


Fig.8. Schematic plot of crossing of various magneto-optical lines in the absence of polaronic coupling. (ref.38).

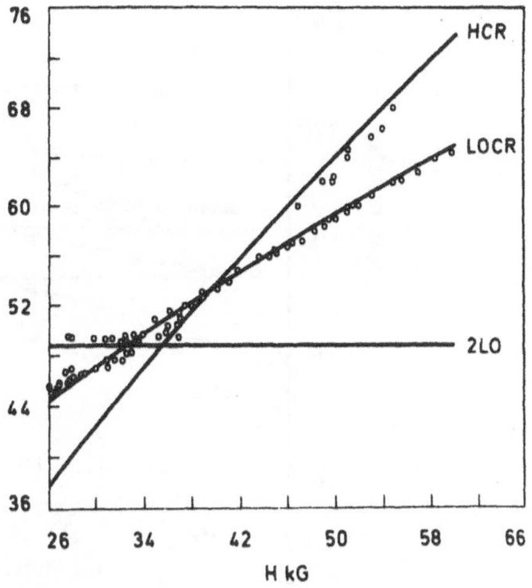


Fig.9. Data points with harmonic resonance (HCR) and phonon assisted resonances (LOCR) shown without interaction. On inclusion of polaronic interaction the intensity of the LOCR branch drops to zero in the cross-over region .(ref.38).

In equation (17) r>1 correspond to harmonics of cyclotron resonance and m>1 corresponds to multiple phonon assistance.The ± signs refer to phonon absorption or emission. Devreese et al (38) have studied the region where the first harmonic of cyclotron resonance(r=2, m=0) and the single LO phonon assisted cyclotron resonance (r=1,m=1) are close to the 2LO energy as shown schematically in figure eight. Polaron pinning effects unlike those observed in the single LO phonon region were reported. The resultant level scheme is shown in figure nine. The middle branch, which away from resonance has the character of the phonon assisted cyclotron resonance (LOCR), loses spectral intensity completely near to resonance.

Similar effects to those outlined for free carriers above can

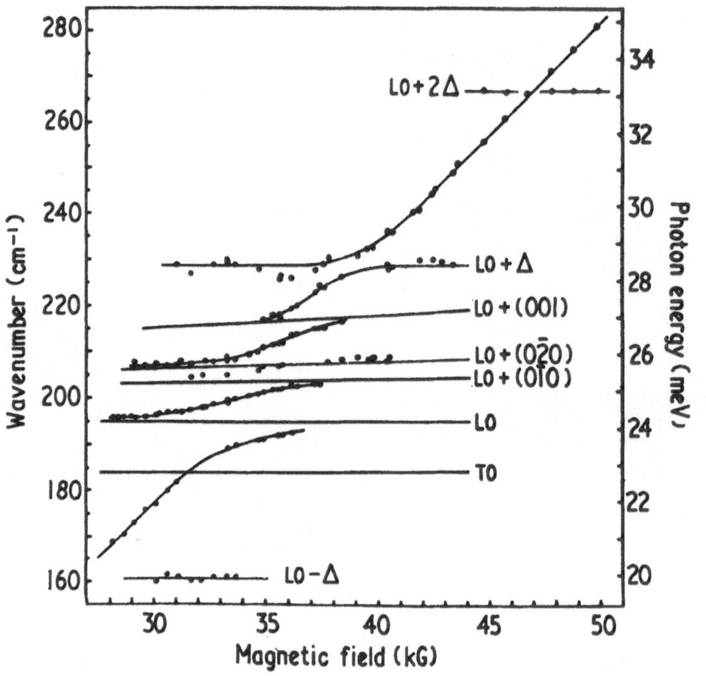


Fig.10. Impurity magneto-absorption in n-InSb showing eight separate branches caused by the different polaronic pinning phenomena described in the text.(ref.40).

be observed with bound electrons. Again, because of its low mass and intrinsic purity most of the initial experiments were performed with n-InSb(39). A particular dramatic series of pinnings was observed (40) when the 1s-$2p_+$ transition (000→010 in the high field nomenclature of Yafet, Keyes and Adams (41))was studied through the region close to the single LO phonon as can be seen from figure ten. Because of the strong absorption a particularly thin sample was used in transmission. Nevertheless the sample was black close to the TO frequency (184 $cm^{-1}$) as can be deduced from the lack of data points in the figure. Two branches are observed either side of the LO phonon frequency (196 $cm^{-1}$) with a frequency offset of slightly greater than 10 $cm^{-1}$; further pinnings are seen when the separation between the $2p_+$ and the $2p_-(0\bar{1}0)$, $3_{d-2}(0\bar{2}0)$ and $2p_0(001)$ respectively become equal to the LO phonon energy. Also seen in figure ten are further pinnings at energies of Δ and 2Δ above and Δ below the LO phonon frequency where Δ = 35 $cm^{-1}$. The offset between the branches at LO + Δ was 4 $cm^{-1}$ whereas that at LO + 2Δ was very small indicating a much stronger coupling to the Δ excitation. Subsequently a splitting at the frequency of Δ itself was detected on the 1s-2p$_0$ transition with an offset of 2 cm$^{-1}$(42) At the time of these experiments the interaction responsible for the pinnings involving Δ was not understood. The probable nature of the interaction will be discussed further in section 4.

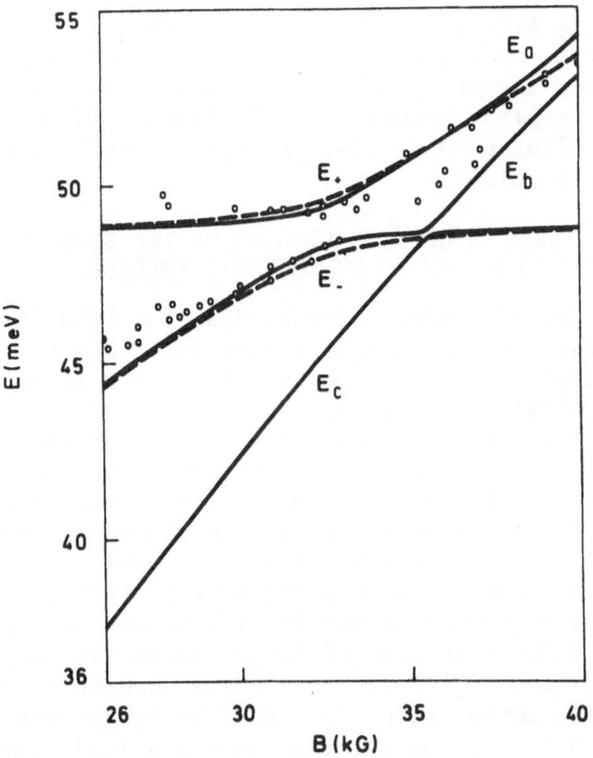


Fig.11. Theoretical curves and experimental points for bound impurity electron in InSb in region of 2-LO energy.(from ref.38).

The various impurity equivalents of the crossings between the HCR and the LOCR in the 2LO phonon region are shown in figure eleven. As was found for the free carrier case, a similar drop-off in intensity is found for the LOCR like excitation (38).

The many features of the many magneto-optical studies of bound and free polarons have been the subject of a number of review articles (1,2,43,44).

Although potentially of great interest because of the modification of both the electronic states and the phonon states by the interface or surface, few results yet exist for the comparable resonant polaron effects which are expected for the quasi-two dimensional electron systems formed by space charge layers in MOS structures or in thin heterojunction structures. Magnetophonon experiments with such systems are discussed in the previous section. A 50% increase in the scattering rate above the LO phonon frequency has been reported for the two-dimensional cyclotron resonance from the electrons in InSb(45). The reason for this increase is not as obvious as for the three dimensional case as the density of states function no longer has the $E^{-\frac{1}{2}}$ dependence on energy. Nevertheless the interpretation put forward is somewhat similar with the broadening arising from enhanced LO phonon emission within the line-width of the corresponding Landau levels. With InAs a splitting of the LO phonon mode has been observed by Raman scattering (46) and attributed to a coupling between the LO phonons and a collective inter-subband excitation mode (ie, the optical equivalent of the electro-phonon effect).

## 4. FAR-INFRARED MAGNETO-OPTICAL STUDIES OF THE BOUND CARRIERS IN n-InSb AS A FUNCTION OF HYDROSTATIC PRESSURE

In the experimental studies discussed so far impurities have been assumed to be hydrogenic with a character described by the effective mass for the nearest band extrema. For the donors in InSb this would seem to be a particularly good approach in view of the extremely small effective mass of the Γ conduction band. However in magnetic fields in excess of about 5T central cell structure becomes apparent on the $1s$-$2p_-$ line in the purest sample, indicating a significant perturbation by the local potential around the donor impurities and, at the highest fields, the magnitude of the chemical shift can correspond to about 10% of the binding energy. This section will discuss how donor states can occur in InSb which are not of Γ character and how the existence of these states can modify the shallow donor spectrum which was previously thought to be described extremely well by simple effective mass theory. The non Γ character of the states can modify the electron-phonon interaction and the associated lattice relaxation effects.

### 4.2 Electron transport experiments

The first indication that donor states might not be associated completely with the Γ conduction band came from Hall and resistance measurements undertaken as a function of hydrostatic pressure and temperature by the group led by Porowski (47,48). The pressure technique developed by this group was particularly simple and well suited to experiments in the confined space provided by a magneto-transport cryostat. The pressure was generated at room temperature in a Be-Cu cell containing a light hydrocarbon as the pressure transmitting medium. The cell is then sealed and cooled to low temperature. Tests have shown that the resultant pressure in the solidified hydrocarbon is uniform to about 1%. Figure 12 shows a pressure cell utilising this technique which has been modified for magneto-optical experiments by adding a sapphire window.

The initial transport experiments were with samples so heavily

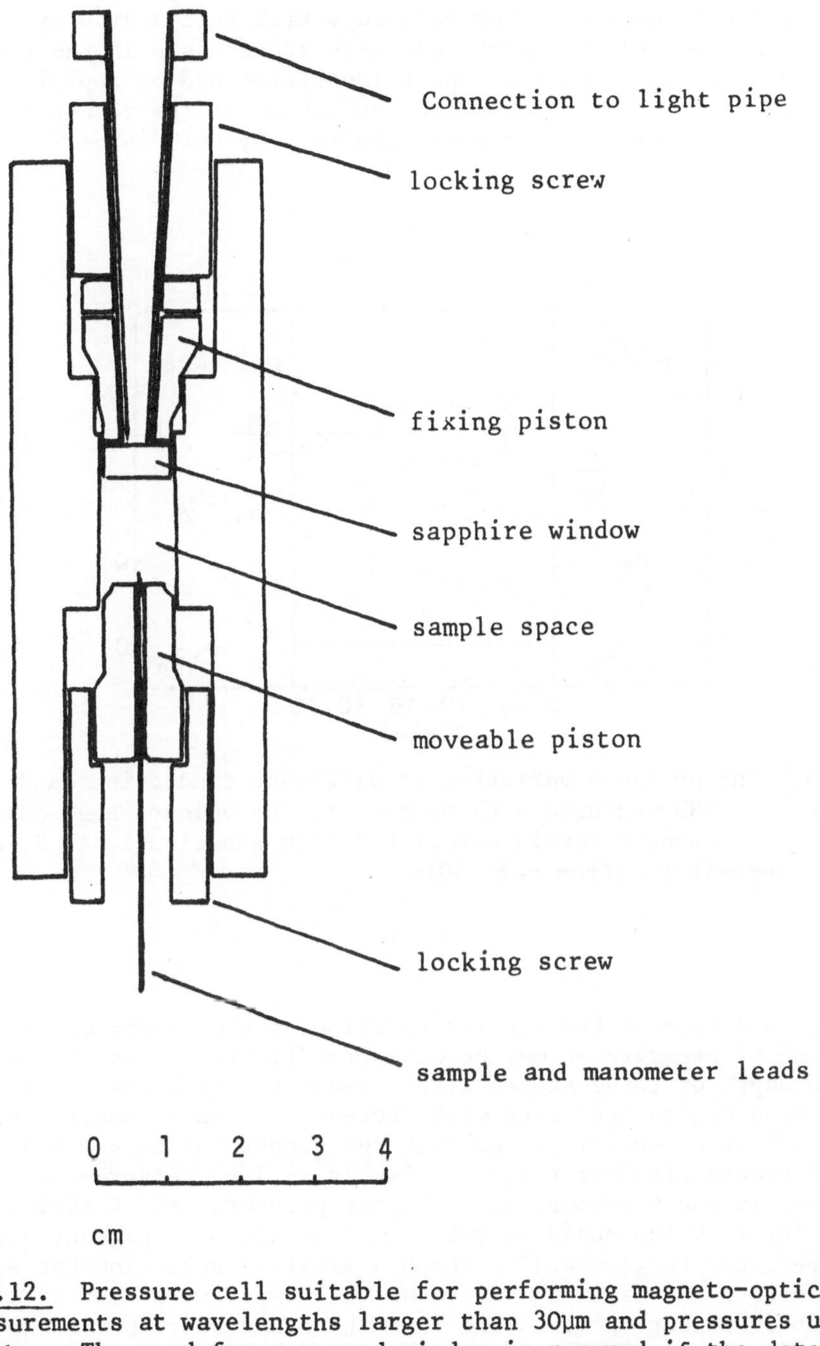


Fig.12. Pressure cell suitable for performing magneto-optical measurements at wavelengths larger than 30μm and pressures up to 20 kbar. The need for a second window is removed if the detector is enclosed in the bomb or if the photo-conductivity from the sample itself is employed as was the case for the experiment shown in figs. 21-27 in this paper.

doped with Te, Se and S (49,50) that the Fermi energy was more than a band-gap above the conduction band edge. With this degree of doping it was possible to detect resonance states deep in the conduction band arising from the dopant impurities and by applying pressure it was possible to study deionisation of the resonance states associated with higher order minima. By this means it was

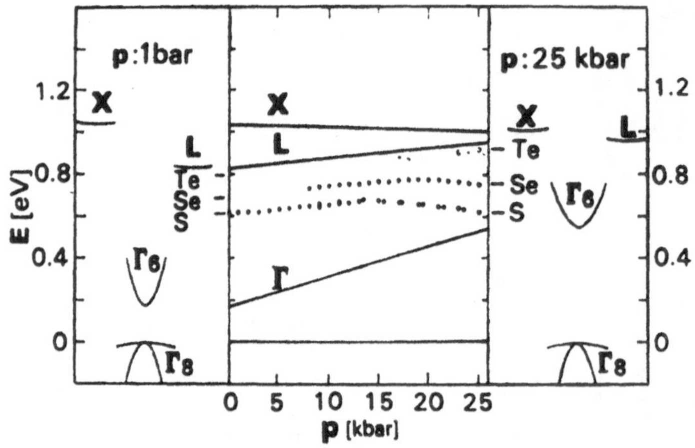


Fig.13. The pressure variation of different conduction band extrema in InSb measured with respect to the valence band edge and of the resonant levels associated with substitutional S,Se and Te impurities (from ref. 50).

possible to determine the precise position of the resonance levels as a function of pressure as may be seen from figure 13. It can be seen that the depth of these states with respect to the higher order conduction band minima increases with decreasing atomic number (increasing electronegativity) and that the slopes for Se and S show distinct breaks with the resonance levels at low pressure following quite closely the L-minima and at higher pressures the X-minima. The results for Te which could only be studied with the highest doping levels were too fragmentary to reach a similar conclusion but are likely to have the same dependence as this impurity is mores likely to be effective-mass like than the other two. The breaks in slope suggest a crossing of two levels from the same impurity, one having predominantly L-character and the other X-character.

With high purity material (ie donor concentrations below $10^{15}$ $cm^{-3}$), deionisation can be found in similar(51) experiments

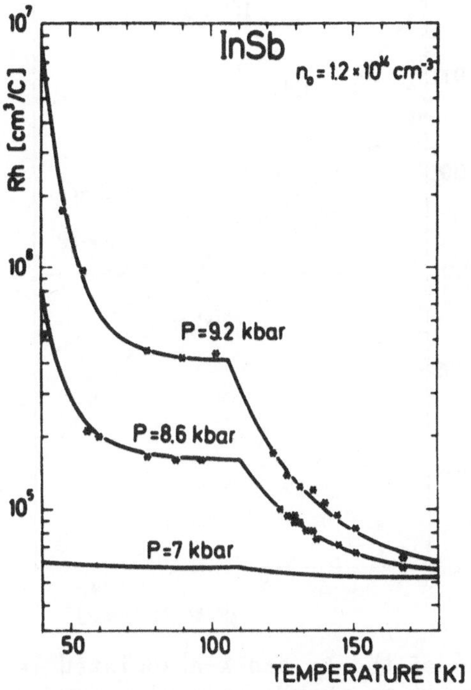


Fig.14. Hall coefficient versus temperature for high-purity n-InSb at various pressures showing deionisation of the L-and X-associated donor states. (ref.51).

except that the two levels lie much closer to the conduction band edge. Fig.14 shows the freeze-out which can be detected onto these levels. These levels arise from a commonly occurring residual donor which, from the position of the levels with respect to those produced by sulphur, selenium and telurium and from a knowledge of the likely contaminants, is believed to be oxygen (52). Figure 15 shows the position of the two levels with respect to the conduction band edge as a function of pressure. It can be seen that these states again move linearly with respect to the bottom of the conduction band with pressure coefficients of -20 and -10.5 meV/kbar

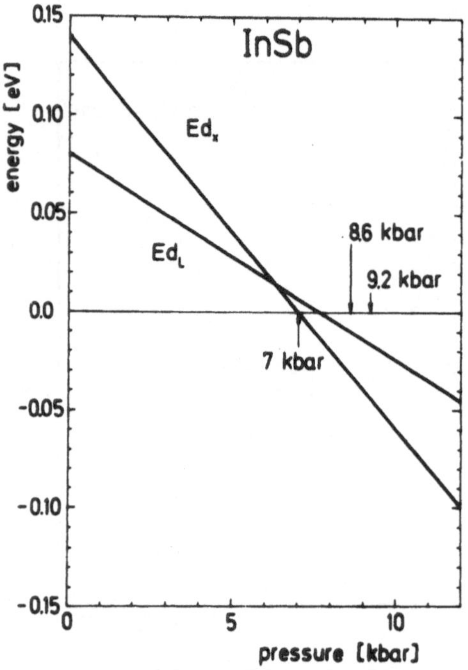


Fig.15. Energies of the L- and X-associated levels measured with respect to the Γ conduction band edge as a function of pressure and zero magnetic field. The pressures at which the results of Fig.14 are taken are indicated by arrows. (ref.51).

and cross into the gap at about 7 and 8 kbars respectively. Again the coefficients are close to those expected for the X- and L-minima: a surprising result in view of the more than 0.5eV separation between the localised states and extrema providing their dominant characters. Nevertheless the separations are far too great for the levels to be described by simple effective mass theory and contributions from many points of the zone must be substantial. The transfer of electrons from the Γ band to the X-like impurities may be inhibited by a large energy barrier (~0.3eV) which appears below 120K due to a strong lattice relaxation around the impurity below this temperature. The time constant for transfer of electrons into or out of this state can become extremely long at low temperatures (eg 270 days at 77 K) and the state can then be considered to be metastable. A magnetic field may be used in a similar manner to pressure to increase the value of the direct gap (dE/dB$\sim$2meV/T)

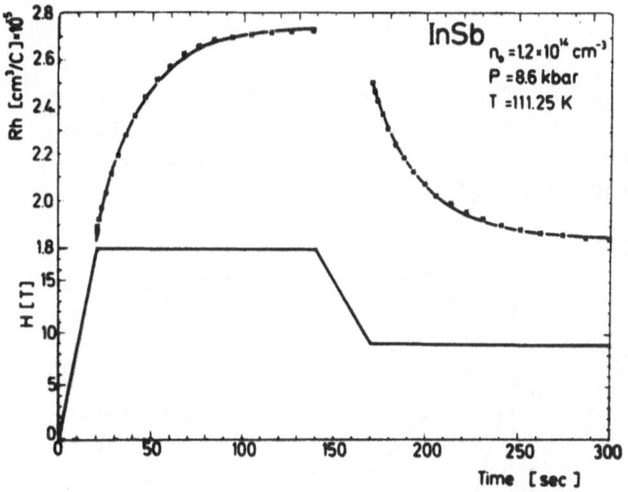


Fig.17. Relaxation time as a function of temperature derived from the type of field cycling shown in figure 16.(ref.61).

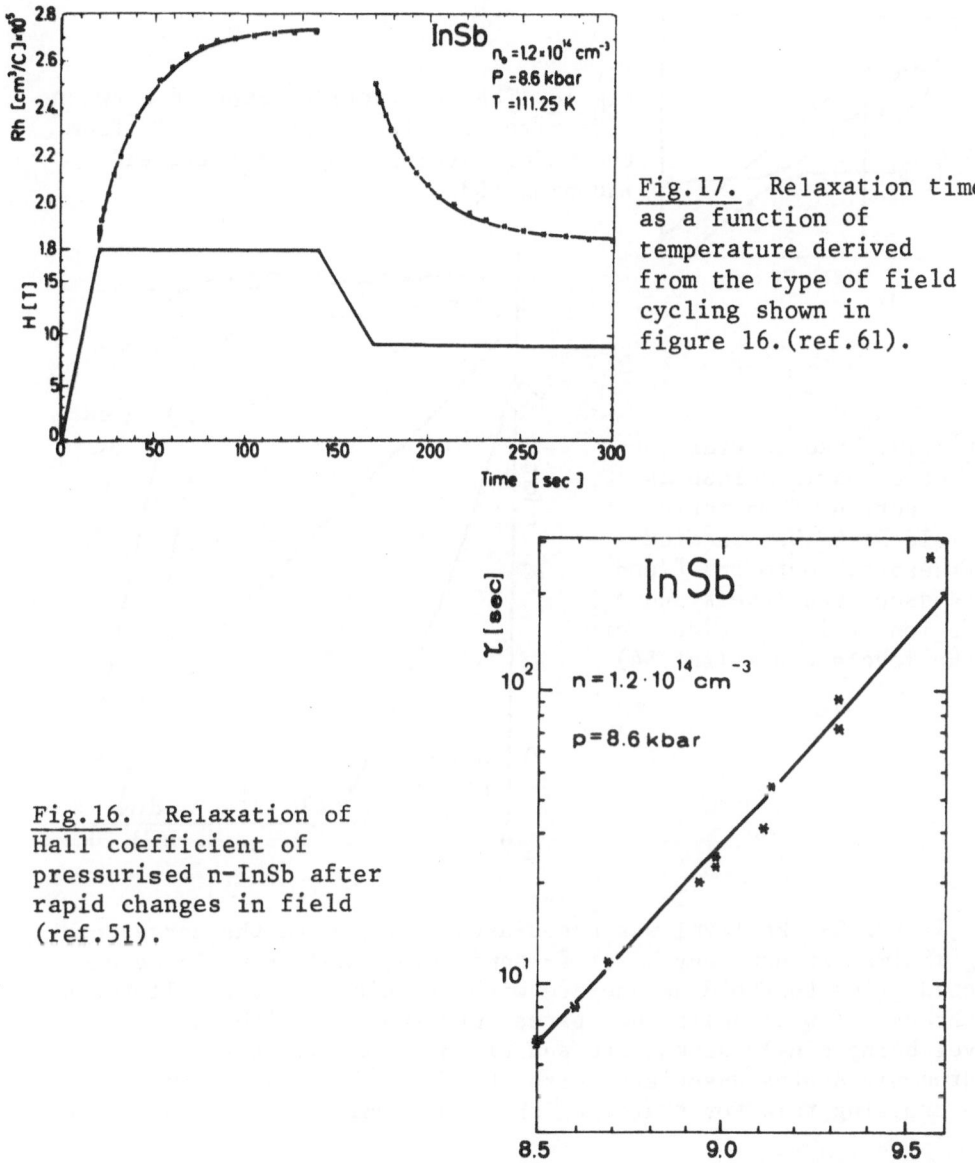

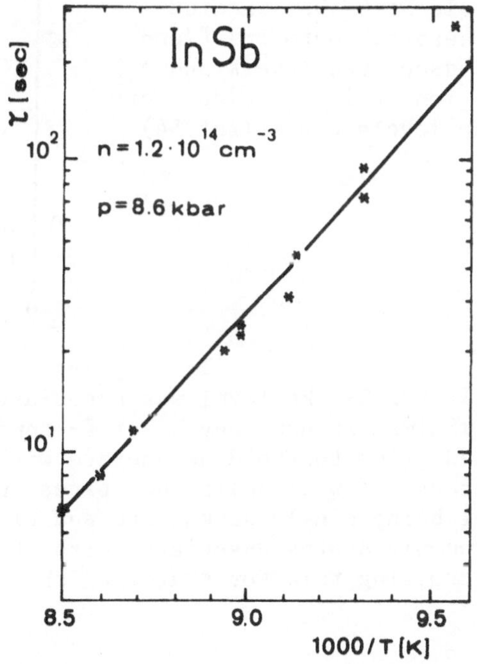


Fig.16. Relaxation of Hall coefficient of pressurised n-InSb after rapid changes in field (ref.51).

and to shift the resonant levels into the forbidden gap. Figure 16 shows the effect of ramping the applied magnetic field between 0 and 18T at 111 K on the population of free carriers at a pressure where the X-like level is lower than the L-level, and figure 17 shows the relaxation time for transfer into the X-states deduced from this type of experiment.

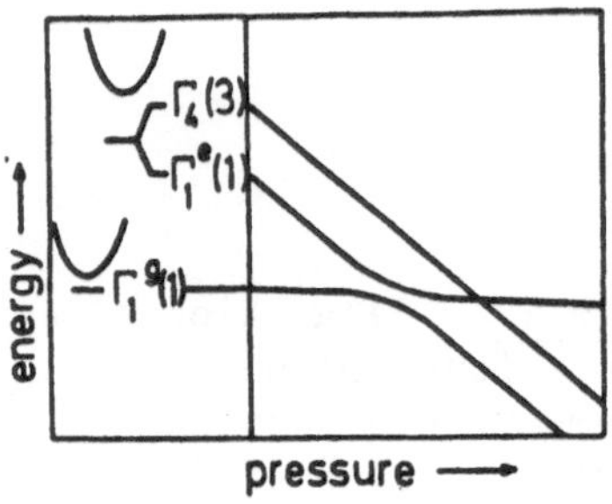


Fig.18. The interaction expected between effective mass-like Γ and L levels from the model developed by Alterelli and Iadonisi (53).

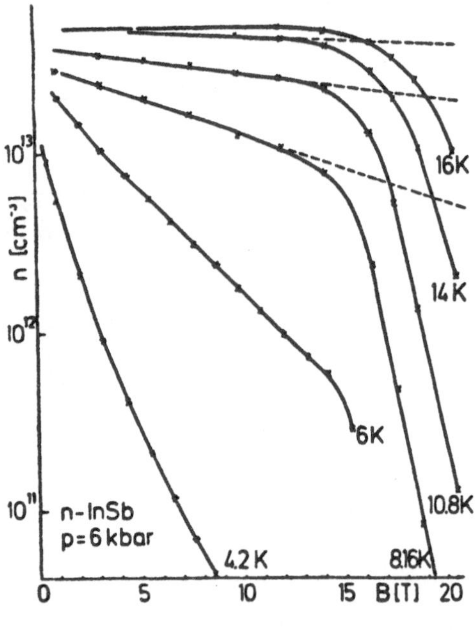


Fig.19. Free carrier concentration in n-InSb as a function of magnetic field at 6 kbar showing freeze-out onto the Γ and L-associated levels and a region of interaction where the levels cross.(ref.54).

If the L-like level was associated solely with the corresponding minima, without any X- or Γ- admixture, then it would be expected to be fourfold degenerate without spin. However valley-orbit interactions will split the states into two sets with the lower level being singly degenerate and of the same symmetry as the hydrogenic states associated with the Γ- minimum. According to the non-crossing rule for states of the same symmetry, an interaction

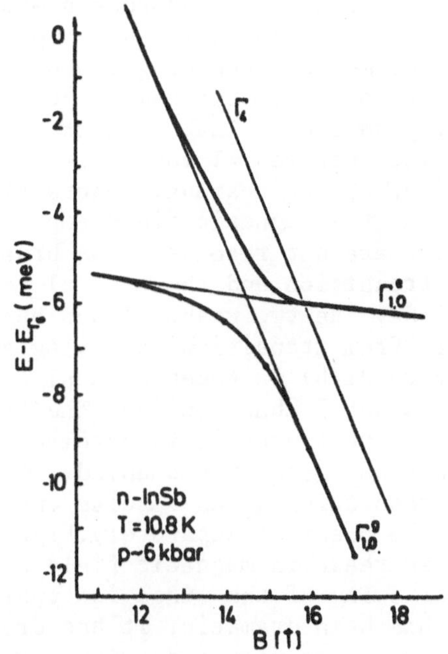


Fig.20. Calculated energies of the L- and Γ-like states in InSb in the region of crossing.(ref.54).

is expected as the states approach one another as indicated in fig.18.(53). In studies of magnetic freeze-out onto the shallow donors an admixture of the L- and Γ- levels was apparent when the two levels crossed (54) as can be seen from figure 19. Analysis suggested the level scheme shown in figure 20 with an offset of 2 meV between the two branches. The extension of the electrical measurements by optical measurements has now permitted greater detail to be discovered about the nature of the interaction and its effects.

4.3 Magneto-optical evidence for an interaction between Γ and L-associated states

In collaboration with the group of Porowski, magneto-optical experiments have been performed in St Andrews aimed at detecting the interaction between the L- and X-associated states and the shallow donors.(55-57). The experimental recording shown in figure 21(a) is typical of the magneto-optical spectrum at zero pressure expected for a high-purity sample of InSb without any intentional doping. The group of lines between 11 and 13 $cm^{-1}$ are 1s-$2p_-$ transitions from the four common residual donors (labelled A,B,C and D in this paper). At the magnetic field employed for this recording (10T), B and C are not resolved. The broader line at 33 $cm^{-1}$ is the 1s-$2p_0$ transition and the central cell structure is not resolved on this line. The two weaker lines between the 1s-$2p_-$ and 1s-$2p_0$ peaks result from transitions starting and terminating on excited states. Figure 21(b) is taken at a slightly lower field but with a pressure of about 5 kbar applied. The lines have narrowed substantially so that donor C is resolved from donor B and, in addition, line A has more than doubled its separation from line B. The increased resolution which results enables the 1s-$2p_0$ line due to donor A to be observed separately. Figure 21(c) shows the effect of a small increase in magnetic field at the same pressure. A further narrowing of the lines is apparent but the effect on donor A has been dramatic. It has dropped in intensity by almost an order of magnitude with respect to the other donors and a further substantial increase in separation from B has occurred. In addition another weak line can be observed on the low frequency side of the 1s-$2p_-$(A) line. The anomalous deepening of the ground state of donor A which can be deduced from increased central cell splitting and the accompanying fall-off in signal intensities can be interpreted as due to the interaction between the Γ and L-levels and the associated admixture of wavefunctions. The fact that the anomalous behaviour is only observed on the deepest of the shallow donors is extremely significant as the transport experiments suggested that oxygen was responsible for the resonance states which could be introduced into the gap by applying pressure (oxygen is the most electronegative of the possible substitutional donors from group IV or group VI). Furthermore there is a strong correlation between the relative strength of donor A in the central cell spectrum and the amount of freeze out than can be induced onto the L- or X-levels. In samples where A is by far the strongest feature in the central cell spectrum before pressure is applied, sample resistances in excess of 100 MΩ can be obtained at pressures in excess of 10 kbar without any magnetic field being applied.

Figures 22 and 23 show the positions of the 1s-$2p_-$,$2p_-$-$3d_{-1}$ and 1s-$2p_0$ lines as a function of pressure at a magnetic field of 11.5T and of field at a pressure 6.1 kbar respectively. The

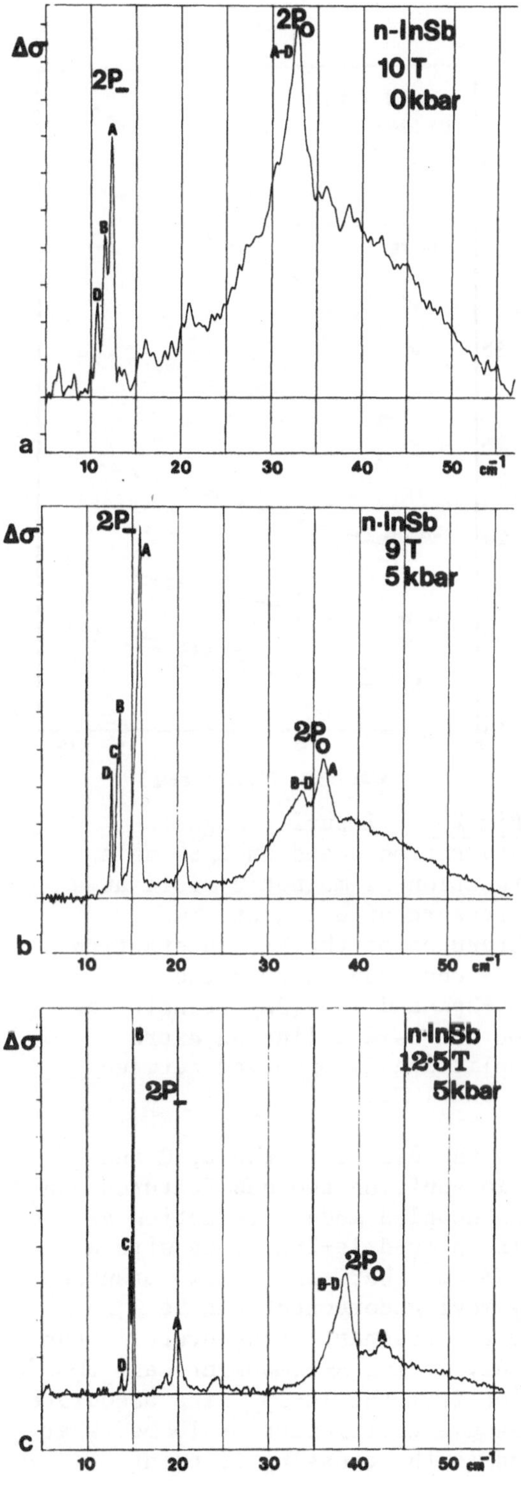


Fig.21 shows the photoconductive spectrum of the donors in InSb in the region of the 1S→2P₋ and 1S→2P₀ lines for three different combinations of field and pressure.

A comparison of (a)-(c) shows a pronounced narrowing of the 1S-2P₋ lines on increasing pressure and field. In(c) the line widths are instrumentally limited on the 1S-2P₋ lines for donors B-D.

(a) is taken at zero pressure and 10T and shows a spectrum similar to those in reference(42) and the separation of donors A and B is 0.8 $cm^{-1}$.

(b) is taken at 4.9 kbar and 9T. The intergrated intensity of donor A is about four times that of B and their separation is 2.5 $cm^{-1}$. As donor A splits off from the remaining donors, it becomes resolved on the broader $1S-2P_o$ line at 35 $cm^{-1}$.

(c) is taken at 4.9 kbar and 12.5T. Donor A now has less than half of the integrated intensity of B and their separation is 5 $cm^{-1}$.

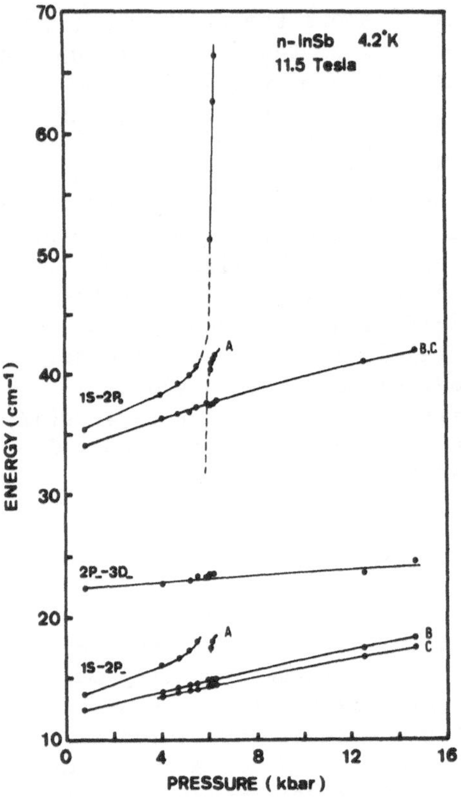


Fig.22. Impurity magneto-absorption peaks in InSb as a function of pressure at 11.5T. At about 6kbar a crossing of the L and Γ associated levels is seen for donor A. The lack of any splitting of the $2p_{-}$-$3d_{-1}$ line shows that hybridisation occurs on the ground state.

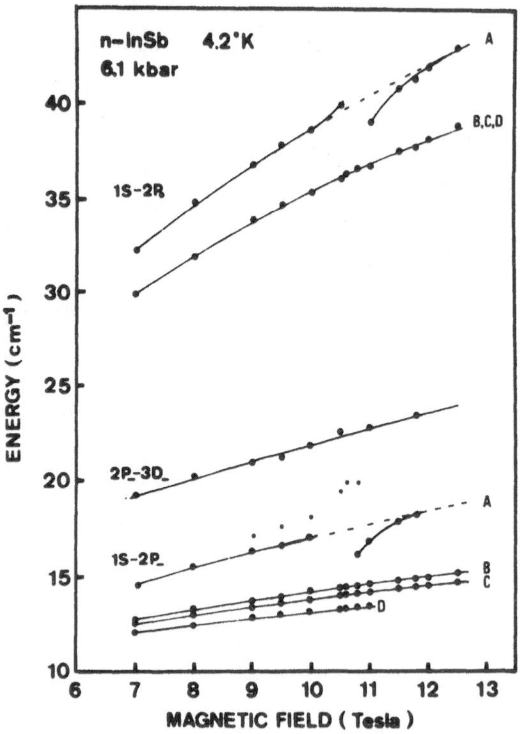


Fig.23. Impurity magneto-absorption peaks in InSb as a function of magnetic field at a pressure of 6.1 kbar. Two branches of the L-Γ interaction are observed on both the $1s$-$2p_{-}$ and $1s - 2p_0$ transitions. On the $1s$-$2p_{-}$ line an extra splitting is observed between 9 and 10T.

anomalous deepening of the A lines with respect to the B, C and D components is clearly displayed. In addition two new features can be seen. Firstly two branches of the coupled mode interaction can be observed very clearly on both the $1s$-$2p_{-}$ and $1s$-$2p_0$ lines with a minimum separation of approximately 3 $cm^{-1}$ between the two branches For the $1s$-$2p_0$ transition a rapidly moving component can be followed to about 30 $cm^{-1}$ away from the resonant cross-over. As can be seen in figure 26(a) such a rapidly moving A-component can also be observed at up to $\sim$30 $cm^{-1}$ separation from the $1s$-$2p_{+}$ line associated with other donors. The observed pressure coefficient of 10 meV/kbar identifies the second level involved in the crossing as being associated with the L-minimum

The offset of 3 $cm^{-1}$ (∿0.4meV) is much less than deduced from the transport measurements presented in the previous section. However the second branch seen in figures 22 and 23 was not detected in the freeze-out analysis. Furthermore it should be noted that in the region of the resonant cross-over the mean separation of the A-component from the B-line is 4 $cm^{-1}$ and that this separation then grows rapidly with increasing pressure or magnetic field. This compares with a separation of 0.8 $cm^{-1}$ at 11T between the same impurities with no pressure applied.(58) This very pronounced deepening of donor with respect to the other donors probably arises from a rapid increase of its chemical shift due to compression of the wavefunction into the strongly attractive local potential for this donor (59) and would certainly contribute to the deepening of the state involved in the freeze-out results shown in figure 20.

## 4.4 Polaron Coupling involving single zone-edge phonon modes

As was shown in figure ten Kaplan and Wallis (40) found resonant modes at energies Δ and 2Δ above the LO phonon energy and Δ below this energy (Δ = 35 $cm^{-1}$). It can be seen from the markedly different offsets for the two crossings that the interaction strength is considerably greater for the +Δ resonance than for the +2Δ resonance. In the work of Kaplan et al (42) similar coupled mode behaviour was found on the 1s-$2p_0$ line. However, in both sets of experiments the nature of the coupling was unclear as the only excitation which could give a resonant mode near this frequency involved single TA phonons whose frequencies are 34 $cm^{-1}$ at the L-point.(60) Two-phonon interactions with both these modes have been detected in magnetophonon experiments (61) but no other single phonon interaction with zone edge phonons has been reported in either magneto-optics or transport experiments, nor would a coupling be expected for electrons associated solely with the Γ minimum. The admixture of Γ and L states invoked to explain the anomalous behaviour of the A component of the central cell spectrum in section 4.2 and 4.3 provides a natural mechanism for appearance of resonant mode behaviour associated with single TA phonons at the L-point. Such an admixture also explains the observation in the experiments of Kaplan and Wallis that the coupling to the Δ modes is stronger than to the 2Δ modes and the fact that coupling to single TA modes at the X-point is not observed although the density of states is greater for these modes as evidenced by the strength of 2TA(X) series in hot electron magnetophonon experiments. The 35 $cm^{-1}$ resonances can be thought of as involving intervalley transfer between localised states mediated by these phonons or as an intervalley resonant polaron. As the A-component of the 1s-$2p_0$ line becomes resolved from the B, C and D lines on the application of pressure, an elegant confirmation of this mechanism for coupling becomes possible. On tuning the A component and the unresolved B-D components successively through the 35 $cm^{-1}$ frequency it is apparent that the resonant coupling can only be found on the A-component. This is consistent with the observations detailed in the previous section which show that the anomalous deepening of the ground state is only found for the A donor and that only L-like states are involved in the resonance. Further evidence that the 35 $cm^{-1}(\Delta)$ mode is indeed a phonon mode is provided by the observation that the resonant energy is pressure independent up to at least 6kbar. If the resonance mode was electronic in character it would be strongly pressure dependent. However, it should be noted that at zero pressure, under the conditions that the $\Delta$ modes were observed on the $1s\text{-}2p_+$ and $1s\text{-}2p_0$ lines, the L-levels were between 40 and 70 meV above the bottom of the $\Gamma$-band, suggesting that some L-admixture is retained for some considerable distance from cross-over. This would also be consistent with the ability to follow the $1s\text{-}2p_0$ (L) branch to more than 30 $cm^{-1}$ from cross-over.

## 4.5 An unidentified mode-crossing

Whereas the pressure experiments provide an explanation for the mechanism for the appearance of $\Delta$-modes in magneto-optical experiments, they have also thrown up at least one further mode-crossing which remains to be identified. This is concerned only with $1s(A)\text{-}2p_-(A)$ transition and can be seen in figure 24 between 9 and 10T. Two branches with a minimum offset of 0.8 $cm^{-1}$ can be observed between these fields. Typical coupled-mode behaviour is found with the lower energy branch losing intensity and the upper increasing in strength with increasing field. The field at which both branches are of equal intensity varies rapidly with pressure and is strongly correlated with the centre position of the subsequent resonant cross-over of the $1s(\Gamma)$ and L states. These features are not inconsistent with the resonance being associated with a crossing of the $2p_-(\Gamma)$ state with the L-state, although the rather weak dependence of the positions of the two branches on field shown in fig. 23 would argue against this interpretation. Another possibility might be a phonon-mediated replica of the $1s\text{-}2p_-$ transition. The very low energy of phonon mode required (0.1meV) would at first sight make this possibility unlikely. However the increasingly local character of the ground state might mix-in a very low energy local mode. Studies of the temperature dependence of the intensities of the components give some credance to the latter possibility. Further experiments with samples having different magnetic field orientations are in progress in an attempt to distinguish between these two possibilities.

## 4.6 Changes in the central cell spectrum produced by the occupancy of the X-like states

It is possible to follow the mangeto-optical spectrum through the region where the X-like and L-like states are nearly degenerate

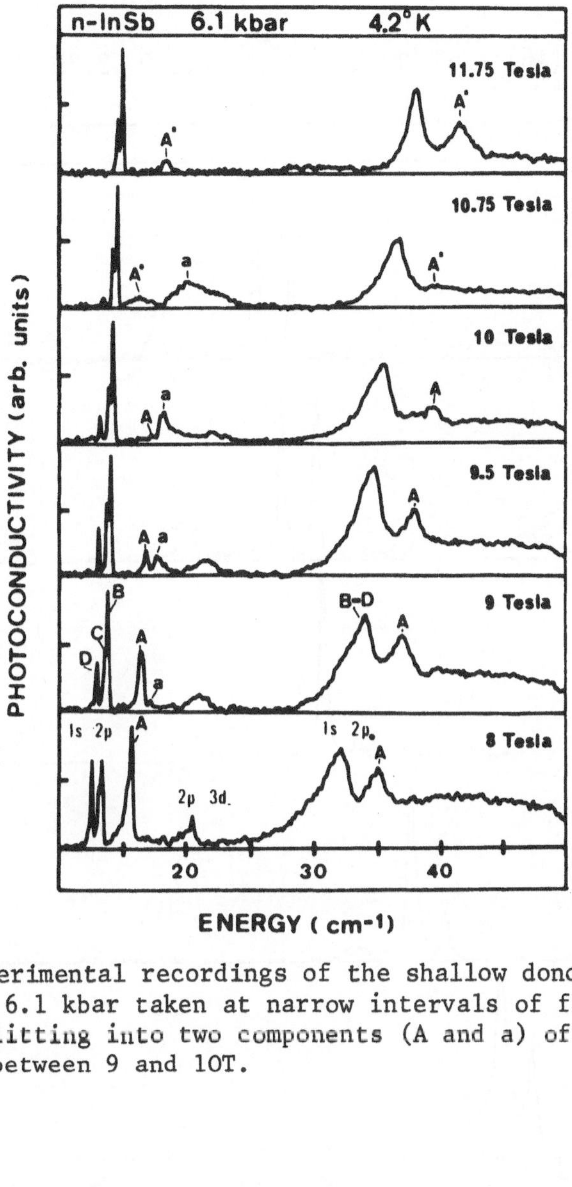


Fig.24. Experimental recordings of the shallow donor spectra in n-InSb at 6.1 kbar taken at narrow intervals of field showing an abrupt splitting into two components (A and a) of the 1s-2p₋ line between 9 and 10T.

provided that sufficient electrons are left in either the L and Γ-like states of donor A or in the other donors (B-D). Figure 25 shows the $1s$-$2p_+$ or impurity-shifted cyclotron resonance (ICR) and the cyclotron resonance (CR) lines at a pressure of 7.63 kbar where the X-level is actually below the L-level as may be seen from figure 15. As the magnetic field is increased from 1.0T to 2.4T the L-component of the ICR line for donor A rapidly separates from the component which is unresolved from the ICR lines from the

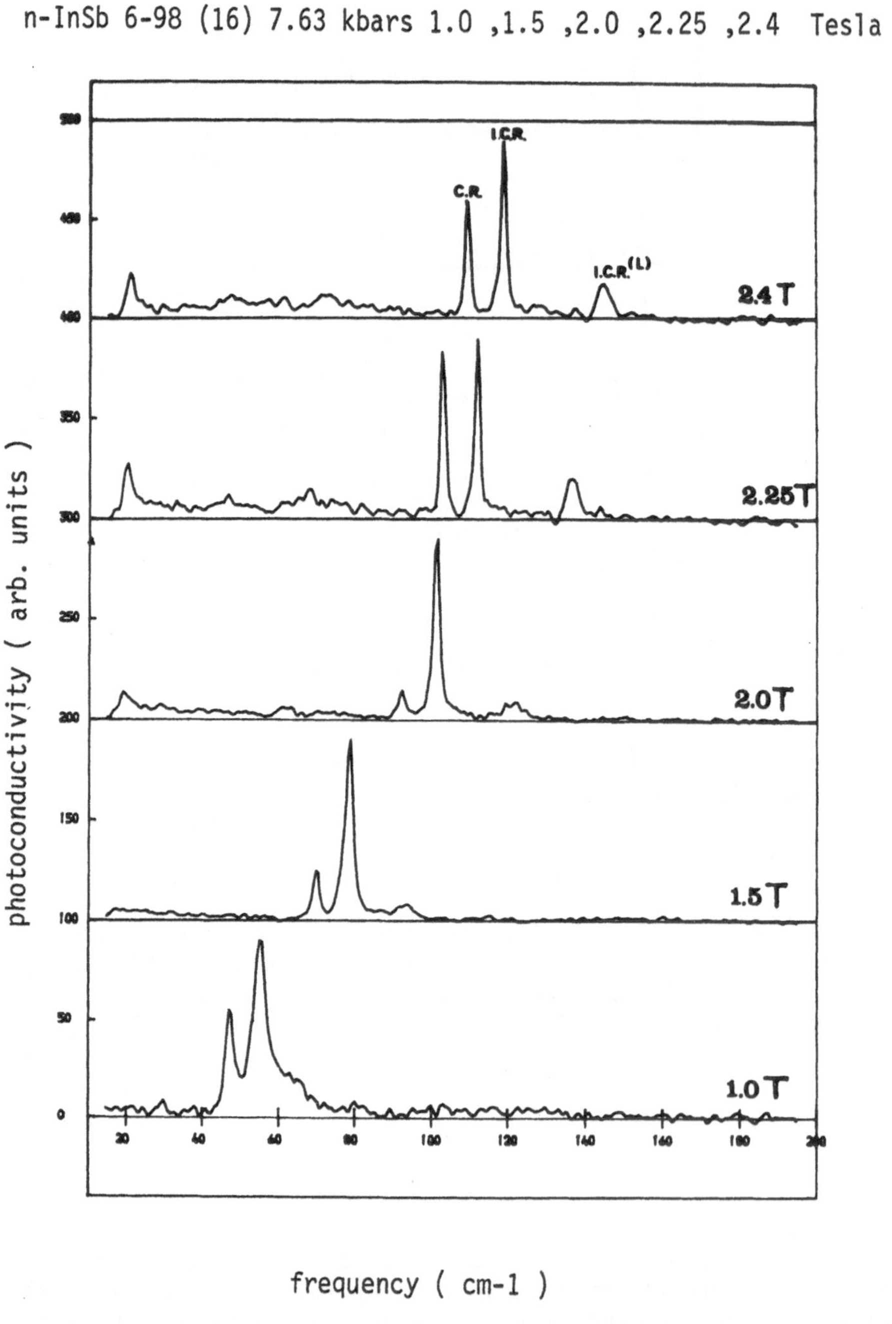


Fig.25. Experimental recordings of the cyclotron resonance (CR) and the impurity-shifted cyclotron resonance or $1s$-$2p_+$ transition (ICR) at a pressure of 7.63 kbar. The L-component of donor A is seen to move rapidly away on increasing the field.

other donors (B-D). As a consequence of the energy barrier due to the large lattice relaxation around the impurity, the population of the X-state is determined by the thermal history of the sample. Thus, if the state is populated at higher temperatures (∿>100K) by choosing a certain combination of pressure and magnetic field, the population at 4.2K can remain practically constant even if there are subsequent changes in the Fermi level with respect to the bottom of the Γ-band. In a magneto-optical experiment the effective number of A-centres

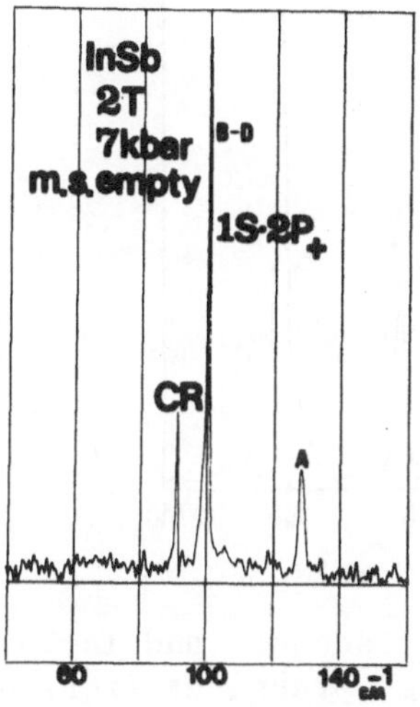


Fig.26(a). Experimental recording of the cyclotron resonance and impurity-shifted cyclotron resonance lines taken when the thermal cycling has been performed to keep the metastable X-state unoccupied. The L-associated component of donor A is clearly visible about 30cm$^{-1}$ above the main ICR line.

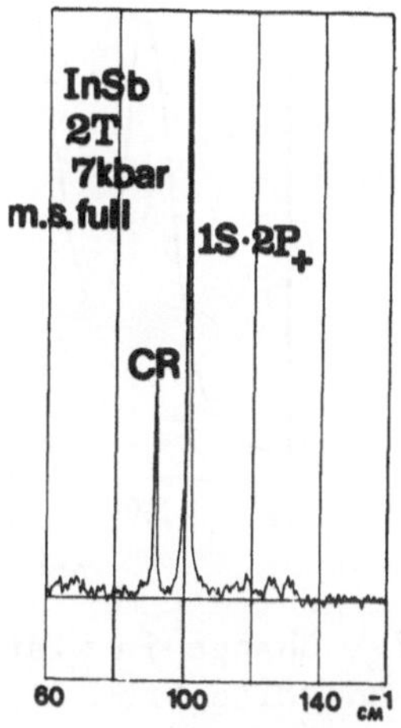


Fig.26(b). The experimental conditions are similar to 26(a) except that the sample has been cooled in such a way that the metastable X-states are almost completely occupied. The L-associated component of donor A has disappeared and the height of the main 1s-2p$_+$ line has decreased with respect to the cyclotron resonance.

appearing in the shallow donor spectrum will be the total number of A centres less those which have electrons occupying the X-states. Figure 26 demonstrates the changes in the 1s-2p$_+$ spectrum which can be brought about by inducing occupancy of the X-states. The pressure and magnetic field have been chosen so that the L-associated states of donor A are about 30 cm$^{-1}$ below the 1s(Γ) levels giving rise to the impurity shifted cyclotron resonance line (1s-2p$_+$) from these states, which can be seen at 130 cm$^{-1}$ in the first recording taken with the X-states unoccupied. As can be seen from the second recording the effect of populating the X-states is to bleach out the L-associated lines entirely and there is also a noticeable decrease in intensity of main ICR line due to the loss of 1s(Γ) component of A.

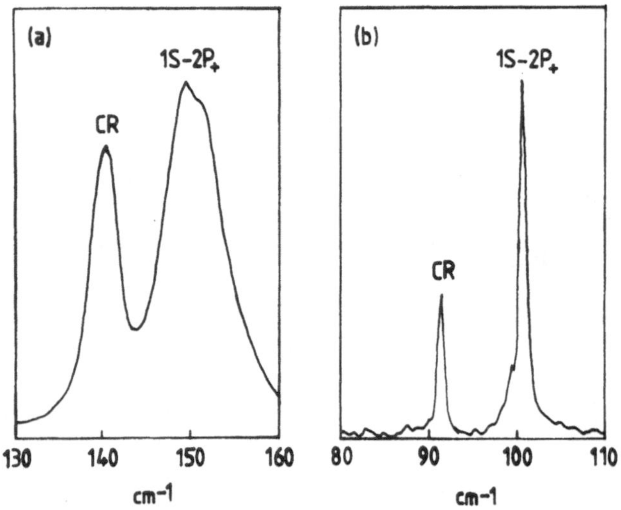


Fig.27. Change in widths of cyclotron resonance and 1s-2p$_+$ lines on going from low pressures to high pressures.(ref.56).

## 4.7 The variation of line width with pressure

One of the intriguing features of the present experiments has been the dramatic narrowing of both the impurity lines and the cyclotron resonance on applying the pressure. The narrowing for the impurity lines was apparent in figure 21 and that of the cyclotron resonance may be seen from figure 27. A possible explanation for this narrowing lies in the correlation effects previously put forward as a mechanism for explaining an unexpected temperature dependence of impurity lines found at very low temperatures.(62) The idea is associated with the tendency for the positively charged ionised donors to be found close to the negatively charged acceptors in compensated material. The resultant dipolar fields have a smaller perturbing effects on the impurities or cyclotron motion than the corresponding number of randomly-distributed monopole fields. However, as the temperature is raised, the electrons on the neutral donors can be excited across onto the previously ionised sites with a resultant increase in broadening. The narrowing found on increasing the pressure can therefore be explained if the energy differences between donor sites located in different regions is increased.

An analysis of the pressure dependence of the observed positions of the cyclotron resonance and impurity lines has been given in a previous publication.(56)

## 5. Conclusion

With all the experiments discussed in this paper InSb has proved to be remarkable in the variety and richness of the results obtained This versatility is a result of the very low effective mass of the conduction band which arises from the narrow direct gap and from the lack of contamination introduced by the growth processes. The low mass enables various resonant polaronic interactions to be observed with ease. Some of the results have been extremely unexpected such as the study of the most shallow of the donor states yet detected for any material which determine the exact location and nature of states which are lying very deeply with respect to the bands that provide their dominant character. The admixture of different bands gives rise to the possibility of new polaronic effects.

## 6. Acknowledgements

The author would like to acknowledge the contributions of his colleagues and collaborators to the investigations described in this paper. In particular the work described in section 4 is the result of a very effective collaboration between the author's group in St Andrews where the experiments involved M Davidson and the group at the High Pressure Institute 'Unipress' of the Polish Academy of Sciences headed by S Porowski. The colloboration has been catalysed by an extended visit from Warsaw to St Andrews by Z Wasilewski supported by a Visiting Fellowship provided by the Science and Engineering Research Council.

REFERENCES


1. P.G. Harper, J.W. Hodby and R.A. Stradling, Reports on Progress in Physics 36, 1 (1973).
2. Y.B. Levinson and E.I. Rashba, Reports on Progress in Physics 36, 1499 (1973).
3. F. Kuchar, E. Fantner and G. Bauer, Phys. Stat. Sol. (a) 24, 513 (1974).
4. J.L. Robert, A. Raymond, R.L. Aulombard and C. Bousquet, Phil. Mag. B42, 1003 (1980).
5. J. Durkham, R.J. Elliott and N.H. March, Rev. Mod. Phys. 40, 812 (1968).
6. H.J. Stocker, H. Levinstein and C.R. Stannard, Phys. Rev. 150, 613 (1966).
7. H.J. Stocker and H. Kaplan, Phys. Rev. 150, 619 (1966).
8. H.J. Stocker, Phys. Rev. Lett. 18, 1197 (1967).
9. S. Komiyama, Proceedings of the NATO Advanced Study Institute on "Physics of Polarons and Excitons in Polar Semiconductors and Ionic Crystals" (1984).
10. A. Onton, Proceedings of the 3rd International Conference on Photoconductivity (Stanford) (1969), p. 329; Phys. Rev. Lett. 22, 288 (1969).
11. E.A. Mazur, Sov. Phys. Semicond. 13, 147 (1979).
12. B.C. Cavenett, Phys. Rev. B5, 3049 (1972).
13. J.R. Hardy, S.D. Smith and W. Taylor, Proceedings of the 6th International Conference on Physics of Semiconductors (Exeter) (1969), p. 521.
14. A.L. Mears, A.R.L. Spray and R.A. Stradling, J. Phys. C1, 1412 (1968).
15. R.J. Nicholas, A.C. Carter, S. Fung, R.A. Stradling, J.C. Portal and C. Houlbert, J. Phys. C13, 5215 (1980).
16. S.M. Puri and T.H. Geballe, Bull. Am. Phys. Soc. 8, 309 (1963); Semicond. and Metals 1, 203 (1966).
17. Y.A. Firsov, V.L. Gurevich, R.V. Parfenev and S.S. Shalyt, Phys. Rev. Lett. 12, 660 (1964).
18. M. Nakayama, J. Phys. Soc. Japan 27, 636 (1969).
19. A.L. Mears, R.A. Stradling and E.K. Inall, J. Phys. C1, 821 (1968).
20. R.J. Nicholas, R.A. Stradling, J.C. Portal and S. Askenazy, J. Phys. C12, 1653 (1979).
21. L. Eaves, R.A. Hoult, R.A. Stradling, R.J. Tidey, J.C. Portal and S. Askenazy, J. Phys. C8, 1034 (1975).
22. D.C. Tsui, T. Englert, A.Y. Cho and A.C. Gossard, Phys. Rev. Lett. 44, 341 (1980).
23. A.S. Barker, J.L. Merz and A.C. Gossard, Phys. Rev. B17, 3181 (1978).
24. S. Komiyama, H. Eyferth and J.P. Kotthaus, Proceedings of the 15th International Conference on the Physics of Semiconductors (Kyoto) (1980), p. 687.

25. E. Yamada and T. Kurosawa, Proceedings of the 9th International Conference on the Physics of Semiconductors (Moscow) (1968), p. 805.
26. R.A. Stradling and R.A. Wood, Solid State Commun. 5, 701 (1968).
27. R.J. Nicholas and R.A. Stradling, J. Phys. C9, 1253 (1976).
28. J.C. Portal, P. Perrier, C. Houlbert, S. Askenazy, R.J. Nicholas and R.A. Stradling, J. Phys. C12, 5121 (1979).
29. R.J. Nicholas and R.A. Stradling, J. Phys. C11, L783 (1978).
30. K. von Klitzing, Solid State Electr. 21, 223 (1978).
31. V.F. Gantmakher and V.N. Zverev, Sov. Phys. JETP 42, 352 (1975); 43, 985 (1976); 44, 1220 (1976); 46, 1223 (1977).
32. T. Instone, L. Eaves, J.C. Portal, C. Houlbert, P. Perrier and S. Askenazy, J. Phys. C10, 1585 (1977).
33. L. Eaves and J.C. Portal, J. Phys. C12, 2809 (1979).
34. R.J. Nicholas and J.C. Portal, Proceedings of the NATO Advanced Study Institute on "Physics of Nonlinear Transport in Semiconductors", ASI series, vol. B52 (1979), p. 255.
35. E.J. Johnson and D.M. Larsen, Phys. Rev. Lett. 16, 655 (1966).
36. F.G. Bass and I.B. Levinson, Sov. Phys. JETP 22, 635 (1966).
37. R.C. Enck, A.S. Saleh and H.Y. Fan, Phys. Rev. 182, 790 (1969).
38. J.T. Devreese, J. De Sitter, E.J. Johnson and K.L. Ngai, Phys. Rev. B17, 3207 (1978).
39. R. Kaplan, Phys. Rev. 181, 1154 (1969).
40. R. Kaplan and R.F. Wallis, Phys. Rev. Lett. 20, 1499 (1968).
41. Y.I. Yafet, R.W. Keyes and E.N. Adams, J. Phys. Chem. Solids 1, 137 (1956).
42. R. Kaplan, R.A. Cooke and R.A. Stradling, Solid State Commun. 26, 741 (1978).
43. B.D. McCombe and R.J. Wagner, Adv. in Electrons and Electron Physics 37, 1 (1975); 38, 1 (1975).
44. C.R. Pidgeon, Handbook on Semicond. 2, 223 (1980).
45. M. Horst, U. Merkt and J.P. Kotthaus, Proceedings of the 4th International Conference on Electronic Properties of 2-D Systems (1981), p. 448.
46. L.Y. Ching, E. Burstein, S. Buchner and H.H. Wieder, Proceedings of the 15th International Conference on the Physics of Semiconductors (1980), p. 951.
47. E. Litwin-Staszewska, S. Porowski and A.S. Filipchenko, Phys. Stat. Sol. (b) 48, 525 (1971).
48. M. Konczykowski, S. Porowski and J. Chroboczek, Proceedings of the 11th International Conference on the Physics of Semiconductors (Warsaw) (1972), p. 1050.
49. E. Litwin-Staszewska, A. Jedrzejczak, S. Porowski and A.S. Filipchenko, Proceedings of the 11th International Conference on the Physics of Semiconductors (Warsaw) (1972), p. 952.
50. L. Konczewicz, E. Litwin-Staszewska and S. Porowski, Proceedings of the 3rd International Conference on Narrow Gap Semiconductors (Warsaw) (1977), p. 211.

51. S. Porowski, L. Konczewicz, M. Konczykowski, R. Aulombard and J.L. Robert, Proceedings of the 15th International Conference on the Physics of Semiconductors (Kyoto) (1980), p. 271.
52. L. Dmowski, M. Konczewicz, R. Piotrzkowski and S. Porowski, Phys. Stat. Sol. (b) 73, 1131 (1976).
53. M. Altarelli and G. Iadonisi, Il Nuovo Cimento 5, 21 (1971).
54. S. Porowski, Proceedings of the 4th International Conference on Narrow Gap Semiconductors (Linz) (1981), p. 420.
55. A.M. Davidson, P. Knowles, P. Makado, R.A. Stradling, S. Porowski and Z. Wasilewski, Solid State Sciences 24, 84 (Springer Verlag, 1981).
56. Z. Wasilewski, A.M. Davidson, P. Knowles, S. Porowski and R.A. Stradling, Proceedings of the 4th International Conference on Narrow Gap Semiconductors (Linz) (1981), p. 183.
57. Z. Wasilewski, A.M. Davidson, R.A. Stradling and S. Porowski, Proceedings of the 16th International Conference on the Physics of Semiconductors (Montpellier) (1982).
58. R. Kaplan, R.A. Cooke, R.A. Stradling and F. Kuchar, Application of High Magnetic Fields in Semiconductor Physics (Oxford), 397 (1978).
59. W. Trzeciakowski and J. Krupski, to be published.
60. D.L. Price, J.M. Rowe and R.M. Nicklow, Phys. Rev. B3, 1268 (1971).
61. R.A. Stradling and R.A. Wood, J. Phys. C3, 2425 (1970).
62. J. Golka, J. Trylski, M.S. Skolnick, R.A. Stradling and Y. Couder, Solid State Commun. 22, 623 (1977).

EXPERIMENTAL STUDY OF HOT-ELECTRONS IN SILVER HALIDES AT CROSSED ELECTRIC AND MAGNETIC FIELDS

Susumu Komiyama

Department of Pure and Applied Sciences
University of Tokyo, Komaba
Meguro-ku, Tokyo / 153 Japan

## 1. INTRODUCTION

In a majority of hot-electron investigations, the electron distribution function in momentum space is assumed to be nearly isotropic.[1] This assumption is justified if the dominating scattering mechanism is of quasielastic character, as in the case of impurity or acoustical phonon scattering. Furthermore, if the electron-electron scattering is significant, a Maxwellian type distribution function characterized by the electron temperature can be assumed.

The hot-electron phenomena observed in silver halides[2-8] are largely different from those phenomena describable in terms of such conventional hot-electron concepts. In ionic crystals such as silver halides, electrons interact strongly with (longitudinal) optical phonons and the emission of the optical phonons by electrons is important in high electric fields. In comparison to the strong influence of the optical phonon emissions, all other scattering mechanisms such as the electron-electron, the impurity and the acoustical phonon scatterings, which usually play a substantial role in familiar hot-electron phenomena, are insignificant; the electron-electron scattering can be completely neglected in ionic crystals because the density of photoexcited electrons is generally extremely low; the impurity and the acoustical phonon scatterings are not very important in pure crystals at low temperatures. In this situation, successive emissions of the optical phonon by electrons are possible and streaming motion of electrons[9] occurs in strong electric fields to yield a highly anisotropic distribution function of electrons in momentum space. A more remarkable electron distribution,[10] including population inversion, is even realized as a consequence of accumulation of electrons into a limited area in momentum space, when magnetic fields of appropriate strength are applied perpendicularly to the electric fields.

In a number of non-ionic crystals also, the interaction between carriers and optical phonons is reasonably strong due to the non-polar optical deformation scattering, and similar phenomena are expected when the crystals are relatively pure and the lattice temperature is low. Actually, these phenomena have been observed not only in silver halides and other ionic crystals,[11] but also in non-polar semiconductors[12-14] such as p-Si and p-Ge. Particularly, in p-Ge, far-infrared emission from accumulated carriers has been observed[15] and the far-infrared amplification[16] seems to have become a realistic subject of experiments. Theories predict further fascinating effects, such as cyclotron resonance amplification,[17-20] superradiance[21] and various types of negative differential conductivities in dc fields.[10,21-24] A brief review of the theoretical aspects of this type of hot-electron phenomena is given in Ref.21. A thorough review of the experimental aspects of the phenomena in both ionic and semiconductor materials is found in a recent paper,[25] where theoretical achievements and predictions are also surveyed.

The intention of these lectures is not to give a general review of the subjects but to present a definite and consistent picture of the phenomena in silver halides through a complete description of the experimental results. This may be worth while because the phenomena in silver halides are supposed to serve as a prototype of similar phenomena in other materials. For this purpose, the high-field galvanomagnetic measurements[2-5] and cyclotron resonance experiments[6-8] on photocarriers in silver halides at helium temperatures are described. The experimental results are compared with theoretical analysis and interpreted consistently. A detailed comparison of the data to Monte Carlo simulation by Kurosawa[21,26,27] is also made to obtain deeper understanding of the phenomena. A possible influence of streaming motion on the polaron self-energy effect is suggested on the basis of the experimental results and the comparison between the experiment and the calculation.

After some experimental notes on measurement techniques in §2, fundamental properties of silver halides is briefly surveyed in §3. Section 4 begins with theoretical analysis of the electron motion in crossed electric and magnetic fields (§4.A). Experimental results are then compared with the calculation (§4.B,C). Peculiar behaviour of trapping lifetime is pointed out in §4.D. Section 5 deals with cyclotron resonance phenomena in intense microwave fields. After giving a semi-qualitative analysis of the expected phenomena (§5.A), experimental results are presented and interpreted (§5.B). In §5.C, interesting aspects of the electron distribution characteristic of the case of intense microwave fields are pointed out on the basis of theoretical work by Kurosawa. Polaron problem in high electric fields is discussed in §6. Related experiments in other materials are briefly mentioned and a short glance is given to theoretical predictions in §7. Appendix provides alternative description of the phenomena in dc fields in terms of quantum mechanics.

## 2. EXPERIMENTAL TECHNIQUES

### A. Galvanomagnetic Measurements

Ionic crystals such as silver halides are excellent insulators at helium temperatures with dark conductivities lower than $10^{-20}(\Omega cm)^{-1}$. The highly insulating charactor is maintained even in high electric fields up to at least $10^4$ V/cm. To study electronic transport properties, carriers are photoexcited by a band gap light. The lifetime of carriers limited by the capture at shallow trap centers is usually short (of order $10^{-11}$psec), and the average distance of the carrier-drift is of the order of $10^{-4}\sim10^{-3}$ cm, which is usually negligibly small in comparison to the size of specimens. Because the trapping lifetime is so short, the density n of photocarriers induced by a photoexcitation of typical intensities is as low as $n\sim10^8/cm^3$. Resultant photoconductivity is therefore low, e.g. of the order of $10^{-8}(\Omega cm)^{-1}$. Because of the low conductivity, together with the difficulty in preparing reliable contacts on these materials, the principle and the method for photocurrent measurements are largely different from those applied to familiar semiconducting materials. In the familiar semiconductors, it is possible to produce ohmic contacts to the material. Electrical currents propagating in the material are thereby able to pass directly into wires leading to current sources. The experimental technique will then be to transmit a known current through the material and to measure the electric-field vector induced in the specimen.

For ionic crystals, however, no workers have succeeded in producing contacts through which electrical currents are properly transmitted at low temperatures, and the application of familiar techniques of semiconductor measurements to the ionic crystals has to be resigned. The low conductivity of the ionic materials permits application of blocking electrodes to these materials.[28] The space charge, which is induced inside the specimen when the electrodes are blocking, does not develop rapidly because of the low conductivity of these materials. Therefore, if pulsed measurements are carried out by use of a short pulse of light to generate carriers, the development of internal polarization fields inside the specimen can be neglected even if electrodes are blocking. In the operation of blocking electrodes, the specimen is regarded as the dielectric in a condenser with the electrodes serving as condenser plates, and carrier motion is detected by charges induced on the electrodes. The experimental technique is then to apply a known electric field to the specimen and to measure the photocurrent vector in the specimen.

Redfield arrangement of blocking electrodes[29] has been widely used for the Hall effect measurements on ionic crystals. The method has contributed greatly to determination of fundamental quantities of the transport phenomena in insulating materials. The Redfield arrangement is, however, not suitable for the measurements at high electric fields. This is because in this method currents are transmitted through one of the electrodes to produce potential gradients in the specimen and the Joule heating of the electrode strictly limits the applicable voltage. The electric field attainable with this arrangement is usually below 100 V/cm at helium temperature.[30] A significant improvement of the arrangement has been introduced by Iye and Kajita[11] to extend the applicable electric fields. With the improved arrangement as shown in Fig. 1a), they could measure Hall effects in pure CdS at 4.2K up to about 3kV/cm. Either of this arrangement or another closely resembling the first one as shown in Fig. 1b) is employed in the experiments[5] described in these lectures. The procedure of measurements will be briefly described in the following.

The first arrangement of electrodes (Fig. 1a) is so designed

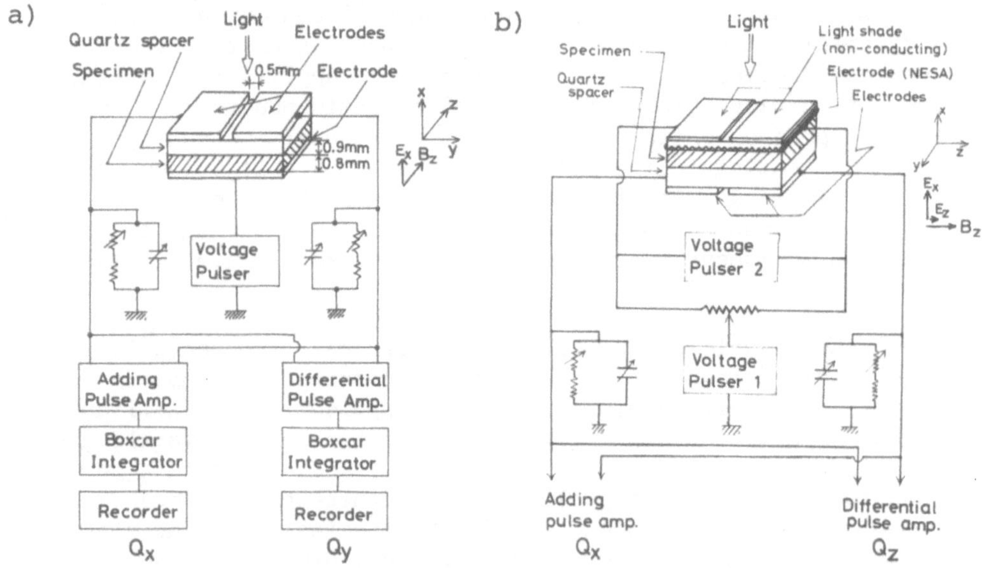


Fig. 1. Experimental arrangements with blocking electrodes for galvanomagnetic measurements. The photocurrent components, $Q_x$ and $Q_y$, are detected in $\vec{E}=(E_x,0,0)$ and $\vec{B}=(0,0,B_z)$ with the arrangement (a), while (b) is for the measurements of $Q_x$ and $Q_z$ in $\vec{E}=(E_x,0,E_z)$ and $\vec{B}=(0,0,B_z)$.

that the dissipative photocurrent $Q_x$ and the Hall photocurrent $Q_y$ are separately detected in the presence of electric field $\vec{E}=(E_x,0,0)$ and magnetic field $\vec{B}=(0,0,B_z)$. The specimens and a quartz slab are sandwiched together between a lower metal-plate electrode and a pair of upper metal-plate electrodes. The specimen is insulated from the upper electrodes by the quartz slab and from the lower electrode by a thin Mylar foil undescribed in the figure. The upper two electrodes are grounded through adequately high resistances, and a voltage pulse is applied to the lower electrodes so that the electric field develops along the x-direction in the specimen. The y-compoennt of the electric field induced by the presence of the gap between the upper electrodes is negligibly small in the specimen because of the spacing of the specimen from the electrodes. This configuration of electrodes includes no electrical parts where current is transmitted and applicable voltages are almost unlimited. In actual measurements the dielectric breakdown of liquid helium environing the electrodes limits the attainable electric field to 30 kV/cm. Carriers are excited in the specimen by a short pulse of light generated either by a xenon flash tube (with duration $\sim 1\mu$ sec). The photocarriers drift under the influence of the electric field until they are captured. Charges proportional to the integrated photocurrent are collected at the upper electrodes, and are measured as a change in potential at the electrodes. If magnetic field is absent, the photocurrent is in the x-direction and the amount of charge collected at either of the electrodes is identical. In the presence of magnetic field $B_z$, the direction of carrier-drift is deflected from the x-direction to yield a component in the y-direction and causes a difference between the amounts of charge at the two electrodes. It can be shown[31] that the sum and the difference of the charges collected at the electrodes are proportional to the components $Q_x$ and $Q_y$, respectively, of the photocurrent vector $(Q_x, Q_y, 0)$:

$$Q_{x,y} = N\tau_t e v_{x,y}, \quad (1)$$

where N is the total number of carriers released by a pulse of light, $\tau_t$ the trapping lifetime, e the unit charge and $v_i$ the drift velocity of carriers in the $i$-direction. The Hall angle $\theta$, the angle between the $\vec{E}$ vector and the $\vec{Q}$ vector, is obtained from the relation

$$\tan\theta = Q_y/Q_x . \quad (2)$$

The second arrangement of electrodes (Fig. 1b) is designed to detect a current $Q_z$ in electric fields $\vec{E}=(E_x, 0, E_z)$ and magnetic fields $\vec{B}=(0, 0, B_z)$. The electrodes consist of a resistance-film plate (NESA coated quartz plate) and a pair of lower metal plates. The specimen is insulated from the resistance-film electrode by a Mylar foil undescribed in the figure. The point of this arrangement is that a current is transmitted in the resistance-film electrode along the z-direction to produce weak electric fields $E_z$ inside the specimen. With this arrangement the difference in the amount of charges collected at the two metal electrodes is proportional to the current compoennt parallel to the magnetic field,

$$Q_z = N\tau_t e v_z . \tag{3}$$

In both the types of electrode arrangements, the strength of electric field $E_x$ is calculated from the applied voltage by taking into account the dielectric constants of the specimen, the quartz plate and the Mylar foil. The measurement is repeated with a time interval ∿600 msec. The sequence in time of pulse operation is shown in Fig. 2. The polarity of the voltages applied to the specimen alternates sequencially to prevent accumulation of space

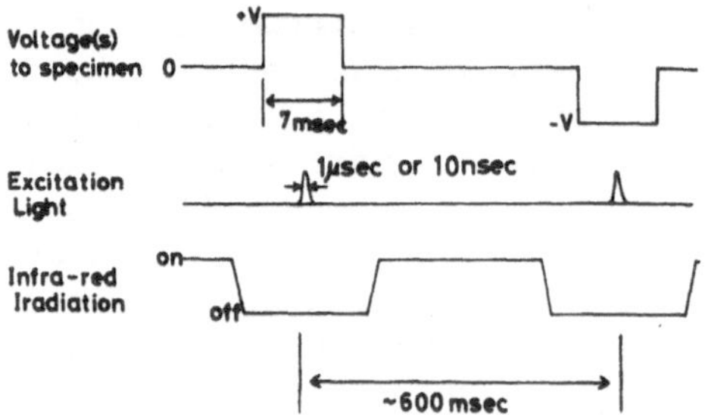


Fig. 2. Sequence in time of the pulse operation for the photocurrent measurements.

charges inside the specimen. Infrared radiation is shed to the specimen to sweep captured carriers out of the trap states and to refresh the specimen after each measurement.

B. Cyclotron Resonance Experiments

The mobility of photoelectrons in the purest crystal used of AgBr at 4.2 K reaches as high a value as $3.5\times10^5 cm^2/Vsec$. With this value of mobility, a sharp absorption line of cyclotron resonance can be obtained in the mm-wave range. Cyclotron resonance experiments on AgBr crystals in linearly polarized strong microwave fields at 35GHz are described in these lectures.

Figure 3 shows a schematic diagram of the experimental apparatus. A standard reflection type spectrometer with a cylindrical sample cavity of the $TE_{112}$ mode is used. A pulsed magnetron with maximum power of about 15 k watt and pulse duration about 200 nsec serves as a strong microwave source. A specimen of typical size of 1×1×1 $mm^3$ is glued onto the end of a thin light guide made of quartz and is held at an appropriate position in the cavity. The microwave absorption induced by the cyclotron resonance of photocarriers is usually less than 0.5%, while the pulsed power from the magnetron fluctuates to ∿7% rms from pulse to pulse. To detect

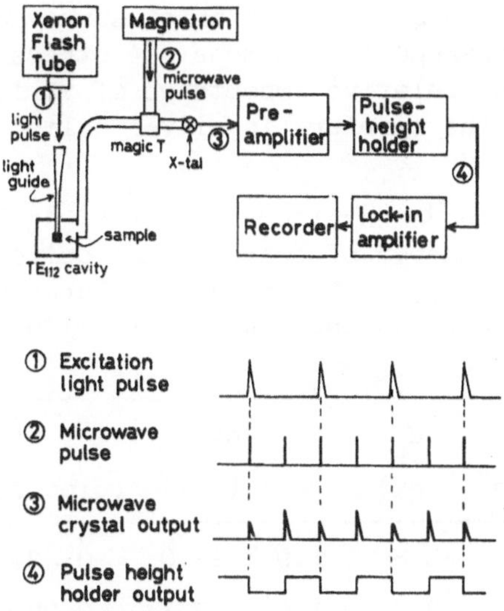


Fig. 3. An experimental setup and the pulse operation for cyclotron resonance experiments at 35GHz.

the small signal compoennt out of the larger fluctuation, a modulation technique is employed. While microwave pulses arrive upon the sample cavity with a repetition rate of about 260 Hz the specimen is alternately illuminated by a zenon flash tube, so that incident pulses of the microwave power suffer absorption alternately (Fig.3). Reflected microwave pulses are detected and the signal component is amplified by a Lock-in amplifier after transformed into a rectangular wave by use of a pulse-height holder. A boxcar integrator serves as the pulse-height holder with averaging time-constant being set to a level much shorter than the aperture duration.

The peak value $E_\omega$ of the microwave field in the specimen, $E_\omega \cos\omega t$, is determined by the following two methods. First, $E_\omega$ is calculated from the value of incident microwave power by taking into account the coupling factor of the cavity, the Q-factor, and the polarization constant of the specimen. Second, $E_\omega$ is derived from the comparison of the resonance lines with those obtained in a transmission-type simple wave-guide system. The values of $E_\omega$ obtained in the two methods agree within an error of ±5%. The absolute accuracy of the $E_\omega$-values is supposed to be ±20%. The relative accuracy at different attenuation levels is $\sim$±2%, being determined by the accuracy of calibrated attenuators in the spectrometer.

## 3. FUNDAMENTAL PROPERTIES OF SILVER HALIDES

Several good reviews[32] are available for the study of fundamental properties of I-VII ionic crystals including silver halides.

Here we will restrict ourselves to a brief survey of the properties significant for the transport phenomena. A stress will be put on the elucidation of the relative importance of various scattering mechanisms acting on the carriers in silver halides.

Table 1. Summary of fundamental properties of AgCl and AgBr. $\hbar\omega_{op}$ denotes the energy of longitudinal optical phonons at 4.2K. $m^*$ or $m_t^*$, $m_l^*$ is the cold polaron mass determined by cyclotron resonance experiments with the free electron mass $m_e$.

| | Band edge Cond. | Valence | $E_g$ (eV) | $\alpha$ | $\hbar\omega_{op}$ (meV) | $m^*$ (electrons) | $m_t^*$ (holes) | $m_l^*$ |
|---|---|---|---|---|---|---|---|---|
| AgCl | Γ | L | 3.3 | 1.9 | 23.0 | $0.43m_e$ | | |
| AgBr | Γ | L | 2.7 | 1.6 | 17.1 | $0.29m_e$ | $1.73m_e$ | $0.79m_e$ |

Several important quantities are summarized in Table 1. The band structures of AgCl and AgBr are similar with each other. The lowest conduction band is of a simple standard form; namely, s-type with a nondegenerate minimum at the point Γ in the Brillouin zone. The top of the valence band is located at the point L. Carriers in either material interact strongly with longitudinal optical phonons. The Fröhlich's coupling constant $\alpha$ for electrons are entered in the table. The effective mass $m^*$ in the table denotes the "cold" polaron mass $m^*(0)$ of electrons determined by cyclotron resonance experiments. There are a number of theoretical calculations of the polaron energy spectrum.[33-35] Among them the Larsen's variational calculation[35] is supposed to provide reasonable measure of the non-parabolicity in the spectrum applicable to high energy ranges. Figure 4 shows the change of effective mass defined by $m^*(\varepsilon) \equiv \hbar^2 k(\partial\varepsilon/\partial k)^{-1}$ with energy, as evaluated from the Larsen's work by Tamura[36] for electrons in AgBr. The result will be used to estimate the average energy of electrons in the cyclotron resonance experiments (§5.B). Holes are self-trapped in AgCl,[37] but they are mobile and make a contribution smaller than that of electrons by a factor 4∼6 to the conduction in AgBr.[38] The effective mass of holes is characterized by anisotropic mass parameters of a spheroidal symmetry.

We will now describe scattering mechanisms acting on the carriers. The Hall mobility of conduction electrons increases monotonically with decreasing the lattice temperature from 300 K to 4.2 K. Values of the mobility of electrons and holes in the crystals used are listed in Table 2. The scattering of carriers at 4.2 K is dominated primarily by impurities. In general the scattering is characterized by the parameters $\tau_{imp}^0$ and p, which are

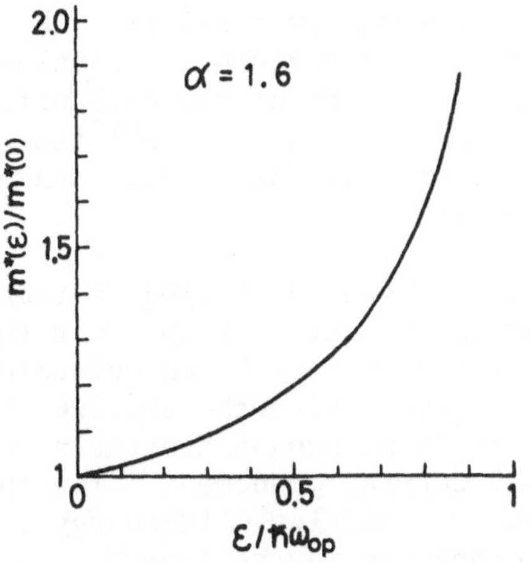


Fig. 4. Polaron mass for α=1.6 (AgBr) as a function of its energy, evaluated by Tamura[36] according to Larsen's theory.[35]

defined by the relation:

$$\tau_{imp}^{-1} = (\tau_{imp}^{0})^{-1}(\varepsilon/kt)^{-p} \quad . \tag{4}$$

Here $\tau_{imp}^{-1}$ is the inverse of the momentum relaxation time of a car-

Table 2. Mobility of electrons and holes in AgCl and AgBr crystals used at 4.2K. The values are determined by Hall effect measurements except for AgBr ZR-3 and C-157', in which the determination is made by cyclotron resonance experiments. The quantities, $\tau_{imp}^{0}$ and p, are defined by Eq. (4).

| | | Electron | | | Hole | | |
|---|---|---|---|---|---|---|---|
| Specimen | | $\mu^0$ (cm²/Vsec) | $\tau_{imp}^0$ (psec) | p | $\mu^0$ (cm²/Vsec) | $\tau_{imp}^0$ (psec) | p |
| AgCl | MO-1 | $2.8 \times 10^4$ | 6.9 | 0 | | | |
| | MG3-2 | $2.0 \times 10^4$ | 5.0 | 0 | | | |
| | M1CB | $3.8 \times 10^4$ | 9.5 | $-\frac{1}{2}$ | | | |
| | M1CD | $1.5 \times 10^4$ | 3.7 | 0 | | | |
| AgBr | ZR3-1 | $1.4 \times 10^5$ | 23 | 0 | $3.5 \times 10^4$ | 20 | 0 |
| | ZR3-2 | $1.0 \times 10^5$ | 16 | 0 | $2.4 \times 10^4$ | 13.5 | 0 |
| | C-157 | $4.1 \times 10^4$ | 6.8 | (0) | | | |
| | ZR3 | $3.5 \times 10^5$ | 56 | (0) | $1 \times 10^5$ | | |
| | C-157' | $1.4 \times 10^4$ | 23 | (0) | | | |

rier with energy ε due to impurity scattering, k is the Boltzmann constant and T the lattice temperature. The values $\tau^{0}_{imp}$ and p determined by detailed measurements of the Hall effects for the crystals,[5] are also listed in the table. In all the crystals except for a crystal designated as M1CB, the energy exponent p is zero, indicating neutral impurities.

That the acoustical phonon scattering of thermal carriers at 4.2 K is insignificant is indicated by the fact that the energy exponent p of the scattering at 4.2 K, as determined by the Hall effect measurements, is zero. Nevertheless, we wish to estimate the strength of the acoustical phonon scattering since it might become significant in hot carrier phenomena. The inverse of the momentum relaxation time of a carrier with energy ε due to acoustical phonon scattering is expressed in the form,[39]

$$\tau_{ac}^{-1} = A\,(2N_{ac} + 1)\varepsilon \;, \tag{5}$$

where A is a constant independent of ε and T, and $N_s = [\exp(\hbar c_s (2m^*\varepsilon)^{1/2}/kT)-1]^{-1}$ is the phonon number with the sound velocity $c_s$ ($2.96\times10^{5}$cm/sec for AgBr[40] ). The constant A is derived from Tamura's work on the temperature dependence of the cyclotron resonance lines[41] to be $A=4.8\times10^{23}/(\text{erg sec})$.[5] The strength of the acoustical phonon scattering is not supposed to be very different between AgCl and AgBr because of the similarity of the band structure.

The strong interaction of carriers with the longitudinal optical phonon (hereafter denoted simply as optical phonon) arises from the polar optical scattering. The scattering rate of a carrier with energy ε is calculated according to a perturbation treatment to yield the form,[42]

$$\tau_{op}^{-1} = 2\alpha\omega_{op}\left(\frac{\varepsilon}{\hbar\omega_{op}}\right)^{-1/2}\left\{N_{op}\,\mathrm{arcsinh}\left(\frac{\varepsilon}{\hbar\omega_{op}}\right)^{1/2} + (N_{op}+1)\,\mathrm{arcsinh}\left(\frac{\varepsilon}{\hbar\omega_{op}}-1\right)^{1/2}\right\} \;, \tag{6}$$

where $\omega_{op}$ is the angular frequency of the optical phonon and $N_{op} = [\exp(\hbar\omega_{op}/kT)-1]^{-1}$ is the phonon number.

We are now able to evaluate the equations (4)-(6) and compare these scattering mechanisms. Figure 5 is a result for electrons in a crystal of AgBr at T=4.2 K. We should note that the overall scattering rate is relatively low in the energy range $\varepsilon<\hbar\omega_{op}$, while it is very high in the range $\varepsilon>\hbar\omega_{op}$ due to the onset of spontaneous emission of optical phonons. The optical phonon scattering is neglected in the range $\varepsilon<\hbar\omega_{op}$, because $N_{op}$ is practically zero at 4.2 K and the phonon absorption is therefore neglected. The carrier-carrier scattering is completely neglected in the transport phenomena in ionic crystals because of the low density of photoexcited carriers; e.g., according to the Conwell-Weiskopf formula,[43] the carrier density of $1\times10^{8}$/cm yields the carrier-carrier

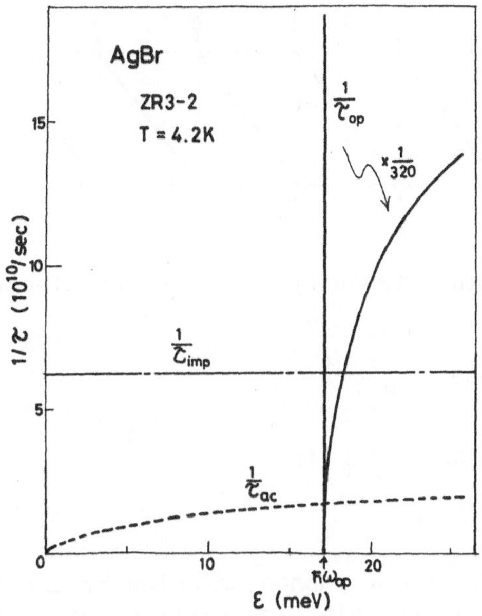


Fig. 5. Scattering probabilities versus carrier energy of an electron in a crystal of AgBr at 4.2 K.

scattering time of order 1 μsec, which is by several orders of magnitude longer than the lifetime of carriers.

## 4. DC ELECTRIC FIELDS

### A. Theoretical Background

In this section we shall describe the motion of eletrons in crossed electric and magnetic fields in the presence of strong interaction with optical phonons. Our treatments will primarily be classical. A complementary description in terms of quantum mechanics is presented in Appendix. We shall begin with the case of zero magnetic field and then introduce the magnetic field, $B_z$, perpendicular to the electric field $E_x$. In the analysis of electron motion we shall assume for electrons an effective mass m* independent of energy.

*Streaming motion.* Let us first assume for simplicity that an electron is free from any scattering mechanism except for the optical phonon emission. If an electric field $E_x$ in the x-direction is applied to the carrier at rest ($v$=0) at time t=0 (Fig.6), the electron is accelerated in the x-direction with the acceleration rate $\dot{v}=eE_x/m^*$ and $v$ increases as $v=(eE_x/m^*)t$. The kinetic energy of electrons, $\varepsilon=(1/2)m^*v^2$, reaches the optical phonon energy $\hbar\omega_{op}$ at time $T^0_{op}$:

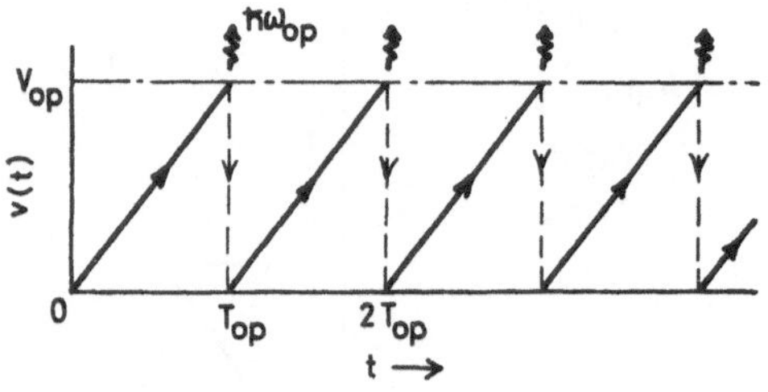


Fig. 6. Ideal streaming motion of an electron.

$$T_{op}^{0} = (2m^{*}\hbar\omega_{op})^{1/2}/(eE_x) \,. \tag{7}$$

The velocity at this moment, $V_{op}$, is

$$V_{op} = (2\hbar\omega_{op}/m^{*})^{1/2} \,. \tag{8}$$

If the probability of optical phonon emission by the electron is infinite at $\varepsilon=\hbar\omega_{op}$ ($v=V_{op}$), the electron emits an optical phonon immediately, and is scattered back to the ground state $v=0$ dissipating all of its kinetic energy. Thereafter the electron is accelerated to $\hbar\omega_{op}$ again and repeats the identical process. We will call this repeated motion the ideal streaming motion. The drift velocity of an ideally streaming electron is

$$v_d^{s} = V_{op}/2 \tag{9}$$

In actual experimental conditions, the streaming motion is possible only when an electron is able to reach $V_{op}$ within the mean free time $\tau$. Thus the condition

$$T_{op}^{0} < \tau \equiv 1/(\bar{\tau}_{imp}^{-1} + \bar{\tau}_{ac}^{-1}) \tag{10}$$

is required, where $\bar{\tau}_{imp}$ and $\bar{\tau}_{ac}$ imply the impurity and acoustical phonon scattering times averaged over $\varepsilon<\hbar\omega_{op}$. This condition sets the lower limit $E_{min}$ to the electric fields required for the streaming motion in a given crystal. On the other hand, the electron should not penetrate significantly into the higher energy range $\varepsilon>\hbar\omega_{op}$. This condition sets the upper limit to E. A rough measure of the maximum field $E_{max}$ will be obtained, for example, from the relation,

$$T_{op}^{0} > \tau_{op}(1.1\times\hbar\omega_{op}) \,, \tag{11}$$

where $\tau_{op}^{-1}(\varepsilon)$ is given by Eq.(6). Streaming motion is expected in the range $E_{min}<E<E_{max}$. The values of $E_{min}$ in the crystals of AgCl and AgBr used range from about 100V/cm to 1kV/cm according to the purity of the crystals. $E_{max}$ is 100kV/cm for AgCl and 70kV/cm for AgBr.

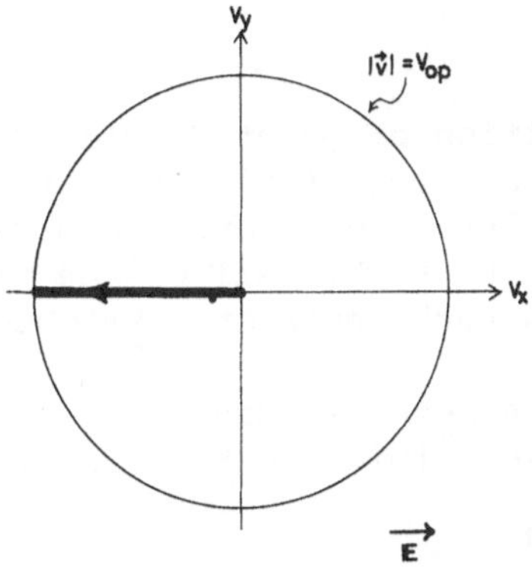


Fig. 7. Velocity distribution of electrons expected for ideal streaming motion.

The velocity distribution of ideally streaming electrons is expressed with a straight line between the points $\vec{v}=0$ and $\vec{v}=(-V_{op}, 0, 0)$, as shown in Fig.7. Each electron passes from the point $\vec{v}=0$ to the point $\vec{v}=(-V_{op}, 0, 0)$ on the trajectory at the constant speed $\dot{v}=eE_x/m^*$.

We wish now to consider the effect of magnetic field $\vec{B}=(0, 0, B_z)$ applied to the electrons ideally streaming in $\vec{E}=(E_x, 0, 0)$. This subject was first treated by Vosilyus and Levinson.[44] The trajectory for streaming motion will be curved due to the Lorentz force. To analyze the motion of electrons precisely, we begin with the classical equation of motion for a free electron;

$$m^* \frac{d}{dt} \vec{v} = -e\vec{E} - e(\vec{v} \times \vec{B}). \quad (12)$$

The equation is immediately solved to yield the following solutions under the initial condition $\vec{v}(0)=(v_x^0, v_y^0, v_z^0)$ :

$$v_x(t) = -V\sin(\omega_c t + \phi) , \quad (13)$$

$$v_y(t) = V\cos(\omega_c t + \phi) - \frac{E_x}{B_z} \quad (14)$$

and $$v_z(t) = v_z^0 , \quad (15)$$

where $\omega_c=eB_z/m^*$ with V and $\phi$ determined by the initial condition; $v_x^0=-V\sin\phi$ and $v_y^0=V\cos\phi-(E_x/B_z)$. The solutions (13) - (15) represent cyclotron oscillations of the electron around the center point at $C=(0, -E_x/B_z, v_z^0)$ on the plane $v_z=v_z^0$ in velocity space. Recalling the optical phonon emission at $\varepsilon=\hbar\omega_{op}$, we find the trajectory for streaming motion in the presence of $B_z$ to be an arc forming a part of the cycloton orbit that passes the point $\vec{v}=0$. The trajectory is therefore determined by a normalized field $\zeta$,

$$\zeta = V_{op}/(E_x/B_z), \tag{16}$$

which specifies the position of point C relative to the sphere $|\vec{v}|=V_{op}$ in velocity space Trajectories for streaming motion on the $v_z=0$ plane at different levels of $\zeta$ are schematically shown in the left half of Fig.8. In the right half of the figure the corresponding trajectories in real space are sketched.

We wish to calculate several quantities determined by the streaming motion. Putting $\vec{v}(0)=0$ in Eqs.(13)-(15), we obtain

$$v_x(t) = -(E_x/B_z)\sin\omega_c t, \tag{17}$$

and $$v_y(t) = (E_x/B_z)(\cos\omega_c t - 1). \tag{18}$$

These are the velocities of a streaming electron starting from $\vec{v}=0$ at t=0. The time $T_{op}$ at which it reaches the state $\varepsilon=\hbar\omega_{op}$ is obtained from $v_x^2(t) + v_y^2(t)= V_{op}^2$ :

$$T_{op} = T_{op}^0 \zeta^{-1} \arccos(1 - \tfrac{1}{2}\zeta^2), \tag{19}$$

where $T_{op}^0$ was defined by Eq.(7). The drift velocities of the streaming electron along the x- and the y-directions are derived by averaging $v_x(t)$ and $v_y(t)$ over time; namely, $v_{dx,dy} = (T_{op})^{-1}\int_0^{T_{op}} v_{x,y}(t)dt$. Elementary calculation yields

$$v_{dx} = -(E_x/B_z)[(1 - \cos\omega_c T_{op})/\omega_c T_{op}] \tag{20}$$

and $$v_{dy} = -(E_x/B_z)[(\omega_c T_{op} - \sin\omega_c T_{op})/\omega_c T_{op}]. \tag{21}$$

Tangent of the Hall angle, $\tan\theta_s \equiv v_{dy}/v_{dx}$, and the magnitude of the drift velocity, $v_{dxy} \equiv (v_{dx}^2 + v_{dy}^2)^{1/2}$, are derived from Eqs.(20) and (21) with Eq.(19):

$$\tan\theta_s = 2\zeta^{-2}\arccos(1 - \tfrac{1}{2}\zeta^2) - (4\zeta^{-2} - 1)^{1/2}, \tag{22}$$

and $$v_{dxy} = V_{op}[\zeta\arccos(1-\tfrac{1}{2}\zeta^2)]^{-1} \times \{\tfrac{1}{4}\zeta^4 + [\arccos(1-\tfrac{1}{2}\zeta^2) - (\zeta^2 - \tfrac{1}{4}\zeta^4)^{1/2}]^2\}^{1/2}. \tag{23}$$

Defining the Hall mobility by $\mu_H^s = \lim_{B_x\to\infty}(B_z^{-1}|v_{dy}/v_{dx}|)$, we obtain from Eqs.(20) and (21)

$$\mu_H^s = (e/m^*)(T_{op}^0/3). \tag{24}$$

Since the drift velocity $v_d^s$ of streaming electrons at $B_z=0$ is $(1/2)V_{op}$, the relation

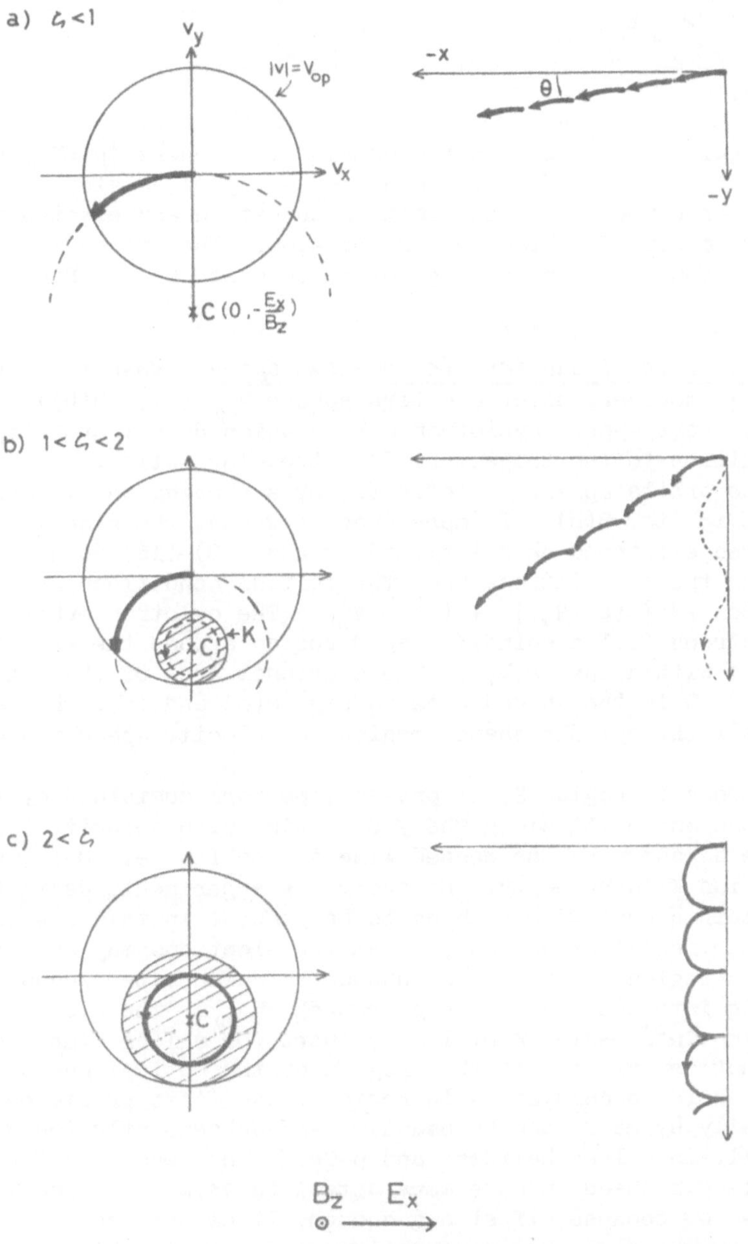


Fig. 8. Different stages of streaming motion at $\vec{B}=(0,0,B_z)$ and $\vec{E}=(E_x,0,0)$, classified by $\zeta \equiv V_{op}(E_x/B_z)$. Electron trajectories in velocity space ($v_z=0$), with the center point $C(0,-E_x/B_z)$ of cyclotron oscillation, are illustrated in the half, and the corresponding trajectories in real space are shown in the right half.

$$v_d^s = (3/2)\mu_H^s E_x \qquad (25)$$

is derived.

When $\zeta<1$, point C is located outside the sphere $|\vec{v}|=V_{op}$ (Fig. 8(a)). In this range of $\zeta$, the kinetic energy of an electron exceeds $\hbar\omega_{op}$ irrespective of the initial condition and electrons are not able to escape from the phonon emission. The unique mode of electron motion is the streaming motion as described in Fig. 8(a).

*Accumulation of carriers in momentum space.* When $\zeta$ exceeds 1 (but not 2), however, point C enters sphere $V_{op}$ (Fig. 8(b)). Accordingly, there appear cyclotron orbits which do not cross sphere $V_{op}$ in addition to the trajectory for streaming motion. The area where these orbits appear is indicated by a shading and is designated as K in Fig. 8(b). In more precise words, the kinetic energies of free electrons as determined by Eqs.(13)-(15) do not reach $\hbar\omega_{op}$ during the oscillations when the initial conditions satisfy the relation $[V + (E_x/B_z)]^2 + (v_z^0)^2 < V_{op}^2$. The cylotron orbits of these electrons fill a spindle-shaped region around the axis (0, $-E_x/B_z$, $v_z$) within sphere $V_{op}$. The sectional area of the region on the plane $v_z=0$ is the shaded area in Fig. 8(b) and (c). Hereafter we will call the spindle-shaped region in velocity space region K.

Electrons in region K, if present, perform sustained cyclotron oscillations and drift along the y-direction with velocity $E_x/B_z$ in real space as shown by the dashed line in the figure. One would expect region K to be empty. However, the experiments definitely see a certain amount of electrons to be present in this region. In actual experimental situations, streaming electrons may occasionally jump into region K via optical phonon emission after adquately penetrating into the higher energy range $\varepsilon>\hbar\omega_{op}$. The electrons once jumping into region K will stay there for a long time since they are not removed out of this region by the optical phonon emission. A carrier accumulation in region K was first predicted theoretically by Maeda and Kurosawa,[10] and subsequently found experimentally in silver halides[2] and p-Ge.[13] The mechanism for accumulation is discussed in some more detail in §4.C. The accumulation is interesting because, if strong enough, it causes population inversion in the distribution function of carriers. Figure 9 shows an example of theoretically-expected energy-distribution-function of heavy holes in Ge.[10] The dashed line in the figure roughly corresponds to the distribution function of streaming carriers. The accumulated carriers introduce the prominent peak at high energies in the distribution function.

The argument so far has been classical. According to the quantum mechanics the energy state of a free electron in the

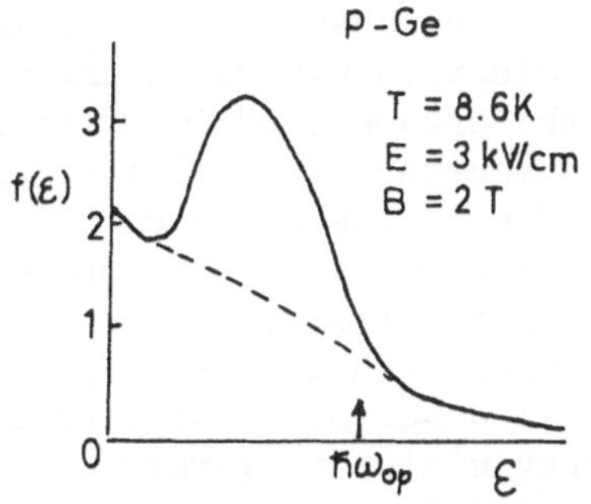


Fig. 9. An energy distribution function of heavy holes calculated by a Monte Carlo simulation.[10] E=3 kV/cm and B=2T correspond to $\zeta$=1.28. Impurity scattering is neglected in the calculation.

crossed fields is quantized into Landau states with the eigenenergy[45] of

$$\varepsilon_{n,v_z,x_0} = (n + \frac{1}{2})\hbar\omega_c + \frac{m^*}{2}v_z^2 + \frac{m^*}{2}(\frac{E_x}{B_z})^2 + eE_x x_0, \qquad (26)$$

where $x_0$ denotes the x-coordinates of the center point of the wave function. The first term corresponds to the rotational kinetic energy relevant to the cyclotron oscillation, the second term to the kinetic energy associated with the free motion in the z-direction, the third term to the translational kinetic energy due to the drift motion in the y-direction, and the last term represents the electrostatic potential energy at the location of the wave function. In the present case (Fig. 8), the quantization effect is neglected for electrons outside region K since the Landau orbits are broken by the phonon emission, but is significant in region K. The Landau states which satisfy the relation,

$$(n + \frac{1}{2})\hbar\omega_c + \frac{m^*}{2}v_z^2 < \frac{m^*}{2}(V_{op} - \frac{E_x}{B_z})^2 , \qquad (27)$$

are left unbroken and form 'Landau cylinders' within region K. Qualitative picture in quantum mechanics and the correspondence to the classical picture are described in Appendix.

*Hot-carrier regime and onset of quantum limit.* When $\zeta$ is even larger than 2, the cyclotron orbit passing the point $|\vec{v}|$=0 does not cross sphere $V_{op}$ and is closed within the sphere (Fig. 8(c)). Streaming motion accordingly becomes impossible and the optical phonon emission plays no longer the essential role in the electron kinetics. If electrons were free from any other scatterings, as has so far been assumed, the average motion of electrons would be reduced to the drift motion along the y-direction with the velocity $(E_x/B_z)$ as shown in the right half of Fig. 8(c). Thus the mode of electron motion would no longer changes with further increasing $B_z$. In reality, however, the electron system in the range $\zeta>2$ passes two different stages due to the presence of impurity scattering. (Though acoustical phonon scatterings will also have similar influence as impurity scatterings on the phenomena, their contribution is less important in the experiments.) The net effect of impurity scattering is to cause a drift of electrons in the x-direction and thus to provide the electrons with energy. Typical processes of the scattering in classical description will be the transitions of an electron to equal-energy states in larger orbits as shown in Fig. 10(a). In the language of quantum mechanics, the scatterings are the transitions of an electron to higher Landau states at a different position in space, as sketched in the right half of Fig. 10(a). By using the notations given by (A2) and (A3) in Appendix the transition is characterized as $\psi_{n,v_z,x_0} \rightarrow \psi_{n',v_z',x_0-\Delta x}$, where $\varepsilon_{n,v_z,x_0} = \varepsilon_{n',v_z',x_0-\Delta x}$ with $n'>n$ and $\Delta x < r_n + r_{n'}$. Thus in this range

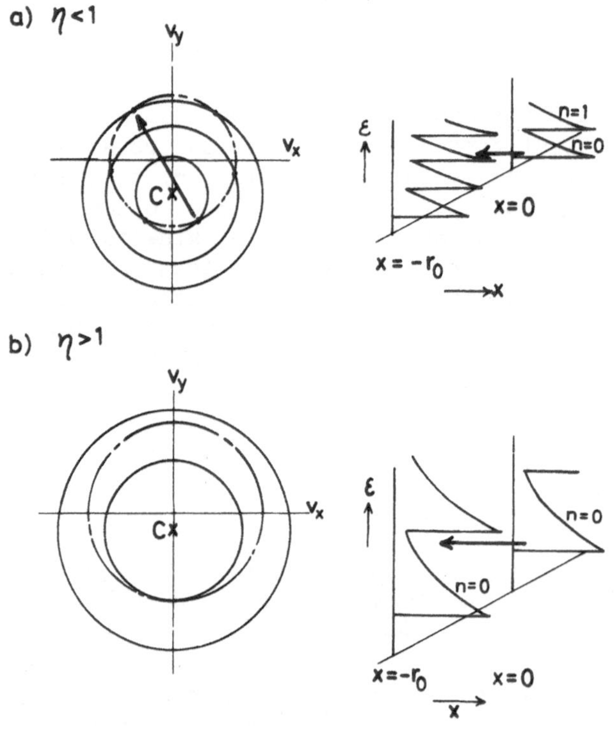


Fig. 10. Two different stages in the range $\zeta>2$. The solid circles and the dot-dash circle in the left half represent the Landau orbits and an equienergy circle in velocity space, respectively. The right half is an alternative description in terms of the density of states of the Landau states at x=0 and $x=r_0$. The slant line indicates the potential gradient $eE_x x$.

of $\zeta$ there still exists a net transfer of electrons to larger cyclotron orbits (or higher Landau states), and the electrons become hot. (Electrons reaching the optical phonon energy will dissipate the kinetic energy by the emission of optical phonons. This also happens for electrons in region K at state (b) in Fig.8. It is suggested that this process does limit the degree of accumulation in p-Ge.) The electron system is *hot* (in the range $\zeta>2$ but $\eta<1$), until the final stage stated below is reached.

When the magnetic field increases further and the relation $\eta>1$ is satisfied, where $\eta$ is defined by

$$\eta \equiv \frac{(\hbar e)^{1/2}}{m^*} \frac{B_z^{3/2}}{E_x}, \tag{28}$$

a quantum limit is achieved. As first pointed out by Kajita and Masumi[46] in experimental work on silver halides, the Landau splitting $\hbar\omega_c$ becomes larger than $eE_x r_0$ when $\eta>1$, where the radius of the lowest Landau state $r_0$ is given by Eq.(A.3). In this quantum-limit situation, transition of an electron initially at the lowest Landau state ($n=0$ and $v_z=0$) to higher Landau states is no longer possible as schematically shown in Fig. 10(b). In the experiments on silver halides, $\hbar\omega_c$ reaches e.g. 2meV at B=5T while kT∿0.36meV at 4.2K. Therefore, the heating of electrons is severely suppressed in this range; electrons at this stage are heated only via transitions to higher $v_z$-states within the lowest Landau state; the probability of such transitions is expected to be relatively low because of the low state-density of the final states as may be seen from the sketch in Fig. 10(b). We may thus classify the phenomena at this final stage as *warm-electron* regime.

We have classified the hot-electron phenomena expected in the crossed fields into four different categories according to the relative strength of $E_x$ and $B_z$. The range of $E_x$ and $B_z$ in Fig. 11 is divided into four regions accroding to the classification for electrons in AgCl. The experiments cover the whole range of $E_x$ and $B_z$, and all the four different stages of the phenomena are clearly observed. A similar classification is indicated for n-InSb in Fig. 12. The *hot-electron* experiments so far reported for n-InSb in the literature[47] are limited to ranges of $E_x$<200V/cm and $B_z$>0.2T (due to the condition $\omega_c$>1), which we classify as for *warm-electron* regime. The same is also true for the vast majority of existing hot-electron investigations on other semiconductors in the crossed fields.

## B. Experimental Results

In this section we shall compare experimental results with the above analysis. In evaluating the derived quantities we shall use the cold polaron mass m*(0) for electrons in silver halides.

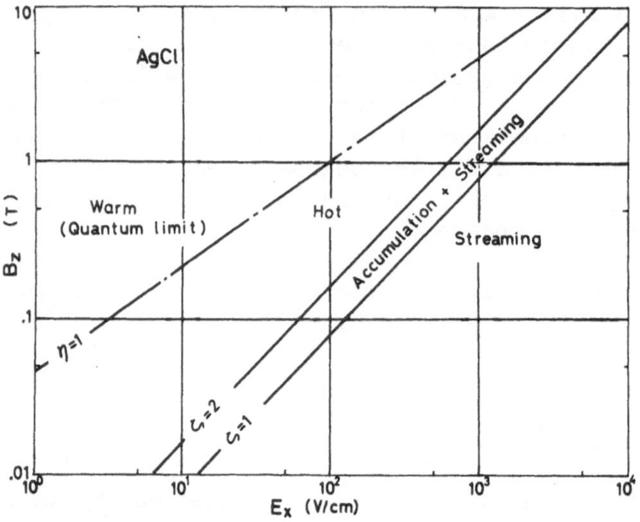


Fig. 11. Classification of hot electron situation in the crossed field condition for electrons in AgCl.

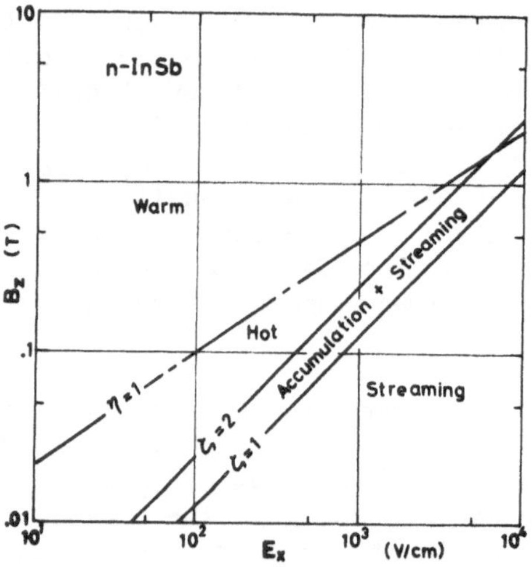


Fig. 12. Equivalent classification for n-InSb.

Streaming motion. Figure 13 shows dependence of the Hall mobility $\mu_H=(1/B_z)(Q_y/Q_x)$, measured at low $B_z$ ($\omega_c$<<1), on $E_x$ for different-purity samples of AgCl and AgBr. The solid line is drawn according to Eq.(24) for electrons, and the arrows indicate the electric fields $E_{min}$ above which streaming motion is expected to appear in respective samples, derived from the relation (10). In each sample, the data points fall close to the solid line above the respective $E_{min}$. The satisfactory agreement of the experimental values of the Hall mobility with the calculation substantially supports the picture of streaming motion in the limit of low magnetic field.

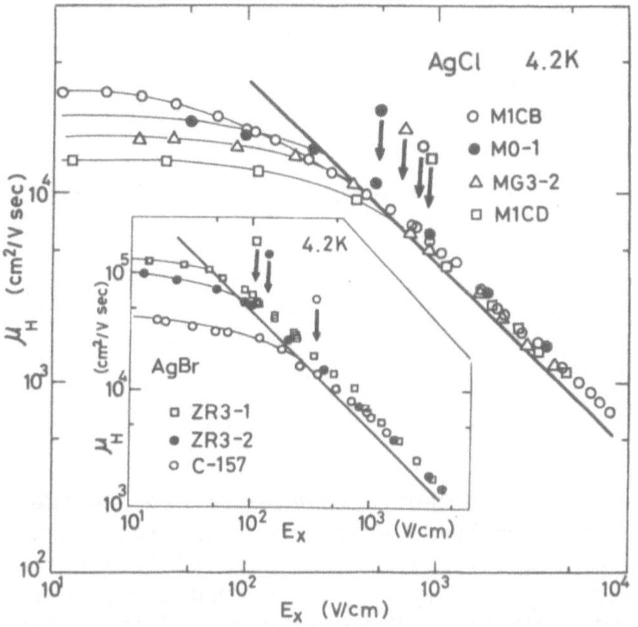


Fig. 13. Hall mobility versus electric field for electrons in several crystals of AgCl. Theoretically expected values for streaming motion (Eq.(24)) are represented by the solid line. The arrow marks $E_{min}$ for each sample. The inset shows similar data for AgBr.

The drift velocity $v_x$ of carreirs cannot be directly measured in the experiemnts, but is calculated from the $\mu_H$ values by using the relation (25). Solid circles in Fig. 14 denote so derived drift velocities $v_d$ in a crystal of AgCl. The solid line indicates the velocity $V_{op}/2$ expected for ideal streaming motion. The dashed line, which departs upwards from the solid line in the range $E_x > 5\times10^3$v/cm, represents theoretical values by Devreese et al.,[48] in which the effect of possible penetration of electrons into higher energy range $v > V_{op}$ in silver halides is included. The experimental values $v_d$ saturate approximately to $V_{op}/2$ above the field of $E_{min}$, and makes certain streaming motion of electrons. Strictly examined, however, the saturation velocity is higher than the value $V_{op}/2$ by a certain amount ($\sim$13%). This result, which is equivalent to the fact that the experimental values of $\mu_H$ are larger than the theoretical values (Eq.(24)) by a certain amount, cannot be directly interpreted as due to carrier penetration into the higher energy range $v > V_{op}$. Instead, it should primarily be attributed to an effect of the non-parabolicity present due to the polaron effect in the energy spectrum of electrons in AgCl. Due to the non-parabolicity, the relations (24) and (25) must be modified and the drift velocity should be differently derived. We have, however, no available theory predicting accurate relationship between $\mu_H$ and $v_d$ for the case of streaming motion of polarons, and the strict

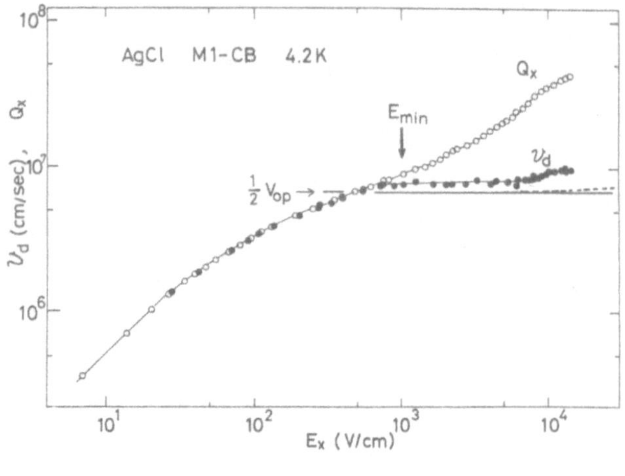


Fig. 14. Plots of photocurrent $Q_x$ (open circles) and the drift velocity (solid circles) versus electric field in a crystal of AgCl. The drift velocity is derived from $\mu_H$ through the relation. $v_d=(3/2)\mu_H E_x$. The dashed line, which deviates from the solid line indicating $V_{op}/2$ in a high $E_x$ range, represents theoretical values by Devreese et al..[48]

discussion about the absolute value of $v_d$ is so far impossible.

Open circles in the figure denote photocurents $Q_x=N\tau_t e v_d$ in arbitrary units. Though the dependence of $Q_x$ on $E_x$ is almost identical to that of $v_d$ in a lower $E_x$-range ($E_x<E_{min}$), the discrepancy is evident in the higher electric field range, where $Q_x$ continues to increase whereas $v_d$ is saturated. It has be confirmed from the measurements of *schubweg*, $\tau_t v_d$, that the increase in $Q_x$ is caused by an increase in the trapping lifetime $\tau_t$ and not by increase in N.[5] Similar behaviour of $Q_x$ is noted in all other crystals of AgCl and AgBr: In every crystal $Q_x$ continues to increase in the high-E range where $v_d$ is saturated due to the occurrence of streaming motion, causing appreciable discrepancy from $v_d$ in the respective range $E_x>E_{min}$. It is thus concluded that the increase in $\tau_t$ is closely connected to the onset of streaming motion. $\tau_t$ also varies with magnetic fields as described in §4.D.

The excellent agreement of the data of $\mu_H$ and $v_d$ with the calculation is not testimony to the needle-like distribution function such as shown in Fig. 7. The quantities, $\mu_H$ and $v_d$, are not very sensitive to delicate shape of the distribution function. In an actual experimental situation impurity (or acoustical phonon) scattering will produce a thin and approximately homogeneous distribution of electrons over the whole sphere $v<V_{op}$. The effect of impurity scattering on the distribution function is, however, not very important in the experiments at high electric fields such that $E>>E_{min}$. What is more important in determining the actual distribution function is a slight but definite penetration of electrons into $\upsilon>V_{op}$, which causes a finite broadening of the needle-like distribution. Later on we will see a realistic broadning expected for electrons in silver halides in theoretical work by Kurosawa (in Figs. 20 and 33). Devreese et al. calculated also the momentum distribution function for electrons in silver halides.[48]

*Carrier accumulation in momentum space.* To see the variation of streaming motion on application of $B_z$, as described in Fig. 8, we measure the photocurrents in the x- and y-directions, $Q_x$ and $Q_y$, as a function of $B_z$. Figures 15 and 16 show the dependence of $Q_x$ and $Q_y$ on $B_z$, respectively, with $E_x$ as a parameter in a crystal of AgCl. The arrows in the figures mark the magnetic fields at which $\zeta=1$ or 2 is satisfied. $Q_x$ at high $E_x$ decreases rapidly with increasing $B_z$ above $\zeta=1$. As will be concluded later, the rapid decrease of $Q_x$ is a combined consequence of (1) the onset of electron accumulation into region K, where average drift velocity of electrons in the x-direction is nearly equal to zero (Fig. 8(b)), and (2) a rapid decrease in $\tau_t$ with $B_z$. The $Q_y$ versus $B_z$ curves at high $E_x$ have maximum values in the range of $B_z$ between $\zeta=1$ and $\zeta=2$. This is also explained qualitatively in terms of electron accumulation into region K, where the drift velocity of electrons in the y-direction is high ($\upsilon_y=E_x/B_z$).

We do not compare the data of $Q_x$ and $Q_y$ directly with Eqs.(20) and (21), because $Q_x$ and $Q_y$ include $\tau_t$ which also varies with $E_x$ and $B_z$. We therefore transform $Q_x$ and $Q_y$ into the tangent of Hall angle,

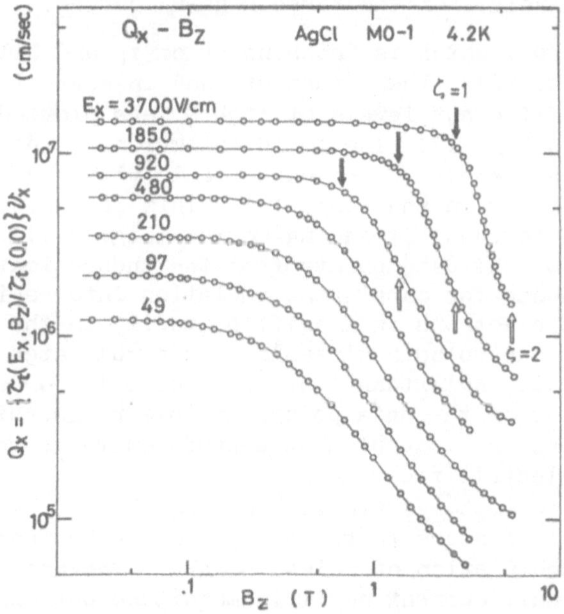


Fig. 15. $Q_x$ versus $B_z$ for a crystal of AgCl, with $E_x$ as a parameter. the black and the white arrows mark the $B_z$ positions at which $\zeta=1$ and $\zeta=2$, respectively.

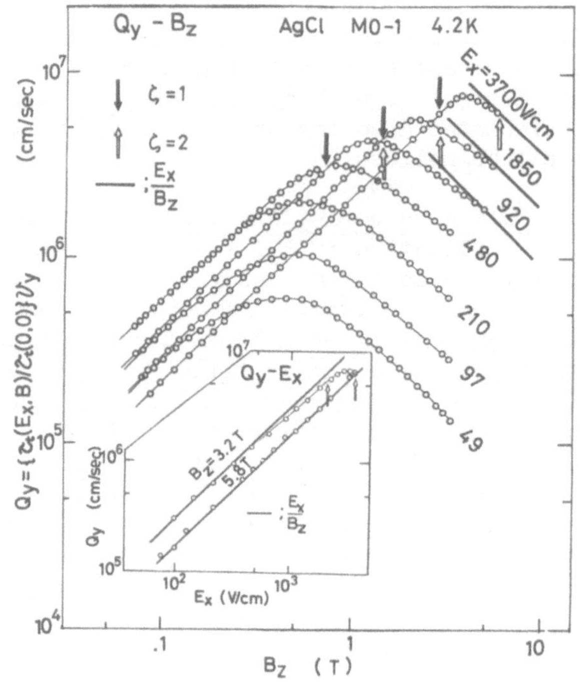


Fig. 16. $Q_y$ versus $B_z$ for the same crystal as for Fig. 15. The black and the white arrows mark the $B_z$ positions of $\zeta$=1 and 2, respectively, for the three high-$E_x$ curves.

$\tan\theta \equiv Q_y/Q_x = v_y/v_x$, which is independent of $\tau_t$ and allows for direct comparison with Eq.(22). The values of tanθ in several crystals of AgCl and AgBr at different levels of high $E_x$ are plotted together as a function of $\zeta$ in Fig. 17. The solid thick line is drawn according to Eq.(22). The agreement of the data points with the theoretical values is satisfactory in the range $\zeta<1$. This fact supports definitely the picture of 'curved' streaming-motion (Fig. 8(a)). The distinct upward departure of the data points from the theoretical line in the range $\zeta>1$ is evidence for carrier accumulation into region K (Fig.8(b)) Since accumulated electrons in K drift primarily in the y-direction, the total current due to both streaming and accumulated electrons yields a larger Hall angle than that expected only for streaming electrons. The scatter of the data points in this range reflects the fact that the degree of accumulation depends on the sample purity and the strength of electric field.

Characteristic features stated above equally apply for both AgCl and AgBr. The only feature that distinguishes AgBr from AgCl is the coexistence of accumulation of holes in AgBr. The accumulation of holes produces a Hall current small in magnitude but opposite in direction to that of electrons. Hence the curves of tanθ versus $\zeta$ in AgBr bend over in the range $\zeta^h>1$, where $\zeta^h$ is defined by Eq.(16) for holes in AgBr. The absence of such a structure in the curves for AgCl is evidence for the fact that the self-trapped states for holes

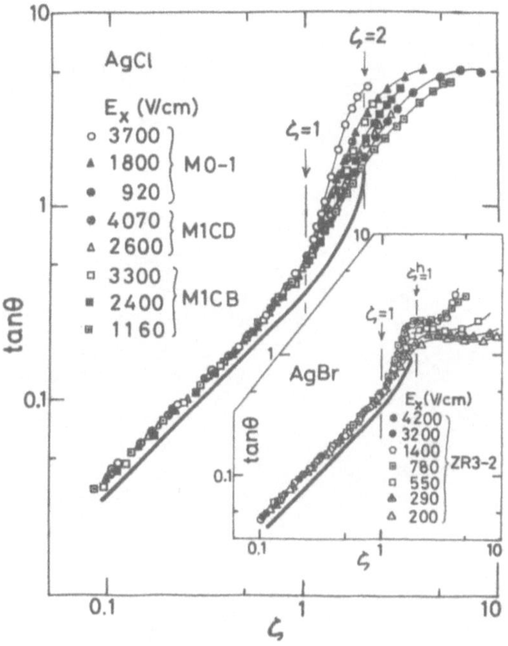


Fig. 17. Tangent of the Hall angle, $\tan\theta \equiv Q_y/Q_x$, versus $\zeta$ for AgCl and AgBr at several levels of high $E_x$. The solid line indicates theoretical values expected for ideal streaming motion without accumulation (Eq.(22)).

in AgCl remain stable even at these high electric fields.

The average values of experimentally derived tanθ in the range $\zeta<1$ are higher than the theoretical values, $\tan\theta_s$, by ~20%. This discrepancy is also ascribed to the non-parabolicity in the polaron energy spectrum. At present, however, there is no theory available for the estimation of the size and the direction of the correction term that should arise from the polaron effect. (We will see in §5 that the averaged cyclotron effective mass $m^*_{sat}$ of a streaming polaron is larger than cold polaron mass $m^*(0)$. If we simply replace $m^*(0)$ by $m^*_{sat}$ in evaluating $V_{op}$, a correction arises in the direction to enlarge the discrepancy. The simple replacement of $m^*$ is thus inadequate.) In addition to the discrepancy in absolute values, the experimental values in the range $\zeta<1$ increase slightly with increasing $E_x$ (by at most 10% in both AgCl and AgBr crystals in the range from 1 to 4kV/cm). This slight increase in tanθ may indicate the effect of streaming electrons penetrating slightly into the higher energy range $\upsilon>V_{op}$.

Kurosawa has made Monte Carlo simulation on electrons in AgBr at 4.2K in the crossed field condition,[21,26,27] including the neutral impurity, the acoustical phonon, and the optical phonon scatterings. To describe the strength and the angular-dependence of the optical

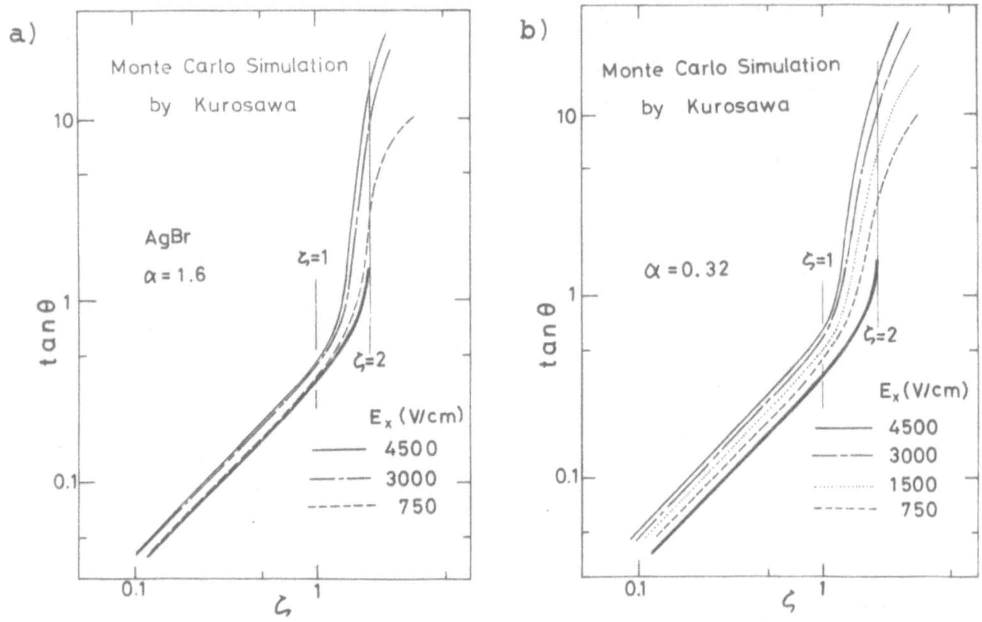


Fig. 18. Theoretical curves of tanθ vs. ζ for AgBr. The calculaiton includes the neutral impurity ( $\tau_{imp}$ =16psec) and the acoustical phonon ($A=4.8\times10^{23}$/erg·sec in Eq.(5) at 4.2K) scatterings. α=1.6 is used for a) while α=0.32 is used for b), for the optical phonon scattering. The solid thick line in either figure indicates the values of Eq.(22).

phonon scattering, the results of perturbation treatment have been used in the calculation. Figures 18(a) and (b) are from unpublished work by Kurosawa,[27] and show tanθ as a function of ζ. The solid thick line, drawn according to Eq.(22) in either figure, serves as a reference for the results of simulation. The data in (a) are for electrons in AgBr with the Fröhlich coupling constant α=1.6. The data in (b) are to demonstrate the effect of reducing the coupling strength by factor 5 with retaining the other parameters unchanged. The theoretical lines of tanθ versus ζ shift upwards with increasing $E_x$ in either case of (a) or (b). The lines are parallel in the range ζ<1. The overall increment of tanθ in the range ζ<1 is smaller for α=1.6 than for α=0.32, indicating the smaller penetration of electrons into $\upsilon>V_{op}$ for the case of stronger coupling. The increment for α=1.6 as measured from 750 to 4500 V/cm is about 18%, which is larger than, but not too different from, the observations. Thus the actual extent of penetration of streaming electrons in silver halides is supposed to be not very different from that expected from the perturbation treatment. However, the experimental results in the range ζ>1 is better simulated by the calculations with the reduced α rather than those with the original α: For instance, the simulated line for 750 V/cm with α=0.32 rises steeply with increasing ζ causing an appreciable discrepancy from the line of Eq.(22) in the range ζ>1 as indeed observed in the experiments, while the corresponding line for

$\alpha$=1.6 scarcely deviates from the values of Eq.(22). This feature is demonstrated more clearly in §4.C, and we will return to the problem again in §6.

A response current $Q_z$ emerges when a small electric field $E_z$ parallel to $B_z$ is applied additionally to the electron system. Another definite evidence for electron accumulation is the $B_z$-dependence of $Q_z$. Figure 19 shows $Q_z$ in a crystal of AgCl as a function

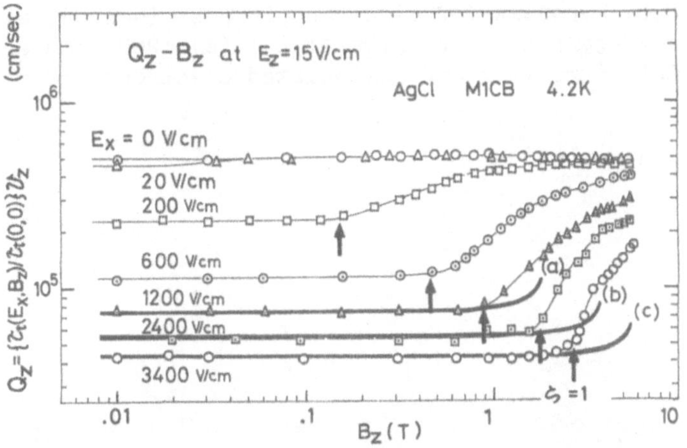


Fig. 19. Response photocurrent $Q_z$ in a crystal of AgCl against the application of a small electric field $E_z$ parallel to $B_z$. The arrow marks for each $E_x$ the magnetic position at which $\zeta$=1. Solid lines (a)∿(c) indicate theoretically expected values when carriers are ideally streaming without causing accumulation for the respective high $E_x$'s.

of $B_z$ at different levels of $E_x$ with $E_z$ fixed to 15 V/cm. The photocurrent $Q_z$ (Eq.(3)) is proportional to the mean free time $\tau$ of the electrons through $v_z=(e\tau/m^*)E_z$. We can therefore examine the change of $\tau$ with $B_z$ by the measurements of $Q_z$. The value of $v_z$ at $E_x=B_z=0$ is deduced from the mobility value at an adequately low $E_x$. In the figure, all the data of $Q_z$ are normalized by $Q_z$ at $E_x$=0 so that the data points represent the quantity $[\tau_t(E_x)/\tau_t(0)]v_z$. (The trapping lifetime $\tau_t(E_x)$ at $E=E_x$ is independent of $B_z$ in the range $\zeta$<1 as is shown in §4.D.) When electrons are streaming, the mean free time $\tau$ for acceleration by $E_z$ is given by $T_{op}/2$, where $T_{op}$ is defined by Eq.(19). Thus the velocity acquired by the streaming electrons is

$$v_z^s = \frac{e}{m^*}\left(\frac{T_{op}}{2}\right)E_z \ . \qquad (29)$$

Solid lines (a)-(c) in the figure indicate the theoretical values of $v_z$ for streaming carriers at the respective $E_x$'s. To enable direct comparison with the data points, the factor $[\tau_t(E_x)/\tau_t(0)]$, as determined by the measurements of $Q_x$ and $v_d$ at $B_z=0$, is taken into account in the presentation of the thoeretical lines. While the agreement of the data points with the theoretical lines is almost perfect in the range $\zeta<1$, the experimental values of $Q_z$ increase rapidly, deviating largely from the theoretical values in the range $\zeta>1$. This is because accumulated carriers with a longer mean free time are introduced in region K to yield the larger current response.

Realistic distributions of electrons on the $v_z=0$ plane at $\zeta=1.83$, as calculated by Kurosawa for electrons in AgBr ($\alpha=1.6$), are shown in Fig. 20 for different two strengths of $E_x$; (a) 1200 V/cm and (b) 3600 V/cm. Both streaming and accumulated electrons are clearly dis-

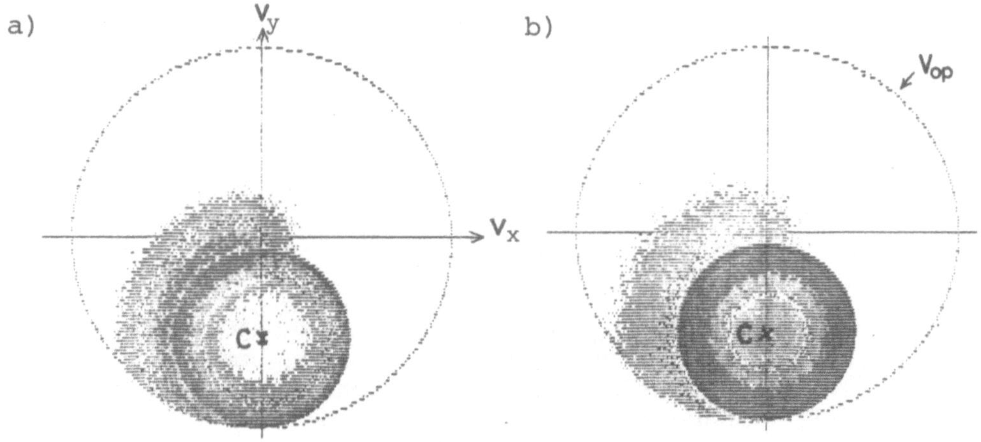


Fig. 20. Electron distribution on the $v_z=0$ plane at $\zeta=1.83$, calculated by Kurosawa for AgBr ($\alpha=1.6$).[21,27] a) is for $E_x$= 1200 V/cm and b) for 3600 V/cm. The same parameters as for Fig. 18 are used for $\tau_{imp}$ and $\tau_{ac}$ in the calculation.

cerned. The distribution function of streaming electrons as represented by an arc in Fig. 8(b) is now subject to broadening due to a slight penetration of the electrons into $v>V_{op}$. With increasing $E_x$, the ratio of electron distribution in the accumulation region increases while the distribution of streaming electrons is further broadened.

*Regime of hot-electrons and onset of quantum-limit.* The finite values of $\tan\theta$ in the range $\zeta>2$ (Fig. 17) are determined primarily by residual impurity scattering. The photocurrents $Q_z$ at $\zeta=2$ (Fig. 19) do not recover the low-field value $Q_z$ (at $E_x=0$), but approach to the value with further increasing $B_z$ in the range $\zeta>2$. These observations indicate that the electrons are still hot in the range $\zeta>2$. Variation in the hot-electron kinetics in the range $\zeta>2$ can be most clearly recognized in the measurements of $\tan\theta$ in a sweep of $E_x$ at fixed high $B_z$'s. Figure 21 is from unpublished data on a crystal of

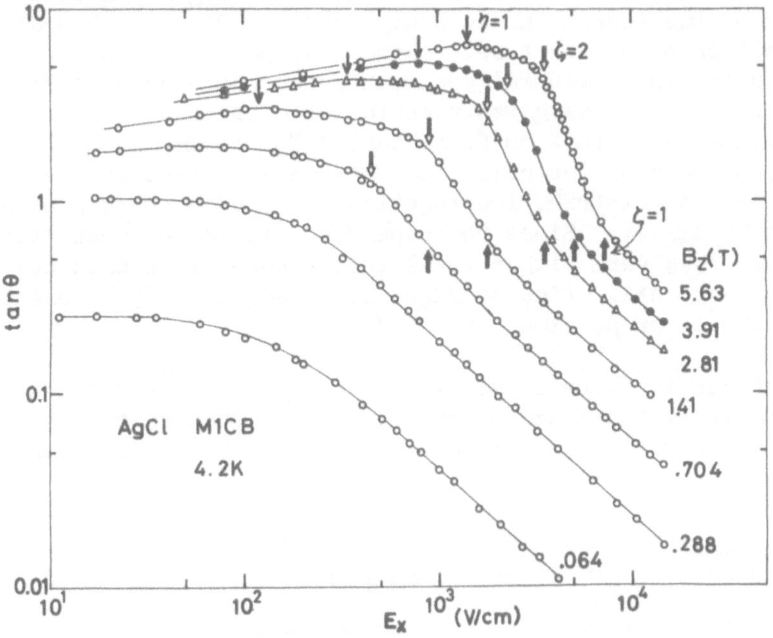


Fig. 21. Curves of $\tan\theta$ versus $E_x$ for a crystal of AgCl at different levels of $B_z$. The $E_x$-positions at which $\zeta=1,2$ and $\eta=1$ are marked for each curve by the arrows.

AgCl.[49] The rapid increase of $\tan\theta$ at high levels of $B_z$ with decreasing $E_x$ in the range $1<\zeta<2$ levels off at $\zeta=2$ because the optical phonon emission is substantially excluded from the kinetics. We wish to focus our attention on the lower $E_x$-range below the arrow $\zeta=2$ (the range $\zeta>2$). The values of $\tan\theta$ gradually increases with decreasing $E_x$ until $\eta=1$ is reached. The values of $\tan\theta$ then begin to decrease with further decreasing $E_x$ below $\eta=1$ (in the range $\eta>1$).

These observations are qualitatively interpreted as follows. In the range between $\eta=1$ and $\zeta=2$, an appreciable distribution of electrons is expected in higher Landau states within region K, because impurity scattering causes the transition of electrons to higher Landau states as pointed out in §4.A. With decreasing $E_x$ within this range the electrons become *less* hot and the average radius of cyclotron orbits of the electrons accordingly decreases. The average displacement per collision of the electrons in the x-direction therefore decreases with decreasing $E_x$. This is the origin for the gradual increase of $\tan\theta$ in the range between $\eta=1$ and $\zeta=2$. When $\eta=1$ is reached, the distribution of electrons is supposed to have been confined practically in the lowest Landau state because the transition of electrons to higher Landau states becomes difficult as described in §4.A. In this case, decrease of $E_x$ causes an increase in the transition probability between the lowest Landau states because of the increase in combined-density of states. (Imagine an electron at the bottom of the lowest Landau state in Fig. 10(b). The probability of the electron being scattered to the adjacent Landau state increases with decreasing the potential gradient because the density of final states increases.) This gives rise to the decrease in $\tan\theta$. The observed dependence of $\tan\theta$ on $E_x$ is $\tan\theta \propto E_x^{0.2}$, which is the measure of the dependence of the collision time on $E_x$. (The exponent 0.2 would reflect the distribution function $f(v_z)$ of electrons within the lowest Landau state.) Since the dependence is rather weak, the electrons become *less* warm within the lowest Landau state with decreasing $E_x$ in this range $\eta>1$. (The average energy acquired by an electron per collision is proportional to $E_x^{1.0}$.)

The change in average energy of electrons in the range $\zeta>2$ can be experimentally confirmed by the $Q_z$-measurements on the same crystal. Figure 22 shows $Q_z$ as a function of $E_x$ with $B_z$ as a parameter.[49] The

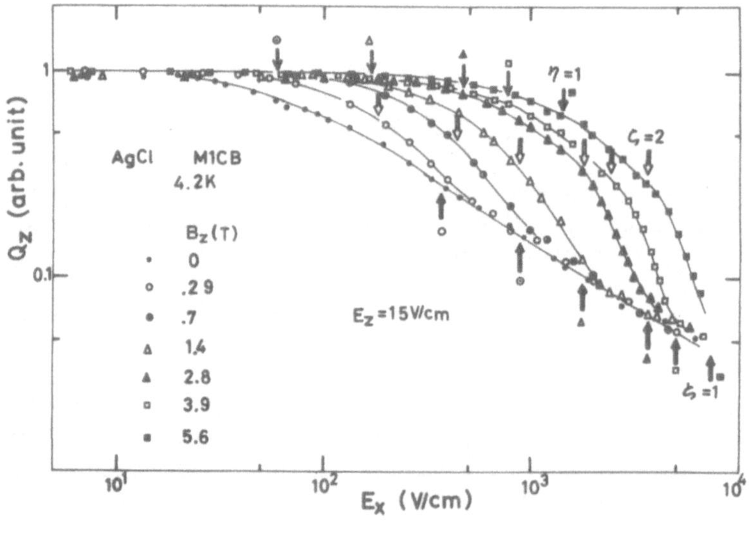


Fig. 22. Curves of $Q_z$ versus $E_x$ at different levels of $B_z$ in the same crystal as that used for Figs. 19 and 21. The $E_x$-positions at which $\zeta$=1,2 and $\eta$=1 are indicated by the arrows for each curve.

condition of the measurements is similar to the measurements of Fig. 19. The results obtained in the range $\zeta<2$ simply reconfirm the earlier results, and we concentrate our attention on the range $\zeta>2$. Since the trapping lifetime $\tau_t$ is independent of $E_x$ or $B_z$ in the range $\zeta>2$ (§4.D), the values of $Q_z$ in this range provide a direct measure of the mean free time of electrons. Moreover, since the energy exponent p of the impurity scattering time (Eq. 4) in this crystal is not zero but p=-1/2, the average energy of electrons is derived from the values of $Q_z$. The increase of $Q_z$ with decreasing $E_x$ below $\zeta=2$ is thus an indication of the decrease of the average energy of electrons in the range $\zeta>2$. We should note that the average energy $\langle\varepsilon\rangle$ of electrons at $\eta=1$ (with high $B_z$) is appreciably higher than the thermal energy; e.g., $\langle\varepsilon\rangle$ at $\eta=1$ with $B_z=5.6$ T is estimated from the $Q_z$-value to be about 2.7 meV, while the average energy of photoexcited electrons in this crystal at $E_x=0$ is about 1 meV. The values of $Q_z$ at $\zeta=2$ are almost independent of $B_z$ above 1.4 T and are smaller than $Q_z$ at $E_x=0$ by a factor about 3.5. This implies $\langle\varepsilon\rangle$ of about 12 meV, which is a reasonable value for electrons at $\zeta=2$ in AgCl with $\hbar\omega_{op}=23$ meV. (It should be mentioned that $\langle\varepsilon\rangle$ in the range $\zeta<2$ is not appreciably higher than this value $\langle\varepsilon\rangle\sim 12$ meV. Roughly speaking, $\langle\varepsilon\rangle$ saturates due to the onset of optical phonon emissoin in the range $\zeta<2$.)

C. Mechanism for Accumulation

We wish to consider the mechanism of accumulation in a similar way to that described in Refs. 10 and 21. Let us denote the numbers of streaming electrons and the electrons in region K by $n_s$ and $n_k$, respectively. We assume that a streaming carrier jumps into region K after the optical phonon emission with a probability $P_k$ and that the electron in region K goes out of this area after some scattering events to join with streaming electrons. By denoting this lifetime by $\tau^k$, we obtain the relation $\partial n_k/\partial t = n_s P_k/T_{op} - n_k/\tau^k$, which yields the ratio of accumulation in the form

$$\frac{n_k}{n_s} = P_k \frac{\tau^k}{T_{op}} \tag{30}$$

in the steady state condition, $\partial n_k/\partial t = 0$. We assume the lifetime $\tau^k$ is limited by a scattering process that removes an electron directly out of this region, and is identical to the average scattering time of electrons in region K. Thus $\tau^k = \bar{\tau}^k_{imp}$, by neglecting the acoustical phonon scattering, where $\bar{\tau}^k_{imp}$ is the average scattering time in region K due to impurities. The probability $P_k$ depends on the extent of penetration of streaming electrons into $v>V_{op}$ (thus depends on the strength of electric field E scaled by the phonon coupling strength). It also depends on the 'volume' of region K (thus on $\zeta$). Kurosawa has recently calculated $P_k$ nuemrically as a function of the normalized strength of E, with $\zeta$ as a parameter.[21]

Experimentally, the ratio $n_k/n_s$ can be derived from the magnitude of discrepancy observed between the experimental values of $\tan\theta$ and the theoretical values $\tan\theta_s$.[5] Figure 23 shows the result in several samples of AgCl at $\zeta=1.4$. The dashed line indicates the theoretical values given by Eq.(30) with the values of $P_k$ calculated for $\alpha=1.9$ by Kurosawa.[21] The experimental values are by orders of magnitude larger than the theoretically expected values. (The very low values

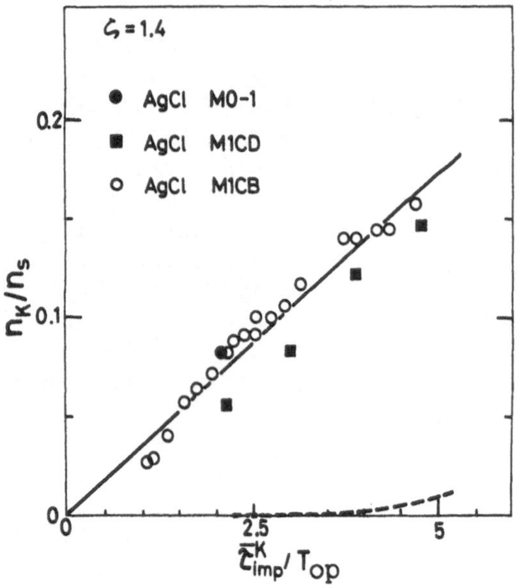


Fig. 23. Ratio of accumulation, $n_k/n_s$, in AgCl. The dashed line is drawn according to Eq.(30) with $\tau^k=\bar{\tau}^k_{imp}$ and theoretical values[21] of $P_k$ by Kurosawa.

of theoretical $n_k/n_s$ are the explicit consequence of the very slight departure of the theoretical lines of tanθ from the reference line Eq.(22) at ζ=1.4 in Fig. 18(a). At higher ζ, appreciable accumulation is derived also from the theory as shown in Fig. 20 for ζ=1.83.) The very low values of the theoretical $n_k/n_s$ manifest the theoretical expectation that the electron optical phonon interaction in silver halides is so strong that it does not permit streaming electrons to penetrate so deeply into $\upsilon>V_{op}$ as to cause appreciable accumulation. One would think the perturbation treatment applied in the theory is inadequate to predict the emission rate of optical phonons in this non-weak coupling case, and would expect a deeper penetration than that given in the theory. However, when one assumes such an adequate penetration as to explain the observed degree of accumulation, one has at the same time to expect an appreciable increase in tanθ with increasing $E_x$ at fixed ζ in the range ζ<1. Such an increase in tanθ is demonstrated in Fig. 18(b), where the weaker coupling constant α=0.32 is used. This is, however, not observed in the experiemnts (Fig. 17). Thus the feature of tanθ observed in the range ζ<1 strongly indicates that the penetration is actually so small as not to cause the accumulation observed in the range ζ>1. These conflicting aspects of the phenomena are not likely to be explained within the simple free-electron picture, and appear to suggest an importance of the polaron self-energy effect on streaming electrons as is discussed in §6.

### D. Trapping Lifetime of Streaming Electrons

Generally, the measurement of photocurrents (Eq. 1) brings information about the quantity $N\tau_t$ in addition to the drift velocity $v$. In ionic crystals, the trap levels usually lie relatively deep (e.g. 36 meV in AgCl and 24 meV in AgBr[50]) below the band edge, and the impact ionization of trapped carriers or the field emission type carrier generation does not take place in the range $E<10^4$ V/cm. Therefore N is kept unchanged in ionic crystals. The photocurrent measurements in ionic crystals thus bring information about the trapping lifetime $\tau_t$.

Masumi has first found that the photocurrents in silver halides increase with electric field E without exhibiting saturation in high E-range.[51] In all the other I-VII ionic crystal ever studied since the work (KBr,[52] KCl,[53] KI,[54] and TlCl[55]), it is the common feature that the photocurrents are not saturated but increase with E in the high E-range where streaming motion is expected. In the II-VI ionic crystal CdS, similar increase of photocurrents is also observed.[11] Several of these experimental results are collected in Fig. 24. Due to the absence of photocurrent saturation, the onset of optical phonon emission in ionic crystals had been questioned for many years. It has been shown in this work on silver halides that the increase in photocurrents is caused by an increase in $\tau_t$ and not by $v_d$. In the work on CdS,[11] the saturation of $v_d$ due to streaming motion is also confirmed

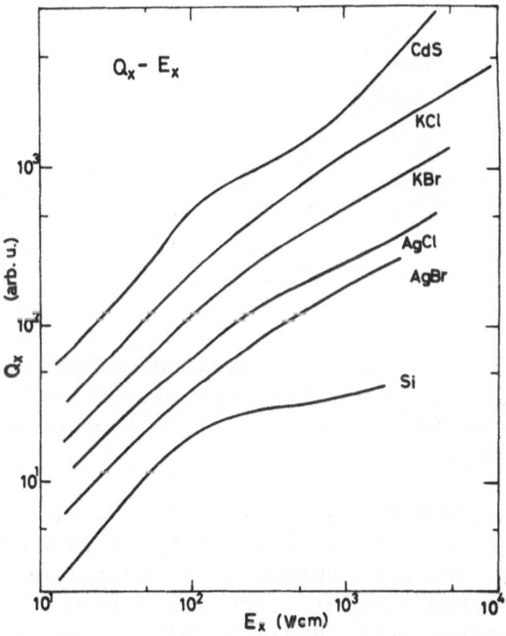


Fig. 24. Photocurrent $Q_x$ versus electric field $E_x$ in various materials at 4.2 K; CdS,[11] KCl,[53] KBr,[52] AgCl,[5] AgBr[51] and Si[12]. Only the photocurrent in Si exhibits saturation, whereas carriers are certainly 'streaming' in all the materials in high $E_x$-ranges.

and an increase of $\tau_t$ is suggested. For all other ionic crystals we do not have any reason to doubt about the occurrence of streaming motion and we should suspect the increase of $\tau_t$ at high electric fields. Since the properties of the trap centers are expected to be largely different among these materials, the increase in $\tau_t$ is not likely to be a charcteristic of particular trap centers in these materials but is expected to be a phenomenon of deeper origin in the ionic material. Unlike the ionic crystals, the photocurrents due to heavy holes in Si do exhibit saturation due to the onset of streaming motion (Fig. 24),[12] and do not provide any indication of increase of $\tau_t$. This fact may reassure the peculiarity of the $\tau_t$-increase in ionic crystals.

In silver halides another distinct aspect is found in the dependence of $\tau_t$ on magnetic field.[5] Figure 25 shows the variation of the

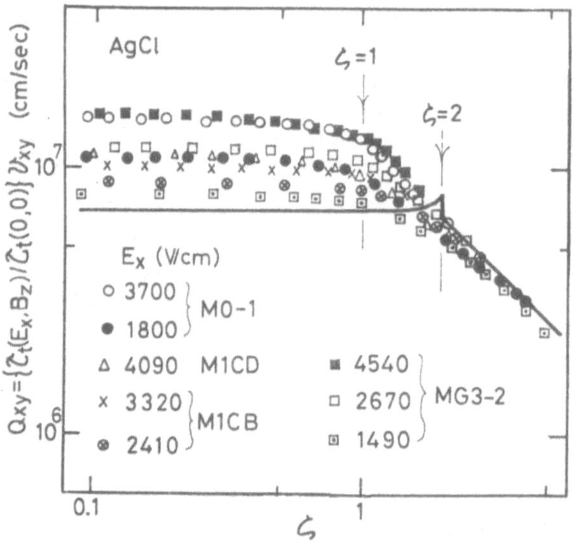


Fig. 25. Variation with $\zeta$ of the magnitude of photocurrents, $Q_{xy} \equiv (Q_x^2 + Q_y^2)^{1/2}$, in several crystals of AgCl. The solid line is drawn according to Eq.(23) in the range $\zeta<2$ and $(E_x/B_z)$ in the range $\zeta>2$.

magnitude of photocurrents, $Q_{xy} = (Q_x^2 + Q_y^2)^{1/2}$, with magnetic field. The measurements are made in a sweep of $B_z$ at fixed high levels of $E_x$. In the figure, the data obtained in several crystals of AgCl are shown together as a function of $\zeta$ converted from $B_z$. The data points represent the quantity $[\tau_t(E_x,B_z)/\tau_t(0,0)]\upsilon_{xy}$ and are compared with a solid line. The solid line indicates the theoretical values $\upsilon_{xy}$ (Eq.(23)) in the range $\zeta<2$ and the values $E_x/B_z$ in the range $\zeta>2$. The discrepancy between the data points and the theoretical line in the range $\zeta<1$ is caused by the relative increase of $\tau_t$; $[\tau_t(E_x,B_z)/\tau_t(0,0)]$. The magnitude of discrepancy is kept almost unchanged at fixed high $E_x$ in the range $\zeta<1$. This indicates that the increased $\tau_t$ at high $E_x$ does not substantially change with $B_z$ in the range $\zeta<1$. However, a rapid decrease of $Q_{xy}$ starts at $\zeta=1$ and all the data points fall close to the theoretical line $(E_x/B_z)$ when $\zeta\simeq2$ is reached. The substantial agreement of the data points with the theoretical line in the range $\zeta>2$ implies that $\tau_t$ is reduced to the low field value $\tau_t(0,0)$ at $\zeta=2$ and no longer changes with further increasing $B_z$ in the range $\zeta>2$. The above findings can be expressed in other words: The increased lifetime $\tau_t$ at high $E_x$ is not affected by the application of $B_z$ when all the electrons are streaming (in the range $\zeta<1$), but begins to decrease with $B_z$ when the number of streaming electrons starts to decrease due to the onset of accumulation $(1<\zeta<2)$, and is finally reduced to the low field value and no longer changes with $B_z$ when streaming electrons disappear $(\zeta>2)$. All the experimental findings thus strongly indicate that the increase of $\tau_t$ is directly triggered by the onset of streaming motion.

It is probably hopeless to expalin the above behaviour of $\tau_t$ within the framework of a simple free-electron picture. It is also difficult to account for the behaviour of $\tau_t$ at $B_z=0$. For example, it has been mentioned in §4.B that the electric field at which the increase of $\tau_t$ occurs coincides substantially with the minimum field $E_{min}$ for the streaming motion in each sample. The field $E_{min}$ differs greatly from sample to sample (ranging, e.g. from 100 V/cm to 900 V/cm). This fact rules out the possibility of $E_x$ acting directly on the capture cross section of the trap centers. Another possibility of the capture rate depending on the average carrier energy is also excluded, since (1) $\tau_t$ does not increase with $E_x$ in the lower $E_x$-range $E_x<E_{min}$ (where the average energy $\langle\varepsilon\rangle$ of the hot electrons increases rapidly with $E_x$), but does increase in the range $E_x>E_{min}$ (where the increase in $\langle\varepsilon\rangle$ is saturated) and (2) $\tau_t$ does not vary with $B_z$ in the higher $B_z$-range $\zeta>2$ (where $\langle\varepsilon\rangle$ decreases significantly with $B_z$) but does decrease with increasing $B_z$ in the range $1<\zeta<2$ (where $\langle\varepsilon\rangle$ is even expected to increase due to the accumulation). This problem also appears to suggest importance of the polaron self-energy effect on streaming electrons as is discussed in §6.

## 5. MICROWAVE FIELDS

The energy-momentum relation of the electrons in ionic crystals deviates from parabolicity with increasing electron energy due to the polaron effect, as is shown in Fig. 4. The change in the effective mass of 'hot' polarons is not so large as to cause significant effects in the Hall effect measurements, but is recognized in cyclotron resonance experiments.[56,57] In the cyclotron resonance work described here,[6-8] the microwave intensity is extended up to ∿3kV/cm, and streaming motion is also realized under the cyclotron resonance condition. It is a significant advantage of the cyclotron resonance experiments on hot polarons that the change in the average energy of electrons and the variation in the scattering mechanism can be examined separately by studying, respectively, the resonance position and the absorption line width. In general an electron system passes through an intermediate hot-electron regime in the course of transition of the electron system from a non-streaming hot-electron regime to the streaming-motion dominated regime with increasing (microwave) electric field. A significant aspect of the extended cyclotron resonance work is the observation of such a transition. In addition, a Monte Carlo simulation by Kurosawa reveals interesting features of streaming motion in strong microwave fields. In this chapter we shall describe and interpret the experimental results semi-qualitatively. We shall also give a glance to the theoretical predictions about the electron motion in strong microwave fields.

## A. Background

Before seeing the experimental results we first wish to speculate on how the hot-electron situation should change with increasing microwave field $E_\omega$ under the cyclotron resonance condition. Figure 26 schematically shows our expectation.

*Hot-electron regime* (Fig. 26(a)). In a low $E_\omega$-range, the average energy of electrons increase with increasing $E_\omega$, but the high-energy tail of the carrier distribution will not reach $\varepsilon=\hbar\omega_{op}$. The momentum

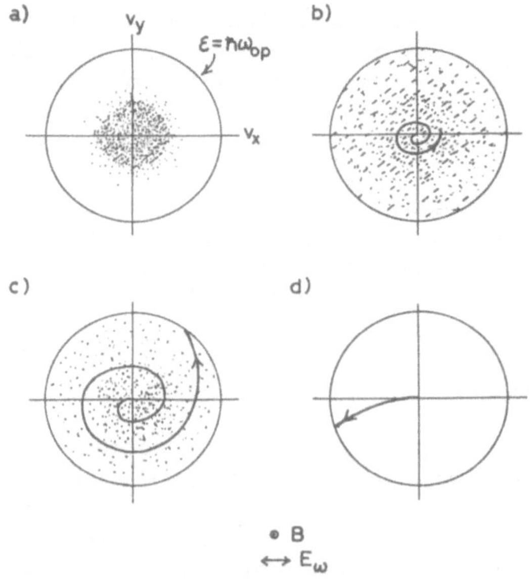


Fig. 26. Sketch of the expected variation with $E_\omega$ in hot-electron kinetics and the electron distribution, under the cyclotron resonance condition. (The pictures are not from simulation calculation.)

relaxation of carriers is domianted by impurity or acoustical phonon scattering, and the energy loss is dominated by the acoustical phonon emission. The phenomenon is amenable to the usual hot-electron concept.

*Pre-streaming regime* (Fig. 26(b)). In an intermediate $E_\omega$-range, the high-energy tail of electron distribution may reach $\varepsilon = \hbar\omega_{op}$ and emission of the optical phonon sets in. The electron distribution is cut off at $\varepsilon = \hbar\omega_{op}$ and the increase in average energy with $E_\omega$ will be quenched. Energy loss of the electron system is dominated by the optical phonon emission while the momentum relaxation will still be dominated by the impurity or acoustical phonon scattering.

*Regime of streaming cyclotron-resonance* (Fig. 26(c)). In a high $E_\omega$-range, it will become possible for electrons to be accelerated directly to $\varepsilon = \hbar\omega_{op}$ before being scattered by impurities or acoustical phonons. Accordingly streamingmotion appears in the cyclotron resonance condition. Both the moemntum loss and the energy loss are dominated by the optical phonon emission. By noting that the cyclotron acceleration in a linearly polarized microwave field, $E_\omega \cos\omega t$, is equivalent to acceleration in the dc-field of $E = E_\omega/2$, one obtains, similarly to Eq.(7),

$$T^{\omega}_{op} = 2(2m^*\hbar\omega_{op})^{1/2} / (eE_\omega)^{-1}, \tag{31}$$

as the travelling time of an electron from $\varepsilon=0$ to $\varepsilon = \hbar\omega_{op}$. The streaming motion is expected in the range $E_\omega > E_\omega^{min}$, where $E_\omega^{min}$ is derived from relation (10) by using $T^{\omega}_{op}$. $E_\omega$ in the experiemnts ranges from 90 to 180V/cm according to the sample purity and the lattice temperature, as indicated by the arrows in Figs. 28 and 29. Noting that the relation $2B/(\Delta B) \simeq \omega\tau$ generally holds for the width ($\Delta B$) of the absorption line with B and $\tau$ being respectively the resonance field and the momentum relaxation time, we obtain

$$(\Delta B)/2 = 2B/(\omega T^{\omega}_{op}) \tag{32}$$

by replacing $\tau$ by $T^{\omega}_{op}/2$, for the half width of the resonance line due to electrons under the streaming cyclotron-resonance.

*Regime of cyclotron streaming-motion* (Fig. 26(d)). In a higher range of $E_\omega$, electrons will reach $\varepsilon = \hbar\omega_{op}$ even within a half period $\pi/\omega$ of the microwave field. The conventional concept of cyclotron resonance no longer applies in the phenomena. This stage, $T^{\omega}_{op} < \pi/\omega$, is reached when $E_\omega$>330 V/cm for electrons in AgBr at $\omega = 2\pi \times 35$GHz. To distinguish the situation from that of the streaming cyclotron-resonance, we call the electron motion at this stage cyclotron streaming-motion.

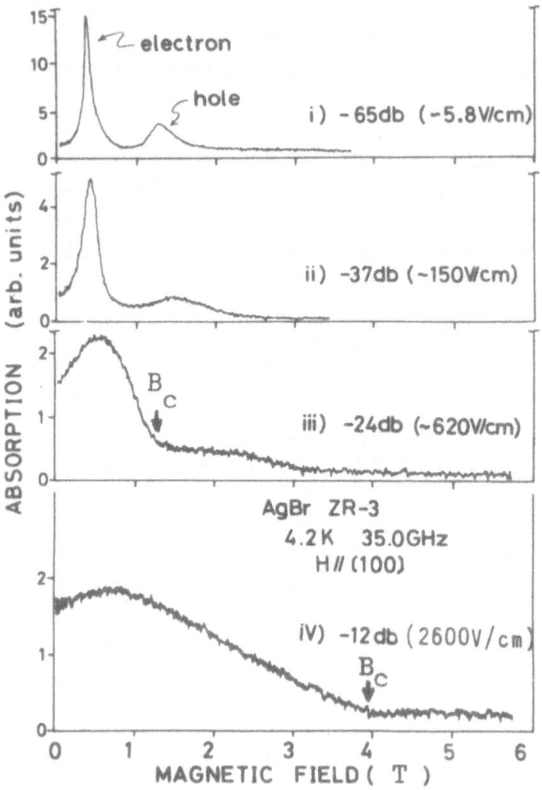


Fig. 27. Variation with $E_\omega$ of the cyclotron resonance line due to photocarriers in AgBr at 4.2K.[6]

B. Experimental Results

The experimental results on photocarriers in AgBr can be substantially interpreted according to the scheme described above. The absorption line varies greatly with increasing $E_\omega$ from 6V/cm to 2.6kV/cm (Fig. 27). The resonance absorption due to holes is well discerned at low $E_\omega$, but it merges into the electron absorption line extending largely towards a higher magnetic field with increasing $E_\omega$. Only absorption due to electrons is considered below. As $E_\omega$ increases, the absorption line gets remarkably broad and the position of the maximum absorption point, $B_p$, shifts to a higher magnetic field. The maximum absorption point $B_p$ is plotted as a function of $E_\omega$ in Fig.28. The half line width $(\Delta B)/2$ is plotted against $E_\omega$ in Fig. 29, where $(\Delta B)/2$ is defined by the field difference between the position of the half maximum point at the higher magnetic field side in the absorption line and the resonance position at the limit of low $E_\omega$. The solid line in the figure indicates the theoretical values Eq.(32) for streaming cyclotron resonance. The data in a lower $E_\omega$-range ($E_\omega$ < 100V/cm) are consistent with the earlier results by Tamura and Masumi.[57] For convenience of description the whole range of $E_\omega$ is divided into four regions: (A) $E_\omega$ < 30V/cm, (B) 30 < $E_\omega$ < 100V/cm,

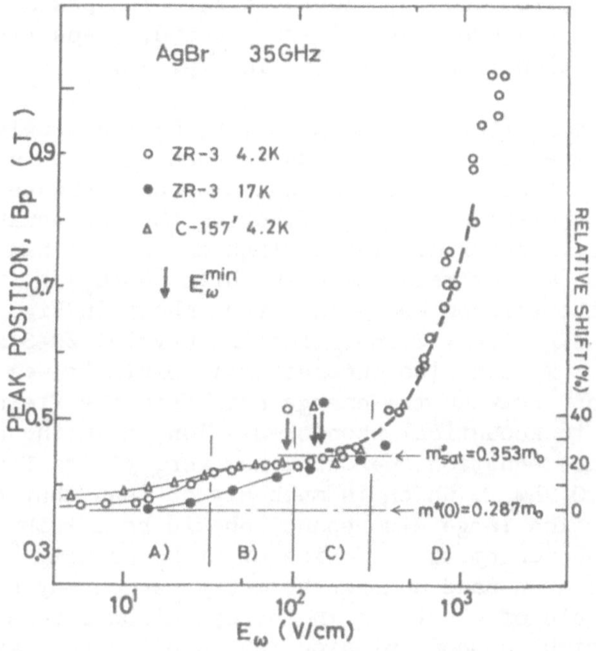


Fig. 28. The maximum absorption point $B_p$ in the cyclotron resonance line versus $E_\omega$.[6,7] The dashed line indicates a result of Monte Carlo calculation.[26]

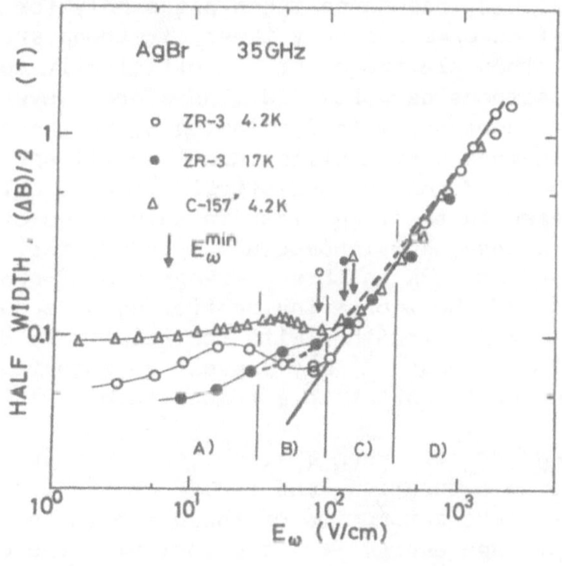


Fig. 29. Half width $(\Delta B)/2$ of the resonance line versus $E_\omega$.[6,7] The solid line indicates the values expected for ideal streaming motion, Eq.(32). The dashed line is a result of Monte Carlo calculation.[26]

(C) $100 < E_\omega < 330$V/cm and (D) 330V/cm $< E_\omega$. The phenomena in each range from (A) to (D) substantially correspond, respectively, to the hot carrier situation from (a) to (d) in Fig. 26.

*Hot-electron regime.* In range (A), $B_p$ shifts slowly to higher B and $(\Delta B)/2$ increases gradually with increasing $E_\omega$ in every experimental condition. The shift of $B_p$ indicates an increase in the average energy of electrons $<\varepsilon>$, reflecting the non-parabolicity in the polaron energy spectrum. The average energy $<\varepsilon>$ can be estimated by the comparison of the relative peak shift $[B_p(E_\omega)/B_p(0)]$ with the calculation of the polaron energy spectrum shown in Fig. 4. The comparison yields, e.g. $<\varepsilon> \simeq 0.2\hbar\omega_{op}$ for the crystal ZR-3 at T = 4.2 K and $E_\omega$ = 10V/cm. We can also theoretically estimate $<\varepsilon>$ by using a balance equation[58] between the energy gain from the field and the eenrgy loss due to acoustical phonon emission, assuming a parabolic band. The balance equation approach, however, yields for the same condition $<\varepsilon> \simeq 0.7\hbar\omega_{op}$, which is much higher than that deduced from the peak shift. The large discrepancy should be attributed either to the optical phonon emission, which might be present already at this $E_\omega$, or to the polaron-band non-parabolicity, which may suppress the rate of energy gain of electrons in the cyclotron resonance condition. Although the optical phonon emission may actually be playing a certain role in cooling the electron system, the effect of the band non-parabolicity is supposed to be more substantial. This is readily expected from the high value of $\omega\tau$ for electrons in the crystal ($\omega\tau^0_{imp} \simeq 12$ for ZR-3) and the probable degree of the non-parabolicity (Fig. 4): At a given magnetic field, resonance takes place only for those electrons whose $m^*(\varepsilon)$ satisfies $\omega_c = \omega$ and only these electrons are selectively heated while the other electrons are not efficiently accelerated: The heating of electrons as a whole is, therefore, severely suppressed. Nevertheless, $<\varepsilon>$ increases with increasing $E_\omega$ in range (A). As $E_\omega$ increases, the electron distribution function will spread to higher energies where the band non-parabolicity is increasingly pronounced. The gradual increase in the line width in this range of $E_\omega$ is interpreted as due to a resultant inhomogeneous broadening. (The line broadening observed at 4.2K in this $E_\omega$-range cannot qualitatively be explained in terms of the increasing scattering by acoustical phonons.) The smaller shift of $B_p$, together with the relatively distinct line broadening, at 17K in range (A) may suggest an importance of the acoustical phonon scattering at this temperature.

*Pre-streaming regime.* When $E_\omega$ increases to cover range (B), the shift of $B_p$ tends to saturation, and $(\Delta B)/2$ stops increasing and even decreases at 4.2K. The saturation of the $B_p$-shift means satuation of the increase in average energy $<\varepsilon>$, and indicates the onset of optical phonon emission. The average energy of electrons $<\varepsilon>$ estimated from the peak shift at $E_\omega$ = 30V/cm and T = 4.2K is about $0.5\hbar\omega_{op}$. This value of $<\varepsilon>$ substantially accounts for the onset of stage (b) of Fig. 26. In this range of $E_\omega$, the average energy $\Delta\varepsilon$ acquired by an electron between successive collisions, $\Delta\varepsilon = (eE_\omega\tau)^2/(8m^*)$, is no longer negligible in comparison to $\hbar\omega_{op}$. If the band non-parabolicity is neglected, $\Delta\epsilon$ would reach $0.3h\omega_{op}$ in the ZR-3 crystal ($\tau \sim 50$psec) at $E_\omega = 50$V/cm and T = 4.2K. This implies that an electron may be subject to a significant change in the effective mass $\Delta m^*$ during the acceleration. The acceleration will be thereby hindered. (The relative phase-shift, $\Delta\phi = 2\tau(\Delta\omega_c)$, between the microwave field and the electron motion caused during the acceleration by the probable mass-change of $\Delta m^* \simeq 0.2m^*$ at $E_\omega = 50$V/cm is estimated to be even larger than $\pi$ when $\omega\tau = 10$!) As $E_\omega$ increases, this situation is gradually relieved by a decrease in $\tau$ caused by the increasing emission of optical phonons, and the phenomena transfer into the regime of streaming cyclotron-resonance in range (C). The line sharpening in range (B) is related to the transitional character of the pre-streaming regime, where the electron system transfers from the non-streaming regime (where the inhomogeneous broadening is significant) to the streaming regime (where collision broadening due to the optical phonon emission is essential). The line sharpening effect is weaker at 17K presumably because collision broadening due to the acoustical phonon scattering is already significant in range (A).

*Regime of streaming cyclotron-resonance.* At $E_\omega \simeq 100$V/cm, the rate of optical phonon emission is expected to be comparable to that of impurity and acoustical phonon scatterings. The value of $\omega\tau$ determined by the combined scattering time $\tau$ is therefore reduced. The value at $E_\omega \simeq 100$V/cm is roughly estimated to be $\omega\tau = 3 \sim 5$ in the present experimental conditions. With this $\omega\tau$, the phase-shift caused by the mass change during the acceleration will not be significant and the acceleration is no longer hindered seriously. This situation permits us to apply the simple analysis of streaming cyclotron-resonance described in the preceding section to the phenomena in range (C). In this range, the line width begins to increase abruptly, while $B_p$ stays almost constant at $B_p \simeq 0.44$T. The data points of $(\Delta B)/2$ for each experimental condition fall close to the solid line (Eq.(32)) calculated with the cold polaron mass $m^*=0.287m_e$ in the respective range $E_\omega > E^{min}$. The excellent agreement is evidence for streaming cyclotron-resonance. The almost constant peak position $B_p$ in this range (C) should be thought to correspond to an effective polaron mass of streaming electrons. Namely, the observed cyclotron mass $m^*_{sat} = eB_p/\hbar\omega \simeq 0.353m_0$, which is larger than the coldpolaron mass $m^*(0) = 0.287m_e$ by 23%, should be regarded as a mass value averaged over the cyclotron-spiral-trajectory from $\epsilon = 0$ to $\epsilon = \hbar\omega_{op}$. In this sense $m^*$ in Eqs.(31) and (32) should be replaced by $m^*_{sat}$ rather than $m^*(0)$. Resultant corrections in $T^{\omega}_{op}$ and $(\Delta B)/2$ are, however, not very large and serious alterations are not necessary for the foregoing arguments.

*Regime of cyclotron streaming-motion.* In range (D), $B_p$ begins to shift drastically to higher B, while $(\Delta B)/2$ still continues to increase according to Eq.(32). In this range of $E_\omega$, an electron passes from $\epsilon = 0$ to $\epsilon = \hbar\omega_{op}$ within a fraction of the half period of the microwave field. The phase-shift caused by the mass change during the acceleration is therefore negligible, and the band non-parabolicity no longer plays a substantial role in the phenomena except that the experiemnts see electrons with the effective mass of $m^*_{sat}$ instead of $m^*(0)$. The drastic shift observed for $B_p$ is connected with cyclotron streaming-motion, in which the ordinary concept of cyclotron resonance breaks down: The shape of absorption lines in this stage represents the variation with B of the number of optical phonons emitted by an electron per half cycle of the microwave field, and the peak position in the absorption line does not directly correspond to the effective mass of electrons. In this situation, acceleration of electrons at a final stage of microwave half-periods, where the microwave field is going to change its sign, is important. Figure 30 schematically elucidates the mechanism of the rapid $B_p$-shift, by comparing velocity variations of an average electron with

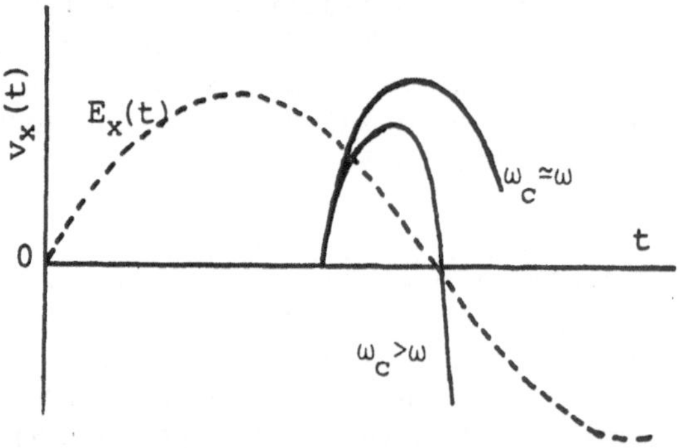


Fig. 30. A schematical explanation of the rapid shift of $B_p$ observed in the regime of cyclotron streaming-motion; the acceleration is more efficient for $\omega_c > \omega$ than for $\omega_c = \omega$ at the final stage of the half circle.

time for two cases of $\omega_c > \omega$ and $\omega_c = \omega$. The electron is more efficiently accelerated for $\omega_c > \omega$ than for $\omega_c = \omega$ at the final stage of the half period, yielding a maximum absorption point at $\omega_c > \omega$. A Monte Carlo simulation by Kurosawa[7,26] reproduces not only the rapid shift of $B_p$ but also the marked increase in $(\Delta B)/2$ as shown by the dashed line, respectively, in Figs.(28) and (29). Moreover, the simulation calculation excellently reproduces the overall line shapes obtained in this range of $E_\omega$.[7,26]

The absorption line in this range of $E_\omega$ is furnished with a loosely defined kink at a higher B position as indicated by the arrow in Fig. 27. The kink position, $B_c$, is plotted as a function of $E_\omega$ in Fig. 31. This cut-off in microwave absorption is interpreted as in the following. The equation of motion of a free electron in a microwave field $\vec{E}_\omega = (E_x, 0, 0)$ and a magnetic field $\vec{B} = (0, 0, B)$, as is obtained by replacing $E_x$ in Eq.(12) by $E_x = E_\omega \cos\omega t$, can be readily solved for arbitrary initial conditions. The motion of an electron is generally expressed by a superposition of oscillatory motions with

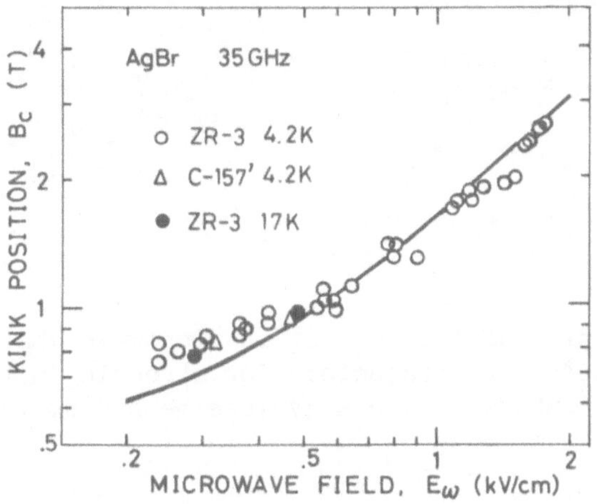


Fig. 31. Plots of the kink position $B_c$ versus $E_\omega$.[8] The solid line indicates the theoretical values Eq.(33), at which streaming motion in the microwave fields disappears.

angular frequencies of $\omega_c$ and $\omega$. Naturally, the kinetic energy of an electron also oscillates. It can be shown[8] that the maximum energy of an electron which passes the point $\vec{v}=0$ in velocity space does not reach $\hbar\omega_{op}$ when $B > B_c^{cal}$, where

$$B_c^{cal} = \frac{E_\omega}{V_{op}} + [(\frac{E_\omega}{V_{op}})^2 + (\frac{m^*\omega}{e})^2]^{1/2}. \tag{33}$$

The solid line in Fig. 31 is drawn according to Eq.(33). The excellent agreement of the data points $B_c$ with the theoretical values $B_c^{cal}$ definitely indicates that the observed cut-off in microwave absorption is caused by the disappearance of streaming motion at the high magnetic fields. The physical implication of Eq.(33) can be easily pointed out by the consideration of the reduced form $B_c^{cal} = 2E_\omega/V_{op}$ of Eq.(33) obtained at the limit of high $E_\omega$. Since $\omega_c \gg \omega$ at $B = B_c$, we may adopt a direct analogy to the dc-field case and describe the electron motion in velocity space by a cyclotron oscillation around point C, which oscillates on the $v_y$-axis as $C=[0, -(E_\omega/B)\cos\omega t, 0]$. Streaming motion clearly disappears when $V_{op} > 2(E_\omega/B)$ as shown in Fig. 32. Thus Eq.(33) corresponds to the condition $\zeta=2$ in the case of microwave field.

## C. Bunching Effects and Higher Harmonic Generation

All the cyclotron resonance data have been reasonably interpreted. A question remains about electron distribution in the regime of cyclotron streaming-motion (Fig. 26(d)). In general trajectories of different electrons at stage (c) or (d) in Fig. 26 are different accord-

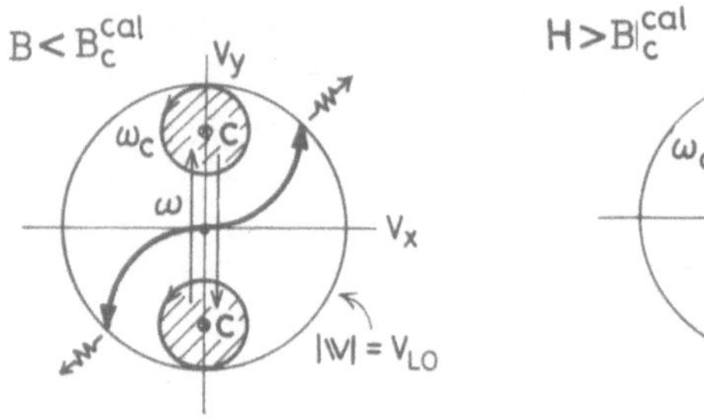

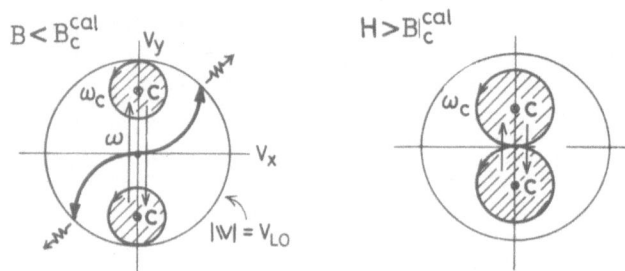

Fig. 32. Explanation of the cut-off in microwave absorption at $B_c^{cal}$. For $B>B_c^{cal}$, the trajectory for streaming motion disappears and the substantial energy-loss mechanism vanishes.

ing to the phase relation of the electron motions to the microwave field. One would, therefore, expect an electron distibution homogeneously spreading over the whole sphere $\varepsilon<\hbar\omega_{op}$, as a consequence of the randomness in the initial condition among different electorns. This is approximately the case for the regime of streaming cyclotron-resonacne (Fig. 26(c)), but not for the regime of cyclotron streaming-motion (Fig. 26(d)). According to the Kurosawa's simulation calculation,[21,26] cyclotron streaming-motion of electrons generally exhibits systematic patterns to yield very remarkable distribution functions. In what follows we shall describe the peculiar features of electron motion which have been predicted theoretically but not been confirmed experimentally.

In general, phonon emission in streaming motion differs from usual scattering events in that the interval of the scatterings is primarily determined by external fields. For example, the motion of electrons is not randomized by the phonon emission in ideal streaming motion in dc-electric fields. In intense microwave fields of range (D), bunching of electrons is even expected. Suppose that two electrons in such an intense microwave field, $E_\omega \sin\omega t$, start from the state $v=0$ at slightly different times $t=0$ and $t=\Delta t$ ($\Delta t<<\pi/\omega$), respectively, in the absence of B. While both the electrons reach $\hbar\omega_{op}$ within the half cycle of microwave field, the acceleration of the latte electron is generally more rapid because the electirc fields that the latter electron experiences are larger. Thus, the time difference $\Delta t'$ between the arrivals of the two electrons at $\hbar\omega_{op}$ is reduced; $\Delta t'<\Delta t$. Such a situation is enhanced by application of small magneti fields B. The work by Kurosawa reveals that, in an appropriate B sucl that $\omega_c/\omega=0.4\sim0.8$, electrons fall into a so called limit cycle and thi moments of phonon emission among different electrons converge, even though they start from random initial conditions. Strikingly enough, the convergence is accomplished practically within a half period of the microwave field! The electrons are accordingly bunched onto a *point* in velocity space in the case of ideal streaming motion, and the bunched carriers travel around the velocity space within the sphere $v<V_{op}$. In actual experimental conditions, a finite distribution of bunched electrons is caused around the focused point by a possible penetration of electrons into $v>V_{op}$. Figure 33 is from the work of

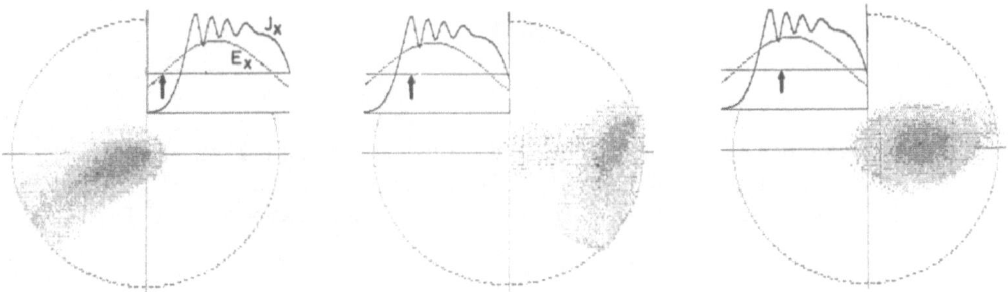


Fig. 33. Sequential change in a half period of the microwave field of the distribution function of bunched electrons on the $v_z=0$ plane at $\omega_c/\omega=0.8$, according to Kurosawa's simulation calculation[26] for electrons in AgBr ($\alpha$=1.6) at T=4.2K and $E_\omega$=1200V/cm with $\omega/2\pi$=35GHz. Parameters used for the scatterings in the calculation are the same as for Fig. 18(a) except that $\tau_{imp}$ =33psec is assumed.

Monte Carlo simulation by Kurosawa for AgBr ($\alpha$=1.6), and shows sequential change of the bunched-electron-distribution within a half period of the microwave field in the condition $E_\omega$=1200 V/cm and $\omega_c/\omega=0.8$. The impurity scattering and the acoustical phonon scattering are also taken into account in the calculation. (They do not have significant influence on the phenomena.) The spread of the electron distribution shown in the figure is the measure of the broadening caused by the penetration of electrons into $v>V_{op}$ at $E_\omega$=1200 V/cm. The dotted and the solid lines in the inset represent, respectively, the microwave field $E_x=E_\omega \sin\omega t$ and the dissipative current $J_x$. The arrows indicate the moments at which the distribution is displayed. Streaming motion of the bunched electrons causes the systematic spikes in $J_x$, each spike corresponding to the arrival of the bunched electrons at $v=V_{op}$.

Another interesting effect expected is a higher harmonic generation. In the regime of cyclotron streaming-motion with high B ($\omega_c>\omega$ but $B<B_c^{cal}$), accumulation of electrons in velocity space also takes place. The accumulation region inthis case travels elliptically in velocity space with the angular frequency $\omega$ within the sphere $v = V_{op}$ as shown in Fig. 34, where the sequential change of the distribution function calculated by Kurosawa[26] for AgBr at $E_\omega$=1200 V/cm and $\omega_c/\omega=3$ is indicated. The accumulated electrons perform cyclotron

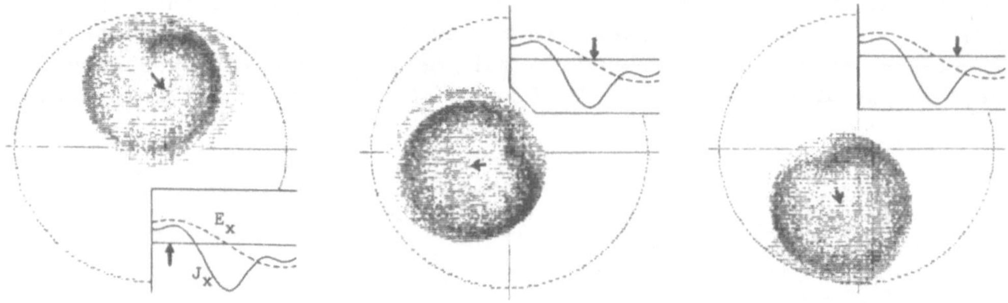


Fig. 34. Sequential change of the distribution function at $\omega_c/\omega=3$ according to Kurosawa's simulation calculation.[26] The parameters used in the calculation are the same as for Fig. 33.

oscillation with $\omega_c$ within the accumulation region. What is essentially different from the accumulation in dc-fields is that the accumulation region is deformed to have asymmetric shape if $\omega_c/\omega$ is an odd integer (larger than 1), as a consequence of the superposition of the eliptic motion with $\omega$ and the cyclotron motion with $\omega_c$. A net dipole moment rotating with $\omega_c$ is accordingly induced, as indicated by the arrows in Fig. 34, and generates a higher harmonic with $\omega_c$. A strong mixing of the third harmonic is clearly discerned in the current $J_x$ shown in the inset of Fig. 34.

Finally I would like to say a few words about the possibility of experimentally exploring these effects. As for a direct detection of the electron distributions as shown in Figs. 33 and 34, it almost falls into the category of fantasy, for experimentalists at the present stage of technology, to see the instantaneous electron distribution and to follow its rapid change within a fraction of the microwave period. However, indirect detection may be possible. When electrons are bunched, the number of phonons emitted by an electron in a half period of the microwave field is generally an integer. The number will increase stepwise as $E_\omega$ increases. Accordingly the absorbed microwave power is expected to exhibit a stepwise increase with increasing $E_\omega$,[26] which may be easily detected experimentally. The detection of the higher harmonics is probably an interesting subject for experiments in semiconductors with a reasonable density of carriers, where the detection of the effect will be relatively easy because a stronger generation is expected.

## 6. POLARON PROBLEM

An electron in ionic crystals is not a simple charged particle but a quasiparticle accompanied by lattice distortion (or virtual optical phonon clouds) around it, the polaron. The question arises on how the quasiparticle character affects the streaming motion and the accumulation of electrons. It has been pointed out in §4.D that the absence of photocurrent saturation in high electric fields, where the drift velocity of photocarriers is expected to be saturated, is a common feature in ionic crystals (Fig. 24). The singular behaviour of the trapping lifetime $\tau_t$ of electrons against $E_x$ and $B_z$, revealed in detail for silver halides (§4.D), is probably not a peculiar characteristic of silver halides but also a common aspect in all other ionic crystals. In addition, the experimental results of $\tan\theta$ (§4.C) have strongly indicated that the electron accumulation occurs without the help of adequate penetration of streaming electrons into $\varepsilon > h\omega_{op}$, which is unaccountable in the simple free-electron picture. These features should be compared to the corresponding phenomena in non-ionic materials: Photocurrent saturation is observed in Si[12] : The carrier accumulation in p-Ge can be substantially accounted for within the simple free-electron picture.[14,25] Hence the peculiar features of the experimental findings in ionic crystals appear to require serious consideration about the polaron problem. The aim of this section is to suggest an importance of the polaron self-energy effect on the phenomena, by providing a tentative explanation of the experimental results.

The polaron picture in the absence of external fields is described as follows. Suppose, first, an electron with an effective mass $m_r^*$ is in a rigid lattice. The energy momentum relation of this electron is sketched by a dashed line in Fig. 35. When one turns on the interaction between the electron and the optical phonons, a polarization field due to lattice distortion is induced around the electron to lower the total energy of the electron-phonon system (the self-energy effect). It will take some finite time for the system to reach an optimum, lowest-energy state. Let us call the optimum state of the electron-phonon system the ideal polaron state. The energy of the ideal polaron is lower than the energy of the rigid-lattice plus one-electron system by an amount $\sim\alpha\hbar\omega_{op}$. The energy momentum relation of the ideal polaron is modified to yield a non-parabolicity $m_p^*(\varepsilon)$ with the cold polaron mass $m_p^*(0) \sim m_r^*(1+\alpha/6)$ at $\varepsilon=0$. The polaron dispersion relation is sketched by a solid line in Fig. 35.

When one applies an infinitesimal electric field to an ideal polaron at the bottom of the band, the polaron will be infinitely slowly accelerated to climb up the polaron dispersion curve acquiring the instantaneous polaron effective mass $m_p^*(\varepsilon)$. The virtual phonon cloud gets thicker and the effective mass gets heavier during acceleration. The polaron will finally reach the state $\varepsilon=\hbar\omega_{op}$, and creates and emits one real phonon. The polaron, immediately returning to $\varepsilon=0$, repeats the same process thereafter. Approximately, this picture of polaron streaming-motion will also apply in finite small

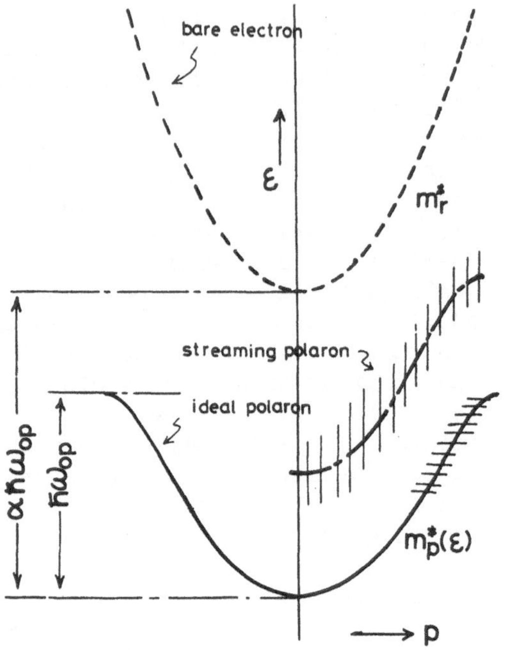


Fig. 35. Sketch of the energy-momentum relations of bare electrons and of ideal polarons, together with a tentative spectrum for streaming polarons in a finite electric field.

electric fields E, if the acceleration is so slow that the lattice distortion and the reaction of the electron properly follow the electron motion. The cyclotron resonance results in the range $E_\omega \lesssim 300$V/cm (§5.B) were interpreted on this scheme. Strictly speaking, however, this picture is no longer valid in finite E because a finite repetition rate of the sudden discontinuous change of the polaron state caused by the phonon emission will make the electron-phonon system incapable of achieving the optimum state at each moment during acceleration. To deal with the question of what should happen to the polaron in this situation, we have no available theory. The argument that follows is a crude speculation of the author.

One possible consequence of applying finite E may be, apart from the lifetime broadening of the energy states $\Delta\varepsilon \sim \hbar/T_{op}$, that the self-energy effect of the electron-phonon system is reduced to cause a rise in the overall energy of the electron-phonon system. The rough measure of the time required for the electron-phonon system to relax into the ideal polaron state would be some periods of the lattice vibration of the optical mode. Therefore, when E is so high that $T_{op} \sim 2\pi/\omega_{op}$ is achieved, the self-energy effect will practically vanish and the polaron states will resemble the states of the rigid-lattice plus one-electron system (apart from the significant uncertainty in energy $\Delta\varepsilon \sim \hbar\omega_{op}$). The electric field E corresponding to $T_{op} = 2\pi/\omega_{op}$

is 9kV/cm for AgBr and 18kV/cm for AgCl. Therefore the states of streaming polarons in the work described in these lectures should be expected to lie somewhere between the ideal polaron states and the states of the rigid-lattice plus one-electron system. The dot-dash line in Fig. 35 indicates a probable dispersion relation for the streaming polaron, with a possible energy-broadening expressed by the thin vertical shading across the line. In contrast to the streaming polaron, the bound polaron state at capture centers, presumably lying well below the bottom of the ideal polaron band with a small Bohr radius of order 30∿50Å, will be less affected by E since the bound state must be primarily determined by the stronger local potential. Hence it is suggested that the upward energy shift of the streaming polaron relative to the bound state causes the observed increase of $\tau_t$ of polarons in the streaming regime.

Let us next consider the polaron accumulation into region K. Generally the problem of self-energy effects on accumulated polarons would be physically more complicated because of the additional presence of magnetic field. We note here only the fact that these polarons do not suffer from sudden discontinuous change of states, being free from phonon emission. This situation will permit the accumulated polarons to achieve 'better' polaron states than those of streaming polarons. Thus the accumulation area K may be located below the streaming polaron band. If this is the case, accumulation of polarons is possible without a significant penetration of streaming polarons into the higher kinetic energy region above $\hbar\omega_{op}$, and the experimental results of $\tan\theta$ can be reasonably interpreted. It is expected that the motion of accumulated polarons is quantized into Landau states with well-defined eigen-energies. Hence one is further tempted to assume that the self-energy effect is even equivalent to that of ideal polarons, expecting the accumulation region in the ideal polaron band as shown with a horizontal hatch in Fig. 35. This hypothesis may lead to trapping lifetimes of the accumulated polarons identical to that of the polarons at E=0, and explains the reduction of $\tau_t(E)$ to $\tau_t(0)$ with increasing B from $\zeta=1$ to $\zeta=2$.

Thus, it is possible to provide tentative explanations for all the conflicting features of the phenomena observed in silver halides. The above discussion has hopefully given a reasonable argument for the possible upward energy-shift of the streaming polarons with respect to the bounded or accumulated polarons. If the arguments have indeed provided some proper insight into reality, polaron theories treating explicitly the self-energy effect at finite E should appear.

Finally it would be appropriate to mention the work by Thonber and Feynman,[59] in which the relation between polaron drift velocity $v$ and the electric field E is theoretically studied with lattice temperature T and the coupling constant $\alpha$ as parameters. The polaron velocity $v(E)$ in this work exhibits saturation with increasing E at low T, similarly to the experimental results. However, the saturation velocity is slightly higher than $V_{op}$ as pointed out by Peeters and Devreese.[60] The factor of two type discrepancy with the experimental value $v \sim V_{op}/2$ is not likely to be attributable to an uncertainty included in the experiemntal value. The theory[59] assumes that a polaron in finite E reaches a steady state $<\dot{p}>=0$, where $<\dot{p}>$ is the expectation value of $\dot{p}$ with the electron momentum p. The theory hence starts from the requirement that the driving force, eE, from the electric field should balance with a resistive force, $<i[p,I]>/\hbar$, from the electron-phonon interaction I, where$[p,I] \equiv pI-Ip$. In reality, however, the polaron repeats the cycle of acceleration and phonon emission, and never does reach the steady state. The steady-state approach would be a proper way to look into the phenomena, if the 'polaron state with a steady velocity $v$' in question is constructed from a proper combination of all the ideal-polaron (or the bare-electron) states that are experienced by the electron during the process of acceleration and phonon emission (or absorption). However, the polaron state in the work is not so constructed. Peeters and Devreese[60] reveal the 'polaron state with $v$' in the work includes at T=0 only the electron state having the velocity $v$. This is clearly inadequate to describe the real situation, and it is now easy to show how the result $v \gtrsim V_{op}$ is derived from the assumption of steady state. The force $<i[p,I]>/\hbar$ coming from phonon scattering is almost zero at $T \sim 0$ when $v<V_{op}$ because of the absence of scattering, and is very strong when $v>V_{op}$ due to the possible emission of phonons. Therefore the polaron state that fulfils the balance equation $eE=<i[p,I]>/\hbar$ at a given finite E must always have a velocity (slightly) higher than $V_{op}$ at low T, independently of the approximation used to calculate $<i[p,I]>/\hbar$. The velocity is insensitive to E when E is not too high because the resistive force by phonon emission is strong. Thus the velocity saturation above $V_{op}$ is a direct consequence of the assumption of a steady state, which is unacceptable from the picture of streaming motion.

## 7. CLOSING REMARKS

Silver halides are not the only materials in which streaming motoin or accumulation of electrons has been observed. Good indications have also been obtained for streaming motion in photocurrent work on Si [12], and even for electron accumulation in work on CdS.[11] In the Hall effect measurements in p-Ge [13,14], a definite evidence for streaming motion and carrier accumulation has been obtained. In this work[14] the ratio of accumulation $n_k/n_s$ has been deduced from the Hall effect data and compared with the theory in a manner similar to that described in these lectures.

A significant aspect of carrier accumulation in p-type materials, in which the light and the heavy hole bands degenerate at the band edge, is the emission of far-infrared radiation. Because of the difference in the effective masses of light and heavy holes, we can realize such a situation in appropriate magnetic fields that accumulation takes place only in the light hole band while heavy holes are streaming.[13,14,16] Since the light-hole accumulation is caused by a redistribution of carriers from the heavy-hole band to the light-hole band, the accumulation is expected to lead to the emission of radiation associated with the transition of accumulated light holes to the heavy-hole band, as first pointed out theoretically by Andronov et al.[16] The emission of radiation via this mechanisms has been observed in p-Ge at 4.2K.[15] The emission work not only opened up the possibility of directly probing the energy distribution function of accumulated carriers, but has also encouraged further work seeking for far-infrared amplification.[16]

The unique aspect of streaming motion and accumulation of hot carriers is the realization of very specific distribution function in the band continuum. As the specific distribution of carriers can directly be manipulated by external fields, the phenomena are furnished with a wide variety of the possibility of further fascinating effects. The far-infrared amplification, as well as the bunching effect and the higher harmonic generation described in §5.C, are examples but do not exhaust the possibilities. A number of theoretical predictions seem now to give a strong impetus to the experiments on semiconductors. However, we do not describe the theoretically expected phenomena here because good review articles to the subject are available. Reviews by Kurosawa[21] and by Al'ber et al.[20] serve as useful general introductions to the theoretical aspects of the expected phenomena. Both experimental and theoretical achievements are summarized and the realizabilities of the theoretical predictions are discussed in Ref. 25. Since the expected phenomena are more or less related to amplifications in the range from mm-waves to the far-infrared, the extention of experimental work to other semiconductors may be of particular interests from the practical point of view in the device physics.

Apart from the novel effects predicted by the theories, an interesting question raised by the experiments in silver halides is the problem of polarons in finite electric (and magnetic) fields. In this aspect the experiments seem to give a strong impetus to the refinement of polaron theories. The key point will be to include the self-energy effect, as pointed out in §4.C,D and §6.

## APPENDIX: QUALITATIVE PICTURE OF STREAMING MOTION AND CARRIER ACCUMULATION IN QUANTUM MECHANICS

We shall consider how streaming motion and onset of carrier accumulation are described in terms of quantum mechanics. The Schrödinger equation for a free electron in the crossed fields $\vec{E}=(E_x,0,0)$ and $\vec{B}=(0,0,B_z)$ is written in the form

$$[\frac{1}{2m^*}(\vec{p} + e\vec{A})^2 + eE_x x]\psi = \varepsilon\psi \;, \tag{A1}$$

where $\vec{p}$ is the momentum operator and $\vec{A}=(0,B_z x,0)$ is the vector potential in the Landau gauge. The equation is exactly solved[45] to yield $\varepsilon$ in the form given by Eq.(26) and

$$\psi_{n,v_z,x_0} = \{\exp[i(\frac{m^*}{\hbar}\frac{E_x}{B_z})y]\}\psi^0_{n,v_z,x_0} \;, \tag{A2}$$

where $\psi^0_{n,v_z,x_0}$ is the eigen function of the n-th Landau state with $v_z$ at $x=x_0$ in the x-coordinates in the absence of electric field. Wave function (A2) represents the superposition of the drift motion along the y-direction with velocity $(E_x/B_z)$ and the cyclotron oscillation of the electron. The measure of spatial extent of the wave function in the x-direction around $x=x_0$ is given by the cyclotron radious

$$r_n = [2\hbar(n + \frac{1}{2})/eB_z]^{1/2} \;. \tag{A3}$$

The correspondence of the quantum mechanical energy (Eq.(26)) to the classical energy will be readily demonstrated. Classical cyclotron orbits Eqs.(13)-(14) are characterized by V, $v_z$ and $x_0$, where $x=x_0 \equiv x(0) - (V/\omega_c)\cos\phi$ is the center axis on the x-coordinates of the motion of an electron with the initial position at x(0). Adding potential energy $\varepsilon_p \equiv e\,x(t)E_x = e[\int_0^t v_x(t)dt + x(0)]E_x$ to the kinetic energy $\varepsilon_k=(1/2)m^*[v_x(t)^2 + v_y(t)^2 + v_z(t)^2]$, we obtain the total energy of an electron in the classical description in the form,

$$\varepsilon_{V,v_z,x_0} = \frac{m^*}{2}V^2 + \frac{m^*}{2}v_z^2 + \frac{m^*}{2}(\frac{E_x}{B_z})^2 + eE_x x_0 \;. \tag{A4}$$

Since $x(t)=(V/\omega_c)\cos(\omega_c t + \phi) + x_0$, the classical cyclotron radius is

$$r_V = V/\omega_c \;. \tag{A5}$$

Equations (A4) and (A5) are identical to Eqs.(26) and (A3), respectively, when V is substituted by

$$V = [\frac{2}{m^*}(n + \frac{1}{2})\hbar\omega_c]^{1/2} \;. \tag{A6}$$

For the discussion below we eliminate n in Eq.(26) by (A3) and obtain

$$\varepsilon_{n,v_z,x_0} = \frac{(eB_z)^2}{2m^*} r_n^2 + \frac{m^*}{2}\left(\frac{E_x}{B_z}\right)^2 + eE_x x_0 \ . \qquad (A7)$$

To express Landau states at $x=x_0$, it is convenient to draw the parabola

$$g_1 = \frac{(eB_z)^2}{2m^*} (x - x_0)^2 + \frac{m^*}{2}\left(\frac{E_x}{B_z}\right)^2 + eE_x x_0 \ , \qquad (A8)$$

as shown in Fig.A. The bottom of the parabola,

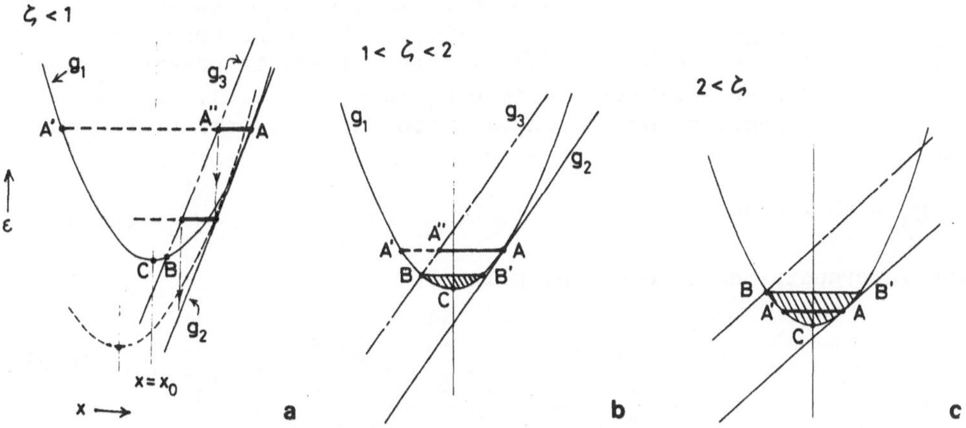


Fig. A. Alternative explanation in terms of quantum mechanics of (a) streaming motion, (b) streaming motion plus accumulation and (c) accumulation. Each stage from (a) to (c) corresponds to the resepctive stage in Fig. 8.

$$C = \left[x_0 \ , \ \frac{m^*}{2}\left(\frac{E_x}{B_z}\right)^2 + eE_x x_0\right] \ , \qquad (A9)$$

corresponds to the center of classical cyclotron orbits as represented by point C in Fig. 8. A Landau state with energy $\varepsilon=\varepsilon_1$ is represented by a horizontal bar at $g_1=\varepsilon_1$ with both sides limited by the parabola. The span of the bar indicates the spatial extent of the Landau state. We will assume the Landau splitting $\hbar\omega_c$ is relatively small and the parabola is filled quasi-continuously with such bars. To include the Landau states at different positions in the x-coordinates, we additionally draw a line

$$g_2 = eE_x x \quad , \tag{A10}$$

which is tangent to the parabola at point A:

$$A = [x_0 + (\frac{E_x}{B_z})/\omega_c \, , \; m^*(\frac{E_x}{B_z})^2 + eE_x x_0] \, . \tag{A11}$$

The Landau states at different positions are expressed by equivalent parabolas obtained by shifting $g_1$ along the line $g_2$. The area above the line $g_2$ is filled with Landau states, while there are no states below the line.

Now we will consider the phonon emission. In the language of quantum mechanics, the optical phonon emission is described by a transition of an electron from a Landau state $\psi_{n,v_z,x_0}$ to another state $\psi_{n',v_z',x_0'}$ with energy $\epsilon_{n',v_z',x_0'} = \epsilon_{n,v_z,x} - \hbar\omega_{op}$. (We will consider for simplicity only the case $v_z = v_z' = 0$.) The transition is possible when the two states have a finite overlapping in space; i.e., $(x_0 - x_0') < (r_n + r_{n'})$. To consider the phonon emission in the scheme of Fig. A, we add a line

$$g_3 = eE_x x + \hbar\omega_{op} \quad , \tag{A12}$$

which intersects the parabola at point B,

$$B = [x_0 - (V_{op} - \frac{E_x}{B_z})\omega_c^{-1} \, , \; \frac{m^*}{2} (V_{op} - \frac{E_x}{B_z})^2 + \frac{m^*}{2}(\frac{E_x}{B_z})^2 + eE_x x_0] \, . \tag{A13}$$

We shall classify the situation by using normalized field $\zeta = V_{op}/(E_x/B_z$

First, when $\zeta < 1$, point B is located on the section C-A, and the energy at point C is larger than $eE_x x_0 + \hbar\omega_{op}$, as shown in Fig. A(a). Therefore electrons at any Landau states are capable of emitting optical phonons. Successive emissions of the phonons, like $\psi_{n,0,x_0} \to \psi_{n',0,x_0'} \to \cdots$, is thus possible and streaming motion occurs. For the streaming motion, the Landau state expressed by the bar A-A′ in $g_1$ and the equivalent states in other parabolas are of primary importance. This is because only these states extend spatially to touch the line $g_2$ and they are the most probable states in accepting electrons after the emission of optical phonons. The Landau state of bar A-A′ is of the Landau index

$$n \simeq \frac{m^*}{2}(\frac{E_x}{B_z})^2 (\hbar\omega_c)^{-1} - \frac{1}{2} \, . \tag{A14}$$

Thus the streaming motion is described by the transition, $\psi_{n,0,x_0} \to \psi_{n,0,x_0 - \Delta x} \to \psi_{n,0,x_0 - 2\Delta x} \to \cdots$, with n given by (A14). An example of the transition is shown in Fig. A(a). From $\varepsilon_{n,0,x_0} - \varepsilon_{n,0,x_0} = \Delta x = \hbar\omega_{op}$, we obtain

$$\Delta x = \hbar\omega_{op}/(eE_x). \tag{A15}$$

By use of (A6) we find the Landau state determined by (A14) corresponds to the cyclotron orbit which passes $v$=0 and provides the trajectory for streaming motion in classical description (Fig. 7(a)).

Second, when 1<ζ<2, point B is located on the section C-A′ and the energy at point C becomes smaller than $eE_x x_0 + \hbar\omega_{op}$. There accordingly appear Landau states below the bar B-B′, in which electrons are incapable of emitting optical phonons. The area in which those states are accommodated is indicated by the shading in Fig. A(b). Accumulation of electrons into this area is possible, similarly, as in the classical description. (The energies of accumulated electrons would seem to be smaller than those of streaming electrons in the present description, because the energy of Landau state A-A′ is obviously higher than that of B-B′. However, the Landau state A-A′ is not a well-defined eigen-state of the system, and the 'energy' of streaming electrons cannot be measured by the eigen-energy of the Landau state A-A′. The energy should rather be measured by the kinetic energy expressed in Fig. A by the average distance of the section A-A" from the line $g_2$, where point A" is the intersection of the line $g_3$ and the bar A-A′. So measured energy of streaming electrons is smaller than the eigen-energies of accumulated electrons.) The Landau state represented by the bar B-B′ has the Landau index n:

$$(n + \frac{1}{2})\hbar\omega_c \simeq \frac{m^*}{2}(V_{op} - \frac{E_x}{B_z})^2 . \tag{A16}$$

By use of (A6), we can immediately find this state to correspond to the cyclotron orbit bounding area K in Fig. 8(b). By expanding the consideration to the case $v_z \neq 0$, we obtain relation (27) for the number of Landau states accommodated within the region for accumulation.

Third, when ζ>2, point B goes to the left hand side of point A′ and the Landau state A-A′, which otherwise provides streaming motion, is included within the shaded area where the phonon emission is impossible (Fig. A(c)). Streaming motion accordingly becomes impossible. By (A6), the Landau states represented by B-B′ and A-A′ correspond to the classical orbit tangent to sphere $V_{op}$ and the orbit passing the point $v$=0, respectively, in Fig. 8(c).

References

1. E.M. Conwell, Solid State Phys. Suppl. 9 (1967).
2. S. Komiyama, T. Masumi and K. Kajita, Proc. 13th Int. Conf.,

Phys. of Semiconductors, Roma 1976 (Tipografia Marves, Roma, 1976), p. 1222.
3. S. Komiyama, T. Masumi and K. Kajita, Phys. Rev. Lett. 42, 600 (1979).
4. S. Komiyama, T. Masumi and K. Kajita, Solid State Commun. 31, 447 (1979).
5. S. Komiyama, T. Masumi and K. Kajita, Phys. Rev. B 20, 5192 (1979).
6. S. Komiyama and T. Masumi, Solid State Commun. 26, 381 (1978).
7. S. Komiyama, T. Masumi and T. Kurosawa, Proc. 14th Int. Conf. Physics of Semiconductors, Edinburgh, 1978 (The Institute of Physics, Bristol, 1978), p. 335.
8. S. Komiyama and T. Masumi, J. Magnetism and Magnetic Materials, 11, 59 (1979).
9. W. Shockley, Bell Syst. Tech. J. 30, 990 (1951).
10. H. Maeda and T. Kurosawa Proc. 11th Int. Conf. Physics of Semiconductors, Warsaw, 1972 (Elsevier, New York, 1972), p. 602.
11. Y. Iye and K. Kajita, Solid State Commun. 17, 957 (1975).
12. K. Kajita, Solid State Commun. 31, 573 (1979).
13. K. Komiyama and R. Spies, Phys. Rev. B23, 6839 (1981).
14. S. Komiyama and R. Spies, Proc. 3rd Int. Conf. Hot Carriers in Semiconductors, Montpellier, 1981, J. de Physique, 42, Suppl. 10, C7-387 (1981).
15. S. Komiyama, Phys. Rev. Lett. 48, 271 (1982).
16. A.A. Andronov, V.A. Kozlov, L.S. Mazov and V.N. Shastin, Sov. Phys. - JETP Lett. 30, 551 (1979).
17. T. Kurosawa, Solid State Phys. 11, 217 (1976); in Japanese.
18. T. Kurosawa, Solid State Commun., 24, 357 (1977).
19. Ya. I. Al'ber, A. A. Andronov, V. A. Valov, V. A. Kozlov and I. R. Ryazantseva, Solid State Commun. 9, 955 (1976).
20. Ya. I. Al'ber, A. A. Andronov, V.A. Valov, V. A. Kozlov, A. M. Lerner and I. R. Ryazantseva, Sov. Phys. - JETP 45. 539 (1977).
21. T. Kurosawa, Proc. 3rd Int. Conf. Hot Carrier in Semiconductors, Montpellier , 1981, J. de Physique 42, Suppl. 10, C7-377 (1981).
22. T. Kurosawa and H. Maeda, J. Phys. Soc. Jpn. 31, 668 (1971).
23. A. A. Andronov, V. A. Valov, V. A. Kozlov and L. S. Mazov, Solid State Commun. 36, 603 (1980).
24. V. L. Gurevich and D. A. Parshin, Solid State Commun. 37, 511 (1981).
25. S. Komiyama, Advances in Physics, 31, 255 (1982).
26. T. Kurosawa, Proc. 15th Int. Conf. Physics of Semiconductors, Kyoto, 1980, J. Phys. Soc. Jpn. 49 (1980) Suppl. A, p. 345.
27. T. Kurosawa, unpublished.
28. See, for example, H. H. Tippins and F. C. Brown, Phys. Rev. 129, 2554 (1963).
29. K. Kobayashi and F. C. Brown, Phys. Rev. 113, (1959).
30. T. Masumi, Phys. Rev. 129, 2564 (1963); ibid, 159, 761 (1967).

31. S. Komiyama, D. Sc. thesis (Tokyo University, 1976) (unpublished).
32. F. C. Brown, in "Polarons and Excitons", (edited by C. G. Kuper and G. D. Whitfield (Oliver and Boyd, London, 1962), and J. W. Hodby, in "Polarons inIonic Crystals and Polar Semiconductors", edited by Devreese (American Elsevier, New York, 1972), p. 389.
33. H. Fröhlich, H. Pelzer and S. Zienau, Phil. Mag. 41, 221 (1950).
34. G. Whitfield and R. Puff, Phys. Rev. 139, A 338 (1965).
35. D.M. Larsen, Phys. Rev. 144, 697 (1966).
36. H. Tamura, unpublished.
37. H. Kanzaki, S. Sakuragi and K. Sakamoto, Solid State Commun. 9, 999 (1971).
38. H. Tamura and T. Masumi, Solid State Commun. 12, 1183 (1973).
39. E. Hanamura, T. Inui and Y. Toyozawa, J. Phys. Soc. Jpn. 17, 666 (1962).
40. D. S. Tannhauser, L. J. Bruner and A. W. Lawson, Phys. Rev. 102, 1976 (1956).
41. H. Tamura, Solid State Commun. 10, 297 (1972).
42. See p. 157 in Ref. 1.
43. E. Conwell and V. F. Weisskopf, Phys. Rev. 77, 388 (1950).
44. I. I. Vosilyus and I. B. Levinson, Soviet Phys. - JETP 23, 1104 (1966), and Soviet Phys. - JETP 25, 672 (1967).
45. W. Zawazki, "Physics of Solids in Intense Magnetic Fields", (edited by E. D. Haidemenakis, Plenum, New York, 1969), p. 307.
46. K. Kajita and T. Masumi, Proceedings of the Twelfth International Conference on the Physics of Semiconductors, Stuttgart, 1974 (Teubner, Stuttgart, 1974), p. 844.
47. See, for example, O. Matsuda and E. Otsuka, Solid State Commun. 26, 925 (1978), see also, N. Kotera, E. Yamada and K. F. Komatsubara, J. Phys. Chem. Solids 33, 1311 (1972).
48. J. T. Devreese, F. Brosens, R. Evrard and E. Katheuser, Proc. 15th Int. Conf. Physics of Semiconductors, Kyoto, 1980, J. Phys. Soc. Jpn. 49 (1980) Suppl. A, p. 341.
49. S. Komiyama and T. Masumi, unpublished.
50. R. S. Van Heyningena nd F. C. Brown, Phys. Rev. 111, 462 (1958), and R. C. Brandt and F. C. Brown, Phys. Rev. 181, 1241 (1969).
51. T. Masumi, Phys. Rev. 129, 2564 (1963); ibid, 159, 761 (1967).
52. M. Mikkor and F. C. Brown, Phys. Rev. 162, 841 (1967).
53. F. Nakazawa and H. Kanzaki, J. Phys. Soc. Jpn. 20, 468 (1965).
54. S. B. Bolte and F. C. Brown, Proc. 3rd Int. Conf. Photoconductivity, 1971 (Pergamon, New York, 1971) p. 139.
55. T. Kawai, K. Kobayashi and H. Fujita, J. Phys. Soc. Jpn. 21, 453 (1966).
56. J. W. Hodby, Solid State Commun. 7, 811 (1969).
57. H. Tamura and T. Masumi, J. Phys. Soc. Jpn. 30, 1763 (1971).
58. H. Kawamura, M. Fukai and Y. Hayashi, J. Phys. Soc. Jpn. 17, 970 (1962).
59. K. K. Thonber and R. P. Feynman, Phys. Rev. B 1, 4099 (1970).
60. D. M. Peeters and J. T. Devreese, Phys. Rev. B 23, 1936 (1981).

DYNAMICAL AND NONLINEAR PROFILES OF POLARONS AND EXCITONS IN PURE AND ULTRAPURE AgCl, AgBr and $AgCl_xBr_{1-x}$

Taizo Masumi

Department of Pure and Applied Sciences
University of Tokyo (Komaba)
3-8-1 Komaba, Meguro-ku, Tokyo 153, Japan

ABSTRACT

This series of lectures starts with a brief description of a number of basic aspects of the theory of polarons. Subsequently experimental problems in the field of polarons are examined. The experimental progress realized in our laboratory over the last decade is reviewed. Our experiments on AgCl, AgBr and $AgCl_xBr_{1-x}$ include i) the investigation of transport phenomena, cyclotron resonance absorption and relaxation mechanisms for polarons, ii) the measurement of optical absorption and luminescence in pure and ultrapure crystals with the analysis in terms of excitons and excitonic molecules.

We emphasize the importance of the dynamical and nonlinear response of polarons and excitons in analysing our experiments.

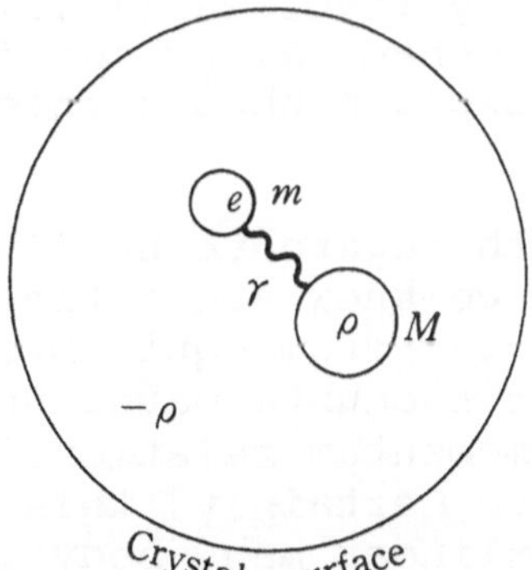


Fig. 1-1. An example of a polaron model in a contemporary style.

## 1. INTRODUCTION

Consider a muscat on a soft surface of jelly cake. By its weight the muscat deforms the surface of the jelly cake (Fig. 1-2). If one gently pushes the muscat (on the jelly cake) it rolls away, accompanied by the deformation underneath as if they form one entity. This is one of the simplest classical pictures of a *"polaron"*. In solid state or condensed matter physics a conduction electron plays the role of the muscat and the ionic crystal or polar semiconductor plays the role of the jelly cake.

The basic description of the large polaron, starting from Landau's concept was introduced by Fröhlich in the earlier period of the 1950's (1950, 1954). It is a typical example of an elementary excitation in polar semiconductors. Although a considerable amount of work has been devoted to the polaron problem (see the books of Kuper and Whitfield 1963, Appel 1968 and Devreese 1972), there still remain several essential problems unsolved both in theory and experiment.

Theoretical physicists usually describe a polaron as an electron in a solid associated with a virtual phonon cloud. The ground state of a polaron was intensively studied by using the perturbation theory (Fröhlich 1950), by the canonical transformation and variational technique related to the "intermediate coupling method" (Lee, Low and Pines 1953), and by using the "path integral method" with a variational principle (Feynman 1955). (For a review, see the books of Kuper and Whitfield 1963 and Appel 1968.)

The problems of excitons, associated with the polaron problem, were reviewed by Elliott (1963), and also by Haken (1963). Dynamical properties of charge carriers in dielectrics such a free polarons were discussed theoretically by Toyozawa (1972). Toyozawa had been interested also in dynamical properties of excitons and naturally in the exciton-phonon interaction (Toyozawa and Hermanson 1968).

With regard to nonlinear properties of polarons, what do we know about the higher energy states of a polaron far from equilibrium? Unfortunately, no one knows the most adequate wavefunction and, as a result, the energy-momentum relation in higher energy states of a polaron. Certainly, this is also one of the most difficult nonlinear many-body problems. Gorkum and Tolpygo (1960) tried to consider a trial wavefunction for a fast polaron but treated the problem by a variational method, whereas Whitfield and Puff (1965) discussed the energy-momentum relation of a polaron at higher energy states near the LO-phonon energy by using the Green function technique. Later, Thornber and Feynman (1970) calculated the impedance of electrons in polar crystals by using the path integral method and obtained numerical results for the field versus velocity curve.

What does a polaron look like, then, for a naive-minded but still curious experimentalist? In the 1950's, we only had a rather preliminary knowledge of the experimental aspects of a polaron, even for the ground state. At the beginning of the 1960's, Professor F.C. Brown and his co-workers (Burnham, Brown and Knox 1960; Brown, Masumi and Tippins 1961; Ascarelli and Brown 1962) started their first experimental attack on the polaron near its ground state using transport measurements, combined with optical and cyclotron resonance techniques.

Experimental studies of the ground state of a polaron by similar but more sophisticated techniques have also been developed particularly at high magnetic fields and reviewed extensively by Hodby (1972).

A series of experimental studies have been also performed in our laboratory, especially concerning the dynamical and nonlinear response of Fröhlich polarons (Masumi 1981). The dynamical characteristics of polarons and excitons have been studied experimentally in particular for AgCl, AgBr and $AgCl_xBr_{1-x}$ in the last decade (Baba and Masumi 1982, Kawahara and Masumi 1980 and Masumi 1981[¶]). Nonlinear response of polarons was first

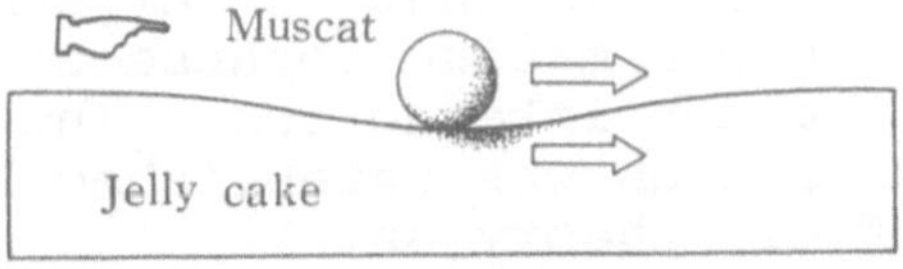


Fig. 1-2. A classical view of a slow polaron.

¶ On the other hand, the transition from a large to a small polaron, e.g. the self-trapping of a hole-polaron, was extensively studied in the formation process of a $V_k$-center starting with a more extended electron-polaron, i.e. a self-trapped exciton in alkali halides and AgCl (Brown 1981).

studied experimentally in AgCl and AgBr by the author (Masumi 1963, 1967). These studies later developed intensively (Komiyama, Kajita and Masumi 1981, Komiyama and Masumi 1978, 1979d, 1979e). In the present paper the dynamical and nonlinear response of ("deformable") polarons and excitons which plays an important role in transport phenomena and optical spectroscopy will be discussed.

We have never expected that our experimental studies would develop in parallel (in space and time) with theoretical studies. The present paper mainly reviews the experimental progress made in the last decade, of physics of Fröhlich polarons and excitons.

## 2. BASIC PROBLEMS IN THE THEORY OF POLARONS

Theoretical physicists have developed and established a set of fundamental theories on polarons (especially of their ground state), by the perturbation theory (Fröhlich 1950), the intermediate coupling method (Lee, Low and Pines 1953) and the path integral method (Feynman 1955). The basic problems of an exciton were described by Elliott (1963) and were also associated with polarons by Haken (1963). Nevertheless, the Schrödinger equation for the polaron has never been solved exactly. Here, we briefly summarize the fundamental problems in the polaron theory.

### 2-1) Basic Problems

First let us consider the conduction electrons moving freely in an ionic crystal. The electrons strongly interact with the LO-phonons, and vice versa, and also with each other. We assume that the crystal can be regarded as a continuous medium because of the long range character of the Coulomb interaction. The basic Hamiltonian for a polaron was established by Fröhlich in terms of quantum field theory as

$$H \equiv H_0 + H_1 \tag{2.1}$$

$$H_0 = -\frac{\hbar^2}{2m^\star}\Delta + \hbar\omega_{LO}\sum_q b_q^+ b_q \tag{2.1a}$$

$$H_1 = \sum_q [V_q b_q e^{iqr_e} + V_q^+ b_q^+ e^{-iqr_e}] \tag{2.1b}$$

Fig. 2-1. Formation processes of a polaron; a typical example of a particle-field problem.

a) A particle and a field both in their ground states and independent from each other ($H_1$=0).

b) Creation of a polaron ($T$=0$K$).

c) Annihilation of a polaron ($T$=0$K$). The lower right inset illustrates the phonon correlation effects.

d) The ground state of a polaron after its formation. It is rather hard to recognize it clearly due to the recoil and phonon correlation effects.

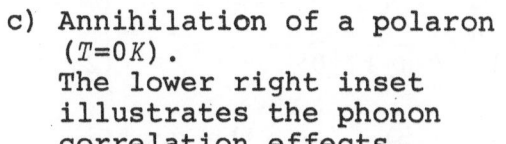

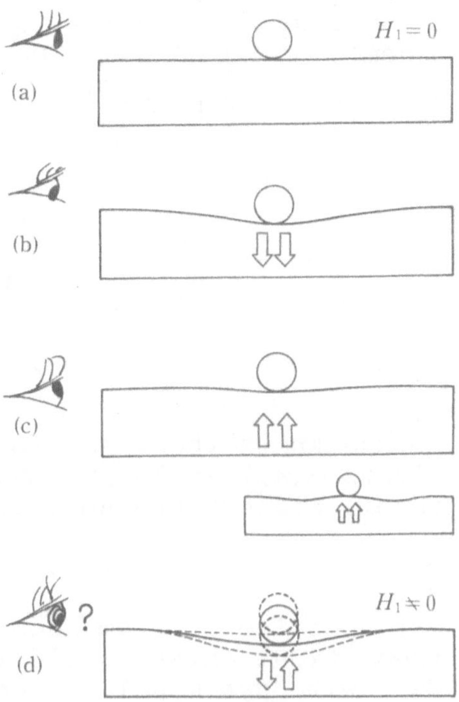


where the first term in $H_0$ of (2.1a) denotes the kinetic energy of an electron with the band mass $m^\star$, $b_q^+$, $b_q$ are the creation and annihilation operators of LO-phonons with the frequency $\omega_{LO}$ and wavevector $q$. $H_1$ represents the electron-LO-phonon interaction with

$$V_q = - \frac{i\hbar\omega_{LO}}{u^{1/2}} (4\pi\alpha)^{1/2} \frac{1}{q} \tag{2.2}$$

$$u \equiv [\frac{2m^\star \omega_{LO}}{\hbar}]^{1/2} \tag{2.3a}$$

$$\alpha \equiv \frac{e^2 u}{2\hbar\omega_{LO}} [\frac{1}{\kappa_\infty} - \frac{1}{\kappa_0}] \tag{2.3b}$$

where $\kappa_0$ and $\kappa_\infty$ are the static and optical dielectric constants, respectively and $\alpha$ is called the coupling constant of the electron in the crystal. The theoretical problem of the polaron theory is now to find the exact solution of the Schrödinger equation

$$\mathcal{H}\Phi = E\Phi \;, \tag{2.4}$$

namely, how to calculate the exact eigenfunction $\Phi$ and eigenvalue $E$. Usually, this problem is treated in a scheme, in which one expands $\Phi$ in terms of the unperturbed Bloch state $\exp[i\vec{k}_0\vec{r}]$ and LO-phonon states $\Pi(b_q^+)^{n_q}/\sqrt{n_q!}|0>$ as follows

$$\Phi(k, \{n_q\}) = \frac{1}{\sqrt{V}} \sum_{\{n_q\}} \Pi\, c(n_q) \times \exp[i(k - \sum_q qn_q)r] \times [(b_q)^{n_q}/\sqrt{n_q!}]|0> \tag{2.5}$$

No one knows the exact form of (2.5), i.e. the coefficients $c(n_q)$ including the phonon correlations with the recoil effects are not rigorously known. Consequently,

$$\Phi(k, \{n_q\}) = ?,\ E(k) = ? \tag{2.6}$$

irrespective of tremendous efforts in a large number of theoretical investigations (Appel 1968). The theory must deal with dynamical and nonlinear many-body problems. The formation processes of a polaron are depicted in Fig.2-1.

## 2-2) Review of Some Theoretical Results Relevant for Experimental Studies

Many theories have been developed to describe the polaron, which is an example of a particle-field problem. Here, we only recall some results, which are important for the analysis of experimental studies, e.g. those obtained mainly by using the second order perturbation method in the weak coupling case ($\alpha \ll 1$) (Fröhlich 1950, 1954).

a) Phonon clouds. The expectation value for the number of virtual phonons $N = \sum_q b_q^+ b_q$ was calculated as follows

$$\begin{aligned} <N> &\equiv {}^{(1)}<k;\ 0|N|k;\ 0>^{(1)} \\ &= \sum_q \frac{|<k-q;\ 1_q|\mathcal{H}_1|k;\ 0_q>|^2}{(\varepsilon_k - \varepsilon_{k-q} - \hbar\omega_{LO})^2} \\ &= \sum_q \frac{|V_q|^2}{(\varepsilon_k - \varepsilon_{k-q} - \hbar\omega_{LO})^2} \\ &= \alpha/2 \end{aligned} \tag{2.7}$$

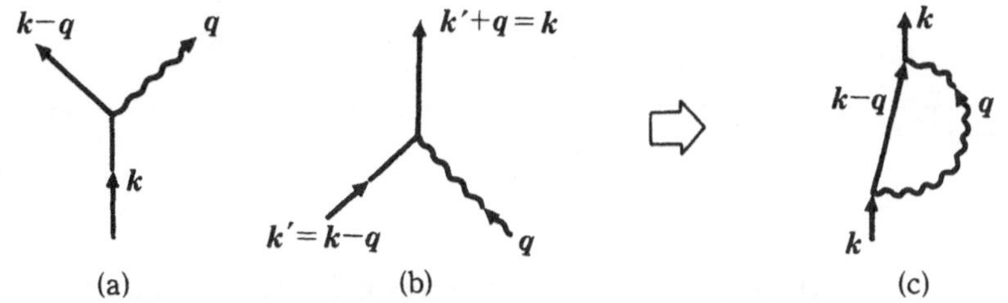


Fig. 2-2. LO-phonon scattering of an electron,
a) LO-phonon emission, b) LO-phonon absorption,
c) Polaron formation as described by the 2nd order term.

b) Self energy and renormalized mass. The corrected eigenvalue $E(k)$ of a polaron in the second order perturbation theory indicated in Fig. 2-2 was calculated as

$$E(k) = -\alpha\hbar\omega_{LO} + \frac{\hbar^2 k^2}{2m^\star}\left(1 - \frac{\alpha}{6}\right) + O(k^4); \qquad (k<<u) \tag{2.8}$$

in the limit of $k<<u$ for "slow electrons". Thus, the first term can be considered to give the self energy and the second term leads to the renormalized polaron mass

$$m_P^\star = \frac{m}{1-\alpha/6}\ ; \quad \frac{m_P^\star}{m^\star} \simeq 1 + \frac{\alpha}{6} \qquad (\alpha<<1) \tag{2.9}$$

in the weak coupling limit ($\alpha<<1$). Theorists usually illustrate the low energy part of the polaron energy spectrum as shown in Fig. 2-3. As the polaron momentum $k$ increases the continuum spectrum is reached.

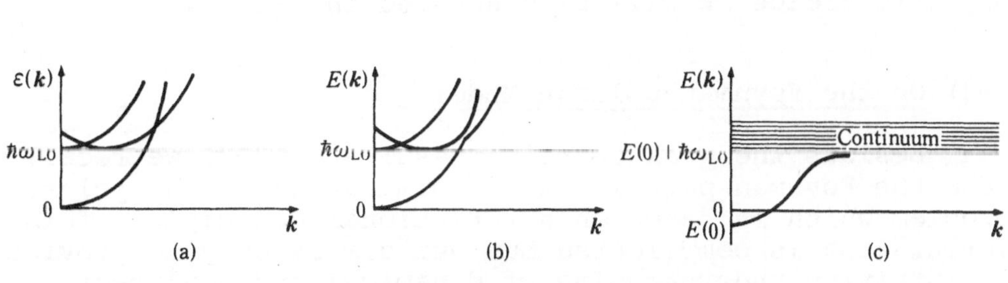


Fig. 2-3. Dispersion relation of a polaron in the low energy part of the energy spectrum.

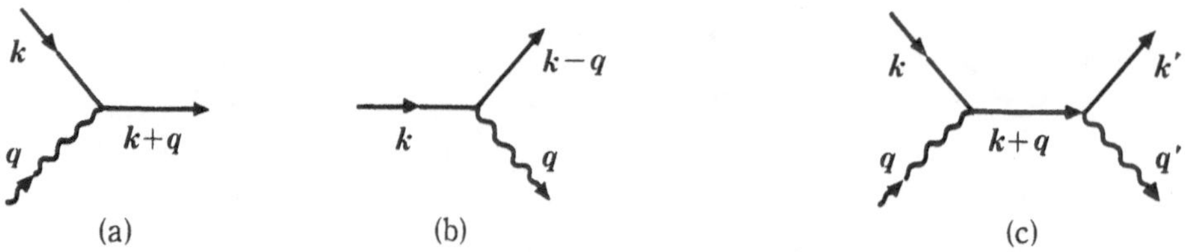


Fig. 2-4. LO-phonon scattering of an electron, a) LO-phonon absorption, b) LO-phonon emission, c) resonance scattering.

c) Damping. At low temperatures, a slow polaron ($k<<u$) can be scattered only by LO-phonon absorption at low electric fields or by resonance scattering, i.e. the successive LO-phonon absorption and emission. An individual process is an inelastic scattering event, while the resonance scattering is an elastic process. The mobility of slow electrons defined by using the momentum relaxation time can be calculated in terms of the scattering time as

$$\mu_d = \mu_H = \frac{e}{2\alpha\omega_{LO}m_P^\star} \cdot f(\alpha)$$

$$\times \left[\exp\left(\frac{\hbar\omega_{LO}}{k_B T}\right) - 1\right] \qquad (2.10)$$

where $f(\alpha) = 1{\sim}5/4$ is a monotonically increasing function obtained on the basis of the intermediate coupling method (Langreth 1967).

In pure crystals at high electric fields, a slow polaron may be accelerated up to the LO-phonon energy $\hbar\omega_{LO}$ with little impurity scattering, eventually to emit an LO-phonon. Such a situation is ideal to realize the streaming motion of hot polarons at crossed electric and magnetic fields as will be discussed in 4-4-c).

## 2-3) On the Feynman-polaron Model

Besides the weak coupling results of 2-2) we recall here the Feynman path integral formulation of the polaron problem which is based on a variational principle. This formulation is complicated but our aim is only to provide an intuitive understanding of dynamical and nonlinear characteristics of a polaron.

Feynman calculates the density matrix for a polaron in the canonical ensemble from the imaginary time dependent correlation function by using his path integral method. This leads to

$$K(r',t';r_0,t_0) = \int_{r(t_0)=r_0}^{r(t')=r'} Dr(t)\, e^{(i/\hbar)\int_{t_0}^{t'} L(\dot{r},r)\,dt}$$
$$= \langle r'|e^{-(i/\hbar)H(t'-t_0)}|r_0\rangle \quad (2.11)$$

with $(t'-t_0) \rightarrow -i\hbar\beta;\ \beta=1/k_B T$,

$$\rho(r',r_0,\beta) = \int_{r(0)=r_0}^{r(\beta)=r'} Dr(u)\, e^{\int_0^{\beta} L(i\dot{r},r)\,du} \quad (2.12)$$

The Hamiltonian corresponding to the Feynman model action is (Kadanoff 1963)

$$H_{FP} = \frac{p_e^2}{2m^*} + \frac{P_M^2}{2M} + \frac{1}{2}\gamma(r_e-R_M)^2 \quad (2.13)$$

Using the inequality

$$\langle e^A\rangle \geq e^{\langle A\rangle} \quad (2.14)$$

an upper bound is obtained for the ground state energy $E_0$ of a polaron in thermal equilibrium. In contrast to other theories, the Feynman theory covers the whole range of coupling constants.

Let us examine the meaning of the Feynman polaron model in more detail. Equation (2.13) describes the motion of a compound system, a "molecule" in which the center of mass moves freely with the effective mass $(m+M)$ but the two particles are bound together to form an harmonic oscillator. This is certainly a simpler description of a polaron than the original form of the Fröhlich Hamiltonian. It constitutes a useful starting point for the calculation of the density matrix in the canonical ensemble by the path integral formulation.

But, what was the original idea of Professor Feynman on his own polaron model? He replaces the electron LO-phonon interaction $H_1$ as illustrated in Fig. 2-5-a) by introducing the fictitious particle bound together with the electron via the spring $\gamma$ to reveal several features

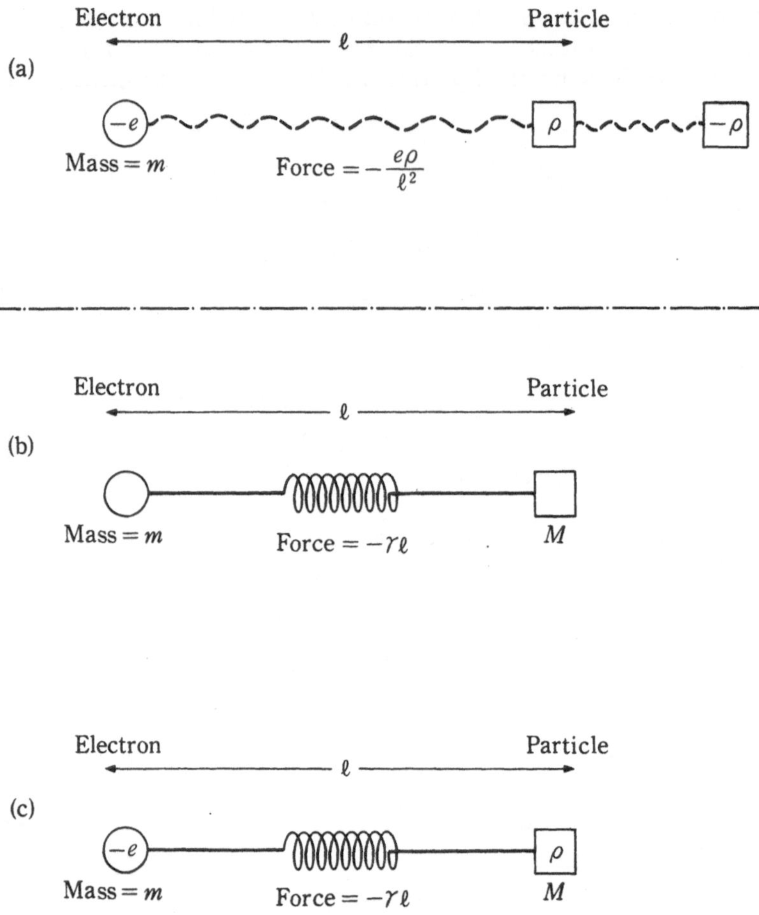


Fig. 2-5. A modification of the Feynman model.
a) The original form of an electron-LO phonon interaction, b) the Feynman-polaron model, c) the supplemented model.

of the particle-field problem as simplified in Fig. 2-5-b)¶. In the Fröhlich Hamiltonian, however, an electron is interacting with a virtual LO-phonon "polarization cloud" illustrated by the charges $\pm\rho$ in Fig. 2-5-a) rather than with the fictitious particle of Fig.

---

¶ Actually, Feynman first adopted the trial action $S_1$, $S_1 = -\frac{1}{2}\int(\frac{dr}{dt})^2\,dt - \frac{C}{2}\iint[r(t)-r(s)]^2\,e^{-w|t-s|}dtds$ instead of the true action $S_0$, in the path integral calculation (Feynman 1955), which is equivalent to choosing the trial Hamiltonian (2.13) (Kadanoff 1963) and the polaron model in Fig. 2-5-b) (Feynman 1972). 2-5-b). Note that the polarization charge $\pm\rho$ has disappeared in the Feynman-polaron model in the process of calculating the density matrix of a single polaron in thermal equilibrium by eliminating all the phonon coordinates in the particle-field problem. However, one sometimes has to recall the existence of these polarization charges. As far as one considers basic properties of a single polaron in thermal equilibrium, the Feynman model successfully reveals most of the relevant features (see Schultz(1963)). However, for the interaction of a polaron with any external system, one better keeps in mind the existence both of the charge of the electron and of the charges corresponding to the induced polarization.

Therefore, we proposed an extension of the Feynman-polaron model shown in Fig. 2-5-c) which becomes effective only for the interaction of an isolated polaron with other systems (Masumi 1975). In our model it is assumed that the induced polarization charge $+\rho$ extends over a limited but finite region (to avoid divergencies) while the polarization charge $-\rho$ appears at the surface of the crystal. The explicit role of the polarization charge $-\rho$ may be disregarded (here) without loss of the main aspects of our model§. The values of $M$ and $\gamma$ in the Feynman-polaron model (Fig. 2-5-b) are not modified in our model. The polarization charge $\rho$ is believed to be of importance to describe the polaron subjected to an external field or interacting with other systems.

The usual approximations for the calculation of the time-dependent correlation function for a polaron still hold.

In principle, the magnitude of $\rho(\alpha)$ may be calculated from the Poisson equation, $-\nabla^2\phi = 4\pi\sigma$ or

---

§ If we explicitly take into account the role of the polarization charge $-\rho$ on the crystal surface, e.g. by adding a second, fictitious, particle with charge $-\rho$ at the surface and connecting the first and second fictitious particles via another spring, this may further improve our description. However, as a first step, it is sufficient to adopt the simple model in Fig. 2-5-c) in order to describe the fundamental dynamical properties of a polaron. We discuss elsewhere the advantages and disadvantages of further modifications of the Feynman-polaron model.

$$\sigma(r,r_{el}) = - \frac{1}{4\pi e} <\Phi|\nabla^2 H_1|\Phi>. \tag{2.15}$$

The total charge $\rho$ is given by

$$\rho = \int \sigma(r,r_{el})\,dr. \tag{2.16}$$

It is complicated to perform the calculation in the path integral formulation but, roughly speaking, (2.16) reduces to

$$\rho(\alpha) = + \left(\frac{\gamma}{e}\right) r_f^{\,3}, \tag{2.17}$$

in the case of the Feynman-polaron model (APPENDIX 1) with the polaron radius

$$r_f = \left[\frac{3\hbar\,(m+M)}{2mM\omega_0}\right]^{1/2} \tag{2.18}$$

and the angular frequency

$$\omega_0 = \left[\frac{\gamma\,(m+M)}{mM}\right]^{1/2} . \tag{2.19}$$

In the classical theory, $\rho$ is given by

$$\rho(\alpha) = \left[\frac{1}{\kappa_\infty} - \frac{1}{\kappa_0}\right] e . \tag{2.20}$$

This charge is valid for a charge in a dielectric medium and is also obtained from the intermediate coupling theory (Pines 1963). The value of $\rho$ never exceeds $e$. Detailed calculations of $\rho$ in the path integral formulation with the variational treatment must eventually lead to the averaged value in Eq. (2.20) after the integration of (2.16) with use of Eq. (2.15).

At first sight, our procedures might seem to be artificial. Actually, however, we are just trying to describe the internal degrees of freedom which are only effective if the polaron interacts with an external system or if it is subjected to an external field. These internal excitations play an important role in the dynamical and nonlinear characteristics of a polaron in quasi- or even non-equilibrium and in the evolution towards equilibrium. In section 6 we will use our model to study the dynamical and nonlinear properties of polarons in interaction with external systems.

## 3. EXPERIMENTAL DIFFICULTIES CONCERNING THE POLARON PROBLEM

There are several experimental difficulties concerning the polaron problem.

First, we must prepare a high quality "jelly cake", i.e. high quality crystals of polar semiconductors or ionic crystals. Here, the term "high quality" means high purity and perfection. To observe a polaron in an ideal state, it is essential to control the creation and annihilation processes of polarons by illuminating a crystal of a wide band gap semiconductor at low temperature and at appropriate wavelength. For this purpose, the two silver halides AgCl and AgBr are the most suitable materials to develop a series of experimental studies of physics of polarons as discussed below.

These are materials of ultra high-purity (e.g. they contain less than $10^{-3}$ p.p.m. of heavy metal impurities and less than 1 p.p.m. of isoelectronic halogen impurities), and also suitable for the creation of an ideal elementary excitation like the polaron by a photon near the visible region.

By now, our knowledge of the physical properties of these materials such as the structure of the conduction band (minimum at the $\Gamma$-point) and the valence band (maxima at the L-points) in the one-electron scheme (Ascarelli and Brown 1962; Tamura and Masumi 1973), the optical band to band transition (Okamoto 1956, Brown, Masumi and Tippins 1961), typical transport phenomena (Masumi, Ahrenkiel and Brown 1965; Burnham, Brown and Knox 1960; Tippins and Brown 1963), and the dispersion relation in the phonon energy spectrum (Vijayaraghavan et al. 1970; Fujii et al. 1977) is very well established, compared to other insulating semiconductors. A summary of fundamental properties of AgCl and AgBr is given in Table 3-1. Therefore, single crystal specimens of zone refined AgCl and AgBr are used as our experimental testing ground of the polaron problem.

On the other hand, there still remain many delicate problems in the chemical (especially photochemical) properties of these famous photosensitive materials (Brown 1976, James 1977). Nevertheless these are less serious in ultra high-purity crystals than in usual materials prepared for photographic purposes. In fact, these silver halides are the best materials for studying the physics of polarons.

Table 3-1; Summary of fundamental properties of $AgC\ell$ and $AgBr$. $E_g$ denotes the indirect band gap at $4.2K$, $e^*/e$ is the effective ionic charge, $\alpha$ is the coupling constant between electrons and LO-phonons, and $\hbar\omega_{LO}$ is the LO-phonon energy at $4.2K$.

| | Band edge Cond. | Valence | $E_g$ (eV) | $e^*/e$ | $\alpha$ | $\hbar\omega_{LO}$ (meV) | $m_p^*/m_e$ (electrons) | $m_t^*/m_e$ (holes) | $m_\ell^*/m_e$ (holes) |
|---|---|---|---|---|---|---|---|---|---|
| $AgC\ell$ | $\Gamma$ | $L$ | 3.3 | 0.78 | 1.9 | 23.0 | 0.41 | | |
| $AgBr$ | $\Gamma$ | $L$ | 2.7 | 0.73 | 1.6 | 17.1 | 0.29 | 1.73 | 0.79 |

(Komiyama, Masumi and Kajita 1979). A correction has been made on the value of $m_p^*/m_e$ (electrons) in $AgC\ell$ according to the recent experimental data (Hodby, Komiyama, Hirano and Masumi, in preparation).

Details on the early experimental studies of polarons in the silver halides have been discussed in the reference (Brown 1972).

## 3-1) Preparation of Pure and Ultrapure Crystals of AgCl, AgBr and $AgCl_xBr_{1-x}$

Preparation of raw material of high purity AgCl and AgBr powder itself requires an enormous amount of effort and patience. These powders are rather popular photosensitive materials. Chemical processes have been developed intensively. Only six kinds of impurities, mostly heavy metals such as iron below 1 p.p.m., were detected by Johnson Mathey Co. by special chemical analysis. Remarkable progress has also been made by Childs at the University of North Carolina over the last two decades (Childs 1977). Kanzaki and Sakuragi (1969, 1970) also made an effort to remove the isoelectronic halogen impurity, mainly Iodine, down to 1 p.p.m., whereas Childs reported his success to reduce it down to $10^{-3}$p.p.m. by a special preparation method.

The physical process involved in the purification of silver halides relies on the fact (as discovered by the workers at the Kodak Research Laboratories on the zone refining in halogen atmosphere) that it is more efficient to remove ferric ion $Fe^{+3}$ than ferrous iron $Fe^{+2}$ in the vacuum or inert gas atmosphere. The purity of the crystal obtained can be examined by spectrochemical analysis, neutron activation, luminescence studies

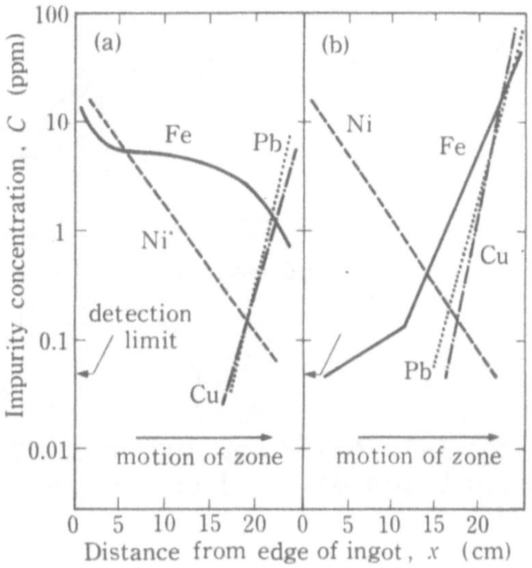


Fig. 3-1. Ultimate Distribution of several kinds of impurities in AgCl ingots zone refined a) in vacuum, and b) in $Cl_2$ atmosphere.

or the measurement of the residual mobility of slow electrons at lower temperatures. As an example, the residual values of the mobility of slow electrons in AgCl of various degrees of purity were tabulated in Table 3-2 and displayed in Fig. 4-2.

We emphasize that it is definitely important for experimentalists to obtain high quality crystals. Anyone who has experience and has spent his precious time and effort to refine and grow the crystals, in a dark room in the case of AgCl, AgBr, $AgCl_xBr_{1-x}$, must greatly appreciate to have reliable crystals available. Only high quality crystals can yield reliable experimental data with which a theory may be compared.

## 3-2) Creation of a Polaron - Optical Absorption

Let us create a polaron in the vicinity of the ground state.

The ground state of the crystal as a whole, as a many-body system, can be realized by placing it at low temperatures and in the dark, in a shielded cryostat. But, the ground state of a polaron can be created only by illuminating the crystal by at least one photon. This is one of the simplest ways to create an elementary excitation in insulators.

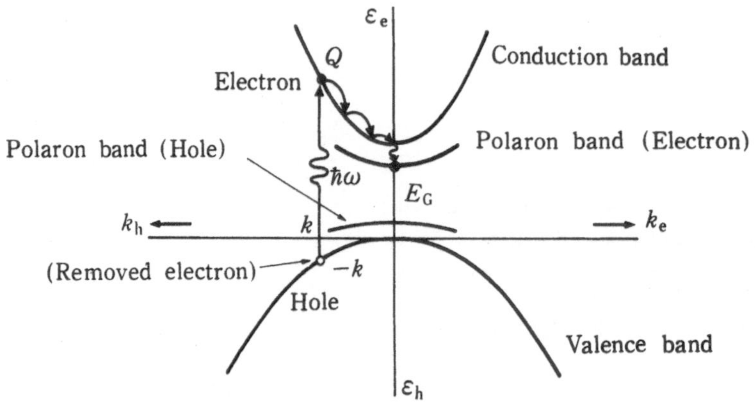


Fig. 3-2. Creation process of a polaron by the direct transition in an ionic crystal.

Ambient temperatures should be as low as possible, the trapping lifetime of photoexcited carriers becomes short as temperature decreases and consequently the transport experiments become difficult. Optical excitation at $\hbar\omega$ beyond the energy band gap $E_G$, must be performed by using light of wavelength $\lambda$ as monochromatic as possible to allow for the analysis of the relaxation mechanisms of carriers after excitation (Fig. 3-2).

In the case of the silver halides, data of optical absorption first indicated the band to band transition which involves an indirect process. Actually, the indirectly allowed transitions were observed in zone-refined AgCl and AgBr crystals with the energy gap of 3.246 eV and 2.6836 eV, respectively (Brown, Masumi and Tippins 1961). For illumination, one can use a Xe flash lamp with appropriate filters or monochromator or even a nitrogen laser-pumped high resolution dye laser with $\Delta\lambda \approx 0.003$ nm. The concentration of photocarriers can be controlled from $10^6$ polarons/cm$^3$, in AgCl appropriate to realize an ideal streaming motion of polarons to $10^{13}$ polarons/cm$^3$ which is sufficient to observe a plasma shift in the cyclotron resonance of polarons in AgBr at high density excitation.

### 3-3) Difficulties in the Conductivity Measurements of High Impedance Materials and Cyclotron Resonance in Special Conditions

Silver halide crystals have the NaCl-type structure and are of highly ionic character, so that they are typical insulators.

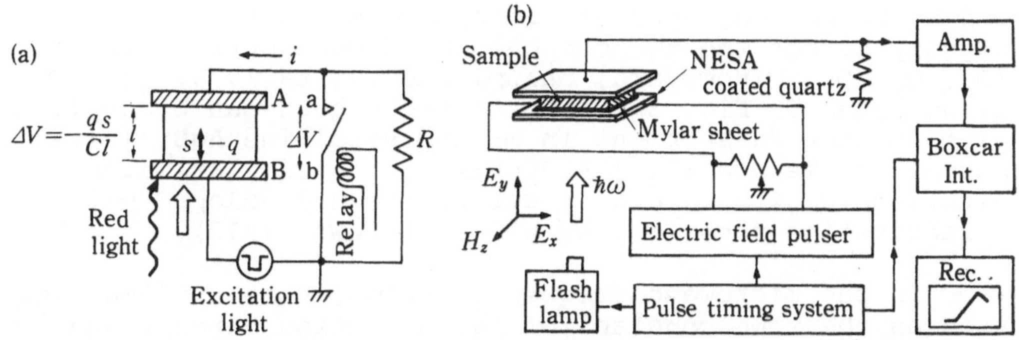


Fig. 3-3.
a) Principles of the transient photoconductivity measurements with blocking electrodes.

b) An apparatus to determine the Hall mobility in transient cases.

As silver halides are wide band gap semiconductors and pure crystals at low temperature become typical insulators with a small number of conduction electrons ($\rho = 10^8 \sim 10^{20}$ Ω.cm), special difficulties in the conductivity measurements of high impedance materials occur associated with the basic problems of sensitivity and time constant of the circuit, the low signal to noise ratio, the non-ohmic properties and instability of the contact electrodes, the build-up of space charge, the problem of leak current, etc.

Accordingly, all of the transport phenomena should be observed by using the fast pulse technique with blocking electrodes. For example, for the Hall effect measurements, we have to apply the pulsed Redfield technique (Kobayashi and Brown 1959), etc. The principles of this technique are illustrated in Fig. 3-3-a),-b) with the time sequence diagram of electric field and exciting light pulses (Fig. 3-4). More refined methods for the transient photoconductivity measurement have been deve-

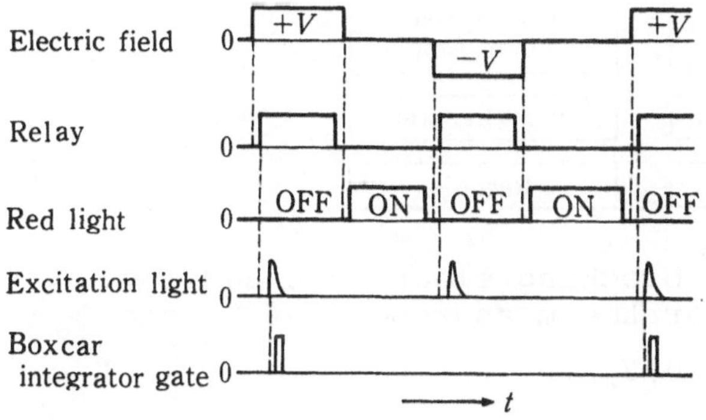


Fig. 3-4. Time sequences of actions of the circuits of various parts in transient photoconductivity measurements.

ped by using two types of blocking electrodes to simultaneously detect the two components of photocurrent vector ($Q_x$, $Q_y$, $Q_z$), (Komiyama, Masumi and Kajita 1979) as illustrated in Fig. 3-5. These techniques can be applied also for the measurement in microcrystalline AgBr (Hirano, Masumi and Takada 1982; APPENDIX 2). A similar method with interdigitated electrodes had been developed with a fabrication technique by Wei and Brown (1973).

In the microwave frequency region, some difficulties such as the time constant of the detection circuit and the build-up of space charge, can be ignored. Apart from the sensitivity problem, one can perform a standard type of experiment of cyclotron resonance of the polaron provided that the crystal is pure enough to fulfil the condition $\omega\tau \gtrsim 1$, where $\omega$ is the angular frequency of the microwave and $\tau$ is the scattering time of the polaron. A more sensitive method called "cross modulation technique" was first applied to ionic crystals by Mikkor, Kanazawa and Brown (1965, 1967) and later intensively developed and extensively applied to many materials by Hodby (1972).

Standard type cyclotron resonance techniques have been further developed by using a magnetron at 35GHz and a cylindrical cavity in the $TE_{112}$-mode at pulsed micro-

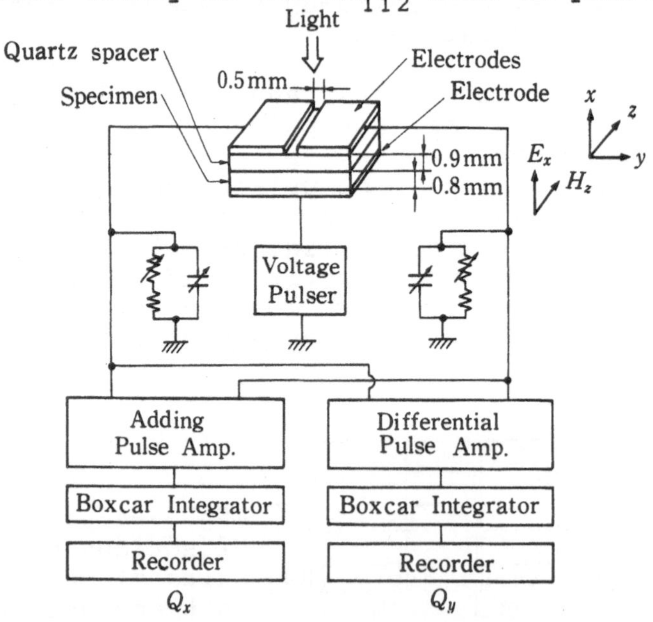


Fig. 3-5. Arrangement of blocking electrodes and the experimental apparatus for the measurement of $Q_x$ and $Q_y$ with $\vec{E}=(E_x,0,0)$ and $\vec{H}=(0,0,H_z)$.

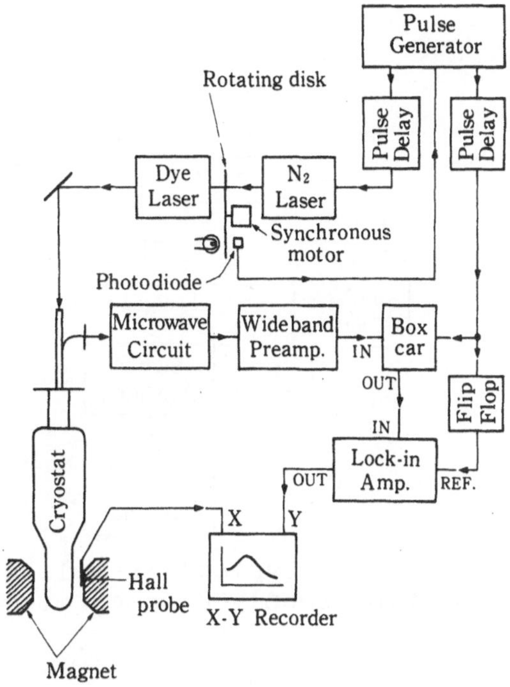


Fig. 3-6. Schematic diagram of an apparatus of the time-resolved cyclotron resonance experiment in the nanosecond region.

wave fields (200nsec) up to 2.6kV/cm to observe the non-linear responses of polarons. Also, a time-resolved cyclotron resonance technique at 34GHz has been developed with a non-resonant waveguide even in the nanosecond region by using a $N_2$-laser pumped dye-laser ($\lambda$=400~465 nm with $\Delta\lambda \simeq 0.05$ nm or 0.3 meV reduced by a beam expander in the optical cavity, output energy 20~100$\mu$J/pulse and pulse width 5nsec) to study the dynamical aspects of the polaron-polaron or polaron-exciton interaction (Fig. 3-6).

## 3-4) Annihilation of Polarons and Excitons - Luminescence

Once an elementary excitation is created from the crystal ground state, it will annihilate some time later. Indeed, an elementary excitation has the complex dispersion relation

$$\tilde{\varepsilon}(\vec{p}) = \varepsilon(\vec{p}) - i\gamma(\vec{p}) \tag{3.1}$$

where p is the momentum of the excitation and there always exists a finite imaginary part $\gamma(\vec{p})$.

The annihilation process of polarons or excitons usually arises through the mutual interaction of the

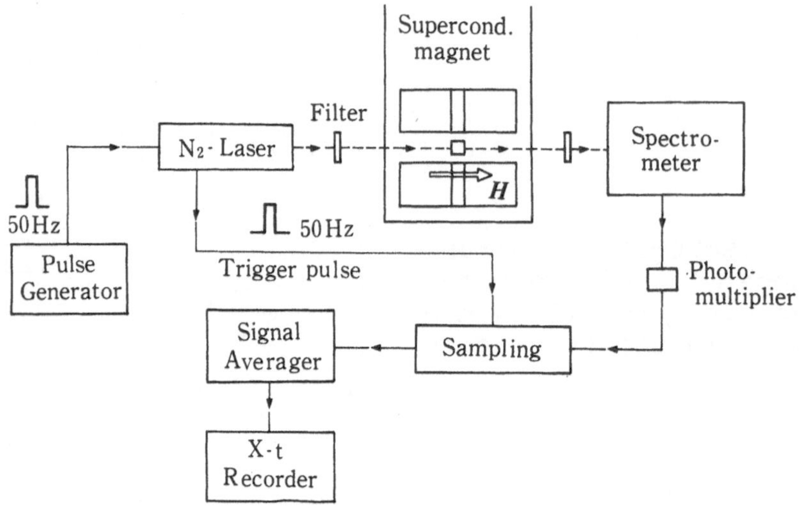


Fig. 3-7. A typical example of the block diagram of the fast time-resolved spectroscopy in the nanosecond region.

electron-polaron and the hole-polaron and/or through the interaction with trapping centers. Under these circumstances, either photons or phonons must be emitted. This results in radiative or non-radiative processes; the former process can be detected by luminescence, e.g. from an exciton, whereas the latter hardly can be detected by the usual method.

Problems usually arise from an ultrashort life-time of free polarons (10∿100p-sec) and free excitons (10∿100n-sec). In this respect, it is essential to prepare ultrapure crystals of silver halides as well as to develop an ultrafast time-resolved spectroscopic technique. Preparation of ultrapure crystals, a "high quality jelly", has been discussed in 3-1). In spectroscopic experiments at high density excitations, a dye laser and/or $N_2$-laser were used. Time resolution of the detection system is normally of the order of 3n-sec. A sampling gate of about 300p-sec can also be used. At the highest excitation and time resolution, a picosecond YAG-laser was used with a pulse duration of 30 p-sec. A typical example of the block diagram of the fast time-resolved spectroscopic apparatus, a set of our vivid eyes, is given in Fig. 3-7 (Baba and Masumi 1983).

## 4. A SERIES OF EXPERIMENTAL STUDIES OF DYNAMICAL AND NONLINEAR CHARACTERISTICS OF POLARONS IN AgCl AND AgBr

### 4-1) Polarons at and in the Vicinity of the Ground State - Cold Polarons

Let us briefly summarize our knowledge of a polaron at and in the vicinity of the ground state.

Studies of the ground state of a polaron have attracted attention of many theoretical physicists and led us, e.g. to Eqs. (2.8) and (2.9). So far, however, nobody knows experimentally how large the self-energy of a polaron is. Rather, for either an electron-polaron or hole-polaron, their effective masses have been best examined by the cyclotron resonance technique at low levels of microwave power. Since the pioneering work by Ascarelli-Brown (1962), an increasing amount of data became available on the effective mass of a cold polaron $m_p^*(0)$ in silver halides (Hodby, 1971, 1972; Tamura and Masumi 1971a, 1973) and even for other materials such as alkali halides as reviewed by Hodby (1971, 1972).

We have also performed standard type experiments of the cyclotron resonance for cold polarons in AgCl and AgBr at 9, 35, 50 GHz and 4.2K-25K (Tamura and Masumi 1971a, 1973). Recently, an even lower value of the electron-polaron mass $m_p^* \simeq (0.415 \pm 0.013)m_e$ has been observed in "purified" AgCl at 139GHz and 1.7K by the cross modulation technique (Hodby, Komiyama, Hirano and Masumi, in preparation). Other data especially in the limit of low levels of microwave power (Komiyama and Masumi 1978) established the cold polaron mass at the ground state $m_p^*(0)$, e.g., $m_p^*(0) = (0.287 \pm 0.003)m_e$ for an electron in AgBr.

The dependence of the peak shifts of the resonance line on both temperature (Hirano and Masumi 1982 for AgCl, Tamura 1972 for AgBr) and microwave power (Tamura and Masumi 1971b, Komiyama and Masumi 1978, 1979d) was also studied. Quite recently, an anomalous temperature dependence of the negative peak shift has been observed for purer AgCl yielding the mass value of cold polarons, $m_p^* \simeq (0.41 \pm 0.01)m_e$, also at 35GHz and around 20K rather than below 10K, see Fig. 4-1-a), (Hirano and Masumi 1983). These results are consistent with those at 139GHz above and will be discussed in section 4-3).

Temperature dependence usually gives a moderately small amount of peak shift and changes of the linewidth,

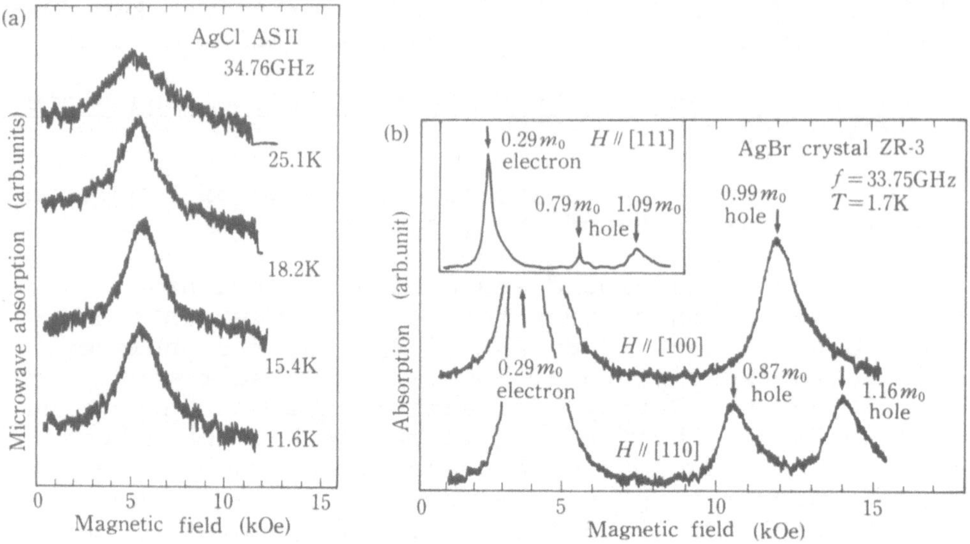


Fig. 4-1. Cyclotron resonance of polarons.
a) electron-polarons in purified AgCl,
b) electron- and hole-polarons in AgBr.

whereas an increase of microwave power yields a remarkable peak shift towards higher magnetic fields. Data of the peak shift and changes of the half width for AgBr (Komiyama and Masumi 1978, 1979d) in the region below 80V/cm indicate, respectively, the existence of the non-parabolicity in the polaron energy spectrum and an inhomogeneous broadening due to the energy-distribution of photo-excited polarons (Tamura and Masumi 1973). It should be noted particularly that the existence of a small concentration of trapping centers influences the observed values of the polaron mass (Komiyama and Masumi 1978). Concerning the data in the higher energy region, we present the discussion of the physics of streaming polarons in section 4-4, and, more in detail, to the preceding paper presented by Dr. Komiyama.

Cyclotron resonance of polarons at higher frequencies and magnetic fields has been intensively studied by Hodby at Oxford (Hodby 1971, 1972; Hodby et al. 1974). He uses a refined cross-modulation technique at 137 and 891GHz with higher resolution and sensitivity to obtain information on the non-parabolicity of the polaron band up to higher energy states. These studies have resulted in interesting data and information which is complimentary to that at lower frequencies. The experiments are still being extended to cover higher magnetic fields and other materials.

Both kinds of data furnish considerable knowledge of the polaron energy spectrum in the vicinity of the ground state and, furthermore, towards the 1 LO-phonon energy region.

## 4-2) Temperature Dependence of the Mobility of Slow Electrons - Slow Polarons

The mobility of slow Fröhlich polarons near the ground state also has attracted attention of theoretical physicists (Low and Pines 1955, Osaka 1961). Experiments on the mobility of slow electrons have been performed rather intensively and supplied considerable knowledge on the scattering mechanisms of slow polarons (Masumi, Ahrenkiel and Brown 1965; Burnham, Brown and Knox 1960). The residual values of the Hall mobility of slow electrons at low temperatures also supplied various information on the sample purity of our AgCl and AgBr specimens (Fig. 4-2-a).

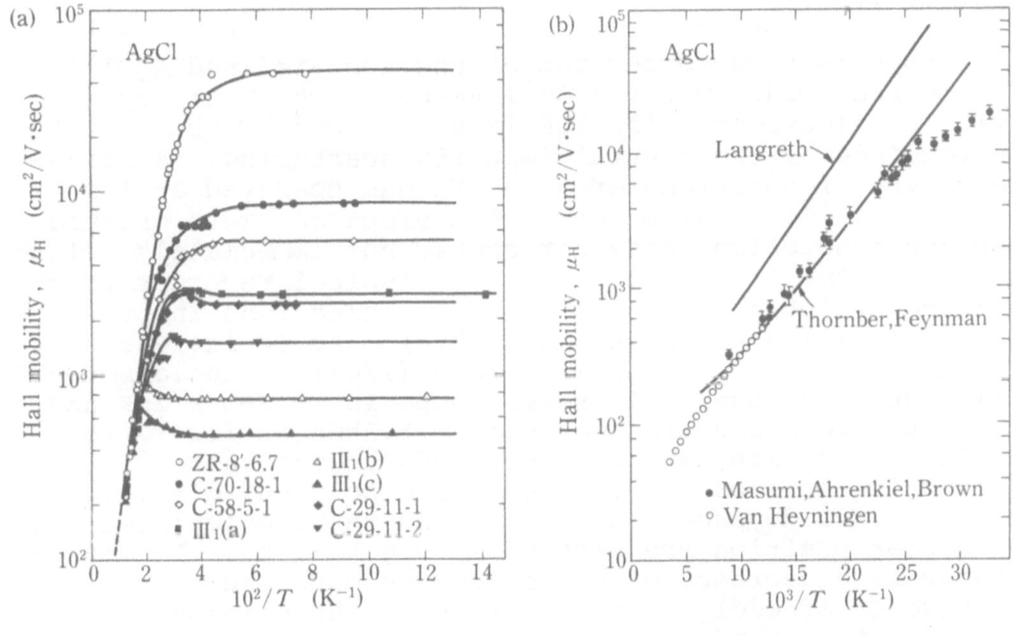


Fig. 4-2. Temperature dependence of the mobility of slow electrons,

a) Hall mobility observed for photoelectrons in various pure and impurity-doped crystals of AgCl. (Refer to Table 3.1).
b) A comparison of experimental and theoretical mobilities.

Data on the slow electron mobility in very pure AgCl crystals may be compared with Eq. (2.10). A comparison between experimental (Masumi, Ahrenkiel and Brown 1965) and theoretical (Thornber and Feynman 1970) mobilities is given in Fig. 4-2-b) (Brown 1972). Note that no arbitrary parameters are employed in this comparison.

All of this knowledge has provided a sound basis for a series of our experimental studies on the dynamical and nonlinear response of Fröhlich polarons.

## 4-3) Relaxation of Photoexcited Polarons due to Hybrid (Resonance) Scattering - "Chilled" Polarons

*"An emergence of oddly cold polarons in pure AgCl at elevated temperatures"* has been clearly recognized in the cyclotron resonance experiment at 35 GHz and 6~30K. This is the first observation of an odd effective cooling phenomenon of the photoexcited electron system by heating the lattice temperature $T_L$ (Hirano and Masumi 1983).

Specimens were cut out of the zone-refined AgCl newly prepared. The residual value of the mobility of slow electrons below 10K is $6.5\times10^4 cm^2/V.sec$; it is determined by the neutral impurity scattering. A standard type of spectrometer at 35GHz was operated at low levels of microwave power. Photoelectrons were created in the conduction band, far above, but immediately relaxed below the 1-LO phonon energy. Typical recorder traces of the cyclotron resonance line for electrons in pure AgCl are illustrated in Fig. 4-3-a). As $T_L$ increases from 6K to 25K, (i) the linewidth $(1/\omega\tau)$ of the absorption curve gradually increases especially above 20K and (ii) the peak position moves from 5.3kOe at low $T_L$ to 5.0kOe at 20~25K, which corresponds to the value of the effective mass of the cold electron polaron, $m^*=0.41m_e$. Temperature dependences of the mass ratio calculated from the peak position are plotted in Fig. 4-3-b). A clearly recognized decrease of the effective mass indicates an emergence of oddly cold electrons in the polaron band. On the other hand, the temperature dependence of the scattering time $\tau$ experimentally reveals a significant contribution of the acoustic phonon scattering (15~25K).

These results suggest to examine the detailed balances of the momentum and energy relaxations of photoexcited polarons in the non-parabolic band. Among several possible mechanisms underlying these peculiar phenomena,

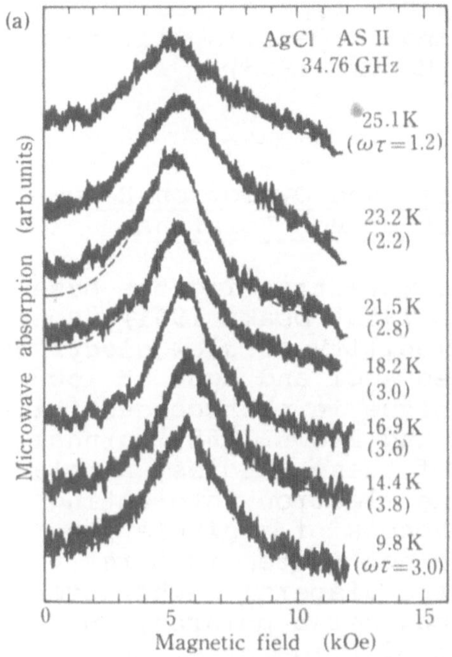


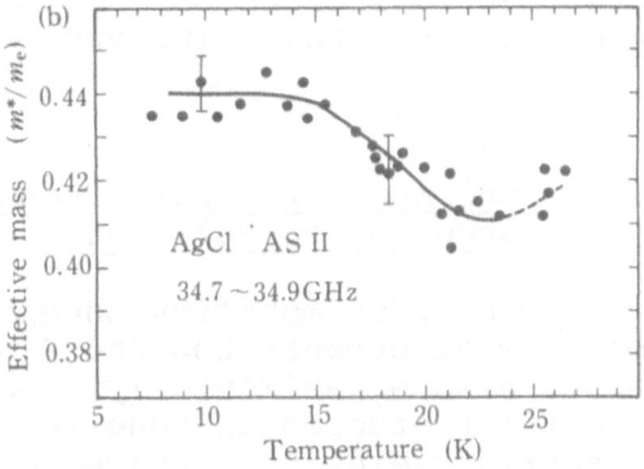


Fig. 4-3-b). Temperature dependence of the effective mass ($m^{\star}/m_e$) of electron-polarons in purer AgCl between 7∿27K.

Fig. 4-3-a). Variation of the cyclotron resonance absorption line of electron-polarons in purer AgCl with temperature (Hirano and Masumi 1983).

we mention the appearance of "*an odd relaxation and distribution of photo-excited but chilled polarons in purified AgCl*" due to e.g. acoustic phonon absorption with LO-phonon emission as illustrated in Fig. 4-4-a),-b)

Fig. 4-4.

a) LA-phonon absorption and LO-phonon emission.

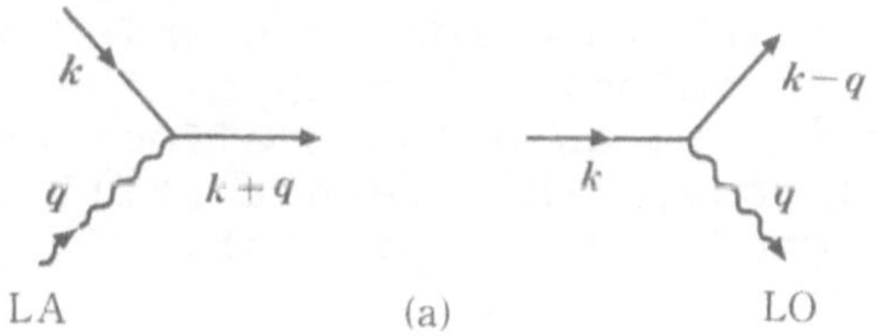


b) Hybrid (resonance) scattering.

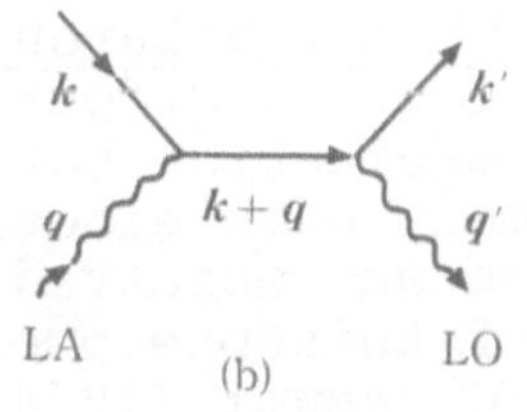

as a possibility here. We call this type of scattering "the Hybrid (Resonance) Scattering". Details will be published elsewhere (Hirano and Masumi 1983).

## 4-4) Nonlinear Transport Phenomena and Cyclotron Resonance due to Polarons - Fast or Hot Polarons

The first mobility theories were treating the case of slow electrons (Low and Pines 1955, Osaka 1961), so that, also experimentally, the mobility of slow electrons was first studied in zone-refined AgCl and AgBr at low electric fields by a pulse technique in photoconductivity measurements (Masumi, Ahrenkiel and Brown 1965; Burnham, Brown and Knox 1960). In most of these studies, a polaron was considered to be a moving electron interacting electrostatically with the LO-phonons of a single frequency $\omega_{LO}(\Gamma)$. Actually, the phonon-system is more versatile with $q$ and sometimes the dispersive character may be essential. We have to consider a polaron, in general, with an LO-phonon cloud of frequencies $\omega_{LO}(q)$ which depends on the wave number $q$.

a) Nonlinear transport phenomena of polarons at high electric fields. Nonlinear transport phenomena at high electric fields were first observed in the transient measurements of photocurrent $Q_x(T,E)$ over a decade and a half ago as illustrated in Fig. 4-5-a) (Masumi 1963, 1967). The $E^{1/2}$-dependence of $Q_x$ has been reexamined with special attention to the Hall mobility measurements at high electric fields shown in Fig. 4-5-b) (Komiyama, Masumi and Kajita 1976). As will be discussed below, it turns out that the mobility of an electron at high electric fields is expected to be a linearly decreasing function of E which gives a saturation of the drift velocity, so that *"the $E^{1/2}$-dependence of $Q_x$"* must be due to an unknown effect on the life-time of fast polarons. This result still remains as *"a peculiar enigma"* even at present.

b) The hot polaron effect in the quantum limit. The hot electron effect in the quantum limit (due to the enhanced excitation between the adjacent Landau levels at extremely high electric fields) characteristic to insulators was observed in the silver halides only. Details of this phenomenon can be found in the work of Kajita and Masumi (1974).

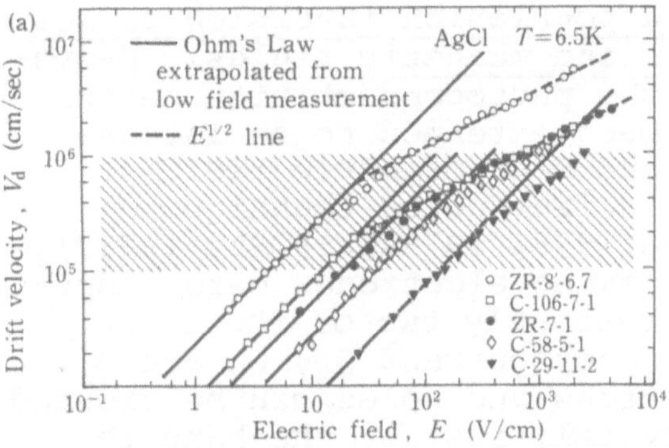


Fig. 4-5-a). Impurity effect of ferrous or ferric ion on the nonlinear photoconductivity of AgCl crystals at $T$=6.5K. Drift velocity $v_d$ for various crystals was calculated using the data in Fig. 1. Hatched zone indicates the region where $v_d$ becomes of the order of the sound velocity $c_s$.

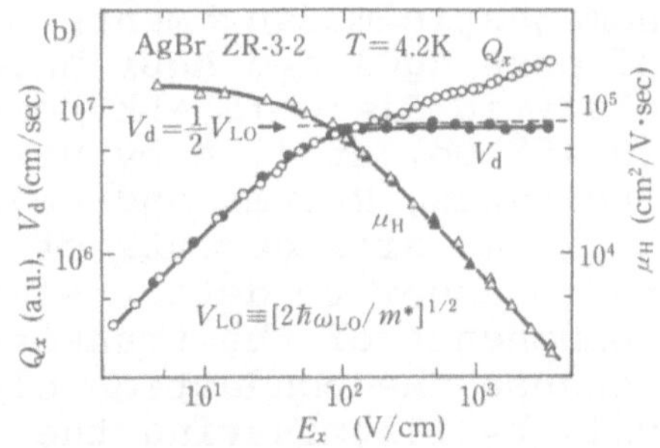


Fig. 4-5-b). $Q_x$, the Hall mobility $\mu_H$, and saturated drift velocity of polarons at high electric fields.

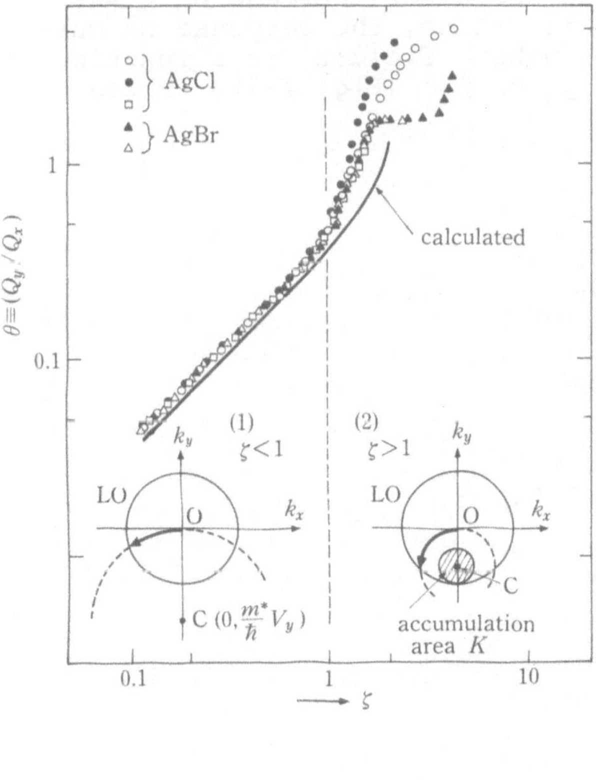


Fig. 4-6. The Hall angle $\zeta\equiv(Q_y/Q_x)$ for AgCl and AgBr as a function of $\zeta$ where $\zeta$ is defined as $\zeta\equiv V_{LO}/V_y$ with the relations $(1/2)m^{\star}V_{LO}{}^2\equiv\hbar\omega_{LO}$ and $V_y=(cE_x/H_z)$. The electron trajectories are indicated for (1) $\zeta<1$ and (2) $\zeta>1$ where, in the latter case, the accumulation area $K$ appears.

c) Streaming motion and population inversion of hot polarons at crossed electric and magnetic fields. Galvanomagnetic measurements of the photoconductivity tensor of pure AgCl and AgBr have been extended to an intense electric field ($E_x \simeq 5kV/cm$) and high magnetic field ($H_z \simeq 58kOe$) at 4.2K by using a fast pulse technique (Komiyama, Masumi and Kajita 1976, 1979a, 1979b, 1979c). Improved arrangements of blocking electrodes were adopted in order to detect simultaneously two of the three components of the transient photocurrent $Q_x$, $Q_y$ and $Q_z$. Because the conduction electrons and holes can be created only by illuminating the crystal, i.e. by the band to band optical excitation, the number of photocarriers can be controlled to be usually $10^7$~$10^9/cm^3$ while, mainly due to the scattering by residual impurities, their mobilities are quite high, $\mu_H \sim 10^4 \sim 10^5 cm^2/V.sec$, so that the effects of applying both the high electric field $(E_x,0,0)$ and the high magnetic field $(0,0,H_z)$ are remarkable. For example, from the simultaneously observed values of photocurrent $(Q_x,Q_y)$, the Hall angle defined by $\theta \equiv (Q_y/Q_x)$ can be plotted as a function of $\zeta \equiv V_{LO}/V_y$, where $(1/2)m^* V_{LO}^2 \equiv \hbar\omega_{LO}$ and $V_y = (cE_x/H_z)$ as illustrated in Fig. 4-6. The significant role of the LO-phonon emission in hot electron kinetics have been revealed in the region $\zeta < 1$ whereas an anomalous increase of $\theta$ was observed in the region $\zeta > 1$. Also, the response of current $Q_z$ with a small perturbing field $E_z$ as a function of $H_z$ behaves anomalously for $\zeta > 1$ (Fig. 4-7). These

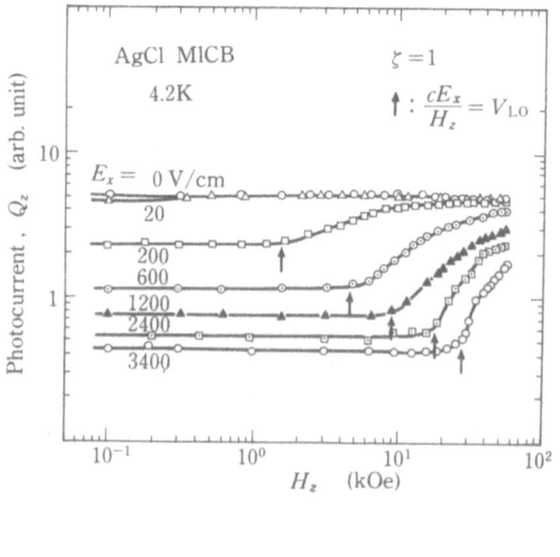


Fig. 4-7. Photocurrent $Q_z$ in the z-direction at crossed fields $\vec{E}=(E_x,0,E_z)$ and $\vec{H}=(0,0,H_z)$. Here the perturbing field $E_z$(15V/cm) is sufficiently low so that $Q_z$ directly reflects the "electron mobility" along the z-direction under the condition $\vec{E}=(E_x,0,0)$ and $\vec{H}=(0,0,H_z)$ An arrow on each $Q_z$-$H_z$ curve indicates the magnetic field position at which $cE_x/H_z=V_{LO}$ (where $(1/2)m^* V_{LO}^2 \equiv \hbar\omega_{LO}$) is fulfilled. All the curves rise steeply just above the indicated field.

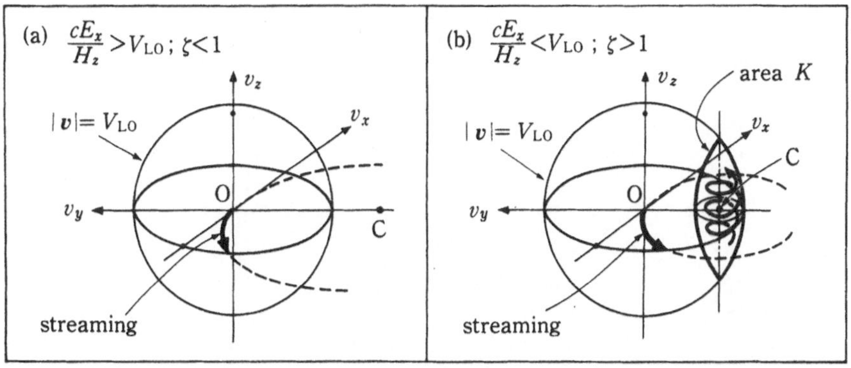


Fig. 4-8. Anomalous distribution of hot electrons in velocity space for two different conditions of the relative strength of crossed fields $\vec{E}=(E_x,0,0)$ and $\vec{H}=(0,0,H_z)$. The point C denotes the center of the cyclotron orbit, $C=(0,-cE_x/H_z,0)$. (a) In the range $cE_x/H_z>V_{LO}$, the point C is located outside the surface $|\vec{v}|=V_{LO}$. The trajectory of a streaming electron is curved by the Lorentz force resulting in an arc around the point C. All the electrons are believed to be distributed on the trajectory of streaming. (b) In the range $cE_x/H_z<V_{LO}$, the point C is located within the surface $|\vec{v}|=V_{LO}$. No serious change is involved in the kinetics of streaming electrons, but, in addition to the trajectory of streaming, there emerges an area $K$ in which a certain amount of electrons are accumulated. An electron in the area $K$ is able to produce a large response current $Q_z$ when a perturbing $E_z$ is applied, by drifting along the z-direction in a helicoidal motion.

experimental results provide, for the first time, conclusive evidences for *"the streaming motion, the anomalous distribution, and the occurrence of population inversion of the polarons"* in the wavenumber $k$-space (Fig. 4-8). The possibility of such an anomalous distribution had been predicted by Maeda and Kurosawa (1972).

d) Cyclotron resonance of polarons at high microwave fields. Similar phenomena can be observed in the cyclotron resonance experiments of polarons in AgBr at high microwave fields. Figures 4-9 illustrate typical traces of cyclotron resonance lines for electrons in AgBr at high levels of microwave power up to 3kV/cm obtained by using a pulsed magnetron at 35GHz. Also shown are data on the dependence of the position of the absorption maximum (Komiyama, Masumi and Kurosawa 1979e). These results, together with those on the half width,

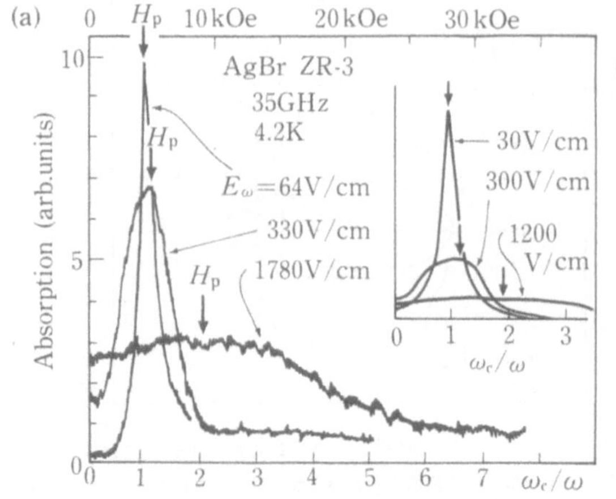


Fig. 4-9-a). Typical recorder traces of resonance spectra for photo-electrons in AgBr at 4.2K. A Monte Carlo calculation reproduces the observed line shape quite well. The magnetic field H=4.42kOe corresponds to $\omega_c/\omega=1$.

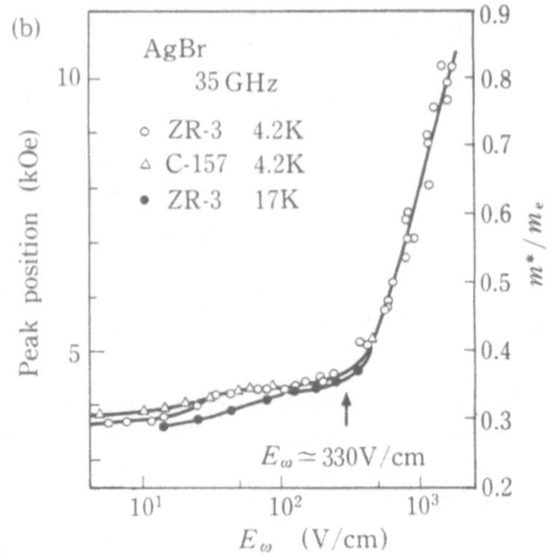


Fig. 4-9-b). Shifts of the peak position with microwave field $E_\omega$.

indicate the non-parabolicity of the polaron energy spectrum and the existence of the streaming motion due to the LO-phonon emission below 330V/cm. Above 330V/cm, an electron is accelerated to the LO-phonon energy within a half cycle of microwave field. *"A new mode of streaming cyclotron motion"* of hot electrons in intense microwave fields appears. Figure 4-10 suggests the changing behaviour of the polaron distribution with increasing microwave power at $\omega \simeq \omega_c$. A Monte Carlo simulation reproduces the observation quite well and predicts *"an electron bunching effect"*. At intense magnetic fields $\omega_c > \omega$, a new kink can be observed in the resonance line as indicated by an arrow in Fig. 4-11 ($\zeta=2$), which reveals that a polaron in streaming cyclotron motion suddenly *escapes the LO-phonon emission* and never contributes to the absorption above a critical magnetic field (Komiyama and Masumi 1979d).

Based on these results, we have partly clarified the dynamical and nonlinear response of polarons in transport phenomena. We note that in these experiments a polaron is associated not only *with virtual phonons* but also *with real phonons*.

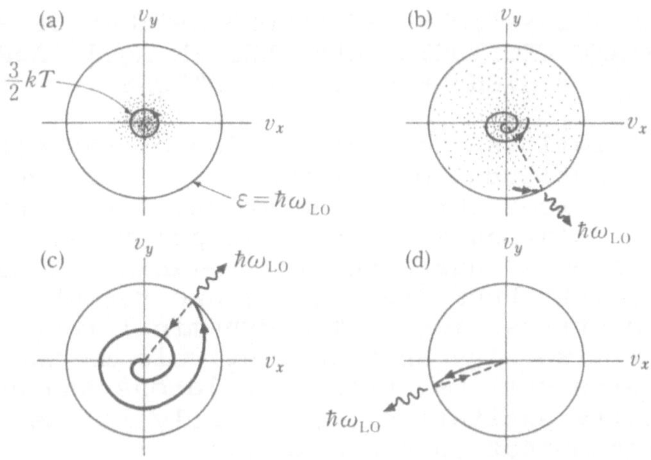


Fig. 4-10. Variation in the electron kinetics and distribution with $E_\omega$ under the resonance condition. (a) Hot electron CR: a small circle denotes $\varepsilon=3kT/2$ at 4.2K; the average energy of the electrons increases with $E_\omega$. (b) Hot electron CR: the electron distribution is cut off at $\varepsilon=\hbar\omega_{LO}$ due to the onset of LO-phonon emissions. (c) Streaming CR: a majority of the electrons perform the streaming cyclotron motion. (d) A new mode of streaming CR: almost all the electrons can be accelerated to $\varepsilon=\hbar\omega_{LO}$ within a half cycle of microwave field.

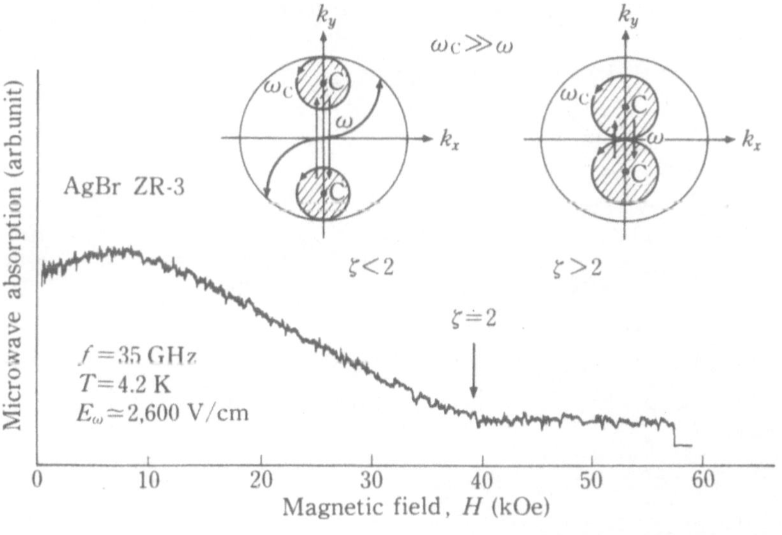


Fig. 4-11. Trajectories of electrons at crossed microwave field and static high magnetic field and a microwave absorption curve in a high magnetic field ($\omega_c>>\omega$) for two cases of $\zeta<2$ and $\zeta>2$. When $\zeta>2$, the trajectory of streaming motion disappears.

## 5. DYNAMICAL RELAXATIONS AND RECOMBINATIONS OF POLARONS, EXCITONS AND EXCITONIC MOLECULES IN AgCl, AgBr AND $AgCl_xBr_{1-x}$ AT HIGH DENSITY EXCITATIONS

In what follows the annihilation of the elementary excitation constituted of the polarons is discussed. The annihilation arises through the mutual interactions of the electron-polarons with the hole-polarons. Here, "the dynamical response" has a two-fold meaning, namely relaxation by the intraband transitions and recombination by the interband transitions. The dynamical response of polarons may also play an important role in the behaviour of many polaron-systems, such as polarons and excitons at high density excitations, especially in their relaxation and recombination processes.

### 5-1) Magneto-Optical Experiments of Absorption, Photo-luminescence due to Polaron and Exciton-Systems at High Magnetic Fields

The purity and degree of perfection of our AgCl and AgBr specimens can be confirmed not only by measuring

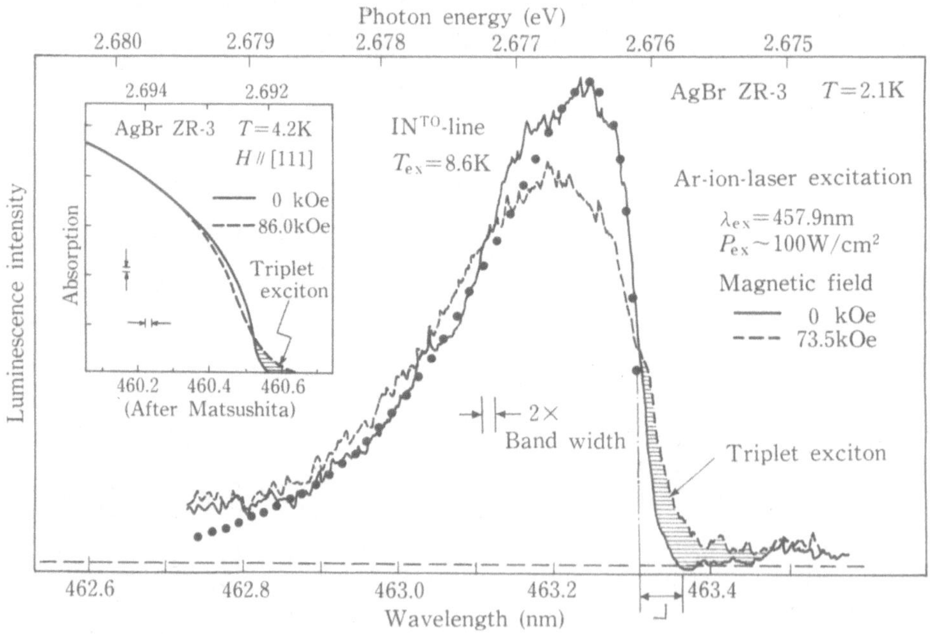


Fig. 5-1. TO-phonon-assisted luminescence lines of indirect free excitons in AgBr ZR-3-3 at 2.1K and at zero and 73.5kOe. The inset indicates the absorption spectra of AgBr ZR-3 at zero and 86.0kOe after Matsushita.

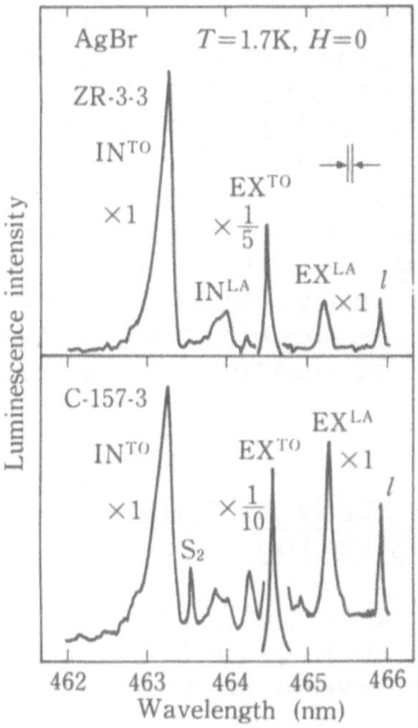


Fig. 5-2. Luminescence spectra of two specimens cut out of different AgBr ingots ZR-3 and C-157, at 1.7K and zero magnetic field.

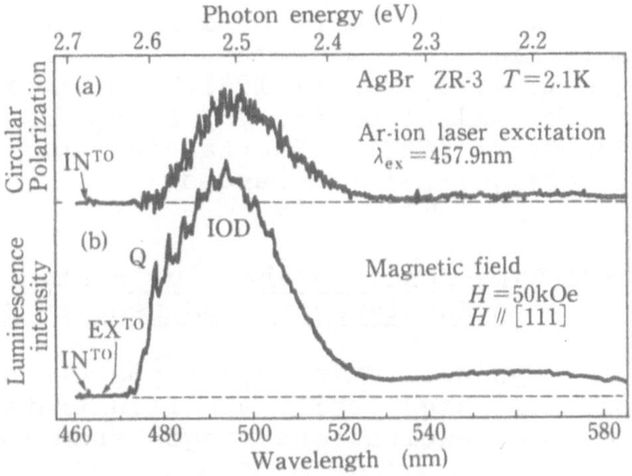


Fig. 5-3. Typical luminescence spectra of AgBr ZR-3-3 at 2.1K and 50kOe under Ar-ion laser excitation (λ=457.9nm) in the cases of detection as
a) circular polarized light and
b) linearly polarized light.

the residual Hall mobility of slow electrons (Masumi, Ahrenkiel and Brown 1965; Burnham, Brown and Knox 1960) but also by observing the optical absorption or more sensitive photoluminescence spectra at low levels of excitations (Baba and Masumi 1979). Thus, the magneto-optical measurements on pure AgBr (Matsushita et al. 1972; Matsushita 1973) have been extended to cover the luminescence study (Fig. 5-1).

As illustrated in Fig. 5-2-a) (Baba and Masumi 1983, Masumi 1981), a narrow luminescence line once ascribed to Bose-Einstein condensed triplet excitons (Fig. 5-2-b)) has never been observed for the purest AgBr specimen so far reported (Baba and Masumi 1979). The electron mobility in this specimen cut out of the ingot ZR-3 was determined by the galvanomagnetic and cyclotron resonance experiments to be about 300,000cm$^2$/V.sec. A new and intrinsic shoulder in luminescence spectra has been found instead on the lower energy edge of the IN$^{TO}$-line for zone refined AgBr crystals ZR-3 at high magnetic fields up to 80kOe (Baba and Masumi 1978). By the analysis based on the data of magneto-absorption (Matsushita 1973) with a set of new values (the exchange energy $\Delta$=0.31meV and the effective $g$-values, $g^c$=0.88 and $g^v_{\parallel}$=0.98 for AgBr (Kurita and Kobayashi 1978)), we ascribed the new shoulder to *"the indirectly allowed triplet excitons"*. This assignment has been supported in part by preliminary results of our recent experiment of circular-polarized luminescence spectra in Fig. 5-3 (Baba and Masumi 1983, Masumi 1981). Quite recently, even a newer set of values on $g^c$, $g^v$, $\Delta$ and $E^b_{ex}$ have been reported (von der Osten 1983). Substantial points of discussion here, however, remain unchanged.

## 5-2) Time-Resolved Photoluminescence Study of Polarons and Excitons at High Density Excitation

Typical luminescence spectra of zone-refined AgBr of ultra-high purity are illustrated in Fig. 5-4. The level of excitations ranges from low levels about

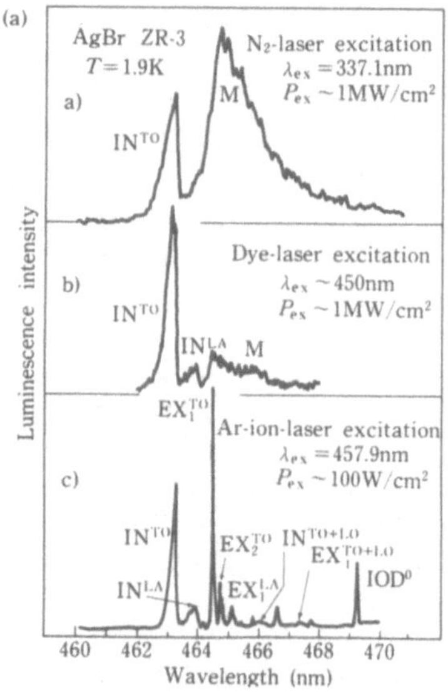


Fig. 5-4-a). Typical luminescence spectra of AgBr ZR-3 at 1.9K under a) pulsed nitrogen laser ($\lambda$=337.1nm), b) pulsed dye laser ($\lambda$=450nm) and c) CW Ar-ion-laser ($\lambda$=457.9nm) excitations.

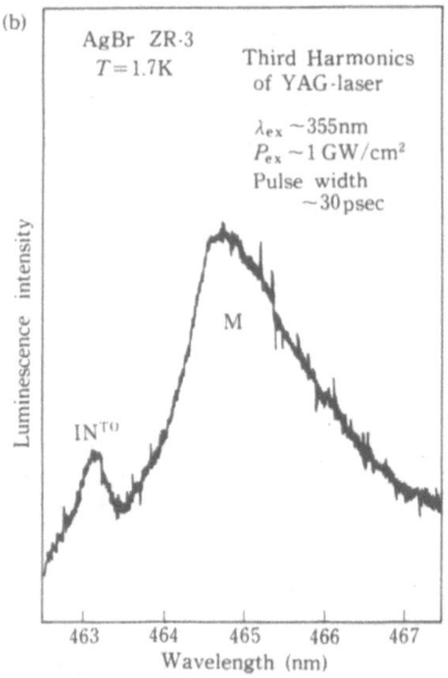


Fig. 5-4-b). Typical luminescence spectra of AgBr ZR-3 at 1.7K and 1GW/cm$^2$ by YAG-laser (3rd harmonics) excitation with pulse duration of about 30psec.

10~100W/cm$^2$ to higher levels of 1 MW/cm$^2$. At high levels of excitations, a new and intrinsic luminescence line-M has been observed in pure AgBr crystals at liquid helium temperatures by a time-resolved spectroscopy technique in the n-sec region (Baba and Masumi 1976, 1977). From the p$^{1.6}$-dependence of luminescence intensity on the excitation power p (Fig. 5-5) and the analysis of the line shape by using Cho's theory (1973), we ascribe the new M-line to the partial recombination emission from *"the free excitonic molecules"* in pure AgBr at high-density excitation.

Quite recently, we have extended our time-resolved luminescence study up to the p-sec region (Baba and Masumi 1983, Masumi 1981) as illustrated in Fig. 5-4-b). The 3rd harmonics of the YAG-laser (Nd-YAG-laser) at 355 nm were used to obtain the excitation power up to 1 GW/cm$^2$ with pulse width of 30 psec. Similar data consistent with those in the n-sec region have also been obtained by the p-sec spectroscopy technique as illustrated in Fig. 5-4-b).

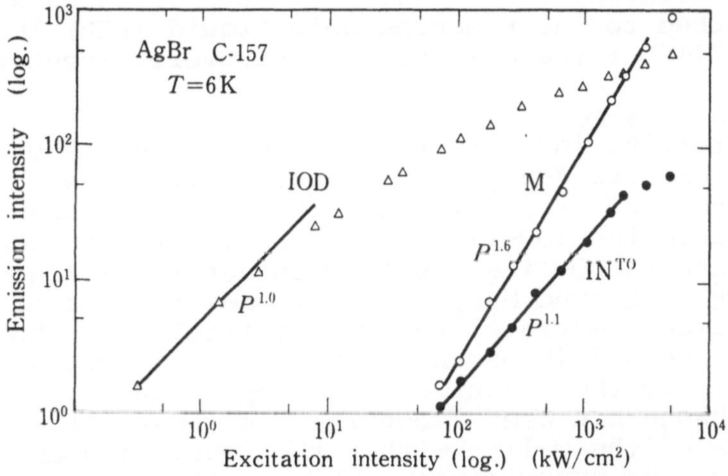


Fig. 5-5. Typical luminescence spectra of AgBr C-157-3 at 6K under c.w. argon-ion-laser (λ=457.9nm) (lower) and pulsed-nitrogen-laser (λ=337.1nm) (upper four) excitations; gate width = 10μs. b) Intensity dependences of the *A*-, *B*- and IOD-lines in AgBr C-157-3 at 5K on the excitation intensity *p* (nitrogen-laser excitation).
• 463.0nm *A*-, IN$^{TO}$-line; ○ 464.7nm *B*-, *M*-line; △ 495.0nm IOD-line.

The binding energy of the excitonic molecules $E^b_{em}$ was estimated to be about (7±1) meV. Theoretically, (Akimoto and Hanamura 1972), it is estimated to be about 1 meV, by using the observed effective mass ratio $\sigma = m_e/m_h \simeq 0.3$ (Tamura and Masumi 1973) and the binding energy of an exciton $E^b_{ex} \simeq 20$ meV (Kanzaki and Sakuragi 1969a, 1969b) for AgBr. The following two effects may account for this discrepancy.

1) The polarizability of the host lattice may enhance the binding energy of the excitonic molecules.
2) The energy states of the excitonic molecule may split into the multiplet due to the many valleys at the L-points for the constituent positive holes (Quattropani, Forney and Bassani 1975).

The second effect may also be important. But, it is even more conceivable that the dynamical character of 4 constituent polarons (2 electron-polarons and 2 hole-polarons) must play a substantial role in the formation of an excitonic molecule as will be discussed later in section 6.

Below 470nm, a rather broad line has been observed, called *"the Q-line"* (Baba and Masumi 1976, 1977, 1979, 1983) as illustrated in Figs. 5-6 and 5-3, which was once ascribed to the Electron-Hole Liquid (EHL) (Hulin et al. 1977) but there is still controversy about its origin.

Our results indicate that 1) the Q-line is characteristic for pure AgBr crystals but 2) it appears even at low levels of excitation and 3) it can be time-resolved into the IOD-line due to isoelectronic iodine impurity and the Q-line itself with several LO-phonon structures (Baba and Masumi 1979). New results of our time-resolved spectra of the Q-line itself are shown in Figs. 5-6 and 5-7 (Baba and Masumi 1983). Furthermore, 4) by applying the magnetic fields up to 50kOe, *no* significant change has been recognized in the position and line shape of the Q-line within the resolution and accuracy of our experiments. There still remain a few unknown but essential points concerning the Q-line.

Another new multiphonon series designated "D-line" has been observed by using a strong UV light from a CW Ar-ion laser (Kleinfeld, Stoltz and von der Osten 1979). Similar structures have been observed also by using a pulsed dye-laser (Baba and Masumi 1983).

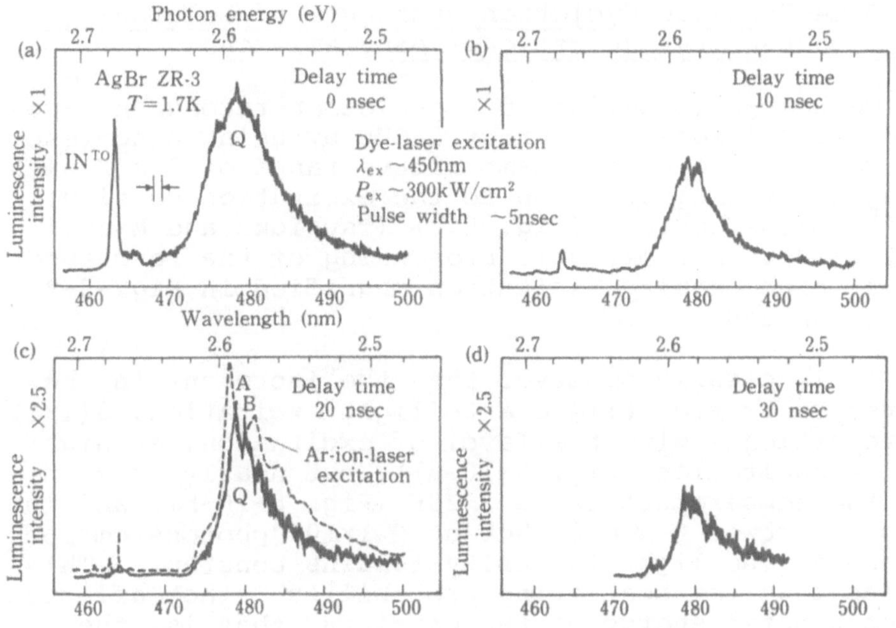


Fig. 5-6. Time-resolved luminescence spectra of AgBr ZR-3 at 1.7K under a pulsed dye-laser excitation $P_{ex} \approx 300kW/cm^2$, $\lambda$=450nm, with pulse width of 5 nsec.

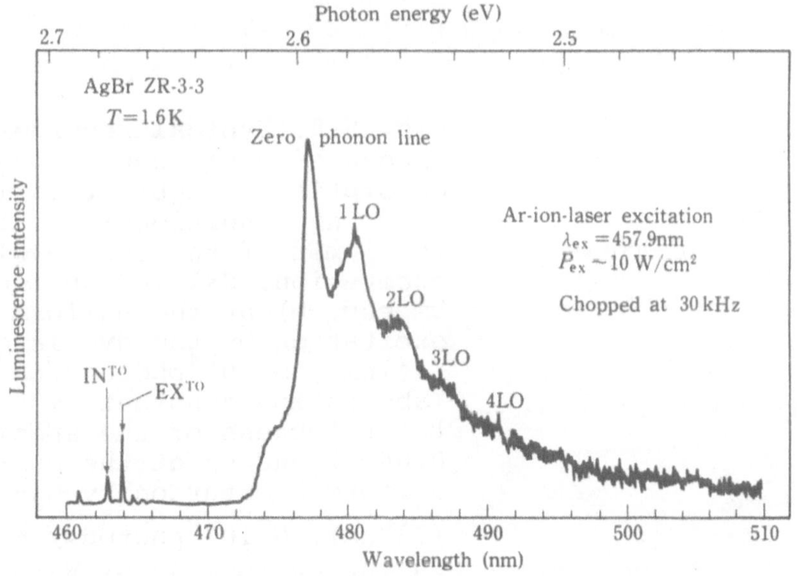


Fig. 5-7. The Q-line spectrum of AgBr ZR-3 at 1.6K excited by an Ar-ion laser mechanically chopped with f=30kHz.

## 5-3) Time-Resolved Cyclotron Resonance of Polarons at High Density Excitation - Many Polarons

We have also performed a cyclotron resonance experiment of polarons in AgBr at 34GHz by using a nonresonant waveguide over the temperature range of 2~23K with pulsed laser excitation up to the excitation density $G \sim 10^{28}$ photons/cm$^3$.sec (Fig. 5.8) (Tsukioka and Masumi 1974, 1980). A remarkable broadening of the resonance line at high density excitation magnified in Fig. 5-9 was recognized.

It is clearly observed that the increment in the inverse relaxation time due to light excitation, $\Delta(1/\tau)$, is proportional with the level of excitation, at high density excitation (Fig. 5-10-a)), but nearly constant over the temperature range 2~23K (Fig. 5-10-b)) and the microwave power range 0~25db at $G>1\times10^{24}$ photons/cm$^3$sec as long as the light intensity remains constant. This implies that the scattering probability is not affected by the kinetic energy of the carriers, that is, the scattering is dominated mainly by neutral centers. The observed values of $\Delta(1/\tau)$ are in fair agreement with those calculated by Erginsoy's formula in the absence of

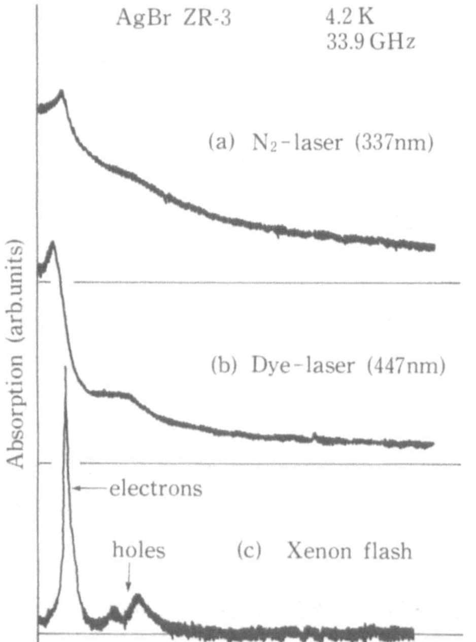


Fig. 5-8. Typical recorder traces of cyclotron resonance absorption in AgBr at various exciting conditions; a) at the limit of the low level excitation, $G<1\times10^{21}$ photons/cm$^3$sec, b) at the uniform excitation by the dye laser (447nm), $G\sim10^{24}$ photons/cm$^3$sec, (absorption constant, $\kappa\cong5$cm$^{-1}$, the thickness of the specimen 0.04cm) and c) at the non-uniform excitation by $N_2$-laser (337nm), $G\sim10^{28}$ photons/cm$^3$sec in the surface layer with the depth of $1/\kappa\cong10^{-4}$cm.

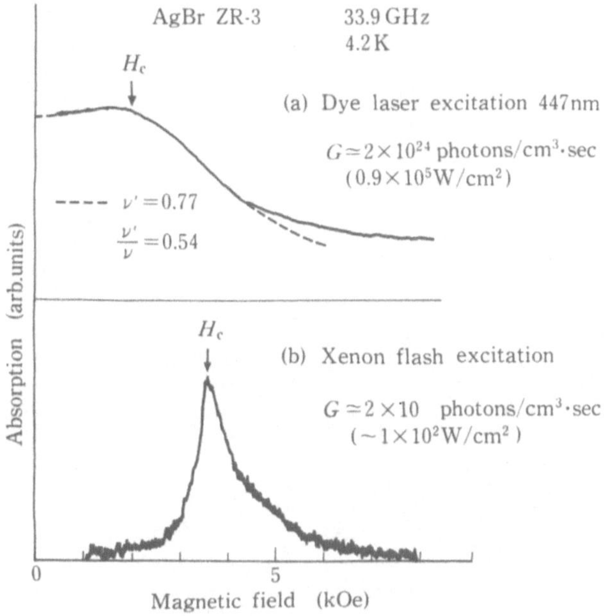


Fig. 5-9. Typical recorder traces of cyclotron resonance absorption due to electrons in AgBr at 4.2K by the dye-laser excitation (447nm) and, in contrast, by a dimmed xenon flash excitation.

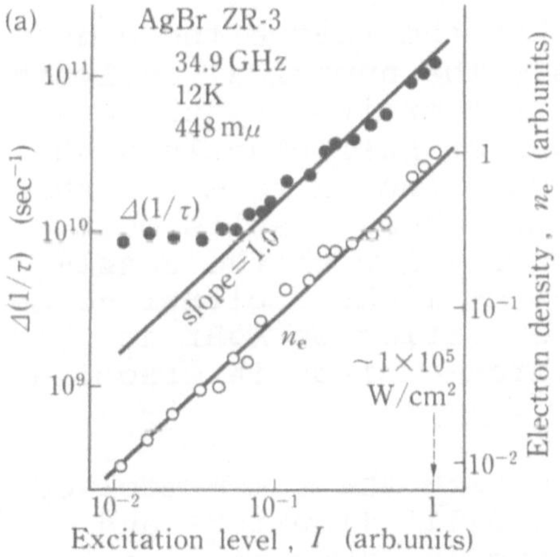


Fig. 5-10-a). The increment of the inverse relaxation time and the density of electrons at 12K as a function of excitation level.

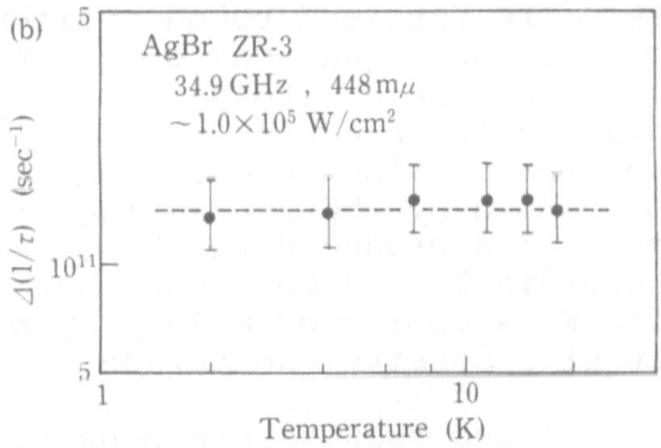


Fig. 5-10-b). The increment of the inverse relaxation time at $I_{max}$ as a function of temperature. Data include the correction factor due to the variation of absorption coefficient with temperature.

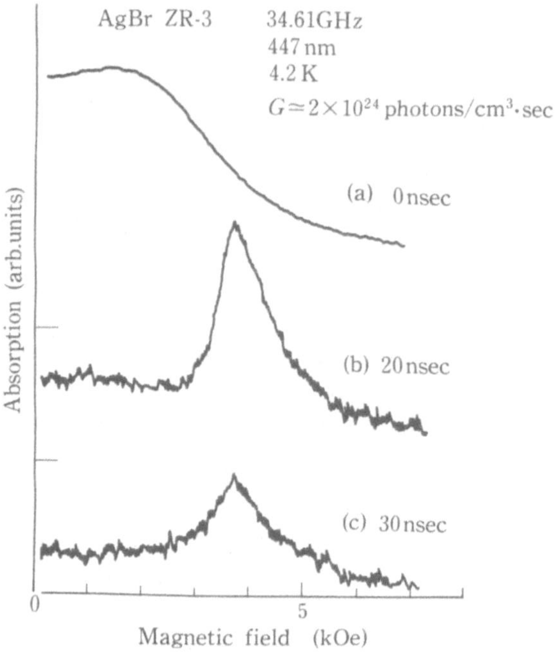


Fig. 5-11. Time-resolved cyclotron resonance spectra at 4.2K, for delay times a) 0nsec, b) 20nsec and c) 30nsec. $G=2\times10^{24}$photons/$cm^3$sec.

the recoil effect and assuming that the scattering centers are free and bound excitons. The numerical analysis of the rate equations of the polaron-exciton system requires the existence of excitons at sufficiently high density to interpret the line broadening in terms of the polaron-exciton interaction. Thus, these results must be caused by the scattering of carriers with the quasi-particles produced in the crystal by light excitation. Accordingly, we conclude that the carrier in AgBr at T<20K is dominated mainly by polaron-exciton interaction at high density excitation.

These results are consistent with the time-resolved cyclotron resonance spectra (Fig. 5-11) (Tsukioka and Masumi 1980) and also time-resolved luminescence spectra (Fig. 5-6) (Baba and Masumi 1980) both in the nsec region. So far, *no* resonance line has been observed due to a new type of elementary excitation. But, the microwave photo-signal at H=0 shows *a hump* at the delay time of $T_d \simeq 15$nsec (Fig. 5-12) (Tsukioka and Masumi 1980), when also *a non-resonant absorption* due to unknown origin was observed at low magnetic field region in the cyclotron resonance spectra (Fig. 5-11-b)).

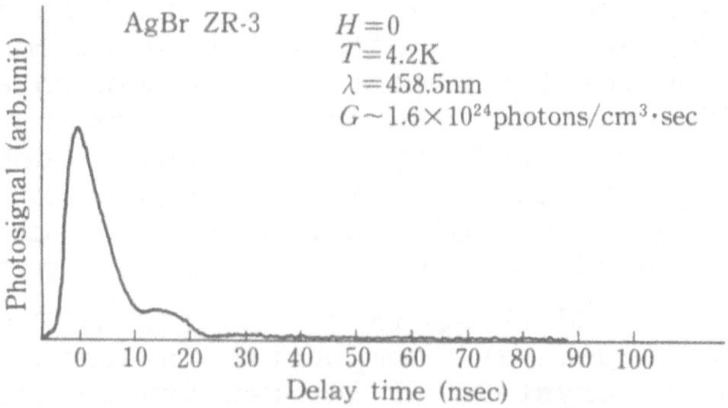


Fig. 5-12. Microwave photosignal at H=0 is shown as a function of the delay time of the sampling gate, $T_d$. $G=1.6\times10^{24}$ photons/$cm^3$.sec.

These results seem to suggest the existence of the e-h droplet, as in optically pumped germanium (Nakamura and Morigaki 1974) which, however, has not been established in the luminescence study of AgBr (Baba and Masumi 1983). Detailed results of further studies on this problem are described in 5-2.

## 5-4) Anomalous Phenomena in Luminescence Spectra in $AgCl_xBr_{1-x}$ at High Density Excitations - Various Types of Polarons and Excitons

New intrinsic lines attributed to *"the partial recombination of the free excitonic molecules"* and *"the Auger process in the two-exciton system"* have been observed, for the first time, in the luminescence spectra of $AgCl_xBr_{1-x}$ mixed crystals, at high density excitation by a time-resolved spectroscopy technique (Kawahara and Masumi 1980).

An exciton in $AgCl_xBr_{1-x}$ mixed crystals is a stable elementary excitation in the free state in the range of $x<0.45$ as well as in pure AgBr, whereas it is self-trapped in the range of $0.45<x$ as well as in pure AgCl, as reported by Kanzaki et al. (1971). Recombination of a free exciton in $AgCl_xBr_{1-x}$ mixed crystals for $x<0.45$ takes place either in the process

$$\text{free exciton} \rightarrow h\nu + TO(L) + nLO(\Gamma) \quad (n=0,1,2,\ldots) \quad (1)$$

or via

$$\text{free exciton} \rightarrow h\nu + nLO(\Gamma) \quad (n=0,1,2,\ldots). \quad (2)$$

The series in process (1) denotes the indirectly allowed transitions for an exciton with the assistance of 1-TO(L)-phonon emission for the momentum conservation and additional n-LO(Γ)-phonons. The series in process (2) is characteristic for mixed crystals in which such a momentum conservation in recombination processes is not necessarily required because of the reduction of translational symmetry. Let us denote the luminescence lines in series (1) by $FE^{TO+nLO}$, and those in series (2) by $FE^{NP+nLO}$. These series are dominant components in luminescence spectra of mixed crystals at low density excitation. What happens in similar situations to excitons in mixed crystals at high density excitation?

Mixed crystals of $AgCl_xBr_{1-x}$ were grown by the Bridgeman technique from melt of pure powder (6-9s) in a $Cl_2$ atmosphere. Samples were prepared carefully with special attention to the surface. The stoichiometry x

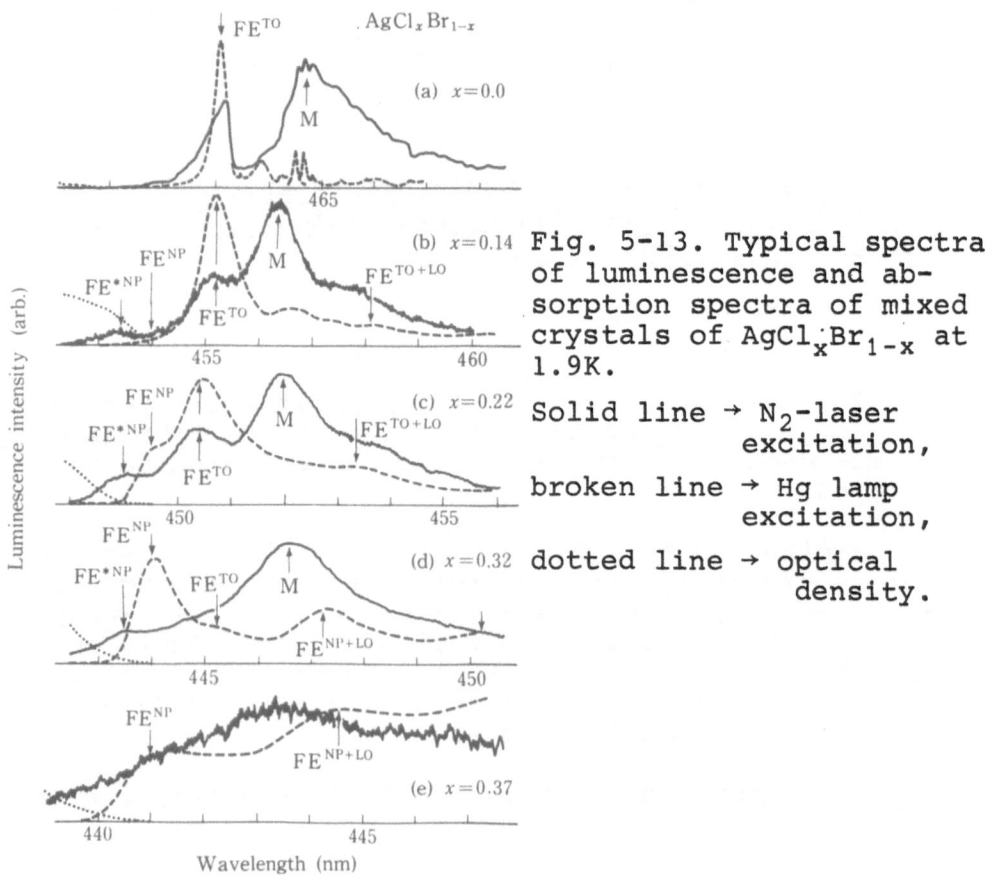

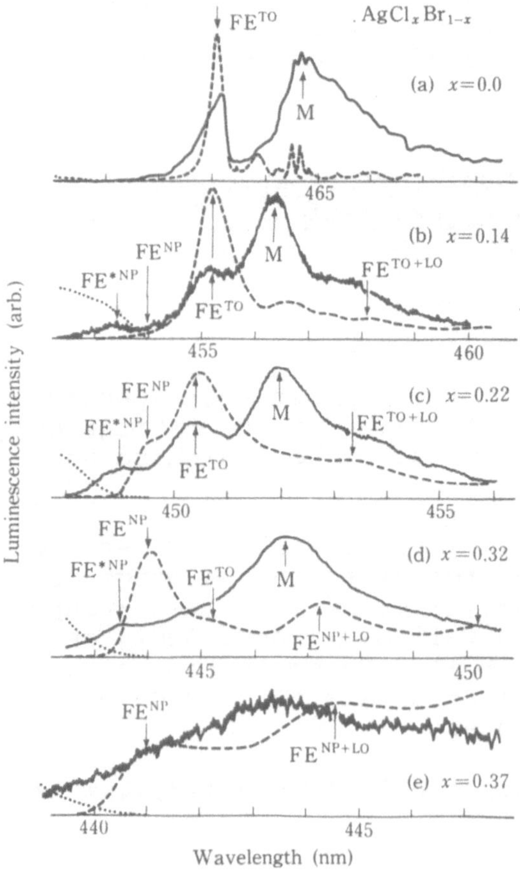


Fig. 5-13. Typical spectra of luminescence and absorption spectra of mixed crystals of $AgCl_xBr_{1-x}$ at 1.9K.

Solid line → $N_2$-laser excitation,

broken line → Hg lamp excitation,

dotted line → optical density.

of each specimen of $AgCl_xBr_{1-x}$ mixed crystal was estimated by observing the position of the optical absorption edge.

Figure 5-13 illustrates typical luminescence spectra of $AgCl_xBr_{1-x}$ over the range $0 \leq x \leq 0.5$ at high levels of excitation and similar data at low levels of excitation together with the optical density spectra, all at 1.9K. At high levels of excitation, new peaks denoted by M were observed, together with the FE-lines with series (1) and (2). Another peculiar feature for $x \neq 0$ is the appearance of a second series of broad lines denoted by $FE^{*NP}$ in the spectra of $AgCl_xBr_{1-x}$ with $x \lesssim 0.2$ at high levels of excitation. Over the range $x \gtrsim 0.5$, including the case of pure AgCl, no effect at high density excitation has been detected so far.

As the excitation power p increases, the intensity of the M-line and the $FE^*$-line increases as $p^{1.5}$ and $p^{1.1}$, respectively, for a mixed crystal with x=0.14 as plotted in Fig. 5-14, whereas the intensity of the $FE^{TO}$-lines is proportional to $p^{1.0}$. The linewidth of the M-line increases as p increases. In the case of mixed crystals, the M-line sometimes overlaps with the series of free exciton lines, so that it is rather difficult to analyze the shape of the M-line. Nevertheless, *no* definite LO-phonon side band can be perceived for the peak of M-lines.

Typical time-resolved luminescence spectra of $AgCl_xBr_{1-x}$ are shown in Fig. 5-15 for the case x=0.14. Here, the luminescence spectra at the highest levels of excitation, obtained immediately after excitation, were observed to change gradually to the spectra at low levels of excitation. Even the M-line at high levels of excitation close to the $FE^{NP+LO}$-line at low levels of excitation in the case of $AgCl_{0.32}Br_{0.68}$ (Fig. 5-13-d)) can be clearly time-resolved with an accuracy better than the resolution power in wavelength. This technique allows one to conclude that the M-line and the $FE^{*NP}$-line in mixed crystals are due to the effect of high density excitation.

First, the characteristics of the M(x)-line for $AgCl_xBr_{1-x}$ described above are quite similar to those of the M-line due to the free excitonic molecules for pure AgBr (Baba and Masumi 1977, Pelant et al. 1977). In pure AgBr(x=0), an excitonic molecule is created in the process

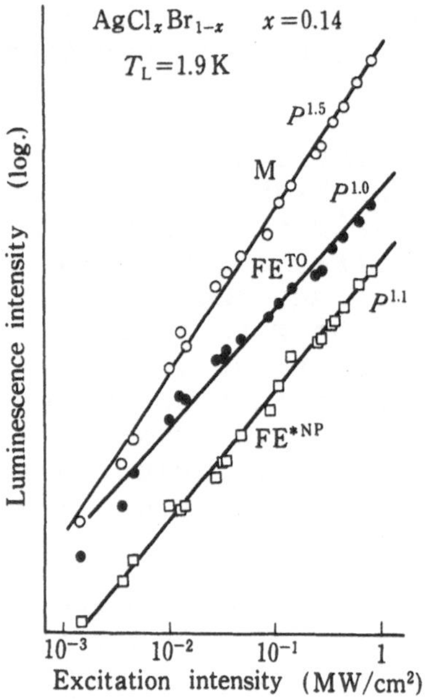


Fig. 5-14. Dependence of the intensity of M-, $FE^{TO}$- and $FE^{\star NP}$-lines on the intensity p for $AgCl_{0.14}Br_{0.86}$
○ M-line: • $FE^{TO}$-line:
□ $FE^{\star NP}$-line

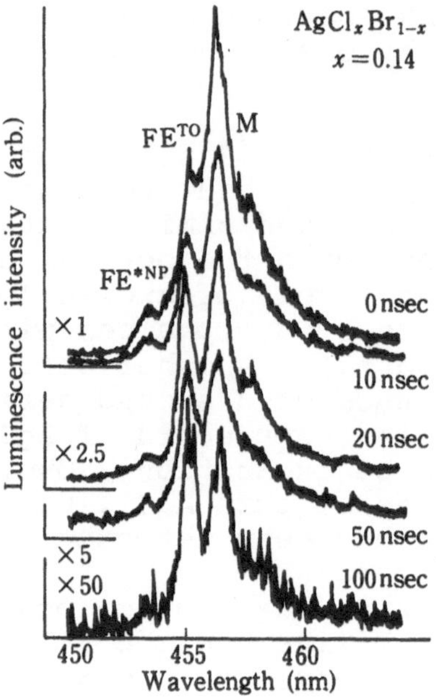


Fig. 5-15. Time-resolved luminescence spectra of $AgCl_{0.14}Br_{0.86}$ at 1.9K under an $N_2$ laser excitation (1MW/cm$^2$)

$$\text{exciton(L)} + \text{exciton(L)} \rightarrow \text{excitonic molecule}(\Gamma \text{ or } X), \tag{3}$$

and annihilates in the process

$$\text{free excitonic molecule}(\Gamma \text{ or } X) \rightarrow \text{exciton(L)} + \hbar\nu + \text{TO(L)}. \tag{4}$$

Actually, in the experiment for pure AgBr crystals by Baba and Masumi (1977), the intensity of the M-line due to free excitonic molecules was proportional to $p^{1.6}$ and the line shape of the M-line fits the curve calculated by using the Cho theory (1973).

Results for $x \neq 0$ in Figs. 5-13 - 5.15 indicate quite similar situations to those for x=o except that the line shape of both the M-line and the FE-lines becomes Gaussian-like for $x \neq 0$. Nevertheless, one can readily recognize that the new M(x)-line tends to the M-line for pure AgBr(x=0) as $x \rightarrow 0$. Thus, we ascribe the M(x)-line to the free excitonic molecules in $AgCl_xBr_{1-x}$ mixed crystals.

Secondly, we have observed another new broad luminescence peak called the $FE^{\star NP}$-line in Fig. 5.13. This peak had never been observed for pure AgBr(x=0) but can be observed *"in $AgCl_xBr_{1-x}$ mixed crystals"* only *"at high density excitation"*. The p-dependence of the $FE^{\star NP}$-line is definitely *"superlinear"*. Also Fig. 5-15 clearly illustrates that the $FE^{\star NP}$-line can be time-resolved as a component at high density excitation. As the position of the $FE^{\star NP}$-line is located just at the indirect absorption edge, a high density excitation effect appears at an energy region higher than the FE-lines which is rather unusual. Also, the shape of the broad line is characteristic for a Boltzmann distribution with temperatures above 30K and the high energy tail of the $FE^{\star NP}$-line even extends into the region of indirect absorption. These results suggest the possibility of a new Auger-like process in the exciton system such as the inelastic collision either between a hot exciton and a hot exciton (5a) or between a hot excitonic molecule and a trapping center (5b). The first possibility is written as

$$\text{free hot exciton} + \text{free hot exciton} \rightarrow \text{exciton} + \hbar\nu^{\star} \tag{5a}$$

Here, the sign ★ denotes an excess gain of energy. The process (5a) can be predominant in the case of high density excitations but so far it has been noted only rarely perhaps due to the lack of relevant experimental data. The second possibility is conceived to be due to the process

$$\text{frustrated excitonic molecule} \rightarrow \text{trapping exciton} + \hbar\nu^{\star} \tag{5b}$$

In the process (5b), the $FE^{\star NP}$-line may be due to the radiative recombination of the free exciton with zero phonons in $AgCl_xBr_{1-x}$, the components of which, in the past, formed an excitonic molecule but were unfortunately dissociated from each other with an unfair energy distribution between its partner-exciton either self-trapped or trapped at a Iodine impurity. *"A frustrated excitonic molecule"* we call this type of molecule.

Finally, the binding energy of a free excitonic molecule, $E^b_{em}(x)$ was studied as a function of x. A peculiar feature here is an unexpected decrease of $E^b_{em}(x)$

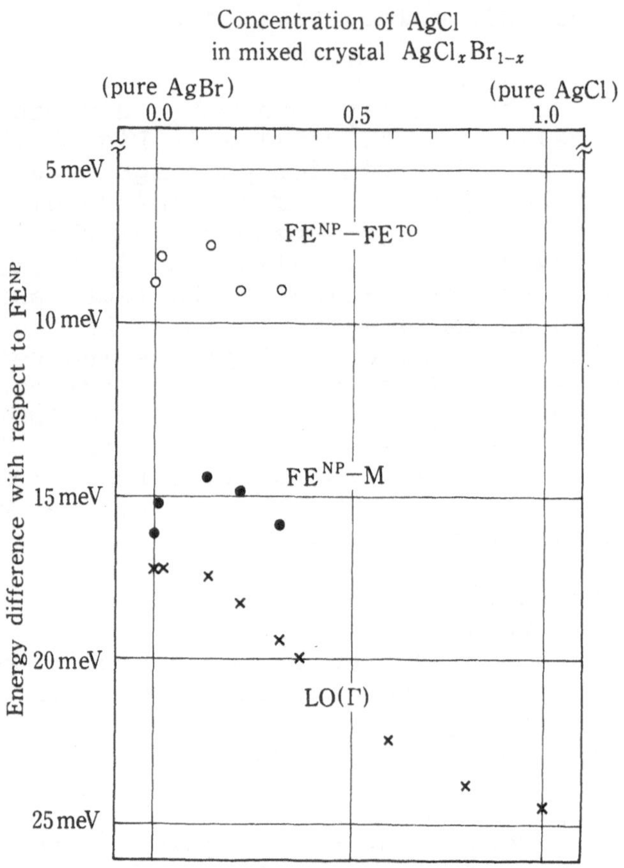


Fig. 5-16. Energy difference between the $FE^{TO}$- and M-lines with respect to the $FE^{NP}$-line.

around $x \lesssim 0.1$ with increasing x and its recovery above $x > 0.2$ as indicated in Fig. 5-16. The fact that the M-line has no higher orders of LO-phonon side-bands in Fig. 5-13 can be explained by the LO-phonon coupling-strength of an excitonic molecule with a quadrupole moment. This coupling is weaker than that for an exciton with only a dipole moment. It is conceivable that a free excitonic molecule is associated with TO(L)- or TA(L)-phonons (besides the momentum conserving TO(L)-phonon) rather than with other types of phonons such as LO(Γ).

We conclude that the new intrinsic luminescence lines attributed to *"the partial recombination of the free excitonic molecules"* and possibly due to *"the Auger process in the two-exciton system"* have been simultaneously observed, for the first time, in $AgCl_xBr_{1-x}$ mixed crystals.

We suggest that the dynamical and nonlinear characteristics of polarons and excitons underlie these experimental results. We shall discuss these problems later in section 6.

## 6. DISCUSSION OF THE DYNAMICAL AND NONLINEAR PROPERTIES OF POLARONS, EXCITONS AND EXCITONIC MOLECULES

With the above mentioned results in mind, we can speculate on the dynamical and nonlinear responses of polarons and excitons (Masumi 1981, 1982).

### 6-1) A Modification of the Feynman-polaron Model

Let us first recall the original idea of the Feynman-polaron model and the modified model (Fig. 6-1) as introduced in 2-3). There, we proposed a modification of the Feynman-polaron model effective only if an isolated polaron interacts with other systems. The fictitious particle occupies a finite volume so that no divergencies related to point charges occur. The values of $M$ and $\gamma$ of a single polaron are not changed by the modification of the model. Note that the polarization charge $\rho$ for the Feynman-polaron model, once eliminated in the process of calculation of the density matrix of a single polaron in thermal equilibrium to eliminate all the phonon coordinates in the particle-field problem, reappears only if the polaron interacts with an external system (Masumi 1979).

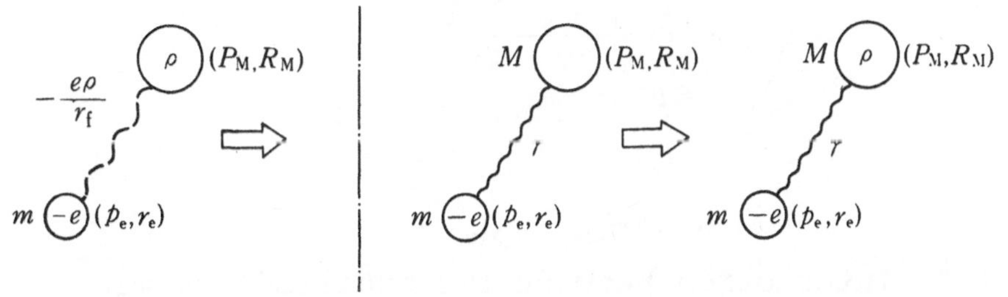


Fig. 6-1. A modification of the Feynman-polaron model. Revival of the polarization charge $\rho$.

6-2) <u>Interaction between Two "Modified" Feynman-polarons</u>

Let us consider the problem of polaron-polaron interaction by considering a sphere consisting of an ionic crystal in which we place two modified Feynman-polarons, as illustrated in Fig. 6-2.

We denote the total Hamiltonian

$$H_{2FKP} = H_0 + H_1 \tag{6.1}$$

where $H_0$ is the unperturbed part for two non-interacting polarons, i.e., the summation of Eq. (2.13) over the $i$-th polaron ($i$-1,2), and $H_1$ denotes the interaction term in space and imaginary time

$$H_1 = \frac{e_1e_2}{R} - \frac{e_2\rho_1}{R-x_1} - \frac{e_1\rho_2}{R+x_2} + \frac{\rho_1\rho_2}{R-x_1+x_2} . \tag{6.2}$$

An elementary exercise in quantum mechanics in the adiabatic approximation in complex time can be performed with $e \equiv e_1 = e_2$, $\rho \equiv \rho_1 = \rho_2$ by expanding Eq. (6.2) for $|R| >> x_1$, $x_2$ and diagonalizing $H_1$ in the standard method of normal mode analysis.

The result yields a decrease of the total energy of the two-polaron system. The total interaction energy of the two modified Feynman-polaron systems to the sixth order of $R^{-1}$ is

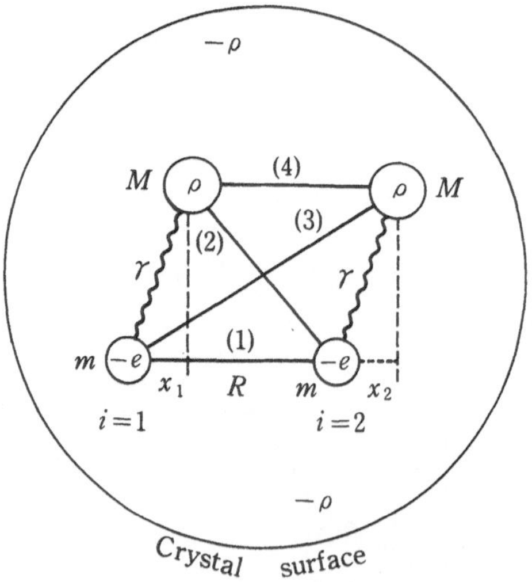


Fig. 6-2. Interaction between two modified Feynman-polarons both placed in a sphere consisting of an ionic crystal.

$$U(R) \simeq \frac{(e-\rho)^2}{R} - \frac{\rho^2(e-\rho)^2}{\gamma R^4}$$

(Screened Coulomb interaction) (Interaction due to the resulting deformation)

$$- \frac{1}{2}\hbar\omega_0\left[\frac{2\rho(e-\rho)}{\gamma R^3} + \frac{\rho^2(e-\rho)^2+\rho^4}{\gamma^2 R^6}\right] .$$

(Interaction due to the coherent polarization) (6.3)

Here, the first term represents the Coulomb interaction between two modified Feynman-polarons screened by the accompanying polarization charge $\rho$ of each other. The second term reflects the lowering of the system energy by the static deformation of both polarons due to the Coulomb interaction. The last term is associated with a van der Waals interaction between dipoles, i.e., "the hetero-polar interaction" between two polarons. These attractive parts are due to the coherence of the virtual LO-phonons associated with two polarons in space and time. As the electron-LO-phonon interaction denoted by the coupling constant $\alpha$ decreases, $M$, $\gamma$ and $\rho$ tend to zero. $U(R)$ given by Eq. (6.3) is reduced to the direct Coulomb repulsion between two bare electrons, whereas, if each electron is completely screened, $\rho \rightarrow e$, it reduces to the van der Waals interaction. Therefore, the modification of the Coulomb interaction expressed by Eq. (6.3) is characteristic for the static softening and dynamical polaron-polaron interaction based on the "modified" Feynman-polaron introduced here. We call Eq. (6.3) "Heteropolar interaction" (Masumi 1975). This is the basic interaction in the two polaron problem in our scheme¶ which is free from the usual theoretical diffi-

¶Actually, the path integral calculation of the density matrix for the coupled Feynman-polaron system and the variational treatment may be necessary to further give the upper bound of the ground state energy $E_0$ on the basis of the "modified" polaron model. Since the Hamiltonian to be used for the trial action contains only harmonic potentials in the diagonalized form, the exact calculation of the path integral can be performed in the adiabatic approximation for imaginary time.

Consequently, a new numerical calculation is requested to obtain $E_0$ by the variational method including $R$ as the parameter. Here, however, we set the path integral formulation and variational calculation aside for future work, and discuss the interaction between two "modified" Feynman-polarons only in the framework of the model Hamiltonian $H_1$ and modified eigenfrequencies $\omega(R)$ as a sort of "conceptual mapping".

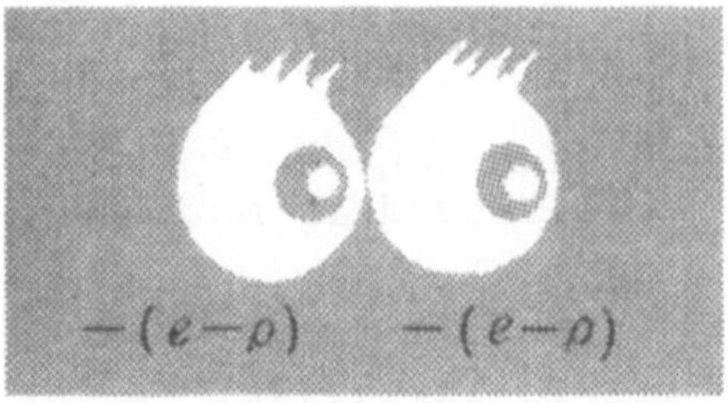


Fig. 6-3. Static and dynamical aspects of a "bipolaron".

culties due to the divergence in higher order perturbation treatments of Coulomb interaction, at least for the polaron-polaron interaction.

As the repulsive force $(e-\rho)^2/R^2$ exceeds the attractive forces at the distance $R$, two polarons are usually independent. However, as they approach each other in the crystal, e.g. at high density excitation or even by chance, the relative weight of the first term compared to the others in the total interaction energy (6.3) decreases. Thus, there might exist a region where the attractive interaction due to the deformation and coherent polarization of two polarons has a real significance as is illustrated in Fig. 6-3. Consequently, there may be a small but nonzero probability that two polarons become bound to each other to form a "bipolaron" which might be metastable or stable.

## 6-3) Polaron Problems in the Formation of an Exciton

Now we turn our attention to the electron-hole or exciton system in ionic crystals.

Suppose the radius of an exciton is $R_{ex}$. In ionic crystals, contrary to non-polar semiconductors, it is significant to examine the relation,

$$R_{ex} \gtrless r_{ef}, r_{hf} \qquad (6.4)$$

where $r_{ef}$, $r_{hf}$ are the radii of an electron-polaron and a hole-polaron. In what follows we limit ourselves to a simplified treatment in terms of the modified Feynman model.

a) Polaronic excitons ($R_{ex} > r_{ef}, r_{hf}$). First we adopt the modified Feynman-polaron model for a hole-polaron as indicated in Fig. 6-4. In analogy to Fig. 6-2, we can visualise the interaction between an electron-polaron and a hole-polaron by Fig. 6-5 at least in the early stages of exciton formation. Further treatments

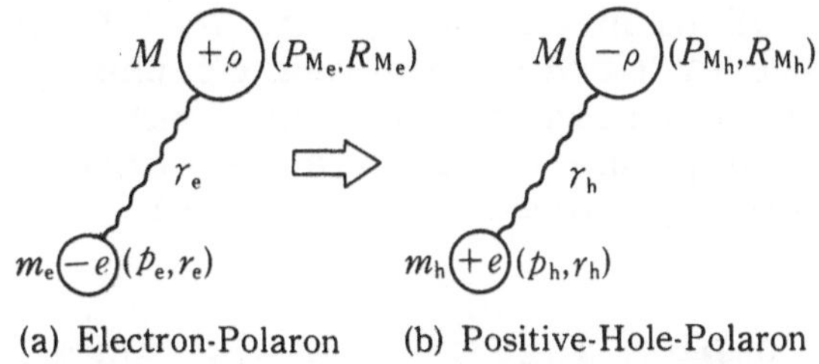


Fig. 6-4. An electron-polaron and a hole-polaron based on the modified Feynman-polaron model.

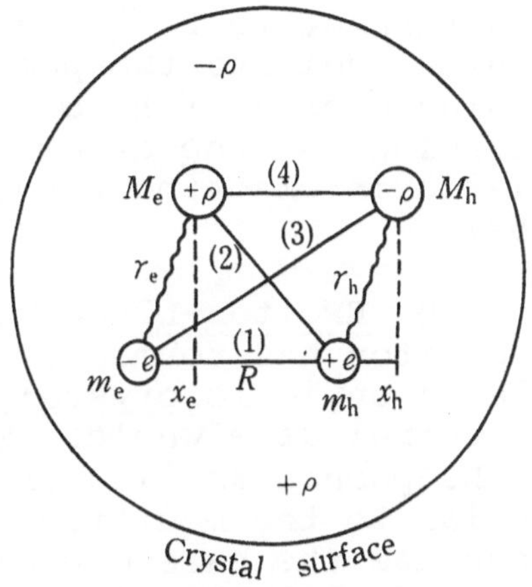


Fig. 6-5. Interaction between an electron-polaron and a hole-polaron based on the modified Feynman-polaron model during the formation of a polaronic exciton.

are slightly complicated because one has to distinguish between the mass values of an electron-polaron and a hole-polaron, $m_e$ and $m_h$, similarly, $\rho_e \neq \rho_h$, $\gamma_e \neq \gamma_h$ and $M_e \neq M_h$. Consequently, $\omega_{0e}$ also differs from $\omega_{0h}$.§ However, if as a first step one ignores all these differences, one may express the total interaction energy of a polaronic exciton system as

$$U(R) \simeq -\frac{(e-\rho)^2}{R} \qquad - \frac{\rho^2(e-\rho)^2}{\gamma' R^4}$$

(Formation energy of an exciton due to the screened Coulomb interaction)

(Static interaction due to the deformation of constituent polarons) (6.5)

$$-\frac{1}{2}\hbar\omega_0' \left[-\frac{2\rho(e-\rho)}{\gamma' R^3} + \frac{\rho^2(e-\rho)^2+\rho^4}{\gamma''^2 R^6}\right]$$

(Dynamic interaction due to the coherent polarizations of constituent polarons) with $\omega_0' \equiv \omega_{0e} \simeq \omega_{0h}$. (6.6)

Note that the second term in the first line and the second term in the square bracket of the second line in (6.3) remain unchanged in going to (6.5) due to their even parity

---

§These points have been studied from the point of view of the exciton-exciton interaction by using appropriate trial wave functions for variational calculations (Adamowski, Bednarek and Suffczynski 1976, 1978, 1979).

with respect to the changes of sign of $(e-\rho)$ and $e$. These results indicate the possibility of an increase of the binding energy of a polaronic exciton due to the static softening and the dynamical response of both an electron-polaron interacting with a hole-polaron (Masumi 1981 - 1982).

b) Exciton-LO-phonon complex ($R_{ex}<r_{ef}, r_{hf}$). Once an exciton has formed from a bare electron and a bare hole in an ionic crystal, one may consider the entity to be neutral as a whole. The exciton now interacts with the LO-phonon as illustrated in Fig. 6-6, in a scheme similar to the modified Feynman-polaron model. Here, we visualise the bare Coulomb interaction between the electron and the hole, i.e. $-e^2/R_{ex}$, by the force constant of a spring $\gamma_{ex}$ and, similarly the correlated virtual phonons induced by the electrons and holes by $\gamma_{LO}^{-1}$. The exciton-LO-phonon interaction is enhanced if

$$\frac{1}{q_{LO'}} \equiv \frac{1}{|q_{ex}|} \simeq R_{ex} \ . \qquad (6.7)$$

For the coupled oscillators in Fig. 6-6, it is reasonable to assume

$$\omega_{ex} \simeq n\omega_{LO'} \ , \qquad (6.8)$$

(n is an integer or the inverse of an integer),

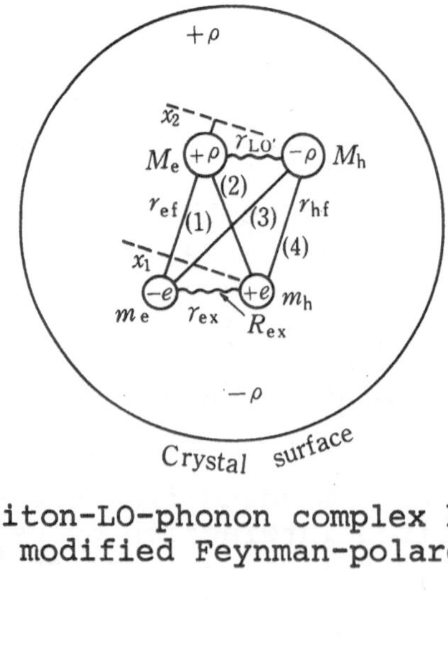


Fig. 6-6. An exciton-LO-phonon complex based on a model analogous to the modified Feynman-polaron model.

For n=1, the interaction energy takes the form:

$$U(r_{ef}, r_{hf}) \simeq -\frac{e^2}{R_{ex}} \qquad -\hbar\omega_{LO'} \langle N_q \rangle_{q \simeq q_{ex}}$$

(Formation energy of a bare exciton) (Self energy of an exciton in ionic crystals)

$$-\frac{1}{2}\hbar\omega_1 \frac{e^2 \rho^2}{\gamma_{ex}\gamma_{LO'} r_{ef}^3 r_{hf}^3}$$

(Dynamic interaction due to coherent polarization of an exciton-LO-phonon complex) (6.9)

where

$$\omega_1^2 = \frac{\omega_{ex}^2 + \omega_{LO'}^2}{2} \pm \sqrt{\frac{4e^2\rho^2}{\mu_{ex} M_{LO'} r_{ef}^3 r_{hf}^3}} \qquad (6.10)$$

with the reduced masses $\mu_{ex}$, $M_{LO'}$. The first and second terms of (6.9) represent the formation energy and the self-energy of an exciton and also lower the total energy.

The main point here is that a neutral entity composed of charged particles interacts with LO-phonons in an ionic crystal (Masumi 1981-1982).

Toyozawa-Hermanson first pointed out the possibilities not only of the formation of an exciton-LO-phonon complex but also the internal excitation of an exciton. Standard theoretical treatments have been devoted to the exciton-LO-phonon interaction by Mahanti-Varma (1970), Toyozawa-Hermanson (1968), and Matsuura (1980).

## 6-4) Polaron Problems in the Formation of an Excitonic Molecule

Polaron effects also play a role in the formation of an excitonic molecule in polar semiconductors or in ionic crystals.

One distinguishes several cases according to the inequality

$$R_{em} \gtrless r_{ef}, r_{hf} \qquad (6.11)$$

where $R_{em}$ is the average inter-polaron distance in an excitonic molecule.

a) <u>Polaronic excitonic molecules</u> ($R_{em}>r_{ef},r_{hf}$). Let us first adopt the modified Feynman-polaron model for all constituent polarons, two electron-polarons and two hole-polarons. One can illustrate an excitonic molecule by the model shown in Fig. 6-7. The interaction energy between two polarons is given by either (6.3) or (6.5) with the approximate conditions

$$\gamma \simeq \gamma' \simeq \gamma'' \quad \omega_0 \simeq \omega_0, \qquad (6.12a,b)$$

One can readily obtain the total interaction energy of a polaronic excitonic molecule $U(R_{em})$ by summing up (6.3) or (6.5) for all constituent polarons. Note that the odd-parity terms with respect to changes of sign of $(e-\rho)$ and $\rho$ cancel each other in part whereas all of the even-parity terms remain in the summation.

The resulting expression is:

$$U(R_{em}) \simeq -\frac{2(e-\rho)^2}{R_{em}} \qquad -\frac{1}{2}\hbar\Omega\frac{(e-\rho)^4}{\beta^2 R_{em}^6}$$

(Static formation energy of an excitonic molecule due to the screened Coulomb interaction) (Dynamic formation energy of an excitonic molecule due to the screened Coulomb interaction)

$$-6\frac{\rho^2(e-\rho)^2}{\gamma R_{em}^4}$$

(Static interaction energy due to the deformation of the constituent polarons in a polaronic excitonic molecule)

$$-\frac{1}{2}\hbar\omega_0\left[-\frac{4\rho(e-\rho)}{\gamma R_{em}^3}+6\frac{\rho^2(e-\rho)^2+\rho^4}{\gamma^2 R_{em}^6}\right]$$

(Dynamic interaction due to the coherent polarizations of constituent polarons in an excitonic polaronic molecule)

(6.13)

Here, the incorporation of polaron effect leads to the third and fourth terms. The first and second terms represent the static and dynamical formation energies of an excitonic molecule from screened electrons and holes with

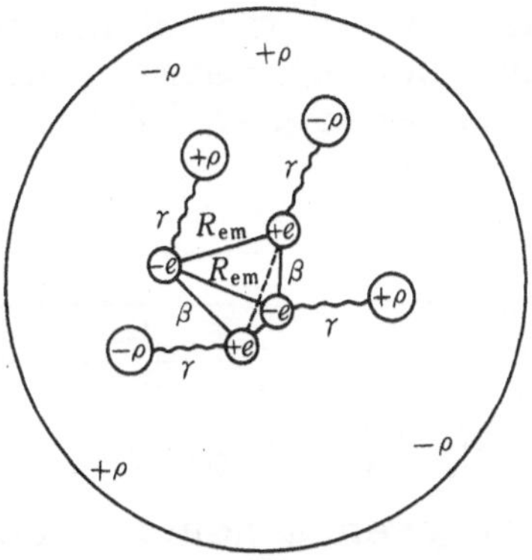


Fig. 6-7. A polaronic excitonic molecule based on the modified Feynman-polaron model.

$$\Omega = \sqrt{\frac{\beta\,(m_e + m_h)}{m_e m_h}}\,. \tag{6.14}$$

Here, $m^{\star}_{ep}$ and $m^{\star}_{hp}$ may be used instead of $m^{\star}_{e}$ and $m^{\star}_{h}$.

We remark that the terms of even-parity with respect to changes of sign of $(e-\rho)$ and $\rho$ in (6.3) and (6.5) are important in the electron-hole system. These terms give rise to nonlinear behaviour as a function of $n$ (the number of polarons) in the many polaron problems whereas they eventually become linear as $n\to\infty$ (Masumi 1981-1982).

b) <u>Excitonic molecule-LO-phonon complex</u> ($R_{em} < r_{ef}, r_{hf}$). Now consider an excitonic molecule-LO-phonon complex. For this complex even the modified Feynman model leads to a rather complicated picture. Therefore we omit drawings here and merely give the result for the total interaction energy for an excitonic molecule-LO-phonon complex as

$$U(\mathrm{r}_{ef}, \mathrm{r}_{hf}) \simeq -\frac{2e^2}{R_{em}} - \frac{1}{2}\hbar\omega_{em}\frac{e^4}{\gamma_{em}{}^2 R_{em}{}^6} - \hbar\omega_{LO''}\langle N_q\rangle_{q\simeq q_{em}}$$

(Static formation energy of a bare excitonic molecule) — (Dynamic formation energy of a bare excitonic molecule) — (Self energy of an excitonic molecule in ionic crystals

$$-3\hbar\omega_2 \frac{e^2\rho^2}{\gamma_{em}\gamma_{LO''}\mathrm{r}_{ef}{}^3\mathrm{r}_{hf}{}^3}$$

(Dynamic interaction energy due to coherent polarization of an excitonic molecule-LO-phonon complex)

(6.15)

where, again, the last two terms describe the polaron effect. The first and second terms represent the static and dynamical formation energies of a bare excitonic molecule. $\omega_2$ is given by

$$\omega_2{}^2 = \frac{\omega_{em}{}^2 + \omega_{LO''}{}^2}{2} \pm \sqrt{\frac{4e^2\rho^2}{\mu_{em} M_{LO''} r_{ef}{}^3 r_{hf}{}^3}} \qquad (6.16)$$

with the reduced masses $\mu_{em}$, $M_{LO'}$.

Remark again that even a neutral complex of excitonic molecules interacts with LO-phonons in ionic crystals.

## 6-5) Polaron Liquid?

As follows from the above discussion the ground state energy of the many polaron system should decrease as the number of polarons increases. However, it is not certain whether the system forms the Polaronic Electron-Hole Liquid (PEHL) or EHL-LO phonon compound. Also for this many body problem it is important to accumulate knowledge on the properties of a single polaron, e.g. to study the polaron radius, etc.

## 6-6) Experimental Situations

Finally, let us consider the experimental situation. From the theoretical point of view it is essential to assume inequalities between $r_{ef}$, $r_{hf}$, $R_{ex}$, $R_{em}$ such as (6.4) and (6.11). But of course our experimental studies do not develop in theoretical order. Nevertheless, we have considerable experimental knowledge on polarons in pure and ultrapure AgCl, AgBr and $AgCl_xBr_{1-x}$ as described in sections 4 and 5. However there also remain difficulties regarding the experimental situation. For example, we have the estimated values of the binding energy of an exciton in AgBr, $E^B_{ex}$, (Kanzaki and Sakuragi 1969a, 1969b, $E^B_{ex} \simeq 20$meV; and more recently, von der Osten 1983, $E^B_{ex} \simeq (32 \pm 5)$meV). On the other hand, the binding energy of an excitonic molecule in AgBr is measured as $E^B_{em} \simeq 7$meV, which is unexpectedly deep. Furthermore we do not know whether a polaron liquid exists. The question is whether or not polarons are underlying these effects. At this moment, nobody knows.

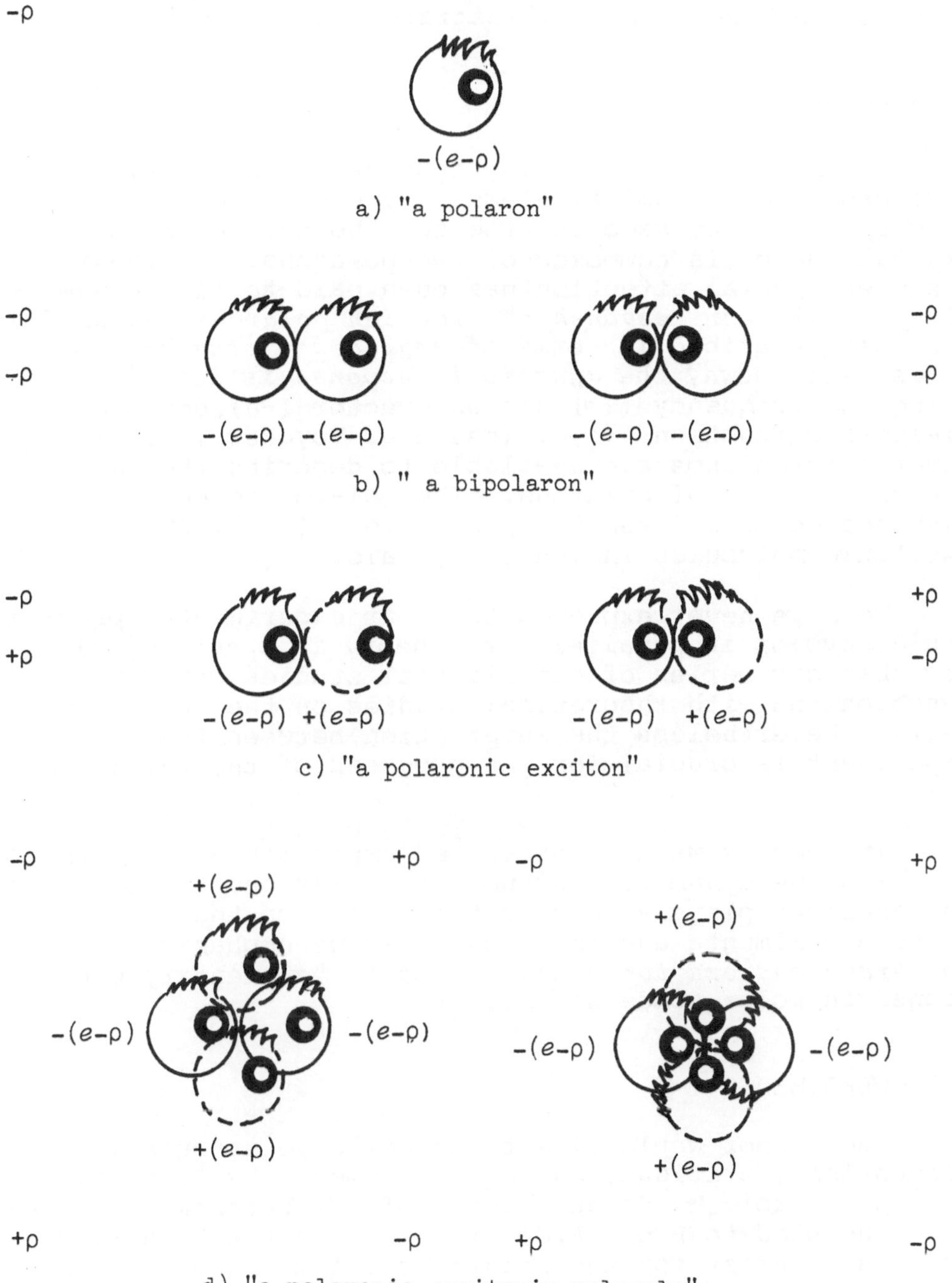


Fig. 6-8. Polarons and polaron-polaron interactions. Only two extreme modes are displayed for a polaronic excitonic molecule in d).

We are convinced that there exist, at least in some of these phenomena, dynamical and nonlinear aspects of polarons and excitons as illustrated in Fig. 6-8.

## 7. CONCLUSION

The Fröhlich polaron is, inherently, a polarizable, deformable and dynamical elementary excitation of a many body system. The same is true for the exciton in ionic crystals which is composed of two polarons. In these lectures special attention has been paid to *"the dynamical and nonlinear response of Fröhlich polarons and excitons"* as described in terms of *real and virtual phonons*. As is well known, the *dynamical response* is described using the frequency ($-\omega$) and wavevector ($-q$) dependent dielectric function $\kappa(q,\omega)$ (see e.g. Toyozawa, 1972). Several formalisms are available to describe the *nonlinear response* of polarons. The polaron concept is also needed to analyse the properties of excitons and excitonic molecules in ionic crystals.

We have never expected that our experimental program could develop in parallel with theory in space and time, nor that our series of experimental studies could be synchronized with theoretical studies on the same subjects. Nevertheless the interaction between theory and experiment is crucial for the progress of the physics of polarons.

In summary we have presented experimental results in which the dynamical and nonlinear response of polarons and excitons plays a fundamental role. On the basis of these experiments and their analysis we emphasize that *"polarons and excitons"* are the basic "elementary excitations" in polar condensed matter.

## ACKNOWLEDGMENTS

The author would like to acknowledge stimulating discussions, critiques and encouragements by Prof. Y. Toyozawa, Prof. W. Sasaki, and Prof. T. Kurosawa. He is also indebted to Prof. F.C. Brown who initially stimulated his interest for the polaron problem and Dr. F. Moser for providing us AgBr crystals used in large parts of these experiments. Sincere thanks are due to all members of his laboratory, especially Drs. K. Kajita, T. Tsukioka, S. Komiyama, T. Baba and Messrs. M. Kawahara, A. Hirano and H. Kubota. We greatly appreciate the continual support of the Cryogenic Center of the University of Tokyo. These works were supported, appropriately in each period, by the Grants-in-Aid for Scientific Research (A) and for Special Project Research, all from the Ministry of Education, Science and Culture, and in part, by the Nishina Memorial Foundation and the Yoshida Foundation for Science and Technology. Several fruitful results have been obtained in the Collaborative Research Project between the University of Oxford and the University of Tokyo, entitled "Studies of Wide Band-Gap Semiconductors" and supported by the Japan Society for the Promotion of Science.

## REFERENCES

Adamowski, J., Bednarek, S., and Suffczynski, M., 1976, The influence of the lattice polarization on the biexciton binding energy, Solid State Commun.,20:785.

Adamowski, J., Bednarek, S., and Suffczynski, M., 1978, Variational wave functions for the biexciton in polar semiconductors, Solid State Commun., 25:89.

Adamowski, J., and Bednarek, S., 1979, Effective exciton-exciton interaction in polar semiconductors, Solid State Electronics, 22:33.

Akimoto, O., and Hanamura, E., 1972, Excitonic molecule. I. Calculation of the binding energy, J. Phys. Soc. Japan, 33:1537.

Appel, J., 1968, Polarons, in:"Solid State Physics", F. Seitz, D. Turnbull, and H. Ehrenreich, ed., Academic Press, New York.

Ascarelli, G., and Brown, F.C., 1962, Cyclotron Resonance in AgBr, Phys. Rev. Letters,9:209.

Baba, T., and Masumi, T., 1976, A new time-resolved luminescence line from excitonic molecules in AgBr, in: Abstracts of Taormina Res. Conf. on the Structure of Matter - "Recent Developments in Optical Spectroscopy of Solids", Taormina.

Baba, T., and Masumi, T., 1977, A new time-resolved luminescence line from excitonic molecules in AgBr, Il Nuovo Cimento, 39B:609.

Baba, T., and Masumi, T., 1978, A new shoulder in luminescence spectra due to the triplet excitons in AgBr, Solid State Commun., 27:1113.

Baba, T., and Masumi, T., 1979, A new shoulder in luminescence spectra due to the triplet excitons in AgBr, in: Proc. 14th Int. Conf. Phys. Semiconductors - Edinburgh 1978, (Inst. Phys. Conf. Ser., no. 43, 1979: Chapt. 24), p. 833.

Baba, T., and Masumi, T., 1983, Luminescence spectra of AgBr at high density excitation (tentative), in preparation.
The authors are indebted to Dr. Aoyagi, Y., Segawa, Y., and Prof. Namba, S., of the Institute of Physical and Chemical Research for giving them an opportunity to use their facilities of psec spectroscopy.

Brown, F.C., Masumi, T., and Tippins, H.H., 1961, Fine structure in the absorption edge of the silver halides, J. Phys. Chem. Solids, 22:101.

Brown, F.C., 1972, Conduction by polarons in ionic crystals, Chapt. 8, in: "Point Defects in Solids", J.H. Crawford and L. Slifkin, ed., Plenum Press, New York and London.

Brown, F.C., 1976, The photographic process, Chapt. 7, in: "Treatise on Solid State Chemistry", Vol. 4-Reactivity of Solids, B. Hannay, ed., Plenum Press, New York and London.

Brown, F.C., 1981, Polarons large and small, in: "Recent developments in condensed matter physics", J.T. Devreese, ed., Plenum Press, New York and London, p. 575.

Burnham, D.C., Brown, F.C., and Knox, R.S., 1960, Electron mobility and scattering processes in AgBr at low temperatures, Phys. Rev., 119:1560.

Childs, C.B., 1977, High purity silver bromide crystals containing less than several parts per billion of iodine, Journal of Crystal Growth, 38:262.

Cho, K., 1973, Emission line shapes of exciton molecules in direct and indirect gap materials, Opt. Commun., 8:412.

Devreese, J.T., ed., 1972, "Polarons in Ionic Crystals and Polar Semiconductors", North-Holland, Amsterdam.

Elliott, R.J., 1963, Theory of excitons - I, in: "Polarons and Excitons", C.G. Kuper and G.D. Whitfield, ed., Oliver and Boyd, Edinburgh and London, p. 269.

Feynman, R.P., 1955, Slow electrons in a polar crystal, Phys. Rev., 97:660.

Feynman, R.P., and Hibbs, A., 1965, Quantum mechanics and path-integrals, McGraw Hill Book Co., New York.

Feynman, R.P., 1972, Statistical Mechanics - A set of lectures, W.A. Benjamin Inc., Reading, Massachusetts.

Fröhlich, H., Pelzer, H., and Zienau, S., 1950, Properties of slow electrons in polar materials, Phil. Mag., 41:221.

Fröhlich, H., 1954, Electrons in lattice fields, Advances in Physics, 3:325.

Fujii, Y., Hoshino, S., Sakuragi, S., Kanzaki, H., Lynn, J.W., and Shirane, G., 1977, Neutron scattering study of the lattice dynamics of AgBr at 4.4K, Phys. Rev. B, 15:358.

Gorkum, Yu.I., and Tolpygo, K.B., 1960, Characteristics of the motion of fast current carriers in ionic crystals, Bull. Acad. Sci. (USSR), Phys. Ser. (English Transl.), 24:91.

Haken, H., 1959, Über den Einflusz von Gitterschwingungen auf Energie und Lebensdauer des Exzitons, Z. Phys., 155:223.

Haken, H., 1963, Theory of Excitons - II, in: "Polarons and Excitons", C.G. Kuper and G.D. Whitfield, ed., Oliver and Boyd, Edinburgh and London, p. 295.

Hirano, A., and Masumi, T., 1983, Cyclotron resonance of polarons in pure AgCl, in preparation.

Hodby, J.W., 1971, Cyclotron resonance of the polaron in the alkali and silver halides III, J. Phys. C: Solid State Phys., 4:L8.

Hodby, J.W., 1972, Experimental study of the electronic transport properties of ionic crystals, in: "Polarons in Ionic Crystals and Polar Semiconductors", J.T. Devreese, ed., North-Holland, Amsterdam, p. 389.

Hodby, J.W., Crowder, J.G., and Bradley, C.C., 1974, Polaron cyclotron resonance in AgBr at microwave and infrared frequencies, J. Phys. C: Solid State Physics, 7:3033.

Hodby, J.W., Komiyama, S., Hirano, A., and Masumi, T., in preparation.

Hulin, D., Mysyrowicz, A., Combescot, M., Pelant, I., and Benoit à la Guillaume, C., 1977, Electron-hole liquid and biexciton pocket in AgBr, Phys. Rev. Letters, 39:1169.

James, T.H., ed., 1977, "The Theory of Photographic Process", 4th ed., Macmillan Publishing Co., Inc., New York.

Kadanoff, L.P., 1963, Boltzmann equations for polarons, Phys. Rev., 130:1364.

Kajita, K., and Masumi, T., 1974, Hot electron effects in silver halides in quantum limit, in: "Proc. 12th Int. Conf. Phys. Semiconductors - Stuttgart 1974, B.G. Teubner, ed., p. 844.

Kanzaki, H., and Sakuragi, S., 1969a, Optical absorption and luminescence of exciton in silver halides containing isoelectronic impurities - part I. AgBr:$I^-$, J. Phys. Soc. Japan, 27:109.

Kanzaki, H., and Sakuragi, S., 1969b, Optical absorption and luminescence of exciton in silver halides containing isoelectronic impurities - part II. AgBr:$Cl^-$ and AgBr, J. Phys. Soc. Japan 29, 924.

Kanzaki, H., and Sakuragi, S., 1970, Optical absorption and luminescence of excitons in silver halides containing isoelectronic impurities - part III. $AgBr:Na^+$ and $AgBr:Li^+$, J. Phys. Soc. Japan, 29:936.
Kanzaki, H., Sakuragi, S., and Sakamoto, K., 1971, Excitons in $AgBr_{1-x}Cl_x$ - Transition of relaxed state between free and self-trapped exciton, Solid State Commun., 9:999.
Kawahara, M., and Masumi, T., 1980, New queues of time-resolved luminescence lines from excitonic molecules in $AgCl_xBr_{1-x}$, Proc. 15th Int. Conf. Phys. Semiconductors, Kyoto 1980, J. Phys. Soc. Japan, 49: Suppl. A, p. 421.
Kleinfeld, TH., Stolz, H., and von der Osten, W., 1979, Solid State Commun., 31:59.
Kobayashi, K., and Brown, F.C., 1959, Hall effect for electrons in silver chloride, Phys. Rev., 113:507.
Komiyama, S., Masumi, T., and Kajita, K., 1976, Anomalous distribution of hot polarons in silver halides at crossed electric and magnetic fields, in: Proc. 13th Int. Conf. Phys. Semiconductors - Rome 1976, Tipografia Marves, Roma, p. 1222.
Komiyama, S., and Masumi, T., 1978, Cyclotron resonance of polarons in AgBr at microwave fields, Solid State Commun., 26:381.
Komiyama, S., Masumi, T., and Kajita, K., 1979a, Definite evidence for population inversion of hot electrons in silver halides, Phys. Rev. Letters, 42:600.
Komiyama, S., Masumi, T., and Kajita, K., 1979b, Definite evidences for the population inversion of hot electrons in silver halides, Solid State Commun., 31:447.
Komiyama, S., Masumi, T., and Kajita, K., 1979c, Streaming motion and population inversion of hot electrons in silver halides at crossed electric and magnetic fields, Phys. Rev. B, 20:5192.
Komiyama, S., and Masumi, T., 1979d, Cyclotron resonance of polarons in AgBr at high microwave fields, J. Magnetism and Magnetic Materials, 11:59.
Komiyama, S., Masumi, T., and Kurosawa, T., 1979e, A new mode of streaming cyclotron motion of hot electrons at intense microwave fields in AgBr, in: Proc. 14th Int. Conf. Phys. Semiconductors - Edinburgh 1978, (Inst. Phys. Conf. Ser. no. 43, 1979: Chapt. 10), p. 335.
Komiyama, S., Kajita, K., and Masumi, T., 1981, Streaming motion and population inversion of hot polarons in silver halides, in: "Recent Developments in Condensed Matter Physics", J.T. Devreese, ed., Plenum Press, New York and London, p. 563.

Kuper, C.G., and Whitfield, G.D., ed., 1963, "Polarons and Excitons", Oliver and Boyd, Edinburgh and London.

Kurita, S., and Kobayashi, K., 1978, Optical absorption of indirect excitons in AgCl and AgBr in high magnetic field, J. Phys. Soc. Japan, 44:1583.

Langreth, D.C., 1967, Polaron mobility at finite temperature, Phys. Rev.,159:717.

Lee, T.D., Low, F.E., Pines, D., 1953, The motion of slow electrons in a polar crystal, Phys. Rev., 90: 297.

Low, F.E., and Pines, D., 1955, Mobility of slow electrons in polar crystals, Phys. Rev. 98:414.

Maeda, H., and Kurosawa, T., 1972, Hot electron population inversion in crossed electric and magnetic fields, in: Proc. 11th Int. Conf. Phys. Semiconductors - Warsaw 1972, Elsevier Publishing Co., New York, p. 602.

Masumi, T., 1963, Mobility of electrons in silver chlorides at high electric field, Phys. Rev., 129: 2564.

Masumi, T., Ahrenkiel, R.K., and Brown, F.C., 1965, Hall mobility of slow electrons in AgCl and the effects of crystal purity, phys. status solidi, 11:163.

Masumi, T., 1967, Nonlinear transport phenomena in AgCl and AgBr at high electric field, Phys. Rev. 159:761.

Masumi, T., 1975, A supplemental improvement of the Feynman-polaron model and the possibility of bipolaron formation, Progr. Theor. Phys., Suppl. 57: 22. The appendix of that article should be replaced by Appendix 1 of the present article.

Masumi, T., 1981, A series of experimental studies into dynamical and nonlinear responses of Fröhlich polarons, in: "Recent Developments in Condensed Matter Physics", J.T. Devreese, ed., Plenum Press, New York and London, p. 543.

Masumi, T., 1981-1982, Profiles of polarons, KOTAI-BUTSURI 16:275, 16:340, 16:559, 17:19 (in Japanese).

Matsushita, M., Kurita, S., Kobayashi, K., and Masumi, T., 1972, Magnetoabsorption of indirect excitons in AgCl, J. Phys. Soc. Japan, 33:1177.

Matsushita, M., 1973, Magnetoabsorption of indirect excitons in AgCl and AgBr, J. Phys. Soc. Japan, 35: 1688

Matsuura, M., 1980, Polaron effects in Wannier excitons, KOTAI-BUTSURI, 15:655 (in Japanese).

Nakamura, A., and Morigaki, K., Transient photoconductivity associated with the formation of electron-hole droplets in optically pumped Germanium, Solid State Commun., 14:41.

Okamoto, Y., 1956, Optische Absorption von Silberchlorid- und Silberbromid-Kristallen, Nachr. Akad. Wiss. Göttingen, IIa, 14:275.

Osaka, Y., 1961, Theory of polaron mobility, Progr.Theor. Phys., 25:517.

Quattropani, A., Forney, J.J., and Bassani, F., 1975, Biexciton in indirect gap semiconductors: Application to GaSe and AgBr, phys. status solidi, 70(b): 497.

Schultz, T.D., 1959, Slow electrons in polar crystals: Self-energy, mass and mobility, Phys. Rev., 116: 526.

Schultz, T.D., 1963, Feynman's path integral method applied to the equilibrium properties of polarons and related problems, in: "Polarons and Excitons", C.G. Kuper and G.D. Whitfield, ed., Oliver and Boyd, Edinburgh and London, p. 71.

Tamura, H., and Masumi, T., 1971a, Effective mass of polarons in AgBr and AgCl, J. Phys. Soc. Japan, 30: 897.

Tamura, H., and Masumi, T., 1971b, Direct evidence for non-parabolicity of the polaron energy spectrum, J. Phys. Soc. Japan, 30:1763.

Tamura, H., 1972, Temperature dependence of polaron cyclotron resonance in AgBr, Solid State Commun., 10:297.

Tamura, H., and Masumi, T., 1973, Cyclotron resonance of positive holes in AgBr, Solid State Commun., 12:1183.

Thornber, K.K., and Feynman, R.P., 1970, Velocity acquired by an electron in a finite electric field in a polar crystal, Phys. Rev. B, 1:4099.

Tippins, H.H., and Brown, F.C., 1963, Magnetoresistance of silver bromide, Phys. Rev., 129:2554.

Toyozawa, Y., and Hermanson, J., 1968, Exciton-phonon bound state: a new quasiparticle, Phys. Rev. Letters, 21:1637.

Toyozawa, Y., 1972, Dynamical properties of charge carriers in dielectrics - Generalization of polaron theory, in: "Polarons in Ionic Crystals and Polar Semiconductors", J.T. Devreese, ed., North-Holland, Amsterdam, p. 1.

Tsukioka, K., and Masumi, T., 1974, Cyclotron resonance of polarons in AgBr under pulsed-laser excitation, Phys. Letters, 49A:185.

Tsukioka, K., and Masumi, T., 1980, Cyclotron resonance of polarons in AgBr at high density excitation, J. Phys. Soc. Japan, 48:1607.

Vijayaraghavan, P.R., Nicklow, R.M., Smith, H.G., and Wilkinson, M.K., 1970, Lattice dynamics of silver chloride, Phys. Rev. B, 1:4819.

von der Osten, W., 1983, Excitons and exciton relaxation in silver halides, in: "Physics of Polarons and Excitons in Polar Semiconductors and Ionic Crystals", J.T. Devreese, ed., Plenum Press, New York and London, p. .

Wei, J.S., and Brown, F.C., 1973, Electron lattice interaction in the silver halides, Photographic Science and Engineering, 17:197.

Whitfield, G., and Puff, R., 1965, Weak coupling theory of the polaron energy-momentum relation, Phys. Rev., 139:A338.

APPENDIX 1

The magnitude of $\rho(\alpha)$ in Fig. 2-5-c) may be obtained in the most simplified way by integrating the averaged value of the polarization charge cloud $\sigma = [-\nabla^2 V_{eq}/4\pi]$ over the polaron radius $r_f$, where $V_{eq}$ is defined as

$$V_{eq} \equiv [V/(-e)] = -\frac{1}{2}\left(\frac{\gamma}{e}\right)(r_e - R_M)^2 . \qquad \text{(A.1)}^{¶}$$

Such a form of potential in the Feynman-polaron model immediately yields the averaged value of

$$\sigma = \frac{3}{4\pi}\left(\frac{\gamma}{e}\right) . \qquad \text{(A.2)}$$

Here, we only seek the self-consistency in the supplemented model Hamiltonian. Then, $\rho(\alpha)$ can be written as

$$\rho(\alpha) = \int_{r_e - R_M \le r_f} \sigma(r_e - R_M)\, d(r_e - R_M)$$

$$= +\left(\frac{\gamma}{e}\right) r_f^3 . \qquad \text{(A.3)}$$

Thus, one may obtain an image of $\rho(\alpha)$ consistent with both the original Feynman-polaron model and the supplemented model in the present work.

---

¶ The notation $V_{eq}$ is identical to $\phi$ in the Poisson equation above Eq. (2.15) of the present text.

# SOME RECENT DEVELOPMENTS IN THE THEORY OF POLARONS

J.T. Devreese

University of Antwerpen (U.I.A. and R.U.C.A.)
Belgium and University of Technology
Eindhoven, the Netherlands

## ABSTRACT

In this paper a review is presented of some of our recent work on the theory of Fröhlich polarons. The following subjects are treated:

- I. On the (Absence of) "Phase Transitions" for Fröhlich Polarons ($\mathcal{H}$=o)
- II. Statistical Properties of Polarons in a Magnetic Field
- III. Polaronic Transport
  - III.1. Ohmic Limit, Boltzmann Equation
  - III.2. The "3/2 kT" Problem in the Low Temperature Polaron Mobility Theory
  - III.3. Velocity Distribution of Polarons in Crossed Electric and Magnetic Fields
- IV. Energy Levels of Bound Polarons

## INTRODUCTION

This paper is based on a series of lectures§ in which I covered both introductory material and recent developments in the physics of polarons.

---

§ Standard notations of polaron theory are used in this paper (see ref. [1])

A short resumé of the material on recent developments in general is mentioned in the Introduction to this volume (page 1).

The subject material of the present paper, as summarized in the abstract, is limited to those recent developments which were realized by myself and my co-workers at the Physics Department of the University of Antwerp. The purpose here is to indicate the main results, insights and possible experimental implications of our work, rather than to provide the technical details which have been (or will be) published in the regular literature.

## I. ON THE (ABSENCE OF) "PHASE TRANSITIONS" FOR FRÖHLICH POLARONS ($\mathcal{H}$=o)

Recently, several authors [2],using [3], have conjectured or claimed that the Fröhlich polaron becomes self-trapped at some critical coupling constant $\alpha_c$. For weak coupling the polaron would be a free quasi particle; at strong coupling it becomes localised. The "transition" would be discontinuous.

These claims about a "transition" of the Fröhlich polaron from a "free" to a "localised" state are obtained from the following considerations. First a "modified" Lee-Low-Pines transformation is performed with the unitary transformation operator

$$U_1 = \exp -\{i\eta \sum_{\vec{k}} \vec{k}.\vec{r}\; a^+_{\vec{k}} a_{\vec{k}}\} \tag{1}$$

where $\eta$ is a variational parameter. In the limit $\eta \to 1$, this is the well-known Lee-Low-Pines transformation. Subsequently, the second Lee-Low-Pines transformation is applied, with the unitary operator

$$U_2 = \exp \sum_{\vec{k}} (a^+_{\vec{k}} f_{\vec{k}} - a_{\vec{k}} f^{\star}_{\vec{k}}) \tag{2}$$

where again $f_{\vec{k}}$ has to be determined variationally. Furthermore, creation and annihilation operators $B^+$ and B are introduced, describing the fluctuations of the electron around its averaged momentum $\vec{p}_o$:

$$p_j = (p_o)_j + \sqrt{\frac{m\hbar\lambda}{2}}\, (B_j + B^+_j) \qquad j=1,2,3 \tag{3a}$$

$$r_j = i \sqrt{\frac{\hbar}{2m\lambda}} \, (B_j - B_j^+) \qquad j=1,2,3 \tag{3b}$$

where the index j denotes cartesian components. Also $\lambda$ and $\vec{p}_o$ are variational parameters. The variational parameters $\eta$, $f_k$, $\vec{p}_o$ and $\lambda$ are then determined by minimizing

$$E = \langle\psi_o|U_2^{-1}U_1^{-1}(H - \frac{\hbar}{2m}\vec{\mu}.\vec{P}_t)U_1U_2|\psi_o\rangle \tag{4}$$

Here, $\vec{\mu}$ is a Lagrange multiplier, to be determined by the condition that the total polaron momentum is zero in the groundstate:

$$\langle\psi_o|U_2^{-1}U_1^{-1} \, \vec{P}_t \, U_1U_2|\psi_o\rangle = o \tag{5a}$$

where $\vec{P}_t$ is the total momentum

$$\vec{P}_t = \vec{p} + \sum_k \hbar\vec{k} \, a_k^+ a_k \tag{5b}$$

However, it can be shown analytically that the condition for zero total momentum implies $\vec{\mu}=o$.

Choosing the phonon vacuum for $|\psi_o\rangle$ one obtains

$$E = \frac{3}{4}\hbar\lambda + \sum_{\vec{k}} V_k f_k \, e^{-\frac{1}{2}k^2(1-\eta)^2\hbar/2m\lambda} \tag{6a}$$

with

$$f_k = -V_k^\star \frac{e^{-\frac{1}{2}k^2(1-\eta)^2\hbar/2m\lambda}}{\hbar\omega + \eta^2\frac{\hbar^2k^2}{2m}} \tag{6b}$$

which is to be minimized with respect to $\lambda$ and $\eta$ in order to obtain $E_o$, the polaron groundstate energy.

These are precisely the expressions for E and $f_k$ which should have been obtained in [3] but due to an algebraic error in [3] the zero-point energy of the electrons was treated incorrectly there. Because of this mistake, also the numerical results in [3] are in error. The numerical values for the polaron groundstate energy obtained by minimizing (6a) are given in [2b].

By minimizing (6a) one finds that the upper bound for the polaron groundstate energy following from the approximations leading to (6a) is given by $E_0 = -\alpha\hbar\omega$ for $0 < \alpha \leq 8.52 \pm 0.01$ and for $\alpha > 8.52$ by values for $E_0$ which coincide with $E_0 = -\alpha\hbar\omega$ at $\alpha = 8.52$ and continuously tend to the asymptotic limit $-\frac{\alpha^2}{3\pi}\hbar\omega$ for $\alpha\to\infty$, while remaining larger than the groundstate energy derived from the Gaussian approximation (and a fortiori larger than the Feynman value) for all $\alpha > 8.52$.

As is well known $E_0 = -\alpha\hbar\omega$ is the second order weak coupling perturbation result, which incorporates translational invariance (again to second order in perturbation theory) and which is also a variational result. Furthermore the asymptotic limit value $E_0 = -\frac{\alpha^2}{3\pi}\hbar\omega$ is a "strong coupling" variational result, obtained using a localised trial wave function.

The authors of ref. [2] incorrectly conclude that the discontinuity of $\frac{\partial E_0}{\partial\alpha}$ at $\alpha = 8.52$ implies that the Fröhlich polaron undergoes a "phase transition" at $\alpha$=8.52 from a "free" to a "localized" state. In ref. [4] we have emphasized that the "phase transitions" reported in [2] are nothing more than artefacts of the approximations involved. Indeed, the claims on self trapping of the Fröhlich polaron in ref. [2] are based on a variational expression for the polaron groundstate energy which is substantially higher in energy than the Feynman 1955 variational expression (Fig. 1a).

It is then obvious that no legitimate claims regarding localisation of Fröhlich polarons can be based on variational results for the groundstate energy, which are (considerably) higher in energy than the best available ones. As discussed in [4] the lowest upper bounds which are available at present for the polaron groundstate energy (based on the Feynman model) and their first and higher derivatives are continuous with respect to $\alpha$ (for all $\alpha$) and therefore show no hint for any localisation phenomena.

In fact it is amazing that, at present considerations are published on self trapping of Fröhlich polarons, based on variational results for $E_0$ obtained in 1978 (such as $E_0 = -\alpha\hbar\omega$ for $\alpha < 8.52$) but which are poor upper bounds compared to the well known Feynman [5] variational result

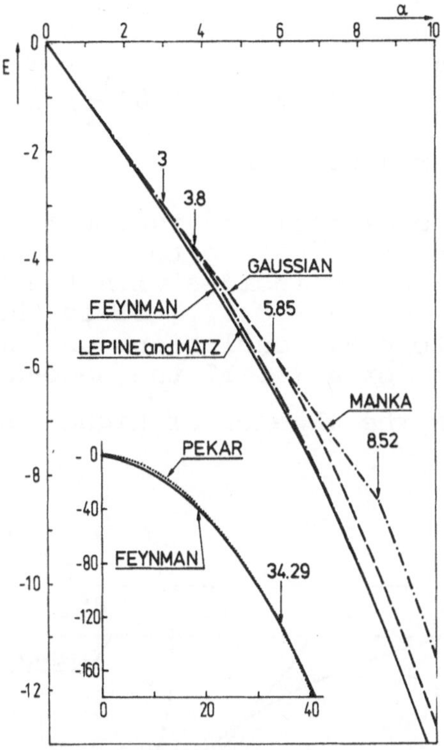


Fig. 1a. Groundstate energy (in units $\hbar\omega$) of the polaron in the theories of Feynman [5] (——), Lépine and Matz [6] (–.–), Mańka [2a] (–..–), the Gaussian model [7] (---), and Pekar [8] (....). In the inset it is shown that the groundstate energy of Pekar's theory is lower than the Feynman result for $\alpha > 34.29$. The arrows in the main figure indicate the position of discontinuities in the first and/or second derivative of E with respect to $\alpha$. (From ref. [4])

obtained almost thirty years ago (1955):

$$E = \frac{3}{4v}(v-w)^2 - \frac{\alpha v}{\sqrt{\pi}} \int_0^\infty \frac{e^{-u}}{[w^2 u + (\frac{v^2-w^2}{v})(1-e^{-uv})]^{1/2}} du \quad (7)$$

which is to be minimized with respect to v and w, leading to the expansions;

$$E_o^F/\hbar\omega = -\alpha - 1.26 \left(\frac{\alpha}{10}\right)^2 - \ldots \text{ as } \alpha \to o \quad (7a)$$

and

$$E_o^F/\hbar\omega = -\frac{\alpha^2}{3\pi} - 3\ln 2 - \frac{3}{4} + O(\frac{1}{\alpha^2}) \text{ as } \alpha\to\infty \qquad (7b)$$

(C = Euler's constant)

It should also be realized that the "best" variational result obtained from (6a) up to $\alpha = 8.52$ corresponds to $\eta$=o, i.e. it simply coincides with the Lee-Low-Pines result of 1952 for $o < \alpha < 8.52$. It is then misleading and illegitimate to draw conclusions on "localisation" from the fact that for $\alpha < 8.52$ the second derivative $\frac{\partial^2 E}{\partial\alpha^2} = o$ i.e. from the absence of higher order terms ($\alpha^2$,

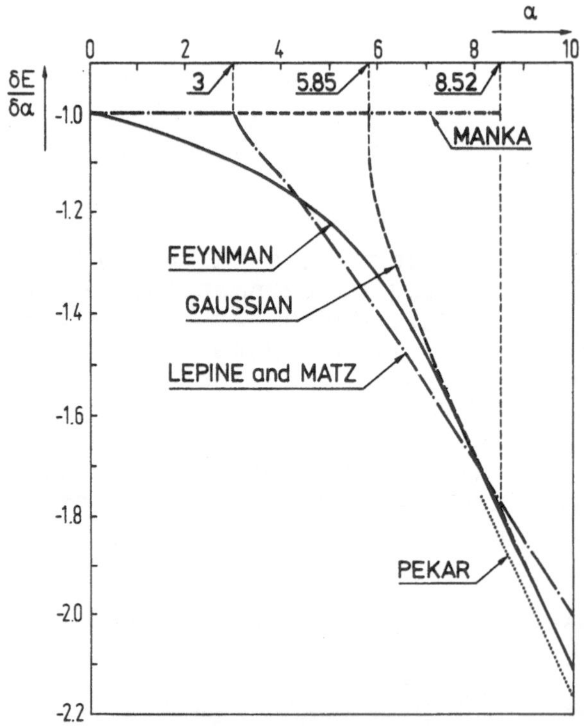


Fig. 1b. First derivative of the polaron groundstate energy with respect to the coupling $\alpha$ for different polaron theories. The asymptotic limit of Pekar (····) is also shown. The approximations of Lépine and Matz (—.—), Gaussian (---), and Mañka (—..—) results in $\partial E/\partial\alpha = -1$ for $\alpha > \alpha_c$ with $\alpha_c$, respectively, equal to 3, 5.85, and 8.52. (From ref. [4])

$\alpha^3$, ...) in (6a) if it is well known from simple perturbation theory, that higher order terms in $\alpha$ exist:

$$E_o^{PT}/\hbar\omega = -\alpha - .0159\,\alpha^2 - A\alpha^3 - \ldots \qquad (8)$$

Our discussion, which shows that no conclusions regarding "localisation of Fröhlich polarons" or "self-trapping" can be derived from the approximations leading to (6a) has implications for several recent studies e.g. for the calculations of ref. [2c]. Explicitly this implies that the first-order phase-transition-like behaviour for Fröhlich surface polarons obtained by the authors of ref. [2c] is an artefact of their approximation which is equivalent to the approximation leading to eq. (6a).

Two further remarks seem in order:

i) The above discussion is limited to the analysis of the Fröhlich Hamiltonian (for the interaction between electrons and polar optical phonons) based on the lowest available variational results for the ground-state energy. Our conclusions are therefore not valid for the "acoustical polaron" (Fröhlich type interaction between the electron and longitudinal acoustical phonons) where the Feynman model suggests localisation indeed. The interaction between the electron and the acoustical phonons is short-ranged (as opposed to the long range electron-LO phonon interaction). The short range nature of the interaction favours localisation of the polaron.

ii) As long as no rigorous results for the groundstate energy of the Fröhlich polaron are derived, of course, the question of self-trapping of the Fröhlich polaron is left open. Nevertheless conclusions regarding localisation drawn from any variational result (and certainly from any variational result other than the one leading to the lowest variational energy) are pointless.

## II. STATISTICAL PROPERTIES OF POLARONS IN A MAGNETIC FIELD; INSTABILITY OF THE POLARON IN A MAGNETIC FIELD

Recently F. Peeters and J.T. Devreese [9] have generalized the Feynman theory for the polaron groundstate to arbitrary magnetic field $\mathcal{H}$ and arbitrary temperature.

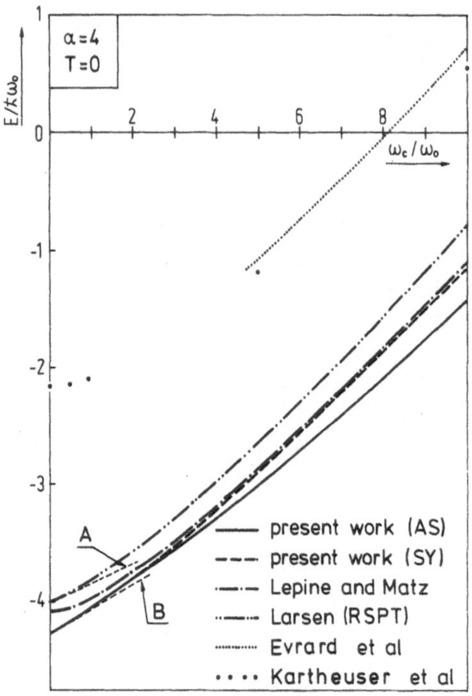


Fig. 2a. Comparison between the polaron groundstate energy of different polaron theories for $\alpha=4$ and zero temperature. We compared the results of ref. [10] (KN), ref. [11] (EKD), ref. [12] (RSPT), ref. [6] (Lépine and Matz), and the present result with a symmetrical (SY) and an anisotropic (AS) effective electron-phonon interaction. (Note that ref. [11] is an asymptotic theory valid for $\mathcal{H}\to\infty$.) (From ref. [9])

In [9] the partition function Z for an ideal polaron gas, expressed as a path integral, is calculated using a conjectured extension of Feynman's variational principle of quantum mechanics§. The trial action introduced in [9] is the generalisation of the Feynman-polaron trial action to the case where a static magnetic field is applied. This trial action has axial symmetry but the

§ No rigorous proof exists of the Feynman variational principle for polarons in a static magnetic field; this problem presents an interesting challenge for theorists.

calculations in [9] are also performed for the case of spherical symmetry.

The free energy for the polaron obtained in [9] is lower than that obtained in any other theory for all coupling strengths $\alpha$, all T and all $\mathcal{H}\neq 0$. This is illustrated in Fig. 2a.

The central result of ref. [9] is the prediction of an instability of the polaron for sufficiently high magnetic field. At a critical magnetic field the electron oscillates so rapidly that it can "escape" from the polaron cloud (at least in the plane perpendicular to the magnetic field). This is seen by the fact that $(\frac{v_\perp}{w_\perp})^2$ undergoes a discontinuous transition at a critical magnetic field (see Fig. 2b) in the quadratic approximation. $v_\perp$ and $w_\perp$ are the generalised expressions, for

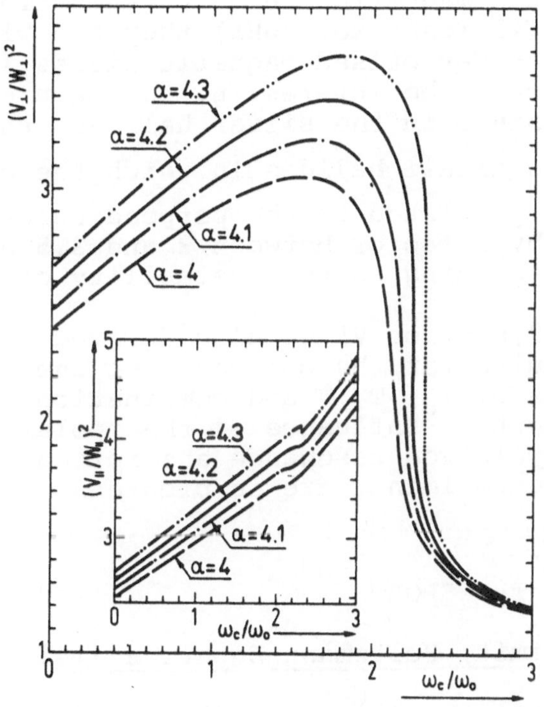


Fig. 2b. Behaviour of $(\frac{v_\perp}{w_\perp})^2$ and $(\frac{v_\parallel}{w_\parallel})^2$, for $\alpha$ = 4, 4.1, 4.2, 4.3. The temperature is equal to zero. (From ref. [9])

nonzero magnetic field, of the standard Feynman variational parameters v and w. As $\mathcal{H}\to 0$ the quantity $(\frac{v_\perp}{w_\perp})^2$ is an approximation to the (more accurate) Feynman polaron mass. Formally the discontinuous transition of $(\frac{v_\perp}{w_\perp})^2$ shows the mathematical structure of a first-order phase transition. It should be emphasized that the discontinuity of the transition presumably is caused by the approximation involved in the calculation of ref. [9]. Nevertheless in view of the accuracy of the Feynman polaron model for the groundstate of the polaron in zero magnetic field, it is expected that the quadratic approximation used in ref. [9] leads to results for the polaron free energy, effective mass, etc. ... which are accurate up to a few percent. Therefore we expect that the physical properties of polarons, which depend on $v_\perp$ and $w_\perp$ undergo a sudden (but continuous) change in a small region of magnetic fields close to the critical magnetic field predicted in ref. [9].

Although the magnetic fields at the "transition" are very high (42 Tesla for AgBr) they are within the reach of present day pulsed magnetic fields [13]. It would in particular be interesting to measure the polaron cyclotron resonance in the silver halides, the thallous halides and the alkali halides in which the parameter $(\frac{v_\perp}{w_\perp})^2$ which is closely related to the perpendicular polaron mass, changes by a factor between 2 and 3.5 at the magnetic field corresponding to the instability of the polaron.

In ref. [9,section V] we also derived numerous analytical results for the polaron free energy for limiting values of $\omega_c$, T, $\mathcal{H}$ and combinations thereof. Calculations on the influence of the "polaron instability" on the cyclotron resonance absorption, for detailed comparison to experiment, are in progress.

## III. POLARONIC TRANSPORT

### III.1. Ohmic Limit; Boltzmann Equation [14]

In ref. [14] the (linearized) Boltzmann equation for polarons has been solved in the Ohmic limit and for small coupling. The solution is rigorous but numerical; only for the asymptotic limit $T\to 0$ an analytical solution was obtained.

Although there exist restrictions on the range of validity of the Boltzmann equation, as applied to polarons, it is certainly applicable for weak coupling, as was shown e.g. by Bogolubov [15]. In the case of weak electron-phonon coupling it is therefore sufficient to calculate the transition rates between different momentum-states of the electron using the first Born approximation, and to solve the resulting Boltzmann equation.

The approach of ref. [14] consists of linearizing the Boltzmann equation with respect to the external electric field; this results in the so-called "Bloch" equation. Laplace-transforming the component of the distribution function linear in the electric field, the problem is then transformed into a matrix equation with a tridiagonal matrix, to be solved numerically. A Maxwellian velocity distribution is assumed at zero electric field.

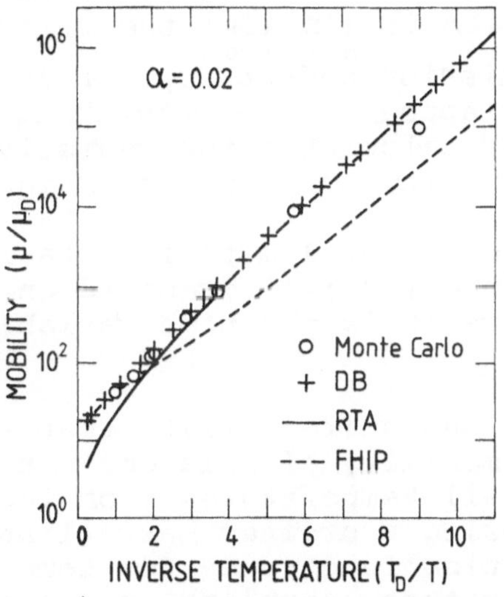


Fig. 3. Mobility of a Fröhlich polaron as a function of inverse temperature for small electron-phonon coupling (+ is taken from ref. [14], —— is the relaxation time approximation, --- corresponds to ref. [16].) This figure is taken from ref. [17].

In the asymptotic limit of low temperature the result of Fröhlich for the polaron mobility is reobtained with the exact solution of the Boltzmann equation.

In Fig. 3 the exact numerical solution for the polaron mobility obtained in [14] is compared to several well known approximative results and with the mobility obtained by solving the Boltzmann equation using the Monte Carlo method.

## III.2. The "3/2 kT Problem" in the Low Temperature Polaron Mobility Theories

In ref. [18] the well known problem of the discrepancy with a factor $\frac{3}{2}\frac{kT}{\hbar\omega}$ between the standard low temperature polaron mobility result $\mu = \frac{e}{2m\alpha\omega\bar{n}}$ as first obtained by Fröhlich, and the corresponding result of Feynman, Hellwarth, Iddings and Platzman (F.H.I.P.) [16] is examined at weak coupling. It is shown explicitly in ref. [18] how the inversion of $\lim_{\alpha\to 0}$ with $\lim_{\omega\to 0}$ in the evaluation of the mobility leads to the above mentioned discrepancy. If one does not consider the correct order of the limits (which is $\lim_{\alpha\to 0}\lim_{\omega\to 0}$) the relative importance of L.O. phonon-emission and-absorption is affected and erroneous results appear. The interchange of limits also occurs if one calculates the resistivity ($\rho \sim \frac{1}{\mu}$) instead of the mobility to first order in $\alpha$.

The correct order of limits is established following Chester and Thellung [19]; it is based on Van Hove's $\lambda^2 t$-limit procedure (t is the time variable and $\lambda^2$ equals $\alpha$ for polarons).

We have also shown (ref. [17]) that the result of F.H.I.P. for the mobility of polarons can be derived for all coupling and all temperatures from the Boltzmann equation, by imposing a drifted Maxwellian electron momentum distribution function. For some values of the parameters (temperature, coupling) the choice of a drifted Maxwellian is inadequate and results i.a. in this discrepancy with a factor $\frac{3}{2}\frac{kT}{\hbar\omega}$. Ref. [20] provides a simple rederivation of the F.H.I.P. and Thornber-Feynman response theory for polarons without the use of path integrals.

## III.3. Velocity Distribution of Polarons in Crossed Electric and Magnetic Fields

In ref. [21] the Boltzmann equation for polarons at finite temperature, in crossed electric and magnetic fields, was solved by F. Brosens and J. Devreese using a procedure, based on analytical methods, which is a generalisation of the "two-circle model" introduced in ref. [22]. This procedure is valid if the population of polarons with kinetic energy $\frac{p^2}{2m} > 2\hbar\omega$ is negligible. This condition is satisfied for the galvano-magnetic measurements on photocarriers in AgCl and AgBr performed by Komiyama, Masumi and Kajita [23]. We refer to the lectures of Professor Komiyama and of Professor Masumi at the present Advanced Study Institute for a detailed description of their measurements. Of particular interest are the "streaming motion" and the "population inversion" of polarons discussed in their lectures.

The two circle method allows to solve the Boltzmann equation at T=o (under the assumption that for almost all electrons $\frac{p^2}{2m} < 2\hbar\omega$) by studying it separately in two distinct geometrical regions ($o < \frac{p^2}{2m} < \hbar\omega$ (region I) and $\hbar\omega < \frac{p^2}{2m} < 2\hbar\omega$ (region II)). In region II the Boltzmann equation reduces to an ordinary linear differential equation from which $f^{II}(\bar{p})$, the polaron distribution function in region II, is easily obtained analytically. The Boltzmann equation in region I (after the introduction in it of $f^{II}(\bar{p})$), also reduces to an ordinary linear differential equation. It is then straightforward to obtain the complete distribution function (by putting $f^{I}(\bar{p}) = f^{II}(\bar{p})$ for $\frac{p^2}{2m} = \hbar\omega$) and the polaron mobility as a function of the electric field.

In ref. [21] this "two-circle method" is generalised to nonzero temperature and to the case of crossed electric and magnetic fields. A crucial step in ref. [21] is a transformation to a system of axes centered at the origin of the cyclotron orbit. Also important is the observation that the "integrated transition rate" for the polaron is almost a constant for $\frac{p^2}{2m} < \hbar\omega$.

For more details of the method we refer to ref. [21]. Some of the results are shown in Figs. 4, 5, 6, 7.

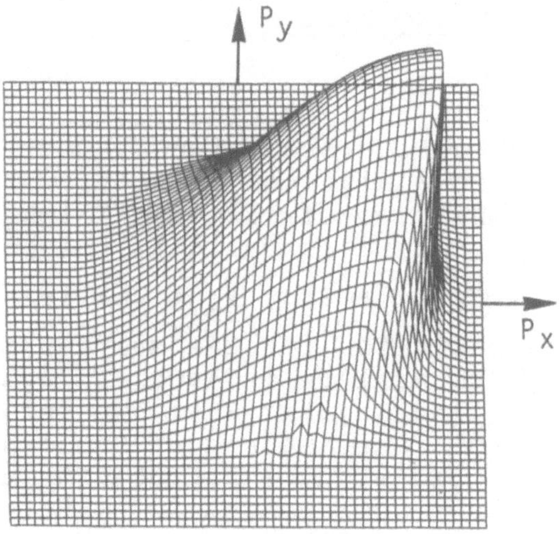


Fig. 4. Calculated polaron distribution for InSb at 77K in an electric field E of 150 V/cm, and without magnetic field B, for $p_z$=o. The x-axis is in the direction of eE, and the z-axis along eB. (From ref. [21])

The complicated distribution function of polarons in crossed electric and magnetic fields is obtained accurately by our method. The results of our calculation of the average polaron velocities and the Hall angle are in close agreement with the measurements of Komiyama, Masumi and Kajita in the regions examined so far.

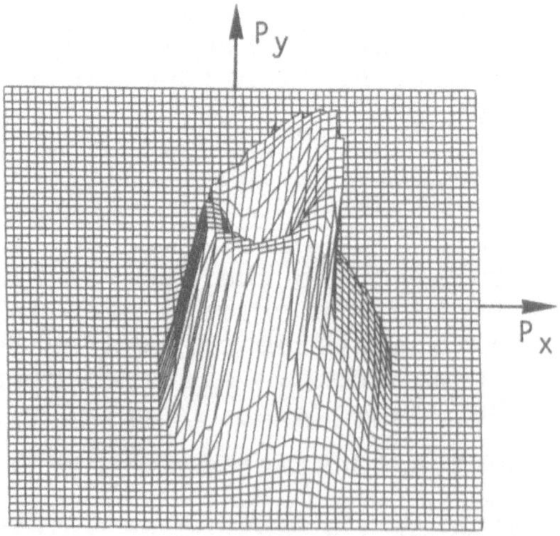


Fig. 5. Same as Fig. 4, except for the magnetic field which is 350 Gauss here. (From ref. [21])

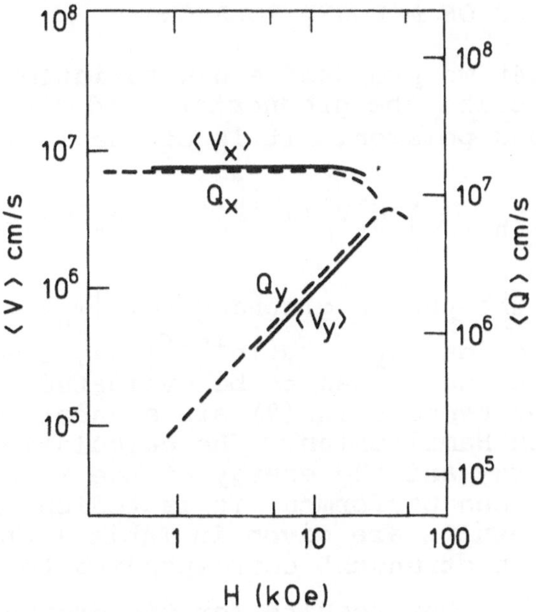


Fig. 6. Calculated values (full lines) of the averaged velocities $(V_x)$ and $(V_y)$ in AgCl at 4.2 K and 3700 V/cm as a function of the magnetic field, superimposed on the experimental currents (dashed lines) $Q_x$ and $Q_y$ from ref. [23]. (From ref. [21])

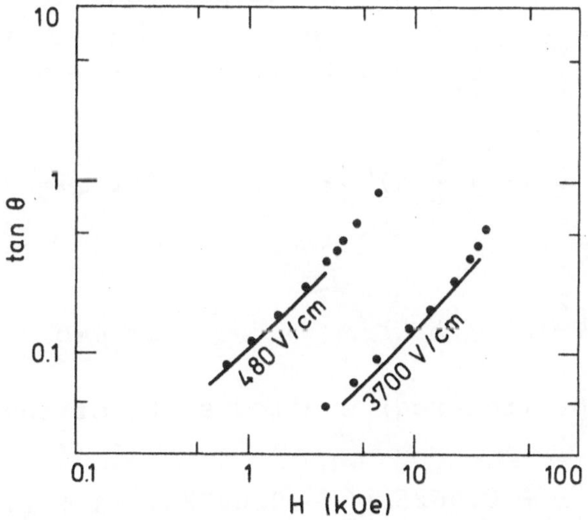


Fig. 7. Calculated Hall angle (full line) as a function of magnetic field in AgCl at 4.2 K, for the electric fields 480 and 3700 V/cm, and compared to the experimental data (points of ref. [23]. (From ref. [21])

## IV. ENERGY LEVELS OF A BOUND POLARON

In ref. [24] we proposed a new variational wave function to describe the groundstate and the excited states of a bound polaron. It is of the form:

$$|\psi\rangle = c|o\rangle|\psi_n\rangle + \sum_{\vec{k}} g_k^{\star} V_k^{\star} (e^{-i\vec{k}.\vec{r}} - \rho_k) a_k^{+} |o\rangle|\psi_n\rangle \qquad (9)$$

$|o\rangle$ stands for the phonon groundstate; $|\psi_n\rangle$ for an electronic wave function, $\rho_k = \langle\psi_n|e^{ik.r}|\psi_n\rangle$, c is a normalisation constant and $g_k$ has to be evaluated variationally. The other symbols in (9) are standard notations for the Fröhlich Hamiltonian. The calculation of the groundstate energy and the energy of "the first excited state (2p)" has been performed; it is tedious but straightforward. The results are given in Table 1 and in Table 2. $\beta$ is the "Coulomb strength" corresponding to the potential energy term $-\frac{\beta}{r}$. Our results for the groundstate energy are close to those obtained by D. Larsen [25]. Both Larsen's and our results are variational. Our results for the 2p-excited bound polaron state are the first which have been obtained for all coupling and for all $\beta$.

The following analytical results have been obtained in ref. [24] (in units $\hbar\omega$):
For the groundstate of the bound polaron:

$$E_{1s} = -\frac{\beta^2}{4} - \frac{5\alpha\beta}{16} - \alpha + \ldots \qquad \text{for } \alpha\to o \text{ and } \beta\to o \qquad (10a)$$

$$E_{1s} = -\frac{1}{4}(\beta + \frac{5}{8}\alpha)^2 + \ldots \qquad \text{for } \alpha\to\infty \qquad (10b)$$

and of course

$$E_{1s} = -\frac{\beta^2}{4} \qquad \text{if } \alpha=o \qquad (10c)$$

For the 2p-type (relaxed) excited state of the bound polaron:

$$E_{2p} = -\alpha - 0.0625\,\beta^2 - 0.097851\,\alpha\beta + \ldots \qquad (11a)$$

for $\alpha\to o$ and $\beta\to o$

$$E_{2p} = -0.03829\,\alpha^2 - 0.0625\,\beta^2 - 0.0978\,\alpha\beta \qquad (11b)$$

if $\alpha\to\infty$

Table 1. Groundstate energy (in units $\hbar\omega$) of the bound polaron for several values of $\alpha$ and $\beta$ as obtained from the present calculation, compared to the variational results of [25]

| | $\beta$ = 6.32 | | $\beta$ = 4.47 | |
|---|---|---|---|---|
| | E (present) | E [25] | E (present) | E [25] |
| 2 | -14.66 | -14.69 | -8.60 | -8.64 |
| 5 | -23.0 | -23.0 | -15.21 | -15.30 |
| 7 | -29.41 | -29.47 | -20.52 | -20.62 |
| 11 | -44.6 | -44.6 | -33.4 | -33.4 |

Table 2. Calculated energy of first excited 2p state (in units $\hbar\omega$) of the bound polaron for several values of $\alpha$ and $\beta$.

| | $E_{2p}$ | | |
|---|---|---|---|
| $\alpha$ | $\beta$=o | $\beta$=1 | $\beta$=2 |
| 1 | -0.7626 | -0.8775 | -1.113 |
| 3 | -1.971 | -2.237 | -2.628 |
| 5 | -3.207 | -3.644 | -4.205 |
| 7 | -4.600 | -5.214 | -5.955 |
| 9 | -6.199 | -6.996 | -7.922 |
| 11 | -8.029 | -9.014 | -10.13 |

Note that $-0.03829\ \alpha^2$ is the correct strong-coupling result for the relaxed excited state of the free polaron. Again, if $\alpha$=o, the result obtained is, of course,

$$E_{2p} = -\frac{\beta^2}{4}\left(\frac{1}{2}\right)^2 \qquad \text{for } \alpha = o \qquad (12)$$

REFERENCES

1. "Polarons and Excitons", eds. C.G. Kuper and G.D. Whitfield (Oliver and Boyd, Edinburgh, 1963); "Polarons in Ionic Crystals and Polar Semiconductors", ed. J.T. Devreese (North Holland, Amsterdam, 1972).
2. a) R. Mańka, Phys. Lett. A67, 311 (1978); Phys. Stat. Sol. (b) 93, 53 (1979).
   b) R. Mańka and M. Suffczynski, J. Phys. C13, 6369 (1980).
   c) E.L. de Bodas and O. Hipólito, Phys. Rev. B27, 6110 (1983).
3. W. Huybrechts, J. Phys. C10, 3761 (1977).
4. F.M. Peeters and J.T. Devreese, Phys. Stat. Sol. (b), 112, 219 (1982).
5. R.P. Feynman, Phys. Rev. 97, 660 (1955).
6. Y. Lépine and D. Matz, Phys. Stat. Sol. (b) 96, 797 (1979).
7. D. Matz and B.C. Burkey, Phys. Rev. B3, 3487 (1971).
8. S.I. Pekar, "Untersuchungen über die Elektronentheorie der Kristalle", (Akademie Verlag, Berlin, 1954).
9. F.M. Peeters and J.T. Devreese, Phys. Rev. B25, 7281 (1982) and B25, 7302 (1982).
10. E. Kartheuser and P. Negrete, Phys. Stat. Sol. (b) 57, 77 (1973).
11. R. Evrard, E. Kartheuser and J.T. Devreese, Phys. Stat. Sol. (b) 41, 431 (1970).
12. D.M. Larsen, in "Polarons in Ionic Crystals and Polar Semiconductors", ed. J. Devreese (North Holland, Amsterdam, 1972), p. 237.
13. N. Miura, G. Kido and S. Chikazumi, Solid State Commun. 18, 885 (1976).
14. J.T. Devreese and F. Brosens, Phys. Stat. Sol. (b) 108, K29 (1981).
15. N.N. Bogolubov, in "Proc. Internat. Symp. Selected Topics in Statistical Mechanics" (Dubna, 1978), p. 3.
16. R.P. Feynman, R. Hellwarth, C. Iddings and P.M. Platzman, Phys. Rev. 127, 1004 (1962).
17. F.M. Peeters and J.T. Devreese (to appear in: "Solid State Physics", eds. F. Seitz and D. Turnbull (Academic Press, New York).

18. F.M. Peeters and J.T. Devreese, Phys. Stat. Sol. (b) 115, 539 (1983).
19. G.V. Chester and A. Thellung, Proc. Phys. Soc. 73, 745 (1959).
20. F.M. Peeters and J.T. Devreese, Phys. Rev. B23, 1936 (1981) and B28 (1983) (to appear).
21. F. Brosens and J.T. Devreese, Solid State Commun. 44, 597 (1982).
22. J.T. Devreese and R. Evrard, Phys. Stat. Sol. (b) 78, 85 (1976).
23. S. Komiyama, T. Masumi and K. Kajita, Phys. Rev. B20, 5192 (1979).
24. J.T. Devreese, R. Evrard, E. Kartheuser and F. Brosens, Solid State Commun. 44, 1435 (1982).
25. F.M. Larsen, Phys. Rev. 187, 1147 (1969).

9. J.M. Kosterlitz and D.J. Thouless, [illegible] (1973).
10. [illegible] Thouless and M. [illegible], [illegible] (1979).
11. [illegible]
12. [illegible] (1979).
13. [illegible] (1981).
14. [illegible] (1980).
15. [illegible]
16. [illegible] (1981).

# TWO APPLICATIONS OF POLARON THEORY : CYCLOTRON RESONANCE AND RELAXATION OF HOT CHARGE CARRIERS

R. Evrard

Institut de Physique B5
Université de Liège
B-4000 Sart Tilman/Liège 1, Belgium

ABSTRACT

Two applications of polaron theory are treated in some detail to give examples of its practical importance. In the first example, it is shown how one can calculate the absorption spectrum for polarons in cyclotron resonance. These theoretical results give a better understanding of polaron effects, like the pinning effect or the phonon assisted-transitions.

The second example deals with the relaxation of hot electrons. The interaction with polar optical phonons is very efficient in relaxing the energy of the hot carriers. It is shown that the final energy distribution reached after the relaxation due to these optical phonons, is far from a Maxwell-Boltzmann distribution. The interaction with the acoustic phonons gives a far slower relaxation, but leads to a Maxwell-Boltzmann distribution at equilibrium, if the inelasticity of the collisions is taken into account.

## I. CYCLOTRON RESONANCE OF POLARONS

### I.1. Introduction

Cyclotron resonance is a standard technique in solid state physics where it is used to determine the band mass of charge carriers. Consider the simple case of a non-degenerate semiconductor with a conduction band having a single isotropic and parabolic minimum at the center of the Brillouin zone. The sample is placed in a constant and uniform magnetic field with magnetic induction $\vec{B}$. Then, it is easily shown that the energy of the electrons in the conduction band is quantized by the magnetic field, at least as far as their motion in the direction perpendicular to this field is concerned. The corresponding quantum levels are called Landau levels. As for the direction along the magnetic field, the electrons behave like free electrons. In short, the possible energies are given by

$$\varepsilon_n(k_z) = (n+\frac{1}{2})\hbar\omega_c + \hbar^2k_z^2/2m \qquad n=0,1,2 \ldots \qquad (1)$$

with

$$\omega_c = \frac{eB}{mc} \qquad (2)$$

in Gauss units (c is the velocity of light). In these relations m and $k_z$ respectively denote the electron band mass and the component of the Bloch wave-vector taken along the direction of the magnetic field which is chosen as z-axis. The notation e is used for the absolute value of the electron charge.

As the electrons behave like harmonic oscillators in the plane perpendicular to $\vec{B}$, it is not surprising that the selection rules valid for harmonic oscillators also apply here. Therefore, the dipolar transitions due to the interaction with the electromagnetic radiation field are allowed only when they occur between adjacent levels, so that

$$\Delta n = \pm 1 . \qquad (3)$$

Thus, the absorption spectrum in the simple case of nearly free electrons consists of a single line, possibly broadened by the collisions with the lattice imperfections or the phonons (later on we will see that the interaction with the longitudinal optical (LO) phonons leads to a somewhat different picture). The frequency at which the absorption takes place is the cyclotron frequency $\omega_c = eB/mc$. Obviously, the absorption curve gives a direct measure of the electron band mass m.

In the same way, one can hope to measure the effective mass of polarons in polar semiconductors or ionic crystals. However, this measurement does not provide a direct determination of the correction to the mass due to the lattice distortion, since the band mass of the "bare" electron (without lattice distortion) is generally not known with enough precision. One could think of increasing the magnetic field to reach a cyclotron frequency larger than the frequency $\omega_0$ of the LO phonons. Then indeed, the speed of rotation of the electrons about the magnetic field would be so large that the lattice distortion would not be able to follow and the electrons would behave as free electrons. Unfortunately to fulfill this condition requires such a high magnetic field that the experiment is feasible in just a few semiconductors. However, the experimental trend goes towards the production of stronger fields, so that new results on the effective mass of polarons could be expected in the near future.

## I.2. The "pinning" effect

Those compounds for which the condition $\omega_c = \omega_0$ is fulfilled at a reasonable value of the magnetic field have a rather weak Fröhlich e-ph interaction, so that perturbation theory can be used to describe the polaron effects. As expected, the first result is that the resonance occurs now at a modified frequency $\omega_c$ such that

$$\omega_c^{\star} = eB/m^{\star}c \tag{4}$$

where

$$m^{\star} = (1 + \alpha/6)m \tag{5}$$

is the second-order perturbation value of the polaron effective mass. As usual, the notation $\alpha$ is used for Fröhlich's coupling constant. This result shows that the slope of the energy of the Landau levels versus the magnetic field is slightly lowered by the e-ph interaction.

However, this result is valid only for the case of a relatively weak magnetic field, such that $\omega_c \ll \omega_0$. When the cyclotron frequency becomes of the same order of magnitude as the phonon frequency, non-degenerate Rayleigh-Schrödinger perturbation theory is no longer valid. Then indeed, one finds two or more states with about the same unperturbed energy. This can be seen in the following way.

With no e-ph interaction, the energy of the whole system (the electron and the phonons) is

$$E = \varepsilon_n(k_z) + \sum_{\vec{q}} n_{\vec{q}} \hbar\omega_0 \tag{6}$$

where the electron energy $\varepsilon_n(k_z)$ has been defined previously (see eq.(1)) and $n_{\vec{q}}$ is the number of LO phonons with wavevector $\vec{q}$. As for the total quasi-momentum along $\vec{B}$, it is given by

$$P_z = \hbar k_z + \sum_{\vec{q}} n_{\vec{q}} \hbar q_z \quad . \tag{7}$$

For instance, consider the case where $P_z=0$. The ground state (no phonon and n=0) as well as the excited states corresponding either to no phonon and n=1 (the first excited Landau level) or to one real phonon and n=0 are represented in fig.1. Let $\vec{q}$ be the phonon wave-vector for this latter states.

Their energy is

$$E_1 = \hbar\omega_o + \frac{1}{2}\hbar\omega_c + \hbar^2 q_z^2/2m \ . \qquad (8)$$

Obviously, these states constitute a continuous spectrum with a threshold at

$$E_{th} = \hbar\omega_o + \frac{1}{2}\hbar\omega_c \ . \qquad (9)$$

The n=1 Landau level enters this continuum when $\omega_c$ becomes equal to $\omega_o$ and the energy levels are then degenerate. Therefore, as claimed above, the usual Rayleigh-Schrödinger theory is no longer valid.

Using Wigner-Brillouin perturbation theory, Johnson and Larsen[1] have shown that the e-ph interaction removes the degeneracy that exists at the crossing point between the n=1, zero-phonon and the n=0, one-phonon states. As can be seen in fig.1, the n=1 Landau level is pushed down, so that it remains below the phonon threshold. This is usually known as the pinning effect.

As for the upper level, it lies slightly above the unperturbed n=1 Landau level. For the same strength of the magnetic field, such that $\omega_c = \omega_o$, other states are also degenerate. For instance, the "n=2, zero-phonon", "n=1, one-phonon" and "n=0, two-phonon" levels give rise to a threefold degeneracy. This case has been studied in detail by Devreese's group and Ngai[2].

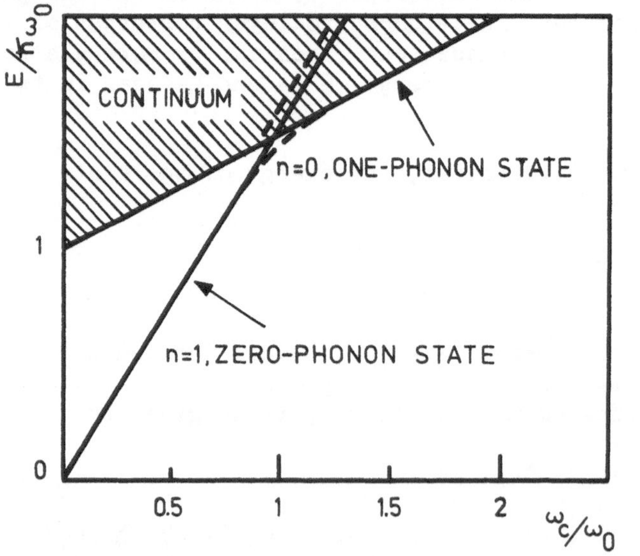


Fig.1. Energy levels of polarons in a magnetic field for $\alpha=0.02$, as predicted by Johnson and Larsen[1]. Full lines: unperturbed levels. Dashed lines: results of Wigner-Brillouin perturbation theory.

Let us make two remarks. First it does not seem possible to observe directly the transitions from n=0 to n=1 by absorption of electromagnetic radiation in the range where $\omega_c$ is close to the LO phonon frequency $\omega_o$. Indeed this corresponds to the reststrahl-frequency region where the lattice is strongly absorbing. However, the pinning effect, can be observed either in transitions where the electrons are excited from a state lying below the conduction band edge (an impurity state or a state in the valence band) or in combined resonances where a spin flip occurs during the transition Then the transition frequency lies beyond the reststrahl region and the lattice is almost transparent to the radiation.

The second remark is the following. The existence of phonon-assisted transitions has been proved theoretically as well as experimentally[3,4]. In one of these transitions, the electron remains in the same Landau state but emits a phonon. Obviously, for the total momentum to be conserved, the electron momentum along the magnetic field changes by $-\hbar q_z$ where $q_z$ is the z-component of the phonon momentum. Therefore, for electrons initially at rest, this phonon-assisted absorption occurs at $\omega_o + \hbar q_z^2/2m$. This shows that there is an absorption band due to phonon assisted transitions with a frequency threshold at $\omega_o$.

Thus, there arises the problem of reconciling the presence of this absorption line with the pinning effect. Indeed, it is hardly credible that the spectrum has three absorption lines, the phonon-assisted transition plus the two lines predicted (and observed) by Johnson and Larsen.

This shows that the determination of the precise shape of the absorption spectrum in cyclotron resonance is a problem of considerable importance and interest. An other reason to study the detailed shape of the spectrum is the following. At weak coupling, the oscillator strength of the phonon-assisted transitions is proportional to the coupling constant. The comparison between the experimental results and the theoretical predictions could give a direct test of the validity of polaron theory and of the value of Fröhlich's coupling constant. This is the motivation of works performed in Antwerp[5] and in Liège[6] . The two approaches are rather similar. The present lecture is based on the latter approach, due to Vigneron, Kartheuser, and myself, since obviously, this is the one I know better.

## I.3. The absorption spectrum

Consider an incident electromagnetic radiation with frequency $\omega$ , propagating along the direction of the magnetic field and having an active circular polarization. The absorption coefficient due to polarons in an initial state $|i\rangle$ is

$$\alpha_i(\omega) = \frac{2\pi^2 e^2 n_e}{\varepsilon^{\frac{1}{2}}(\omega) m^2 c\omega} \sum_f |\langle f|A^+|i\rangle|^2 \delta(E_f - E_i - \hbar\omega) \tag{10}$$

where $n_e$ is the carrier density and $\varepsilon(\omega)$, the dielectric constant, i.e. the square of the index of refraction of the crystal, measured at the frequency $\omega$. The sum runs over all the possible final states $|f\rangle$. Here $|i\rangle$ and $|f\rangle$ denote polaron states with respective energies $E_i$ and $E_f$. Due to the e-ph interaction they differ from the unperturbed states.

The operator $A^+$ and its hermitian conjugate A are rising and lowering operators for transitions between Landau levels. Therefore, they are such that

$$A^+|n\rangle = (n+1)^{\frac{1}{2}}|n\rangle \ , \tag{11a}$$

$$A\ |n\rangle = n^{\frac{1}{2}}\ |n-1\rangle \ . \tag{11b}$$

The effects of the temperature are treated simply in taking the statistical average of $\alpha_i(\omega)$, so that the total absorption is described by the absorption coefficient

$$\alpha(\omega) = \langle \alpha_i(\omega) \rangle \tag{12a}$$

$$= \sum_i \exp(-\beta E_i)\alpha_i(\omega) / \sum_i \exp(-\beta E_i). \tag{12b}$$

where

$$\beta = 1/kT. \tag{13}$$

It is straightforward to show that the calculation of this absorption coefficient reduces to that of the function $G(\omega)$ defined as

$$G(\omega) = \langle G_i(\omega) \rangle \tag{14a}$$

$$= \sum_i \exp(-\beta E_i) G_i(\omega) / \sum_i \exp(-\beta E_i) \ , \tag{14b}$$

with

$$G_i(\omega) = \lim_{\eta \to 0} \langle i|A(H - E_i - \hbar\omega - i\eta)^{-1} A^+|i\rangle. \tag{15}$$

Indeed, comparing eq.(10) with eq.(14) and (15) shows that

$$\alpha(\omega) = \frac{2\pi e^2\ n_e}{\varepsilon^{\frac{1}{2}}(\omega) m^2 c\omega} \operatorname{Im} G(\omega) \ . \tag{16}$$

As pointed out earlier, the states $|i\rangle$ with energy $E_i$ are eigenstates of the full Hamiltonian describing the electron and phonons interacting together. For the same reasons, in eq.(14), H denotes this full Hamiltonian. However, as the coupling is weak in most of the systems of practical interest, the idea coming first to mind is to expand the function $G(\omega)$ in powers of Fröhlich's coupling constant $\alpha$. The drawback of using a perturbation expansion for the absorption spectrum is the following : among other effects, the e-ph interaction shifts and broadens the line of cyclotron resonance. It is clear that the perturbation, which is the difference between the perturbed spectrum and the unperturbed $\delta$-function like spectrum is not a small correction. Therefore, there is no hope that the method gives convergent results. A first improvement is obtained in introducing a natural linewidth $\gamma$. This leads to replace the $\delta$-function in eq.(10) by a Lorentz function

$$L(E_f-E_i-\hbar\omega,\gamma) = \frac{\gamma}{2\pi} \frac{1}{(E_f-E_i-\hbar\omega)^2+\gamma^2/4} . \tag{17}$$

This line width is used to take phenomenologically into account the effects of scattering mechanisms other than the collisions with the polar LO phonons, e.g. scattering by impurities, defects and acoustic phonons. This changes the expression (15) of $G_i(\omega)$ into

$$G_i(\omega) = \langle i|A(H-E_i-\hbar\omega-i\gamma/2)^{-1}A^+|i\rangle . \tag{18}$$

For the case of no electron-phonon (e-ph) interaction, one obtains immediately

$$G_i^0(\omega) = -(n+1)/\hbar(\omega-\omega_c+ i\gamma/2\hbar) , \tag{19}$$

where n is the quantum number for the Landau level occupied by the electron in the unperturbed state corresponding to $|i\rangle$ . The statistical average gives

$$G^0(\omega) = -\rho/\hbar(\omega-\omega_c + i\gamma/2\hbar) \tag{20}$$

with

$$\rho = \sum_n (n+1)\exp(-\beta n\hbar\omega_c)/\sum_n \exp(-\beta n\hbar\omega_c) . \tag{21}$$

However, the improvement brought about by the introduction of the line width $\gamma$ is not yet sufficient when the shift and the broadening due to the interaction with the polar LO phonons are large compared to this initial line width $\gamma$. Then indeed, the correction is about as large as the unperturbed spectrum itself.

To avoid this difficulty, we write

$$G(\omega) = -\rho/\hbar\left[\omega-\omega_c + Q(\omega) + i\gamma/2\hbar\right] , \tag{22}$$

therefore adding a term $Q(\omega)$ in the denominator to take the effects of the interaction into account. Clearly, $Q(\omega)$ goes to zero if the interaction is switched off. If $Q(\omega)$ was a constant independent of $\omega$, its meaning would be quite simple. Its real and imaginary parts would respectively represent the shift and broadening of the cyclotron resonance due to the interaction with the LO phonons. Then, it could be calculated by the Breit-Wigner theory. However, as we will show later on a complete calculation of $Q(\omega)$ up to second order of perturbation (the first order correction is zero) leads to an expression which actually depends on $\omega$. This dependence is essential to account for the phonon-assisted transitions.

The calculations can be performed in the following way : As we are mainly interested in semiconductors, like the III-V and II-VI compounds, which are weakly ionic, we can use a perturbation expansion for $Q(\omega)$. Far from any resonance, $Q(\omega)$ is small compared to the rest of the denominator of eq.(22). Then, the expression (22) of $G(\omega)$ can be expanded to get, at order linear in $Q(\omega)$,

$$G(\omega) = G^{o}(\omega) + \frac{\partial G^{o}(\omega)}{\partial \omega} Q(\omega) \ , \qquad (23)$$

where $G^{o}(\omega)$ is the function defined earlier (see eq.(20)), which gives the absorption when the e-ph interaction is zero.

On the other hand, it is possible to obtain the expansion of $G(\omega)$ in powers of the coupling constant $\alpha$. Up to the linear order in $\alpha$(second order of perturbation), one has

$$G(\omega) = G_{o}(\omega) + G_{1}(\omega) \ . \qquad (24)$$

The calculations leading to (24) are straightforward but lengthy. To report them in detail here seems of no interest. The reader interested in these details is referred to the research papers by ourselves[6] or by Prof. Devreese and his group[5]

Comparing eq.(24) with eq.(23) gives the second-order expression of $Q(\omega)$ (linear in $\alpha$). As explained above, the expansion (24) of $G(\omega)$ is not convergent for values of the frequency of the incident electromagnetic radiation near the cyclotron resonance. However the form (22) used for the absorption function $G(\omega)$ leads us to think that the expression derived from (24) for $Q(\omega)$ remains valid by analytic continuation even near the resonance.

The theory described here has been applied to the study of the absorption spectrum in different frequency ranges and for different values of the magnetic field $\vec{B}$. The first interesting case is when the magnetic field B is such that the cyclotron frequency $\omega_c$ is close to $\omega_o$. The results for this case are shown in fig.2, where the absorption coefficient is plotted versus the frequency of the incident radiation for different values of the

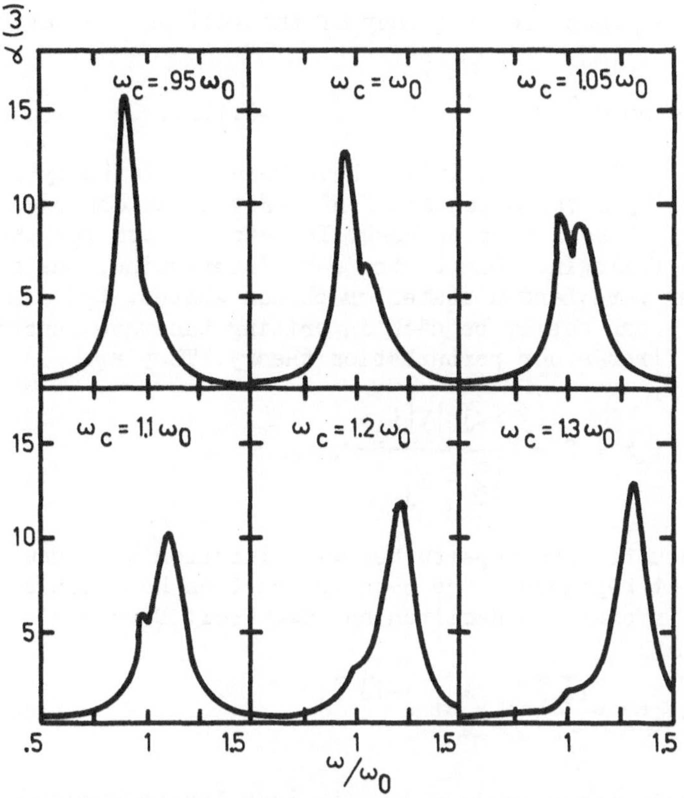


Fig. 2. Absorption coefficient $\alpha(\omega)$, in arbitrary units, versus the frequency of the incident electromagnetic radiation. The absorption spectra are shown for different values of the magnetic induction B, corresponding to the values of $\omega_c$=eB/mc indicated in the figure. We have used $\alpha$=0.02 and $\gamma=0.2\hbar\omega_o$.

cyclotron frequency $\omega_c$=eB/mc. The values $\alpha$=0.02 and $\gamma/2=0.1\ \hbar\omega_o$ have been used in the numerical computations. The pinning effect is clearly seen. However, the picture emerging from our results is more complete than that obtained by Johnson and Larsen since it gives not only the position of the peaks, but also their shape and their oscillator strength. The system behaves more or less like two coupled oscillators, one with fixed frequency $\omega_o$, the other with a frequency $\omega_c$ that can be changed from below $\omega_o$ to above it. In fact, the frequency $\omega_o$ is the threshold of a continuum of possible excitations and, when $\omega_c$ becomes larger than $\omega_o$ the pinned state and the absorption band located at $\omega_o$ merge in a single band so that the former can no longer be seen as an individual line.

Another interesting case is when phonon-assisted transitions can occur, i.e. when the frequency of the incident radiation satisfies

$$\omega = \omega_o + n\omega_c \ . \qquad (n=0,1,2,\ldots)$$

These phonon-assisted transitions have been predicted by Bass and Levinson[3] and by Enck, Saleh and Fan[4] as well as observed for the first time by these latter authors. The explanation for their existence is the following. Due to the e-ph interaction, the states are no longer pure Landau states or phonon states, but a mixture of both. This can easily be seen in writing the wave functions as given by first-order perturbation theory. They are

$$|i\rangle = |i_o\rangle + \sum_j \frac{|j_o\rangle\langle j_o|V|i_o\rangle}{E_i^o - E_j^o} , \qquad (25)$$

where $|i_o\rangle$ denotes the unperturbed wave function with energy $E_i^o$. As for V, it represents the e-ph interaction Hamiltonian (Fröhlich's interaction Hamiltonian. See Prof. Devreese's lectures)

$$V = \sum_{\vec{q}} (V_q a_{\vec{q}} e^{i\vec{q}.\vec{r}} + V_q^{\star} a_{\vec{q}}^{+} e^{-i\vec{q}.\vec{r}}) \ . \qquad (26)$$

This shows that the polaron state $|i\rangle$ is a linear superposition of the unperturbed states; some of them contain one phonon, since, in the matrix element of V in (25), a creation operator $a_q^+$ is applied to $|i_o\rangle$. Moreover, due to the presence of the factor $\exp(-i\vec{q}.\vec{r})$ which multiplies the creation operator, the electron is no longer in a pure Landau state, but in a mixture of all these Landau states. This changes the selection rules, so that new transitions are allowed. They involve the emission of a phonon with transitions of the electron to any final Landau state. Therefore, the possible transition frequencies are

$$\omega = \omega_o + n \ \omega_c \qquad (n=0,1,2 \ldots) \qquad (27)$$

as announced above.

The absorption in the region of phonon-assisted transitions, obtained with our theory, is shown in fig.3. As the intensity of the lines is proportional to the coupling constant, the comparison with experiment should provide a direct test of polaron theory and of the value used for α. However, the only case where phonon-assisted transitions have been observed is InSb [7]. In this case, the theoretical results compare qualitatively well with experiment[8] . However, the coupling is too weak for the phonon-assisted transitions to be intense enough, so that a quantitative comparison

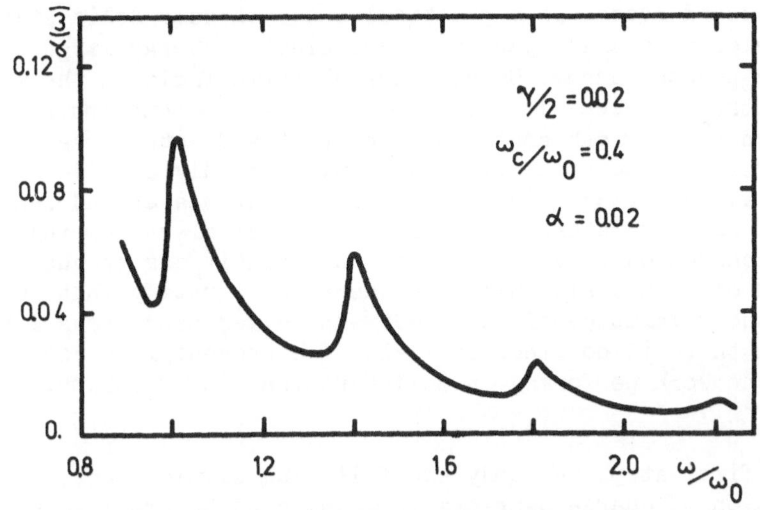


Fig. 3. Absorption coefficient $\alpha(\omega)$, in arbitrary units, in the region of the spectrum where contributions of phonon-assisted transitions are possible. The value of B is such that $\omega_c = 0.4\ \omega_o$. The values $\alpha = 0.02$ and $\gamma = 0.04$ have been used in the calculations.

is not possible.

Experimental results on crystals with larger values of the coupling constant would be particularly interesting, since they would provide a quantitative test of the theory.

## II. RELAXATION OF HOT ELECTRONS DUE TO POLAR LO PHONONS

### II.1. Introduction

The study of relaxation of hot electrons is interesting for two main reasons. The first one is of fundamental importance. Indeed, if one knows that, in the absence of an electric field or after this latter has been switched off, the electrons relax to a Maxwell-Boltzmann distribution (for a non-degenerate electron gas), the details of this relaxation are not yet clear. There remains many important questions, like : How good is the relaxation-time approximation ? What are the respective contributions of the different possible relaxation mechanisms? Are these contributions additive ?

The second reason is a practical one : In some solid-state devices, electrons are injected with a kinetic energy large compared to the phonon energy. Usually the electric field in the active region of these devices is not zero. However, in some instances, this region has a width small compared to the electron mean free path. Therefore, the relaxation takes place mainly outside this region, i.e. in a place where the field is far weaker and can be neglected. Another interesting case is that of electrons injected into the conduction band by band-to-band transitions produced by absorption of light. All these considerations justify that we describe the relaxation of hot electrons in semiconductors for the case where there is no electric field. The present lectures are based on the work performed in collaboration with Ph. Lambin and J. Schmit.[10]

In a first step, we study the following simple model. The concentration of charge carriers is assumed to be small enough for the electron-electron interactions to be negligible. The interaction with the phonons has two components : A Fröhlich interaction with the polar LO phonons and a deformation potential interaction with the longitudinal acoustic (LA) phonons. The interaction with impurities and defects has not yet been introduced in our calculations. The collisions with these impurities and defects are usually assumed to be elastic;with this assumption, their effect could easily be accounted for by introducing a suitable relaxation time, in a way similar to the case of the elastic contribution of the collisions with the LA phonons, which is described later in this paper.

We restrict the study to the case of a band with a single isotropic, parabolic valley at the center of the Brillouin zone. Therefore, intervalley scattering is supposed to be negligible or nonexistent. This has for consequence that the optical phonons are far more efficient than the acoustic ones in relaxing the energy of the hot electrons, at least when emission or absorption of optical phonons are possible. This can be shown in the following way. As an example, consider the emission of a phonon with wavevector $\vec{q}$ and frequency $\omega(q)$. Conservation of momentum and energy implies that

$$\hbar^2|\vec{k}+\vec{q}|^2/2m = \hbar^2k^2/2m + \hbar\omega(q) \tag{28}$$

where $\vec{k}$ is the electron Bloch vector after emission and, consequently, $\hbar^2k^2/2m$ is its final energy. From eq.(28), one immediately obtains the energy lost by emission of the phonon $\vec{q}$ :

$$\hbar\omega(q)= \hbar^2(q^2+ 2\vec{k}.\vec{q})/2m \, . \tag{29}$$

For emission of optical phonons, this energy loss is of the order of 30 meV. For acoustic phonons,

$$\omega(q) = qv_s \tag{30}$$

where $v_s$ is the sound velocity. As we will see later on, the possible values of q are rather small, so that the linear dispersion (30) can be used. Then, eq.(29) gives

$$q = 2mv_s/\hbar - 2k\cos\theta \ . \tag{31}$$

where $\theta$ is the angle between $\vec{k}$ and $\vec{q}$. This shows that the momentum of the emitted phonon and therefore, also its energy $\hbar q v_s$, are maximum when $\cos\theta=-1$. Then

$$q = 2mv_s/\hbar + 2k \tag{32}$$

and the maximum energy lost by the electron during emission of an acoustic phonon with wave-vector $\vec{q}$ is

$$\hbar q v_s = 2mv_s^2 + 2\,\hbar k\, v_s \ . \tag{33}$$

This energy loss is about 1 or 2 meV for electrons with initial energy of 100 meV and becomes of the order of 5 to 10 meV for an initial energy of 1 eV. Note that these results are obtained in the most favourable conditions, when the phonon is emitted in the forward direction. Emissions in other directions lead to a lower energy transfer. Of course, the exact figures depend on the characteristic properties of the crystal under consideration, such as the speed of sound and the band mass.

In conclusion, the optical phonons are far more efficient than the acoustic phonons in relaxing the energy of the charge carriers, specially when there is a strong interaction with these optical phonons. This is the case for the polar semiconductors where the interaction is a Fröhlich interaction. On the contrary, the acoustic phonons are rather efficient in relaxing the direction of the electron momentum.

For these reasons, the relaxation of hot electrons interacting only with polar LO phonons will be first described. The effects of elastic collisions with the LA phonons will then be briefly discussed. We will come to the inclusion of the effects of inelasticity in the collisions with the LA phonons only at the end of these lecture notes.

## II.2. Effects of the interaction with the polar LO phonons

The same simple model as before is used : the band is assumed to have a single isotropic and parabolic minimum at the centre of the Brillouin zone. We restrict ourselves to the cases of a low electron density and a weak interaction with the phonons.

Then, Boltzmann's equation applies. In the absence of electric field or after this latter has been switched off, this equation becomes

$$\frac{\partial f(\vec{k},t)}{\partial t} = -W(k)f(\vec{k},t)+\int d^3k'W(\vec{k}',\vec{k})f(\vec{k}',t), \tag{34}$$

where $f(\vec{k},t)$ is the probability of occupation of the state $\vec{k}$ and $W(\vec{k}',\vec{k})$ is the probability of transition from $\vec{k}'$ to $\vec{k}$, per unit time and unit volume of the $\vec{k}'$ space. As for $W(k)$, it denotes the total probability per unit time for an electron to leave the state $\vec{k}$, so that

$$W(k) = \int d^3k'W(\vec{k},\vec{k}') . \tag{35}$$

For weakly ionic crystals, like the III-V or the II-VI compounds, the Born approximation can be used to calculate the probability of transition $W(\vec{k},\vec{k}')$ due to emission or absorption of one polar LO phonon. One obtains[9]

$$W(\vec{k}',\vec{k})=\alpha\pi^{-1}h\omega_o^2(h/2m\omega_o)^{\frac{1}{2}}\Big[N\delta(\frac{\hbar^2k^2}{2m}-\frac{\hbar^2k'^2}{2m}-\hbar\omega_o)+$$
$$+(N+1)\delta(\frac{\hbar^2k'^2}{2m}-\frac{\hbar^2k^2}{2m}-\hbar\omega_o)\Big] / |\vec{k}-\vec{k}'|^2 . \tag{36}$$

In this expression, N denotes the statistical population of phonons with energy $\hbar\omega_o$, i.e.

$$N = [\exp(\beta\hbar\omega_o)-1]^{-1}.$$

The $\delta$-functions in eq.(36) come from the conservation of energy Obviously, the first term between the brackets of eq.(36) is due to phonon absorption and the second, to phonon emission.

One easily gets by integration

$$W(k)= 2\alpha\omega_o\varepsilon^{\frac{1}{2}}[N\ \mathrm{Arsh}\varepsilon^{\frac{1}{2}}+ (N+1)\ \Theta\ (\varepsilon-1)\mathrm{Arch}\varepsilon^{\frac{1}{2}}] , \tag{38}$$

where $\varepsilon$ is the electron energy expressed in units $\hbar\omega_o$ :

$$\varepsilon = \hbar k^2/2m\omega_o . \tag{39}$$

The function $\Theta$ in the right-hand member of eq.(38) is the Heaviside cutoff function defined as

$$\Theta(x) = 0 \quad \text{for} \quad x < 0 , \tag{40a}$$

$$\Theta(x) = 1 \quad \text{for} \quad x > 1 . \tag{40b}$$

This means that the second member between brackets in eq.(38) is zero if

$$\hbar k^2/2m\omega_o < 1 , \tag{41}$$

which can also be written as

$$\hbar^2 k^2/2m < h\omega_o . \tag{42}$$

This is what one expects, since this term is due to emission of phonons.

An expansion in Legendre polynomials is appropriate to the isotropic situation which prevails here. This expansion consists of writing

$$f(\vec{k},t) = \sum_{\ell=o}^{\infty} g_\ell(\varepsilon,t)P_\ell(u) \tag{43}$$

with

$$\varepsilon = \hbar k^2/2m\omega_o \tag{44a}$$

and

$$u = \cos\theta \tag{44b}$$

where $\theta$ is the angle between $\vec{k}$ and the direction of the z-axis, chosen in a arbitrary way. The notation $P_\ell(u)$ is used for the Legendre Polynomial of order $\ell$.

The denominator of eq.(36) can also be expanded in Legendre Polynomials in the following way :

$$|\vec{k}'-\vec{k}|^{-2} = (k'^2+k^2-2kk'\cos\Phi)^{-1}$$

$$= (2kk')^{-1}\left(\frac{k^2+k'^2}{2kk'} - \cos\Phi\right)$$

$$= (2kk')^{-1}\sum_{\ell=o}^{\infty}(2\ell+1)Q_\ell\left(\frac{k^2+k'^2}{2kk'}\right)P_\ell(\cos\Phi) \tag{45}$$

where $\Phi$ is the angle between $\vec{k}$ and $\vec{k}'$ and the $Q_\ell$'s are Legendre functions of the second kind. Using

$$\varepsilon = \hbar k^2/2m\omega_o \tag{46a}$$

and

$$\varepsilon' = \hbar k'^2/2m\omega_o \tag{46b}$$

leads to

$$(\vec{k}'-\vec{k})^{-2} = \frac{\hbar/2m\omega_o}{2\sqrt{\varepsilon\varepsilon'}} \sum_{\ell=0}^{\infty} (2\ell+1) Q_\ell\left(\frac{\varepsilon+\varepsilon'}{2\sqrt{\varepsilon\varepsilon'}}\right) P_\ell(\cos\Phi). \tag{47}$$

Now, use is made of the addition theorem :

$$P_\ell(\cos\Phi) = P_\ell(\cos\theta)P_\ell(\cos\theta') + 2\sum_{n=1}^{\infty} (-1)^n P_\ell^{-n}(\cos\theta) P_\ell^n(\cos\theta')\cos n\varphi, \tag{48}$$

where the $P_\ell^n$ 's are associated Legendre functions. The meaning of the notations $\theta$, $\theta'$ and $\varphi$ can be found in fig.4.

In the integral term of the Boltzmann equation, the integral over the angles can easily be performed. It is of the type

$$\int_0^\pi P_\ell(\cos\theta')\sin\theta' d\theta' \int_0^{2\pi} |\vec{k}-\vec{k}'|^{-2} d\varphi \ .$$

$$= 2\pi \frac{\hbar/2m\omega_o}{\sqrt{\varepsilon\varepsilon'}} Q_\ell\left(\frac{\varepsilon+\varepsilon'}{2\sqrt{\varepsilon+\varepsilon'}}\right) P_\ell(\cos\theta) \tag{49}$$

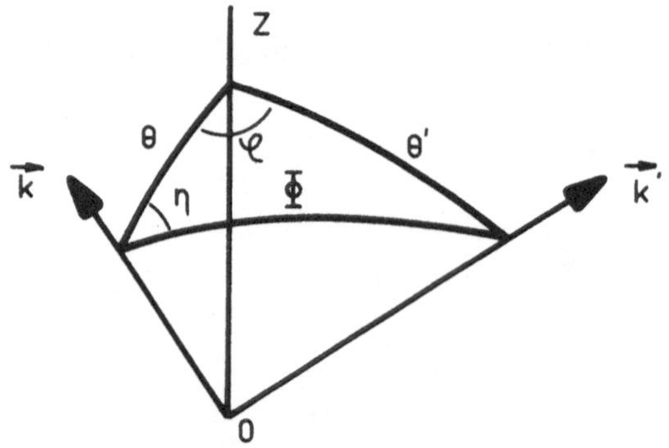


Fig. 4. This figure gives the meaning of the notations used for the angles between $\vec{k}$, $\vec{k}'$ and the z-axis.

To obtain the result of eq.(49), we used the fact that

$$\int_0^{2\pi} \cos n\varphi \, d\varphi = 0 \qquad (n \neq 0), \tag{50}$$

and also that

$$\int_{-1}^{1} P_\ell(u) P_{\ell'}(u) du = 2\delta_{\ell\ell'}/(2\ell+1) \tag{51}$$

where $\delta_{\ell\ell'}$ is the usual Kronecker symbol.

The integration over the modulus of $\vec{k}'$ is straightforward, due to the presence of the δ-functions in the integrand. It leads to the following form for the Boltzmann equation :

$$\sum_{\ell=0}^{\infty} P_\ell(\cos\theta) \left[ \frac{\partial g_\ell(\varepsilon,t)}{\partial t} - A_\ell(\varepsilon)\, \Theta\,(\varepsilon-1) g_\ell(\varepsilon-1,t) + B(\varepsilon) g_\ell(\varepsilon,t) - C_\ell(\varepsilon) g_\ell(\varepsilon+1,t) \right] = 0 \tag{52}$$

with

$$A_\ell(\varepsilon) = \alpha N \omega_o \varepsilon^{-\frac{1}{2}}\, Q_\ell\left(\frac{2\varepsilon-1}{2\sqrt{\varepsilon(\varepsilon-1)}}\right) , \tag{53a}$$

$$B(\varepsilon) = 2\alpha\omega_o \varepsilon^{-\frac{1}{2}} \left[ N \,\mathrm{Arsh}\varepsilon^{\frac{1}{2}} + (N+1)\, H\,(\varepsilon-1) \mathrm{Arch}\varepsilon^{\frac{1}{2}} \right] , \tag{53b}$$

$$C_\ell(\varepsilon) = \alpha(N+1)\omega_o \varepsilon^{-\frac{1}{2}}\, Q_\ell\left(\frac{2\varepsilon+1}{2\sqrt{\varepsilon(\varepsilon+1)}}\right). \tag{53c}$$

This shows that the different Legendre components of the distribution function relax independently. The component $g_\ell$ of order $\ell$ relaxes according to the equation

$$\frac{\partial g_\ell(\varepsilon,t)}{\partial t} = A_\ell(\varepsilon)\, \Theta\,(\varepsilon-1) g_\ell(\varepsilon-1,t) - B(\varepsilon) g_\ell(\varepsilon,t) - C_\ell(\varepsilon) g_\ell(\varepsilon+1,t) \tag{54}$$

This equation is no longer an integral equation, but a matrix equation. Indeed, if we introduce x and n such that

$$\varepsilon = n+x \ , \ 0<x<1 \ , \ n=0,1,2,\ldots, \tag{55}$$

then, the values of $g_\ell(\varepsilon,t)$ can be considered as the components $h_n^\ell(x,t)$ of a vector. They are defined by the following equation

$$h_n^\ell(x,t) = g_\ell(n+x,t), \qquad n = 0,1,2 \ldots \tag{56}$$

These components obey the equation

$$\frac{\partial h_n^\ell(x,t)}{\partial t} = - \sum_{m=0}^{\infty} w_{n,m}^\ell(x) h_m^\ell(x,t) . \tag{57}$$

The coefficients $w_{n,m}^\ell(x)$ are the elements of a tridiagonal matrix $w^\ell(x)$. More precisely

$$w_{n,n-1}^\ell(x) = -A_\ell(n+x) , \qquad n > 1 , \tag{58a}$$

$$w_{n,n}^\ell(x) = B\ (n+x) , \tag{58b}$$

$$w_{n,n+1}^\ell(x) = -C_\ell(n+x) . \tag{58c}$$

All the other matrix elements of $w^\ell(x)$ are zero.

The solution of eq.(57) is easily found. It is

$$h_n^\ell(x,t) = \sum_i c_i(x)\ b_{i,n}^\ell(x)\ \exp(-\lambda_i^\ell(x)t) \tag{59}$$

where $\lambda_i^\ell(x)$ and $b_i^\ell(x)$ denote respectively the eigenvalues and eigenvectors of the matrix $w^\ell(x)$, i.e.

$$\sum_m w_{n,m}^\ell(x)\ b_{i,m}^\ell(x) = \lambda_i^\ell(x)\ b_{i,n}^\ell(x) . \tag{60}$$

The coefficients $c_i(x)$ are to be determined from the initial value of the distribution function. Indeed, taking $t=o$ in (59) leads to

$$h_n^\ell(x,o) = \sum_i c_i(x)\ b_{i,n}^\ell(x) . \tag{61}$$

Standard numerical techniques can be used to find the eigenvalues and eigenvectors of $w^\ell(x)$. On the other hand, as the vectors $b_i^\ell(x)$ form a complete basis, the coefficients $c_i$'s are easily computed in practice. Fig.5 gives the results for the time evolution of the $g_0$ component of a distribution which initially is a displaced Maxwellian :

$$f(\vec{k},o) = C\ \exp(-\beta_0 \hbar^2 |\vec{k}-\vec{k}_0|^2/2m). \tag{62}$$

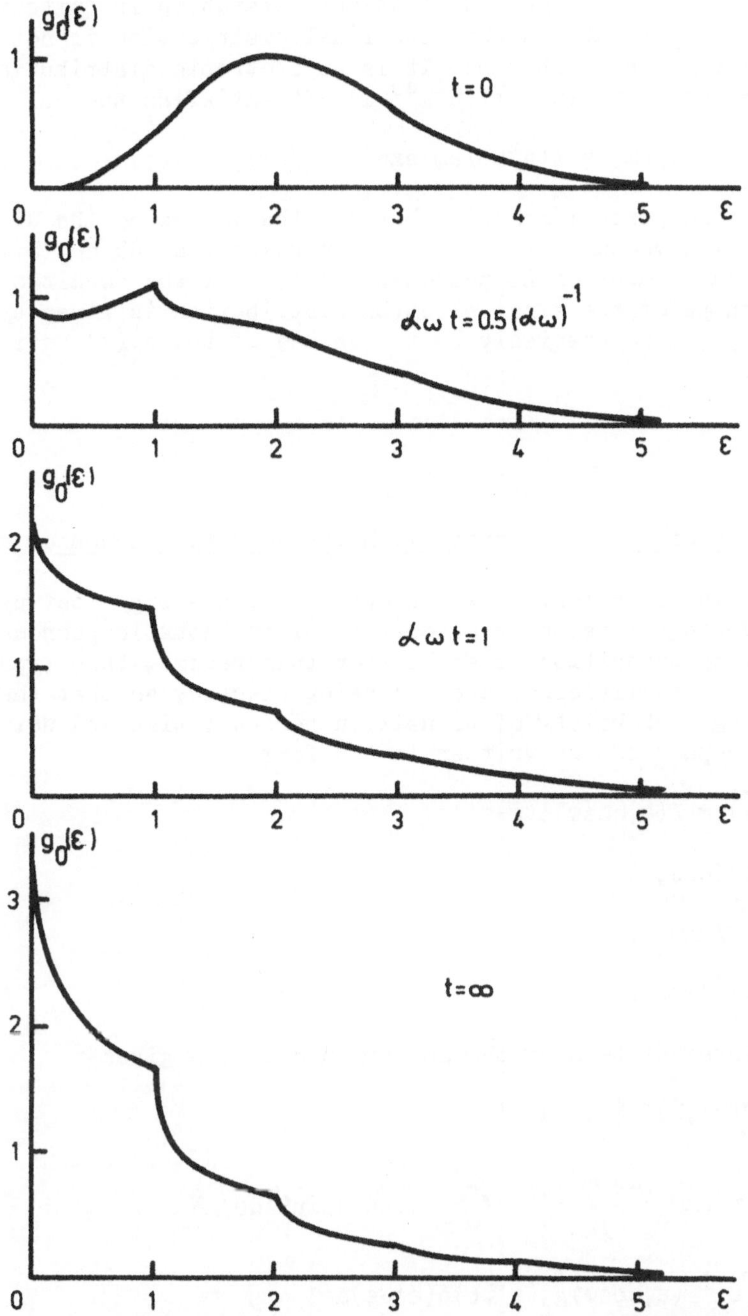


Fig.5. Time evolution of the $g_0$ component of a hot-electron distribution which initially is a displaced Maxwellian. The characteristic time $(\alpha\omega_0)^{-1}$ is typically from $10^{-12}$s to $10^{-13}$s. The value $\varepsilon=1$ corresponds to an electron energy equal to $\hbar\omega_0$. The temperature is given by $KT=\hbar\omega_0$.

From this figure, it is obvious that the relaxation is quite fast. What is more important, is that the final distribution is not a Maxwell-Boltzmann distribution. It is an isotropic distribution, so that one can write it as $f(\hbar^2k^2/2m)$. It satisfies the rule

$$f(\hbar^2k^2/2m + h\omega_o) = f(\hbar^2k^2/2m)\exp(-\hbar\omega_o/KT), \tag{63}$$

but there is no general relation between the values of the distribution function measured for electron energies that do not differ by an integral number of LO phonons. Apart from the condition (63), the exact shape of the final electron distribution is given by its initial shape, more precisely by the values of the $c_i(x)$ defined in eq.(61).

## II.3. Effects of the interaction with the acoustic phonons

In the introduction, it was shown that the energy lost by the electrons during emission or absorption of an acoustic phonon for an intravalley transition is small. For this reason, this type of scattering is usually considered as being elastic, so that the corresponding probability of transition per unit time and unit volume of k-space can be written in the form

$$W(\vec{k}',\vec{k}) = F(\varepsilon,\cos\Phi)\delta(\varepsilon-\varepsilon') , \tag{64}$$

where, as before,

$$\varepsilon = \hbar k^2/2m\omega_o , \tag{65a}$$

$$\varepsilon' = \hbar k'^2/2m\omega_o . \tag{65b}$$

Then, the integral term of the Boltzmann equation gives

$$\begin{aligned} I &\equiv \int W(\vec{k}',\vec{k})f(\vec{k}',t)d^3k' \\ &= \frac{1}{2}(\hbar/2m\omega_o)^{-\frac{3}{2}} \sum_{\ell=0}^{\infty} \int_0^{2\pi}d\varphi\int_0^{\pi}P_\ell(\cos\theta')\sin\theta' d\theta' \times \\ &\times \int_0^{\infty}\varepsilon'^{\frac{1}{2}}F(\varepsilon,\cos\Phi)g_\ell(\varepsilon',t)\delta(\varepsilon'-\varepsilon)d\varepsilon' \\ &= \frac{1}{2}(\hbar/2m\omega_o)^{-\frac{3}{2}}\varepsilon^{\frac{1}{2}} \sum_{\ell=0}^{\infty} g_\ell(\varepsilon,t)\int_0^{2\pi}d\varphi\int_0^{\pi}F(\varepsilon,\cos\Phi)P_\ell(\cos\theta')\sin\theta' d\theta' . \end{aligned} \tag{66}$$

This integral is more easily calculated in taking $\vec{k}$ as polar axis for the integration in spherical coordinates. Then one obtains

$$I = \frac{1}{2}(\hbar/2m\omega_o)^{-\frac{3}{2}} \sum_{\ell=0}^{\infty} \varepsilon^{\frac{1}{2}} g_\ell(\varepsilon,t) \int_0^{2\pi} d\eta \int_0^{\pi} F(\varepsilon,\cos\Phi) P_\ell(\cos\theta') \sin\Phi d\Phi. \tag{67}$$

The meaning of the angles appearing in this eq.(67) can be found in fig.4. Obviously

$$\cos\theta' = \cos\theta\cos\Phi + \sin\theta\sin\Phi\cos\eta \; . \tag{68}$$

As in the case of the polar optical phonons, the addition theorem is used to express $P_\ell(\cos\theta')$ (see eq.(48)). Again, the terms containing a factor $\cos n\eta$ give zero by integration. Therefore

$$I = \pi(\hbar/2m\omega_o)^{-\frac{3}{2}} \sum_{\ell=0}^{\infty} \varepsilon^{\frac{1}{2}} g_\ell(\varepsilon,t) P_\ell(\cos\theta) \int_{-1}^{+1} F(\varepsilon,u) P_\ell(u) du, \tag{69}$$

where the notation $u=\cos\Phi$ has been introduced.

The total probability of transition per unit time is calculated in the same way. It is

$$W(k) = \int W(\vec{k},\vec{k}')d^3k'$$
$$= \pi(\hbar/2m\omega_o)^{-\frac{3}{2}} \varepsilon^{\frac{1}{2}} \int_{-1}^{+1} F(\varepsilon,u)du. \tag{70}$$

Gathering all these results leads to introduce a relaxation time $\tau_\ell(\varepsilon)$ depending on the energy, for each Legendre component. These relaxation times are defined by

$$\tau_\ell^{-1}(\varepsilon)=\pi(\hbar/2m\omega_o)^{-\frac{3}{2}} \varepsilon^{\frac{1}{2}} \int_{-1}^{1} F(\varepsilon,u) \left[1-P_\ell(u)\right] du \; . \tag{71}$$

Consider the case where the electrons interact with both polar LO phonons and acoustic phonons, and where the scattering due to these latter is elastic. Then, the only change in the formalism described in the preceding section is that the diagonal matrix element $w^\ell_{n,n}(x)$ defined by eq.(58b) is now replaced by

$$w^\ell_{n,n}(x) = B\,(n+x) + \tau_\ell^{-1}(n+x). \tag{72}$$

As a consequence, the relaxation becomes in general faster. However, as $P_o(u)=1$ , eq.(71) shows that

$$\tau_o^{-1}(\varepsilon) = 0 \; . \tag{73}$$

Therefore, the elastic scattering mechanisms don't contribute to the relaxation of the isotropic part of the distribution of the hot carriers. This is quite natural, since elastic interactions change only the angular distribution and not the energy distribution.

The numerical results for the type of relaxation described here will be given in a forthcoming paper[11]. The main conclusion is the following. Even if the relaxation is faster (at least for $\ell \neq 0$) when the interaction with the LA phonons is taken into account, the shape of the Legendre components remain qualitatively the same. In particular, the discontinuities in the slope remain and the equilibrium distribution reached after relaxation is in general not a Maxwell-Boltzmann distribution.

To understand how, in actual situations, the relaxation leads to a Maxwell-Boltzmann distribution for a non-degenerate electron gas, it is essential to include inelastic mechanisms. In these lectures notes, the effects of the energy transfer during emission or absorption of LA phonons is briefly described.

The Boltzmann equation no longer reduces to a matrix equation, so that more complicated methods must be used. Here the solution is obtained by the following iterative procedure. Symbolically, the Boltzmann equation can be written as

$$\frac{\partial f(\vec{k},t)}{\partial t} = -Lf(\vec{k},t) \ , \tag{74}$$

where L is an integral operator, the definition of which is easily obtained by comparing eq.(74) with the initial Boltzmann equation.

The formal solution of (74) is

$$f(\vec{k},t) = e^{-L(t-t_o)} f(\vec{k},t_o) \ . \tag{75}$$

Expanding the exponential in a power series, leads to

$$f(\vec{k},t) = \left[1+ \sum_{n=1}^{\infty} \frac{(-1)^n}{n!} L^n (t-t_o)^n\right] f(\vec{k},t_o) \ . \tag{76}$$

It is clear that the larger the time interval $t-t_o$, the slower the convergence of the expansion (76). It is not possible to include many terms in the calculation of eq.(76), since the n-th power of the operator L implies a convolution of n-1 integrals. In practice, the expansion of eq.(76) has been restricted to four terms. To extend the results to a longer time interval, we reiterate the process, applying eq.(76) as many times as required to reach a stationary value for f(k,t). The results are shown in fig.6. They show that the relaxation involves two different time scales. First, there is a rapid relaxation involving LO phonons and the elastic part of the scattering by the LA phonons. This relaxation goes on till the distribution function becomes isotropic (if it was not

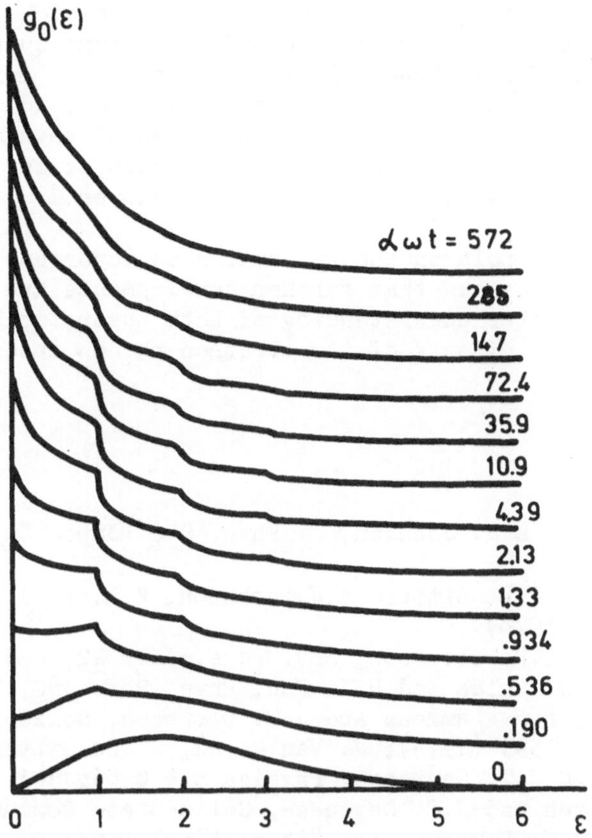


Fig. 6. Time evolution of the $g_0$ component of a hot-electron distribution (arbitrary units). The values of the parameters correspond to GaAs at 420 K. The effects of the interaction with the polar LO phonons and with the LA phonons are included and the energy transfer to the LA phonons is taken into account.

such at the beginning)and such that

$$f(\hbar^2k^2/2m+h\omega_o) = f(\hbar^2k^2/2m)\exp(-\hbar\omega_o/kT). \tag{75}$$

There is also a far slower relaxation, due to the inelasticity of the scattering by the acoustic phonons. This second relaxation goes on for a longer time till the distribution becomes the equilibrium Maxwell-Boltzmann distribution.

## CONCLUSION

In these lectures, we have shown on two examples, that the interaction between the charge carriers and the polar optical phonons play an important role in practical applications. On the other hand, from the fundamental point of view, some questions remain unanswered. For instance, is the concept of an effective mass still valid when the polaron kinetic energy is close to the energy of the LO phonons and if yes, what is the value of this mass ? What would be the effects of a stronger coupling, such as for the silver halides, on the results described in these lectures ?

This shows that polaron theory remains an important chapter of solid-state physics and that further developments are still needed to have a better understanding of both theoretical and practical aspects of the effects of the electron-phonon interactions in solids.

REFERENCES

1. D.M. Larsen and E.J. Johnson, J. Phys. Soc. Japan Suppl.21, 443 (1966).
2. J.T. Devreese, J.De Sitter, E.J. Johnson, K.L. Ngai, Phys. Rev. B17, 3207 (1978).
3. F.G. Bass and I.B. Levinson, Sov. Phys.-JETP 22, 635 (1966).
4. R.C. Enck, A.S. Saleh and H.Y. Fan, Phys. Rev. 182, 790 (1969).
5. J. Van Royen, L.F. Lemmens and J.T. Devreese, Solid-State Commun. 15, 591 (1974); J. Van Royen, J. De Sitter, L.F. Lemmens and J.T. Devreese, Physica B & C 81, 101 (1977); J. Van Royen and J.T. Devreese, Solid-State Commun. 40, 947 (1981) ; J.T. Devreese in: "Theoretical Aspects and New Developments in Magneto-optics", ed. by J.T. Devreese (Plenum, 1980) p.217.
6. J.P. Vigneron, R. Evrard and E. Kartheuser, Phys. Rev. B 18, 6930 (1978).
7. M.H. Weiler, R.L. Aggarwal and B. Lax, Solid-State Commun. 14, 299 (1974).
8. J.P. Vigneron, Solid-State Commun. 32, 595 (1979).
9. J.T. Devreese and R. Evrard, Phys. Status Solidi b 78, 85 (1976).
10. R. Evrard, Ph. Lambin and J. Schmit in : "Proc. of the 16th Int. Conf. on the Phys. of Semiconductors", Montpellier 1982.
11. Ph. Lambin, R. Evrard and J. Schmit, to be published.

# COUPLED PLASMON - POLAR PHONON MODES IN SEMICONDUCTORS: SPATIAL DISPERSION AND OTHER PROPERTIES

Wolfgang Richter

I.Physikalisches Institut der RWTH Aachen
Sommerfeldstraße, D-5100 Aachen,FR Germany

## 1. INTRODUCTION

It is the purpose of this paper to present a survey on the characteristic properties of the collective excitations of the free electron gas. These excitations are termed plasmons. They are in contrast to excitations of single electrons, so called single particle excitations (SPE). Classically the plasmon is described by a synchronous motion of all free electrons against the positive background of the ions. At infinite wavelength,i.e. wavevector K=0,this motion is solely determined by Coulomb forces and the free electron gas oscillates with the plasma frequency $\Omega_p^2 = n\, e^2/m\varepsilon_0$ against the positive ionic background. For finite wavelength, i.e. $K \neq 0$, the electron gas no longer oscillates homogeneously, but density gradients build up. In addition to the motion caused by Coulomb forces then diffusion type motion will occur and the eigenfrequency of the plasmon will change with wavevector: $\Omega(K)$. This is what is termed spatial dispersion.

The behaviour of plasmons can be most conveniently studied in semiconductors, where, by doping, the free carrier concentration can be changed about several orders of magnitude. Nearly all semiconductors, with the exception of the diamond-structure ones (C,Si,Ge,α-Sn),exhibit in addition IR-active (polar)phonons. Their longitudinal modes couple strongly via the longitudinal macroscopic electric field to the plasmons. Thus coupled plasmon-LO-phonon (PLP) modes are the collective excitations of the system polar phonons and free carriers.

The coupling of plasmons and longitudinal polar phonons was first investigated theoretically by Varga /1/ for long wavelength modes, i.e. wavevector K=O.This situation is to an excellent approximation fullfilled in 1-photon experiments (reflectivity, transmission) where $K \simeq 10^{-3}\pi/a \simeq 0$, i.e. very small compared to the size of the Brillouin-Zone. However, in all other physical situations (transport, screening, inelastic scattering) the restriction K=O has not to be obeyed, and the spatial dispersion of the PLP-modes has to be taken into account. The important wavevector range for this dispersion mechanism is approximately a few percent of the maximum wavevector in the Brillouin-Zone. In this range the phonon component of the PLP-modes does not contribute to the spatial dispersion, since the short range ion-ion interactions still play no role at these relatively long wavelengths. The spatial dispersion of the PLP-modes is only caused by the free carrier contribution.

Experimentally inelastic light scattering (Raman scattering) turns out to be the ideal probe for this wavevector range. A similar range can be covered by inelastic scattering of low energy electrons. However, besides having a much lower resolution (in $\Omega$ as well as K), the penetration depth of electrons (30Å) is much smaller than for photons (visible:>1000Å) and thus electron scattering probes specifically surface excitations. Light scattering on the other hand is first of all sensitive to volume excitations. With the availibility of many suitable laser lines within the last years a number of Raman scattering investigations have been performed in order to study the spatial dispersion of plasmon-LO-phonon (PLP) modes in semiconductors /2-11/.

From the theoretical side several different approximations and results have been given for the spatial dispersion of the PLP-modes /12-16/. While in the limit of small wavevector they all give identical results, for larger K the solutions are not just different but contradictory. Fig. 1 sketches the problem. In the simplest case of one polar phonon, two coupled PLP-modes are observed. Their dispersion is usually assumed to be quadratic, i.e. $\propto K^2$ at small K values. At larger wavevectors, on the other hand, only one longitudinal mode, the lattice dynamical LO-phonon is observed. It will be the purpose of this paper the elucidate the intermediate region between the well established K=O PLP-modes and the pure

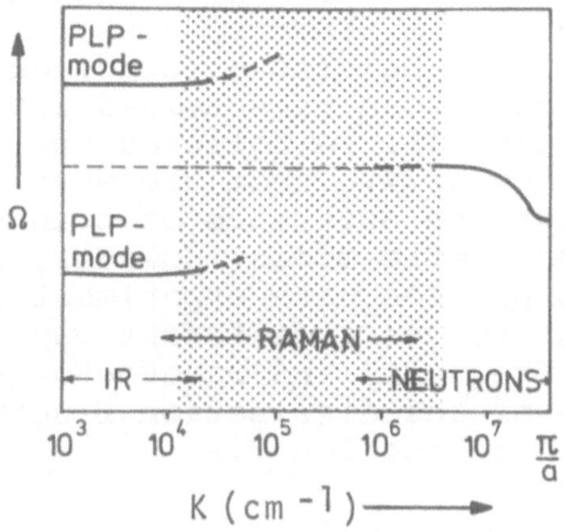


Fig. 1. Dispersion of plasmon-LO-phonon modes (note the logarithmic K scale).

LO-phonon at large wavevectors.

For this purpose we will first generally look for solutions of transverse and longitudinal modes accompanied by an electric field (polaritons). In order to describe the spatial dispersion, the free carrier electric susceptibility $\chi_{FC}(\Omega,K)$ turns out to be the central function. Different models for $\chi_{FC}(\Omega,K)$ are considered (chapter 3) and compared with the experimental results (chapter 4,5). The good agreement enables us to give a general discussion of the spatial dispersion in the intermediate region as sketched in Fig. 1.

## 2. MODE-EQUATIONS

Plasmors and phonons are first of all elementary excitations, that is approximate eigenstates of the microscopic system hamiltonian. However, their nature is collective rather than quasiparticle. This implies that macroscopically observable quantities are connected with them. Since we are discussing charge density oscillations, coupled via the electric field to polar phonons, such a macroscopic quantity is the electric field. This means that the coupled modes must be also electromagnetic modes and that their frequencies can be obtained from Maxwell equations. This approach will be followed in the next section (2.1), where we arrive from the wave equation at certain conditions for the existence of transversal and longitudinal electromagnetic modes. However, for the general discussion and comparison with experiment it turns out to be more convenient to use a different way to define the mode spectrum in frequency and wavevector space. This is done with the use of the dissipation-fluctuation theorem, which is discussed in section 2.2.

## 2.1 Wave-Equation

We start from Maxwell-equations

$$\frac{1}{\mu_0}\nabla\times\underline{B} = \varepsilon_0\dot{\underline{E}} + \dot{\underline{P}} + \underline{j} + \nabla\times\underline{M} \tag{1}$$

$$\nabla\times\underline{E} = -\dot{\underline{B}} \tag{2}$$

$$\varepsilon_0\nabla\underline{E} = \rho - \nabla\underline{P} \tag{3}$$

$$\nabla\underline{B} = 0 \tag{4}$$

together with the material equations

$$\underline{P} = \varepsilon_0\chi\,\underline{E} \quad \text{and} \quad \underline{j} = \sigma\underline{E}\,. \tag{5}$$

We look for plane wave solutions of the form

$$\psi = \psi_0\exp[i(\underline{K}\underline{r} - \Omega t)] \tag{6}$$

for all quantities in equs.(1) to (5). Inserting the plane wave ansatz into Maxwell equations one obtains

$$\frac{1}{\mu_0}i\underline{K}\times\underline{B} = -i\Omega(\varepsilon_0\underline{E} + \underline{P}) + \underline{j} + i\underline{K}\times\underline{M} \tag{7}$$

$$\underline{K}\times\underline{E} = \Omega\underline{B} \tag{8}$$

$$i\varepsilon_0\underline{K}\cdot\underline{E} = \rho - i\underline{K}\cdot\underline{P} \tag{9}$$

$$\underline{K}\cdot\underline{B} = 0 \tag{10}$$

the so-called fourier transformed formulation of the Maxwell-equations. Eliminating now B from the first two equations we obtain the wave equation

$$\underline{K} \times (\underline{K} \times \underline{E}) = -\Omega^2\mu_0(\varepsilon_0\underline{E}+\underline{P})-i\Omega\mu_0 j+\Omega\mu_0\underline{K}\times\underline{M} . \qquad (11)$$

By using the material equations (5) and assuming unmagnetic media (M=O) we arrive at

$$\underline{K} \times (\underline{K} \times \underline{E}) = -\Omega^2\mu_0\varepsilon_0\underline{\underline{\varepsilon}}\underline{E} \qquad (12)$$

where the dielectric function ε is given by

$$\underline{\underline{\varepsilon}} = \underline{\underline{1}} + \underline{\underline{\chi}} - \frac{i\,\underline{\underline{\sigma}}}{\varepsilon_0\Omega} \qquad (13)$$

Now in order to project the transverse and longitudinal solutions out of (12) all quantities are splitted into their longitudinal and transverse components (e.g.)

$$\underline{E} = \underline{E}_T + \underline{E}_L \quad \text{with} \quad \underline{K}\times\underline{E}_L \equiv 0 \quad \text{and} \quad \underline{K}\cdot\underline{E}_T \equiv 0 \qquad (14)$$

The left side of (12) then yields the identity

$$\underline{K}\times(\underline{K}\times\underline{E})=K^2\,\underline{E}_L-K^2\underline{E}_T-K^2\underline{E}_L = -K^2\underline{E}_T \qquad (15)$$

and thus (12) can be rewritten as

$$-K^2\,\underline{E}_T = -\,\Omega^2\mu_0\varepsilon_0\underline{\underline{\varepsilon}}\,(\underline{E}_T+\underline{E}_L) . \qquad (16)$$

Since the above equation has to be valid for all $\underline{E}$ it follows for longitudinal modes

$$0 = \varepsilon_L(\Omega,K) \qquad (17)$$

and for transverse modes

$$K^2 = \Omega^2\mu_0\varepsilon_0\varepsilon_T(\Omega,K) \qquad (18)$$

or with $\mu_0\varepsilon_0=1/c^2$

$$K^2c^2/\Omega^2 = \varepsilon_T\,(\Omega,K), \qquad (19)$$

where we have now also explicitly written the Ω and K dependence of ε. If we assume a scalar dielectric function at K=O (isotropic materials) now at finite K there are two components due to nonlocal interactions, the transverse and the longitudinal dielectric function. In general they will be different.However, in the limit of small K

$$\lim_{K\to 0} \varepsilon_T(\Omega,K) = \lim_{K\to 0} \varepsilon_L(\Omega,K) \tag{20}$$

and the difference vanishes. Since we are going to use the transverse dielectric function only for wavevectors nearly equal to zero, we drop the indices T and L and write

$$\varepsilon_L(\Omega,K)=\varepsilon(\Omega,K) \quad \text{and} \quad \varepsilon_T(\Omega,K)=\varepsilon(\Omega,0) \ . \tag{21}$$

The conditions (17) and (19) then finally read

$$\text{longitudinal:} \quad \varepsilon(\Omega,K) = 0 \tag{22}$$

$$\text{transverse} \ : \quad K^2c^2/\Omega^2-\varepsilon(\Omega,0) = 0 \quad . \tag{23}$$

It is now possible to calculate the eigenfrequencies of the modes from (22), (23) if $\varepsilon(\Omega,K)$ is known. Since in general $\varepsilon(\Omega,K)$ will be complex, the solutions will be too.

The modes described by these two equations are often termed polaritons, in addition also with a prefix,depending on the main contribution to $\varepsilon(\Omega,K)$: phonon-polariton, exciton-polariton, plasmon-polariton (plasmariton) ect.. Quite often, however, the term polariton is exclusively used for the transverse modes only.

## 2.2 Dissipation-Fluctuation Theorem

A different way to obtain information on the mode spectrum of a system is to study its fluctuation spectrum in $\Omega$ and K. The dissipation-fluctuation (DF) theorem serves to calculate the fluctuation spectrum. It connects in thermal equilibrium the power spectrum of fluctuations of a quantity X with the imaginary part of a linear response function T:

$$\langle XX^*\rangle = \frac{\hbar}{\pi}\,[n(\Omega)+1]\ \mathrm{Im}T \ , \tag{24}$$

where $n(\Omega)$ is the Bose-function. T relates the linear response of the system in the quantity X under the influence of an external force F:

$$X = T \cdot F \tag{25}$$

In our case we may take for X the electric field $\underline{E}$ associated with the modes. An external force which induces an electric field inside the sample (response) is then given by an external (fictitious) polarisation, $\underline{P}_{ext}$. The total polarisation inside the sample is

$$\underline{P}_{tot} = \underline{P}_{ind} + \underline{P}_{ext} \tag{26}$$

$$= \varepsilon_0 \cdot \chi \cdot \underline{E} + \underline{P}_{ext} \quad . \tag{27}$$

From the first two Maxwell equations (1), (2) together with (27) one obtains then in a similar way as above in 2.1/17,18/:

$$(K^2 - \frac{\Omega^2}{c^2} \varepsilon) \; \underline{E} = \frac{\Omega^2}{c^2} \cdot \frac{\underline{P}_{ext}}{\varepsilon_0} - \frac{1}{\varepsilon} \; \underline{K}(\underline{K} \cdot \underline{P}_{ext}/\varepsilon_0) \; . \tag{28}$$

Thus a linear response relation between $\underline{E}$ and $\underline{P}_{ext}$ can be written as

$$\underline{E} = \frac{\varepsilon \frac{\Omega^2}{c^2} - \underline{K}\,\underline{K}}{\varepsilon(K^2 - \frac{\Omega^2}{c^2} \varepsilon)\varepsilon_0} \; \underline{P}_{ext} \quad . \tag{29}$$

We now specify for the longitudinal and transverse case. Putting $\underline{K}$ for example into the z-direction

$$\underline{K} = K \cdot (001) \tag{30}$$

the longitudinal case

$$\underline{P}_{ext} = P_{ext} \cdot (001) \tag{31}$$

gives from (29) the longitudinal response relation

$$E = - \frac{P_{ext}}{\varepsilon_L \varepsilon_0} \quad . \tag{32}$$

In the transverse case

$$\underline{P}_{ext} = P_{ext} \cdot (100) \tag{33}$$

the dyadic product in (29) vanishes and we obtain for the transverse response relation

$$E = \frac{\Omega^2/c^2}{(K^2 - \Omega^2 \varepsilon_T/c^2)\varepsilon_0} \; P_{ext} \quad . \tag{34}$$

Comparing with (25) the linear response functions are obtained as

$$T_{\text{longitudinal}} = \frac{-1}{\varepsilon_L \varepsilon_o} \qquad (35)$$

$$T_{\text{transverse}} = \frac{1}{(K^2 \cdot c^2/\Omega^2 - \varepsilon_T)\varepsilon_o} \; . \qquad (36)$$

The DF-theorem, (26), then reads

$$\langle EE^* \rangle_{\text{longitudinal}} = \frac{\hbar}{\pi\varepsilon_o}[n(\Omega)+1] \cdot \text{Im}\, \frac{1}{-\varepsilon(\Omega,K)} \qquad (37)$$

$$\langle EE^* \rangle_{\text{transverse}} = \frac{\hbar}{\pi\varepsilon_o}[n(\Omega)+1] \cdot \text{Im}\, \frac{1}{K^2c^2/\Omega^2 - \varepsilon(\Omega,O)} \, , \qquad (38)$$

where in addition we have used the approximations expressed in (21).

From the above equations the whole transverse and longitudinal fluctuation spectrum in $\underline{E}$ can be calculated if the dielectric function is known. The modes are then obtained through the maxima in $\langle EE^* \rangle$. From the similar structure of (37), (38) with (22), (23) one can guess already at this point that the definitions given through the maxima of $\langle EE^* \rangle$ for the modes are equivalent to the ones given in the last section. This will be discussed below.

## 2.3 Comparison of Both Approaches

From the preceding sections we are left with the conditions (22), (37) for longitudinal modes and with equations (23), (38) for transverse modes. Splitting now $\varepsilon(\Omega,K)$ into its real and imaginary part

$$\varepsilon(\Omega,K) = \varepsilon_1(\Omega,K) + i\varepsilon_2(\Omega,K) \qquad (39)$$

and inserting into the imaginary part of equ. (38) we obtain

$$\text{Im}\, \frac{1}{K^2c^2/\Omega^2 - \varepsilon(\Omega,O)} = \text{Im}\, \frac{(Kc/\Omega)^2 - \varepsilon_1(\Omega,O) + i\varepsilon_2(\Omega,O)}{|(Kc/\Omega)^2 - \varepsilon_1(\Omega,O) - i\varepsilon_2(\Omega,O)|^2}$$

$$= \frac{\varepsilon_2(\Omega,0)}{|Kc/\Omega)^2-\varepsilon(\Omega,0)|^2} \tag{40}$$

and similarly from (37)

$$\mathrm{Im}\,\frac{-1}{\varepsilon(\Omega,K)} = \frac{\varepsilon_2(\Omega,K)}{|\varepsilon(\Omega,K)|^2} \tag{41}$$

Thus we see that equs. (22), (23) and the definition via the maxima of $\langle EE^*\rangle$, given in equs. (37) and (38), are equivalent. It should be pointed out, however, that equs. (37) and (38) do not contain only the maxima but the complete fluctuation spectra, and thus give the whole story. This is why differences occur in the practical use of both approaches.

In solving (22) and (23) we will in general get complex solutions for $\Omega$ and K since $\varepsilon$ is complex:

$$\Omega = \Omega_1 + i\Omega_2,\ K = K_1 + iK_2. \tag{42}$$

This corresponds to the temporal and spatial damping of the modes. In order to plot dispersion curves one usually solves (22) or (23) by assuming $\Omega$ or K real and the other quantity to be complex. The solutions look quite different as can be seen in the example of Fig. 2a,b. Which situation is appropiate is determined by the experiment, which imposes certain additional conditions on $\Omega$ and K. The situation in Fig. 2a for example would correspond to a light scattering experiment in a transparent crystal(wavevectors well defined),which measures then the temporal damping of the modes in question. Fig. 2b would be adequate for a reflectivity measurement,where the frequency is determined by the spectral apparatus and this time the spatial damping is relevant.

In plotting now, for the same example, the fluctuation spectrum, as shown in Fig. 2c, the physical situation with respect to dispersion and damping in the $\Omega$-K plane is clearly demonstrated. The appropiate experimental situation is then imposed afterwards upon the fluctuation spectrum by convoluting it with corresponding resolution functions (in K or $\Omega$ or both) of the experiment under consideration. In the remaining part of the paper, where we discuss only the longitudinal modes of the system free carriers - polar phonons, always the fluctuation spectrum will be used.

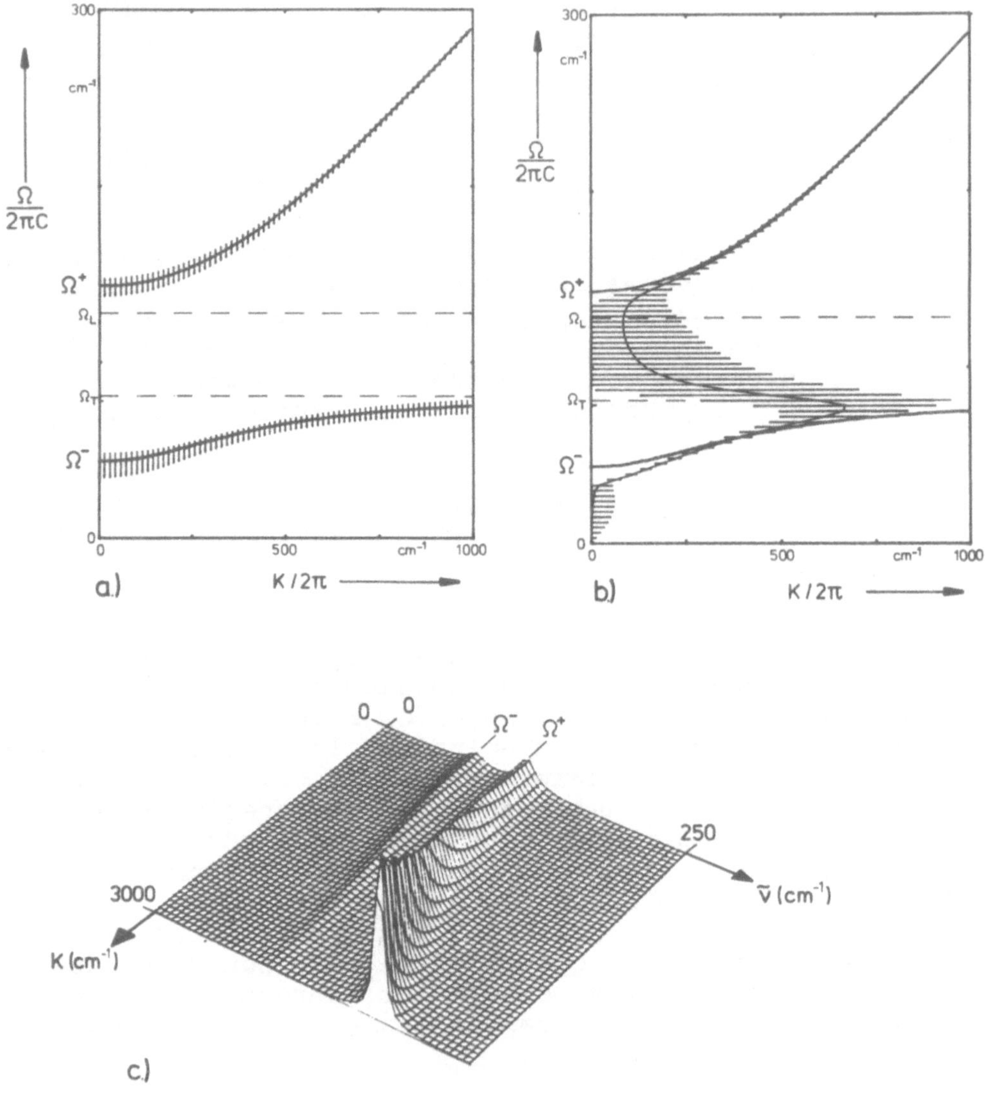


Fig. 2. Different representations of spatial dispersion (example: GaAs, $n=10^{+17}cm^{-3}$, transverse polariton):
(a) K real, Ω complex with halfwidth of imaginary part indicated by bars;
(b) Ω real, K complex with halfwidth of imaginary part indicated by bars;
(c) fluctuations of <EE*>.

## 3. DIELECTRIC FUNCTION

The central quantity for the coupled plasmon-phonon modes is the dielectric function, as seen in the previous chapter. In doing experiments on these modes one measures the fluctuation spectrum in certain regions of the $\Omega$-K space and thus, in principle, one is able to determine the dielectric function. However, the much simpler way to analyze experimental data is to start from theoretical models for the dielectric function and compare the calculated predictions directly with the experiment. It is the purpose of this chapter to illustrate some of the features of these models.

Since the materials in question are semiconductors with polar phonons and a variable concentration of free carriers, we make the following ansatz for the dielectric function:

$$\varepsilon(\Omega,K) = 1 + \chi_{VE} + \chi_{PH}(\Omega) + \chi_{FC}(\Omega,K) \qquad (43)$$

$\chi_{VE}$ is the contribution from all interband transitions and is taken to be constant in the frequency range of the PLP-modes, because the transition frequencies are usually much higher. In this approximation one also abreviates $\varepsilon_\infty=1+\chi_{VE}$.

The phonon contribution $\chi_{PH}$ is excellent described by a harmonic oscillator

$$\chi_{PH}(\Omega) = \frac{\Delta\chi\cdot\Omega_{TO}^2}{\Omega_{TO}^2-\Omega^2-i\Omega\Omega_{\tau p}} , \qquad (44)$$

which gives a frequency,but not wavevector dependent contribution to $\varepsilon$. Equ. (44) has been verified in many IR-experiments and with respect to the K independence by Raman scattering.

The free carrier contribution finally has to provide the K dependence of the dielectric function and is the cause for the spatial dispersion of the PLP-modes.In the following we discuss two different models for the wavevector and frequency dependent free carrier susceptibility.

Two kinds of interactions have to be taken into account for $\chi_{FC}$. The first are coherent interactions within the electron gas and the second incoherent interactions between the electron and its environment. The coherent processes are described in a selfconsistent field approximation by Lindhard's dielectric function /19/. In order to take the incoherent processes into account a modification of the Lindhard expression has been proposed by Mermin /20/. A numerically simpler approach is the so-called hydrodynamical theory, which is treated in section 3.2.

### 3.1 Lindhard (Mermin)

The Lindhard expression is discussed intensively in the literature /21/. It is derived in RPA approximation from the quotient of a selfconsistent potential V, which acts on the electrons. This potential is the sum of an external potential $V_a$ plus the potential $V_i$ induced by $V_a$ in the electron gas:

$$V = V_a + V_i \tag{45}$$

By definition the susceptibility is then

$$\chi = - V_i/V_a \,. \tag{46}$$

The result of the calculation is:

$$\chi_{FC}(\Omega,K) = \lim_{\eta \to 0} \tilde{\chi}_{FC}(\Omega,K) \tag{47}$$

with

$$\tilde{\chi}_{FC}(\Omega,K) = \frac{e^2}{\varepsilon_0 K^2 (2\pi)^3} \int \frac{f(k+K)-f(k)}{E(k+K)-E(k)-\hbar\Omega-i\eta} \, d^3k \,, \tag{48}$$

where f is the occupation function for the energy states E(k). The transitions, over which is integrated, and the notation can be seen in Fig. 3. $\eta$ is at this point only a mathematical parameter which is taken to zero afterwards. Later on it will describe damping of the electron gas to an external system (impurities, acoustical phonons ect.) and will be equated to $\hbar\Omega_{\tau e}$ .

Taking now Fermi statistic for f, assuming isotropic and parabolic energy bands and transforming into spherical coordinates one obtains from (48)

$$\tilde{\chi}_{FC}(\Omega,K) = -2\pi C \int_0^\infty k^2 F^{-1} dk \int_{-1}^{+1} \left(\frac{dx}{\alpha_- + i\eta} - \frac{dx}{\alpha_+ + i\eta}\right) \tag{49}$$

with $F = \exp[\beta(\hbar^2 k^2/2m - E_F) + 1]$, $\beta = 1/k_B \cdot T$

and $\alpha_\mp = \frac{\hbar^2}{2m}(\mp K^2 - 2kKx) + \Omega$

and C is the prefactor in front of the integral in (48)

$$C = e^2/\varepsilon_o K^2 (2\pi)^3 .$$

Splitting finally (49) into real and imaginary part we obtain:

$$\mathrm{Im}[\tilde{\chi}_{FC}(\Omega,K)] = -2\pi C \int_0^\infty k^2 F^{-1} dk \int_{-1}^{+1} \left(\frac{-\eta dx}{\alpha_-^2 + \eta^2} - \frac{-\eta dx}{\alpha_+^2 + \eta^2}\right) \tag{50}$$

and

$$\mathrm{Re}[\tilde{\chi}_{FC}(\Omega,K)] = -2\pi C \int_0^\infty k^2 F^{-1} dk \int_{-1}^{+1} \left(\frac{\alpha_- dx}{\alpha_-^2 + \eta^2} - \frac{\alpha_+ dx}{\alpha_+^2 + \eta^2}\right) . \tag{51}$$

In the case of zero temperature analytical solutions can be easily found for (50), (51), because the occupation function is then given by the step function. The well known expressions are for example quoted in /21/. The imaginary part is exactly confined to $\Omega$,K values within the shaded area of Fig.3. T=O is an excellent approximation whenever $E_F \gg k_B \cdot T$, that is for low temperatures and high carrier concentrations.

For the case T≠O the imaginary part still can be given analytically

$$\mathrm{Im}[\chi_{FC}(\Omega,K)] = \lim_{\eta\to 0} \mathrm{Im}[\tilde{\chi}_{FC}(\Omega,K)] = \frac{me^2 k_B^2}{4\pi\varepsilon_o \hbar^4} \cdot \frac{T}{K} \ln\left(\frac{a + e^{-\gamma_-^2}}{a + e^{-\gamma_+^2}}\right) \tag{52}$$

with $a = e^{-\beta E_F}$ and $\gamma_\pm^2 = \frac{\beta\hbar^2}{2m}\left(\frac{m\Omega}{\hbar K} \pm \frac{K}{2}\right)^2$.

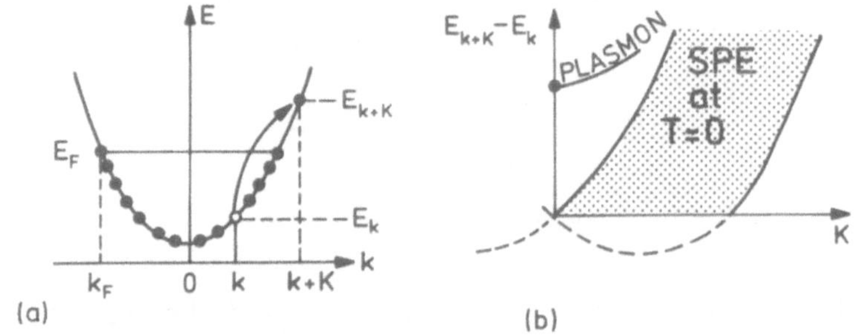


Fig. 3. Single particle excitations at T=0.
(a) transitions; (b) region in Ω-K space were transitions are allowed.

The imaginary part now is no longer confined to the shaded area of Fig.3, but extends into the neighbouring Ω,K regions. For the real part of $\chi_{FC}(\Omega,K)$ no analytical solutions can be given. It has to be obtained either by a Kramers-Kronig transformation from (52) or by direct numerical integration of (51).

In the above expression for $\chi_{FC}(\Omega,K)$ no coupling to external excitations is considered. As a consequence the damping (linewidth) of the plasmon will be entirely caused by the electron gas itself. For the description of the experimental situation, however, damping processes like impurity scattering have to be taken into account. Thus in order to compare later on with experiment, it turns out to be necessary to introduce at least a phenomenological damping constant $\tau_e=1/\Omega_{\tau e}$ for the free carrier system. The simplest way to do so, would be to identify the parameter $\eta$ in (48) with $\hbar\Omega_{\tau e}$

$$\eta = \hbar\Omega_{\tau_e} \tag{53}$$

and to avoid the limes in (47). However, in order to conserve particle number, it was shown by Mermin /20/, that the susceptibility should have the following form:

$$\chi_{FC}(\Omega,K) = \frac{(1+i\Omega_{\tau e}/\Omega)\cdot\tilde{\chi}_{FC}(\Omega,K)}{1+(i\Omega_{\tau e}/\Omega)\cdot\tilde{\chi}_{FC}(\Omega,K)/\tilde{\chi}_{FC}(0,K)} , \tag{54}$$

where for $\tilde{\chi}_{FC}(\Omega,K)$ the appropriate Lindhard expression for T=O or T≠O have to be inserted and the whole expression is then splitted into its real and imaginary part. The result is of course a considerable numerical expenditure, since in general three integrations have to be performed.

Results calculated for the dielectric function with the Lindhard expression (50), (51) and the Lindhard-Mermin extension (54) are shown and discussed in section 3.3.

## 3.2 Hydrodynamic Theory

The hydrodynamic theory /22/ starts from the one particle density matrix $\rho(\underline{r}_1,\underline{r}_2,t)$ for free electrons. The equation of motion reads /10,23/:

$$\dot{\rho} + \frac{i\hbar}{2m}(\Delta_1-\Delta_2)\rho = \frac{1}{\hbar}[V(\underline{r}_1) - V(\underline{r}_2)]\rho \tag{55}$$

From this equation a hierarchy of balance equations can be deduced:

$$e\dot{n} + \mathrm{div}\underline{J} = 0 \tag{56}$$

$$\underline{\dot{J}} + \mathrm{Div}\,\underline{\underline{\Sigma}} = -\frac{e}{m}n\nabla V \tag{57}$$

with $n(\underline{r})=\rho(\underline{r},\underline{r})$; $J(\underline{r}) = \frac{ie\hbar}{2m}(\nabla_1-\nabla_2)\rho|_{\underline{r}_1=\underline{r}_2}$; (58)

and $\underline{\underline{\Sigma}} = \frac{e\hbar^2}{4m^2}(\nabla_1-\nabla_2)(\nabla_2-\nabla_1)\rho|_{\underline{r}_1=\underline{r}_2=\underline{r}}$ (59)

Instead of proceeding in the hierarchy with an equation for the tensor $\underline{\underline{\Sigma}}$ we terminate with (57) by the following approximations: (i) $\underline{\underline{\Sigma}}$ is given the value that emerges from a linearized solution of (56), (ii) $n\nabla V$ is linearized as $n\nabla V=n_0\nabla V+n\underline{F}$, where $\underline{F}$ is an external stochastic field which is the source of incoherent damping. The term $n\underline{F}$ may be replaced by a relaxation term of the form $\Omega_\tau\underline{J}$ /24/. Equation (57) then takes the "hydrodynamic" form

$$\underline{\dot{J}} + \Omega_{\tau e}\underline{J} + \mathrm{Div}\underline{\underline{\Sigma}} = -\frac{e}{m}n_0\nabla V\ . \tag{60}$$

which is a generalized form of the Drude equation

$$\underline{\dot{J}} + \Omega_{\tau e}\underline{J} = \varepsilon_0\Omega_p^2\underline{E} \tag{61}$$

The rather complicated term Div $\underline{\underline{\Sigma}}$ in (60) is a linear operator applied to n(r). It takes into account coherent processes within the electron gas. Equation (61) leads to a susceptibility of the form

$$\chi_{FC}(\Omega,K) = \frac{\Omega_p^2}{D(K,\Omega)-\Omega^2-i\Omega\Omega_{\tau e}} \tag{62}$$

with

$$D(\Omega,K) = \frac{\hbar^2}{4m^2} \frac{\displaystyle\int \frac{[f(\underline{k}+\underline{K})-f(\underline{k})](2\underline{k}\underline{K}+K^2)^2 d\underline{k}}{2\underline{k}\underline{K}+K^2-2m\Omega/\hbar}}{\displaystyle\int \frac{[f(\underline{k}'+\underline{K})-f(\underline{k}')]d\underline{k}'}{2\underline{k}'\underline{K}+K^2-2m\Omega/\hbar}} \;. \tag{63}$$

In (62), (63) the same approximations as in the Lindhard dielectric function are used. However, in contrast to the latter the coherent interactions (D) and the incoherent interactions ($\Omega_{\tau e}$) appear in different terms and thus are analytically decoupled. This gives the possibility to recast (63) in a much simpler form for certain approximations. In addition one can, at least principally, apply a more detailed microscopic treatment in order to describe the incoherent interaction with external systems ($\Omega_\tau$).

For K=0 and D(Ω,K)=0 the classical Drude result

$$\chi_{FC}(\Omega,0) = \frac{\Omega_p^2}{-\Omega^2-i\Omega\Omega_{\tau e}} \tag{64}$$

emerges from (62). In other limiting cases the processes effective in D(Ω,K) can be identified as diffusion or as so-called Landau-damping due to single particle decay. For the conditions of optical interactions $\Omega \gg Kv_F$ and $K \ll k_f$ one finds:

$$D(\Omega,K) = \frac{3}{5}\frac{\hbar^2}{m^2} \int \frac{K^2 d^3K}{\cosh^2 \frac{E-E_F}{2k_BT}} \Big/ \int \frac{d^3K}{\cosh^2 \frac{E-E_F}{2k_BT}} \tag{65}$$

For the degenerate case $E_F \gg k_B \cdot T$ this reduces to

$$D(\Omega,K) = \frac{3}{5} v_F^2 K^2 \quad . \tag{66}$$

It should be noted that by adding a diffusion type term $D_0 \mathrm{grad}_\rho$ to the Drude equation, (61), the result (62) with (66) is easily obtained. However, the prefactor $3v_F^2/5$ in (66) has to be obtained by comparison with the T=O Lindhard expression /11,25/. We also point out,that (66) does not imply a quadratic plasmon dispersion, since it appears in the denominator of (62).

### 3.3 Comparison of the Different Models for $\chi_{FC}$

In order to compare the different $\chi_{FC}(\Omega,K)$ we give in the following figures plots of Imε, Reε and Im 1/ε obtained from Lindhard, (52), Lindhard-Mermin, (54) and the hydrodynamic theory, (62) with (65). The contributions from valence electrons and phonons where added according to (43,44) and numerical parameters appropiate for GaAs were chosen.

From a first glance at Figs. 4...7 the results of all models look quite similar. But qualitative and quantitative differences will show up during the course of this paper. Starting with Imε (Fig.4), we recognize first of all the K-independent phonon contribution which proceeds parallel to the K-axis. The free carrier term, $\chi_{FC}$, is largest at Ω=K=O and decreases with increasing K and Ω.A general broadening of the free carrier contribution is observed with increasing damping (hydrodynamic theory $\tau_1$→hydrodynamic theory $\tau_2$ and Lindhard→Lindhard-Mermin). For equal τ, however, the hydrodynamic $\chi_{FC}$ shows much sharper structure than the Lindhard based $\chi_{FC}$. Similar features are observed for the Reε (Fig.5). In Figs. 6 and 7, finally, Im -1/ε is plotted as 3-dimensional plot (Fig.6) and with lines of equal height in Im -1/ε (Fig. 7). Differences between the different models occur again with respect to the sharpness of the structures shown. The general features are the following. At small K two quite pronounced fluctuation maxima are observed, which correspond to the two coupled PLP-modes $\Omega^-,\Omega^+$ (in the literature also often termed $L_-,L_+$), whose frequencies at K=O, in excellent agreement with experiment, are given by:

$$\Omega_\pm^2 = \frac{\Omega_{LO}^2+\Omega_p^{*2}}{2} \pm \sqrt{\frac{(\Omega_{LO}^2+\Omega_p^{*2})^2}{4} - \Omega_p^{*2}\,\Omega_{TO}^2} \tag{67}$$

with $\Omega_p^* = \Omega_p/\varepsilon_\infty{}^{1/2}$

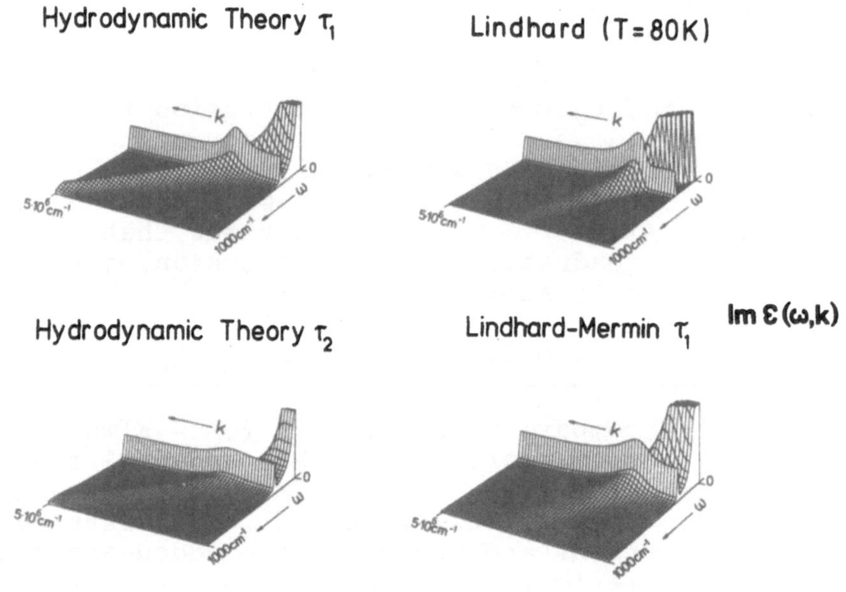


Fig. 4. Imaginary part of the dielectric function, $\text{Im}\,\varepsilon(\Omega,K)$, as a function of $\Omega$ and K. Left side hydrodynamic theory with two different relaxation times ($\tau_2<\tau_1$). Right side Lindhard theory($T\neq0,\tau=\infty$) and Lindhard-Mermin($T\neq0,\tau_1$). GaAs parameters was used.

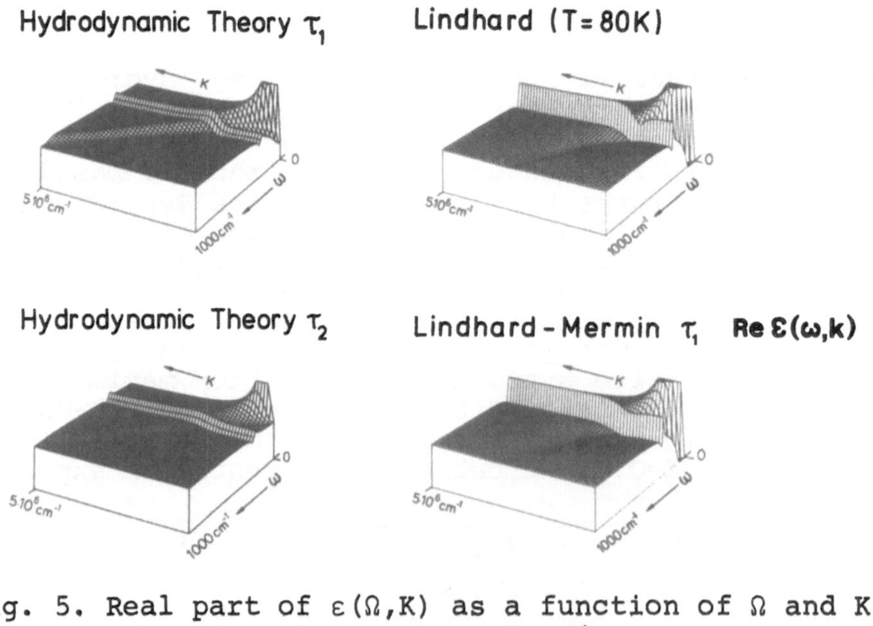


Fig. 5. Real part of $\varepsilon(\Omega,K)$ as a function of $\Omega$ and K. Otherwise the same as Fig. 4.

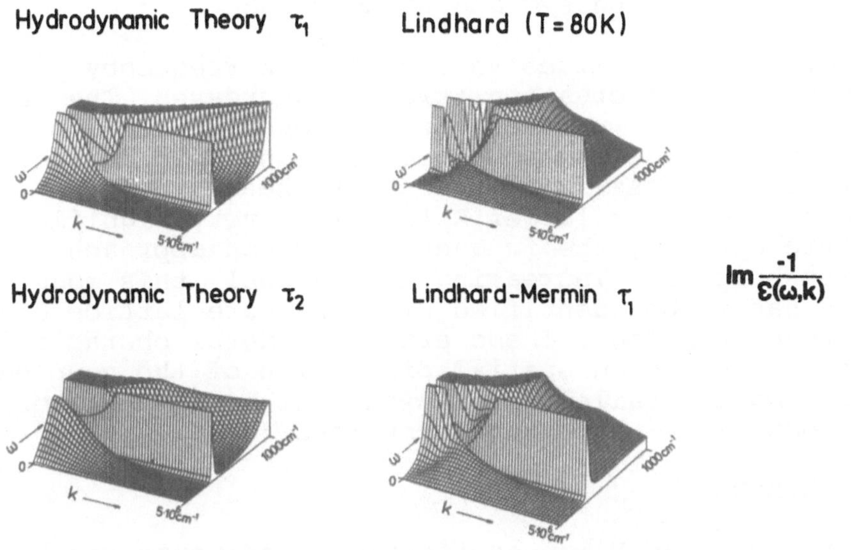


Fig. 6.: Im $-1/\varepsilon(\Omega,K)$ as a function of $\Omega$ and K. Otherwise the same as Fig. 4.

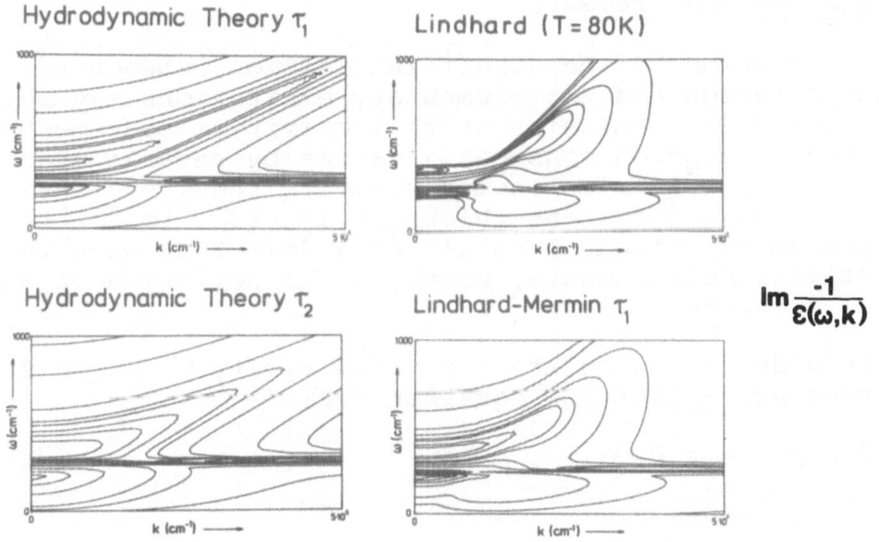


Fig. 7.: Same as Fig. 6, but plotted two dimensional with lines of equal height i.e. Im $-1/\varepsilon(\Omega,K)$=constant.

a result, which directly can be obtained from (43) with (44), (64) by taking $\varepsilon=0$ and using $\Omega_{\tau e}=0$.

For larger K values an increase in frequency is observed and the fluctuation strength decreases. The reason is that the single particle excitation (SPE) region is approached and the modes become more and more damped. In the SPE region itself, the modes are heavily overdamped and smeared out. At larger K finally a very significant structure appears, the frequency of which approaches $\Omega_{LO}$ from below for increasing K. Obviously this mode for large K has to be identified with the pure lattice dynamical LO-phonon mode. These are the general characteristics of the expected spatial dispersion of the coupled PLP-modes. A more detailed discussion will be given later after comparison with experiment (chapter 5).

## 4. EXPERIMENT

Infrared and Raman scattering experiments can be performed in order to determine the spatial dispersion. In infrared experiments essentially the mode frequencies at K=0 are determined, while in Raman scattering the scattering wave vector and thus K can be varied up to approximately a few percent of the Brillouin zone.

### 4.1 Infrared Measurements

Information on the direct interaction between infrared radiation and the elementary excitations may be obtained from transmission or reflectivity measurements. Since the absorption constant $\alpha$ varies by several orders of magnitude, transmission measurements require a series of samples with different thickness in order to fulfill the optimum experimental condition $\alpha \cdot d=1$. Thus usually reflectivity measurements, which can be performed on one sample, are made.

In order to calculate the reflectivity we start from the transverse polariton equation (23)

$$K^2c^2/\Omega^2 = \varepsilon(\Omega,0) \tag{23}$$

Since

$$K^2c^2/\Omega^2 = K^2/K_o^2 = \tilde{n}^2 \tag{68}$$

($K_o$=vacuum wave vector, $\tilde{n}$ = complex refractive index: $\tilde{n} = n + i\kappa$) we obtain from (23)

$$\tilde{n}^2 = (n+i\kappa)^2 = \varepsilon(\Omega,0) \tag{69}$$

The reflectivity for a semiinfinite geometry is then

$$R = \frac{(n-1)^2 + \kappa^2}{(n+1)^2 + \kappa^2} \ . \tag{70}$$

For other geometries (e.g. layers) multiple reflections and interference effects may have to be taken additionally into account. For a given model for $\varepsilon(\Omega,0)$ and a given $\Omega$ it is thus possible to calculate R with the help of (69) and (70) and to adjust it to the experimental data. The dielectric function used in the fit procedure then gives via the maxima of (37) the PLP-mode frequencies. A different possibility is to extract the dielectric function from R directly via a Kramers-Kronig transformation /26/ and apply (37). Since the first approach is usually more convenient, the latter is only taken, whenever it is not possible to obtain a good fit to the experimental data.

Fig. 8 gives two examples for reflectivity measurement and fit. The spectra were taken with a Fourier transform spectrometer, the usual type of instrument for far infrared measurements /27/. It turns out to be neccessary to take data over a wide spectral range in order to determine $\chi_{VE}$ in (43) very accurate, since this sets the accuracy for the determination of $\Omega_{\pm}$. The mode frequencies, obtained by fitting with equs. (43), (44) and (64), are indicated in the figures.

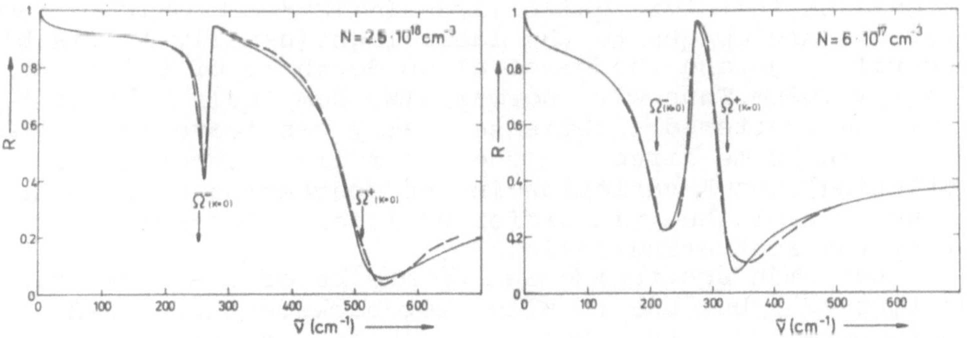


Fig. 8. Infrared reflectivity spectra (solid line)of two GaAs samples with different electron concentrations. Dashed line: fit with equs. (43), (44) and (64).

## 4.2 Raman Scattering

Like in any scattering experiment additional degrees of experimental freedom, as compared to an infrared measurement, are opened, because now two waves are involved in the interaction. Energy and wave vector conservation require

$$\hbar\omega_i - \hbar\omega_s = \hbar\Omega \tag{71}$$

$$\hbar\underline{k}_i - \hbar\underline{k}_s = \hbar\underline{K} \tag{72}$$

for the creation of an elementary excitation (Stokes process) in the scattering process. Since

$$|\underline{k}_{i,s}| = n_{i,s}\Omega/c \tag{73}$$

we obtain from (71), (72) for the transfered scattering vector K:

$$|K| = (\omega_i^2 \cdot n_i^2 + \omega_s^2\, n_s^2 - 2\omega_i\omega_s n_i n_s \cos\varphi)^{1/2}/c \tag{74}$$

This function is plotted in Fig. 9 with values appropiate for GaAs. Other diamond or zincblende-type semiconductors give similar plots. For a given scattering experiment, i.e. laser frequency and scattering angle fixed, the analyzing monochromator scans along such a line in the $\Omega$-K-plane. It should be noted, however, that scattering angles other than 180° (backscattering) are hard to realize in most semiconductors,since,because of their small energy gaps the absorption constants are due to interband transitions and therefore quite large ($\alpha=10^{+4}...10^{+5}cm^{-1}$): the materials are opaque to the laser light(usually in visible spectral range)and the penetration depth is of the order of a few 100nm.Thus with normal size samples(thickness $\simeq$ 0.5mm)the scattered light essentially can leave the sample only in the same direction the laser light entered(back scattering).The K-variation in the scattering experiment is thus established via different laser frequencies and not by the scattering angle.

The Raman scattering set up may be of the conventional type /28/,but has to allow for backscattering and low temperatures. As compared to phonon scattering,the plasmon scattering intensities are usually much lower (approximately 0.1 counts per second and mW laser power).Thus excellent stray light rejection must be obtained and a photomultiplier with a small number of darkcounts used.

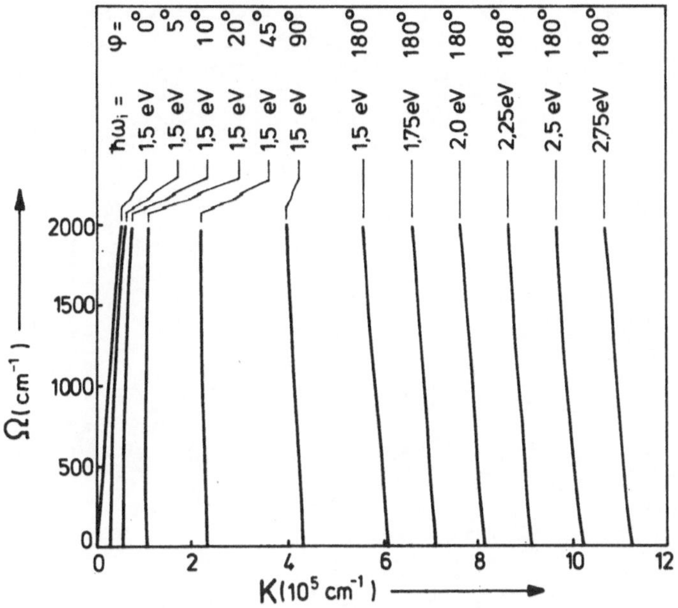


Fig. 9. Scan curves in a light scattering experiment for different scattering angles and photon energies. Calculated with (74) and GaAs parameters.

The differential Raman cross section for scattering by charge density fluctuations may be written as /18/

$$\frac{d^2\sigma}{d\theta d\Omega} = C \cdot K^2 \cdot \{1+n(\Omega)\} \cdot \mathrm{Im}\, \frac{-1}{\varepsilon(\Omega,K)} \qquad (75)$$

There θ is the solid angle over which the scattered light is gathered and C contains the scattering matrix element, which in general will also depend on exciting frequency $\omega_i$ (resonant Raman scattering). The last term in (75) reveals that the scattering experiment determines in a very direct way the fluctuation spectrum.

A typical set of Raman spectra obtained with different laser lines on the same sample is shown in Fig.10. The observed scattering peaks for the two coupled modes shift with excitation frequency $\omega_i$. Since $\omega_i$ determines K, equ. (74), this plot gives direct evidence for the spatial dispersion of the coupled plasmon-LO phonon modes. Additional peaks seen, are at the frequency positions of the TO- and LO-phonon. The TO-phonon does not couple into the carrier system. It appears only weakly because it is not allowed by the K=0 selection rules /29/, applied to the (100)-surfaces used here. The

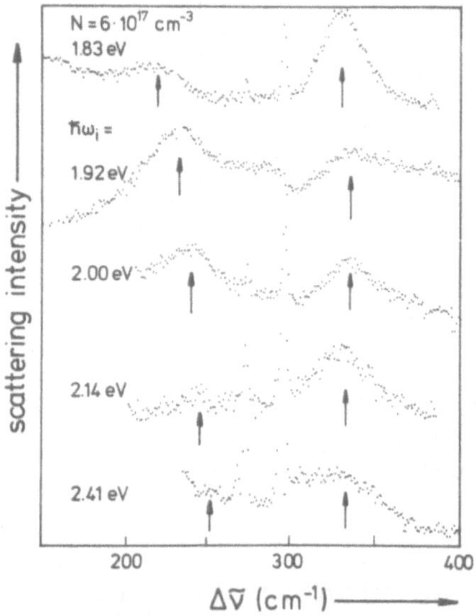


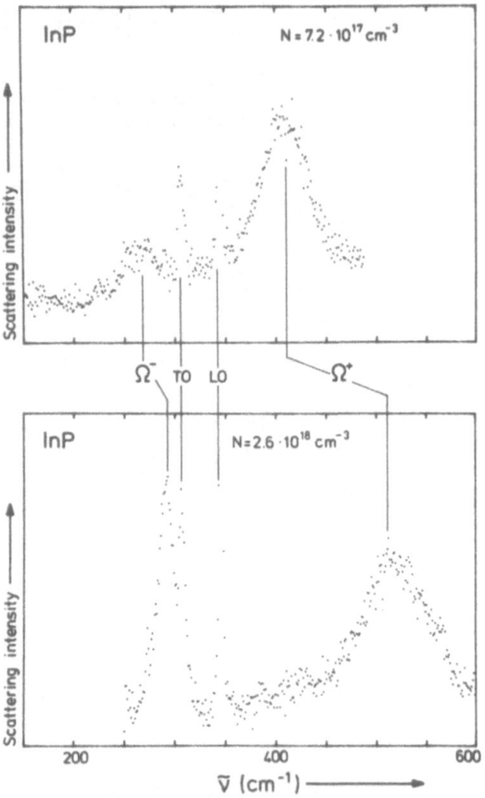


Fig. 10. (Left) Raman spectra of PLP-modes in GaAs for different laser photon energies. Arrows indicate positions of PLP-modes.

Fig. 11. (Right) Raman spectra of PLP-modes in InP for two different carrier concentrations ($\hbar\omega_i$=2.18eV).

LO-phonon scattering peak, which is unexpected, originates from a carrier-free depletion layer at the surface (10...30nm) and is in addition allowed by the selection rules. Another possibility for a mechanism contributing to this peak will be discussed later. The change of the PLP-mode spectra with carrier concentration is demonstrated in Fig. 11 with the example of InP. With increasing carrier concentration the PLP-mode frequencies increase. While the $\Omega^-$-mode moves close to the TO-phonon and is now more phonon-like with narrow linewidth, the $\Omega^+$-mode moves away from the phonons and becomes more plasmon-like /30/. At still higher carrier concentrations the $\Omega^-$-mode corresponds then to a longitudinal motion of the atoms where the additional Coulomb force is screened by the free carriers and thus the eigenfrequency is $\Omega_{TO}$. The $\Omega^+$ represents then a pure plasmon motion.

More experimental results will be shown and discussed in the next chapter, where we compare them with calculations.

## 5. ANALYSIS AND DISCUSSION

In the previous section it was shown already, that the IR-reflectivity data can be excellently described by $\varepsilon(\Omega,0)$. This is independent of the models used here for $\chi_{FC}(\Omega,K)$, since in the limes $K\to 0$ in every case the Drude result is reproduced. Thus, it remains to test $\chi_{FC}$ with the Raman data.

### 5.1 Raman spectra

In Figs. 12-14 we give a couple of examples for the comparison of scattering spectra with calculations from (75) and (43) using the different models for $\chi_{FC}$. Usually there remains only a scaling factor as fit parameter. Absolute cross section measurements are not performed and, on the other hand, are also complex to calculate. All other free carrier parameters ($\tau$,n,m) are at least roughly known from other experiments (Hall effect, IR-reflectivity) and allow therefore little variation for adjustment of the calculations to the measured spectra. However, a final adjustment can be used to obtain more accurate values for the parameters in question.

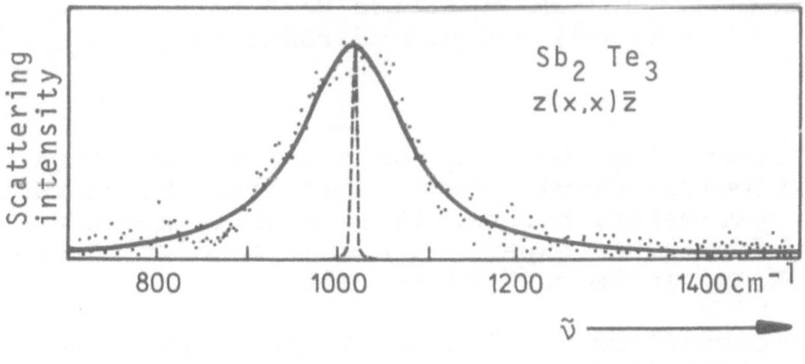


Fig. 12. PLP-mode Raman spectrum of $Sb_2Te_3$. Theoretical fits: (a) hydrodynamical theory, equ. (75), (43), (62), (65) and $\Omega_{\tau e}$=110cm$^{-1}$ (solid line); (b) Lindhard, equs. (75), (43), (52) (dashed line).

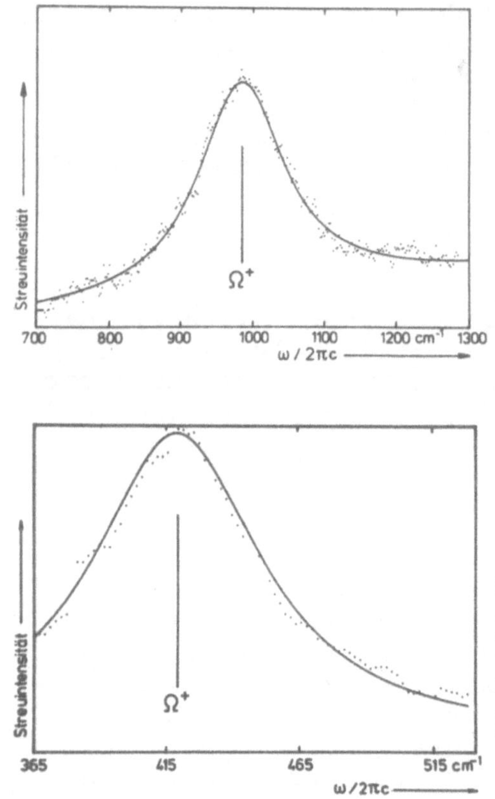


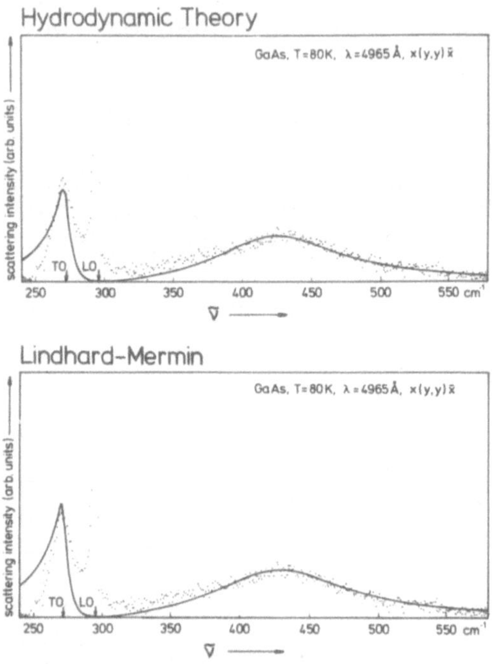


Fig. 13. (Left) PLP-mode($\Omega^+$) Raman spectrum of InP fitted with the hydrodynamic theory,equs.(75),(43),(62),(65) and $\Omega_{\tau e}$=160 and 150 cm$^{-1}$.

Fig. 14. (Right) PLP-mode Raman spectrum of GaAs ($n=1.3\ 10^{18}$cm$^{-3}$) fitted with Lindhard-Mermin ($\Omega_{\tau e}$=90cm$^{-1}$) and hydrodynamic theory($\Omega_{\tau e}$=130cm$^{-1}$).

This is especially the case for n and m. In contrast, the phenomenological damping term $\tau$ has more the character of a fit parameter, because it is usually known with less accuracy from other experiments and in addition depends also on K as can be seen in Fig.15.

The comparison in Figs. 12-14 shows that the models for $\chi_{FC}$, which include a damping term, are quite able to describe the data, in contrast to the Lindhard $\chi_{FC}(\tau=\infty)$ (Fig.12). The latter yields a linewidth much smaller than the experiment. The reason is that at the carrier concentration discussed here, carrier scattering by ionized impurities or intrinsic defects is the dominant free carrier scattering mechanism and determines the plasmon lifetime. Such a process of course cannot be described by the Lindhard susceptibility (47),(48) which includes only interactions within the electron gas. It should be pointed out, however, that the values obtained for $\tau$ from the hydrodynamic $\chi_{FC}$ and the Lindhard-Mermin $\chi_{FC}$ are quite different (Fig.14). This obviously comes from the fact that the hydrodynamic theory, in the simple diffusion approximation used here, has to compensate for the other carrier mechanism (Landau damping and higher order gradients) by a larger $\tau$. Of course, a better approximation than (65) or (66) may be taken for the hydrodynamic theory.

The small linewidths obtained from Lindhard $\chi_{FC}$ can be seen also in Fig. 6, which shows very narrow (in frequency) PLP-mode fluctuations outside the SPE-region at small K. Inside the SPE region, however, the results are similar as in the other three calculations of Fig. 6. Our scattering measurements, on the other hand, hardly cross the SPE boarder. Scattering experiments at larger K are hampered by the non-availability of appropiate laser lines (UV) and the low penetration depth at these laser frequencies, which is of the order of the depletion layer thickness. Therefore an experimental test deep inside the SPE region is not possible at present via light scattering. Neutron scattering also offers no alternative, since, even at high flux reactors, the intensity limited K resolution is much less than required for such a verification.

## 5.2 Dispersion

A set of experimental dispersion curves is shown in Fig. 15 for GaAs with four different carrier concentrations. The frequency and wavevector position of the scattering peaks is given together with their linewidths. A comparison between measured dispersion and calculated dispersion is given in Fig. 16. Despite the large differences in linewidth, all models give essentially the same spatial dispersion. The differences lie within the accuracy of measurement. The set of dispersion curves of Fig. 15 is shown in Fig. 17, compared with the calculated dispersion from the hydrodynamic theory. No fit parameter is involved in these calculations and also those for Fig. 16. Thus, the agreement has to be considered as excellent. Differences between experiment and calculation occur for the lower concentration samples (Fig. 17). For the $\Omega^-$-mode this might be due to a larger uncertainty in the measurement (low scattering intensities, because of a large depletion

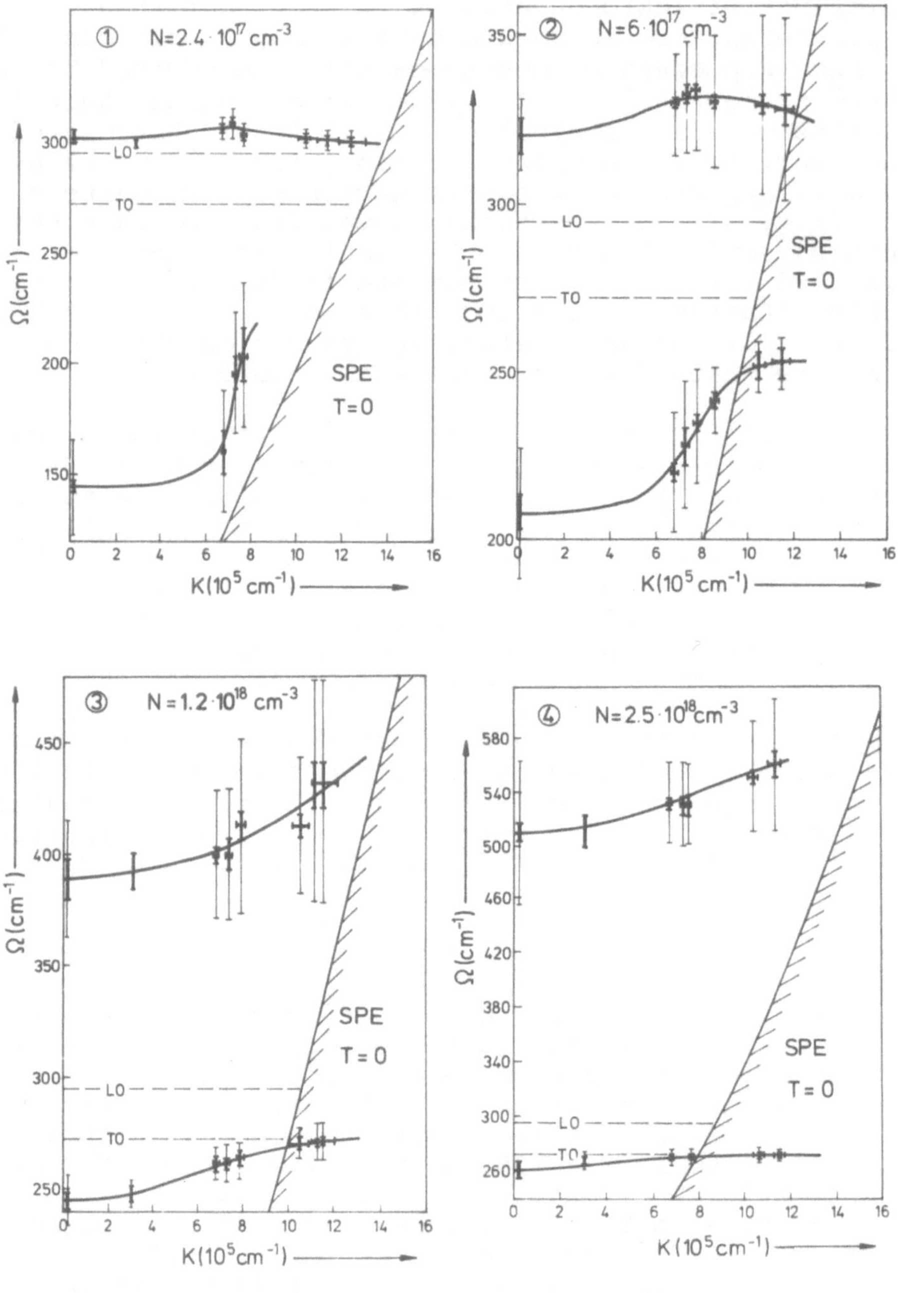


Fig.15. Experimental spatial dispersion of PLP-modes for four GaAs samples with different carrier concentrations at T=80K. Solid line: guide line for the eye. Thick solid bars: uncertainties in Ω and K. Thin solid lines: halfwidth of scattering peaks

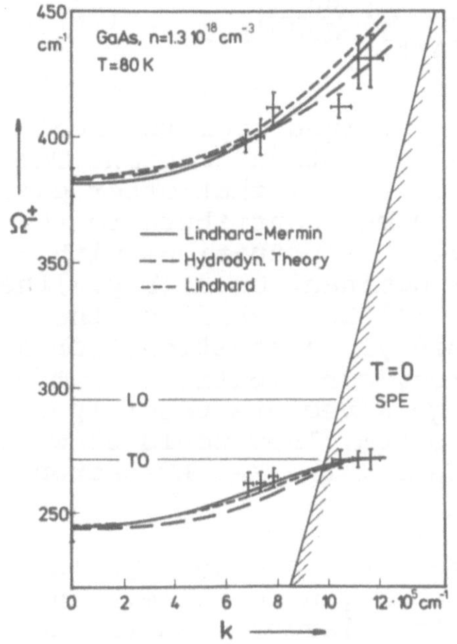


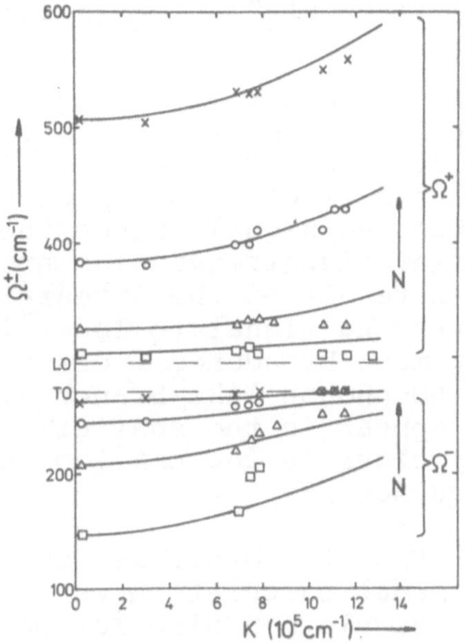


Fig. 16. (Left) Dispersion of PLP-modes for one carrier concentration in GaAs. Crosses: measurement; lines: different model calculations.

Fig. 17. (Right) Concentration dependence of PLP-mode dispersion in GaAs. N(in $10^{+17}$cm$^{-3}$)=2.4 (□), 6.0 (Δ), 12.0 (O), 25.0 (x). Solid lines: hydrodynamic theory.

layer). For the $\Omega^+$-mode, on the other hand, where the experimental points bend off at larger K-vectors, the discrepancy can be partly explained by the lower K-resolution one has at these wavevectors. This comes about, because the samples show stronger absorption at the laser frequencies, neccessary for producing these K vectors (Fig.9). The refractive index $\tilde{n}$ then has a larger imaginary part and the wavevector transfer in the scattering experiment is less well defined. Such a smearing out of K-vectors can be taken into account by integrating the cross section over a Lorentzian distribution of K-vectors /11,31,32/:

$$\frac{d\sigma}{d\theta d\Omega}\Big|_{\Omega,<K>} = \frac{\pi}{\alpha}\int_{-\infty}^{+\infty}\frac{dK}{(K-<K>)^2+(\Delta K)^2}\cdot\frac{d\sigma^2}{d\theta d\Omega}\Big|_{K,\Omega} \qquad (76)$$

where $<K>$ denotes the wavevector obtained from the real part of $\tilde{n}$ from (74) and $\alpha$ is the absorption constant. The effect of a larger range of K vectors is, that other parts from the fluctuation spectrum also may contribute to the scattering signal. Since the small K fluctuations with smaller eigenfrequencies are the dominant ones, (Fig.6), the final result of the integration will be a shift of the scattering signal to lower frequency. In addition, for a very much broadened K distribution, contributions from the LO-phonon like branch at large K vectors possibly may appear in the scattering spectrum. They would show up very close (a few $cm^{-1}$) to, but below, the K=0 LO-phonon frequency /35/.

From the otherwise good agreement between experiment and model calculation we feel encouraged to believe in the calculation also for wavevectors larger than the measured ones. Thus, the following picture for the PLP-mode spectrum and spatial dispersion is obtained (Fig.18): first of all, the modes get heavily damped when entering the SPE-region. For very large K-vectors at the right side of the SPE-range a phonon like-mode appears, which approaches the LO-phonon frequency from below and is then finally the pure lattice dynamical LO-phonon. Between this and the $\Omega^-$mode at small K-vectors there is a deep minimum in fluctuation strength, and it seems not to be realistic to connect them to a single dispersion curve/33,34/. Moreover, in Fig. 18b we see that the $\Omega^-$-mode is essentially repelled from the higher frequency range if one follows the maxima in $Im-1/\varepsilon$.

Within the SPE-region it seems difficult to define mode frequencies via the fluctuation maxima, since different choices, because of the weak structures, are possible. The dashed line in Fig.18b gives the results of one of the earlier calculations of Devreese et al. /35, 36/ referring to this subject. It is obtained by using the Lindhard (T=0) free carrier susceptibility and taking $Re(\varepsilon)=0$. Because of the dominating imaginary part, however, the corresponding maxima in $Im-1/\varepsilon$ are not very pronounced within the SPE-range. Thus, in concluding this discussion, one can state, that the term coupled plasmon-LO-phonon modes should not be stressed too much within the single particle excitation region. Finally, we should also point out, that some of the features discussed, have been calculated already by Lemmens et al. /35,36/.

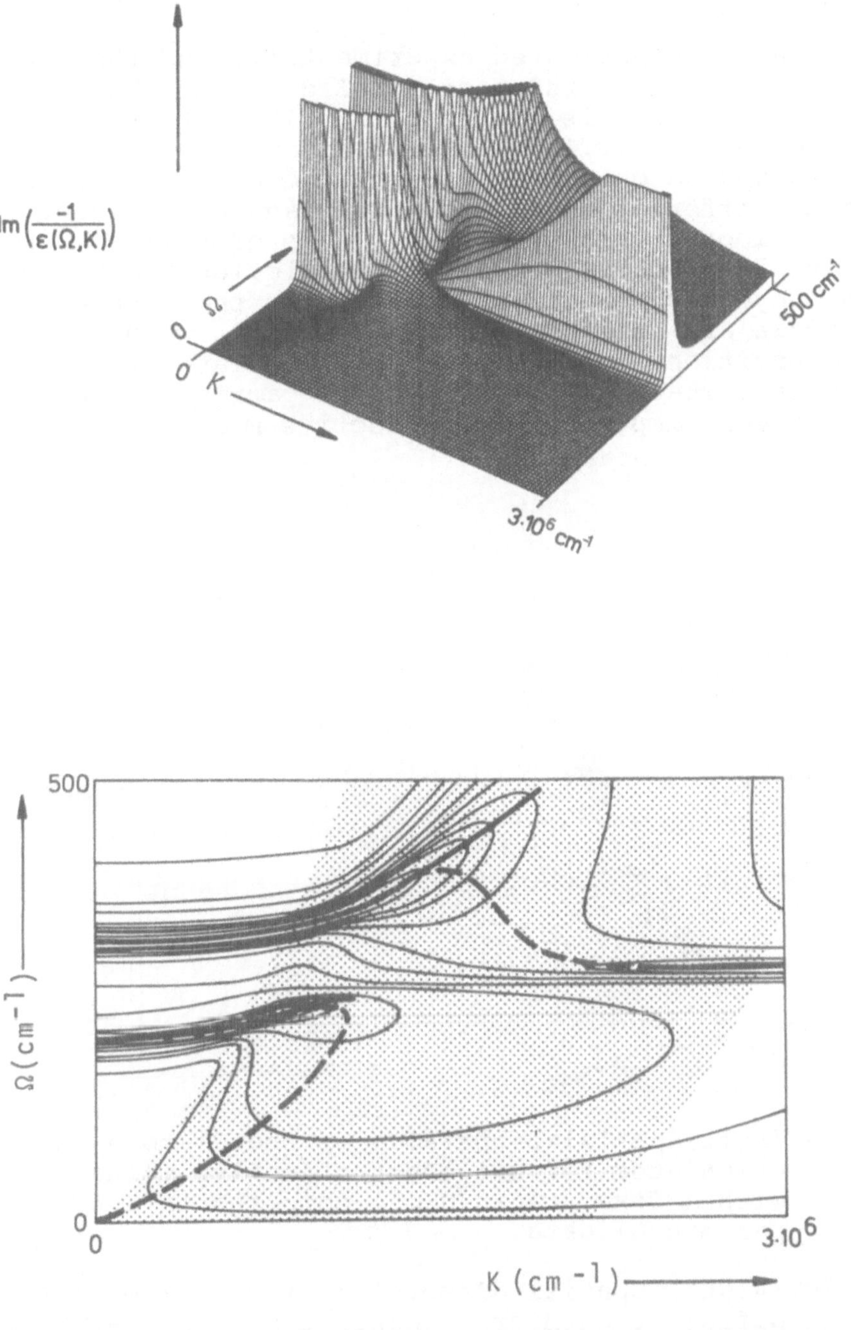


Fig.18. PLP-mode fluctuation spectrum, Im$-1/\varepsilon(\Omega,K)$, calculated with GaAs parameter, $N=5\cdot10^{-17}cm^{-3}$ and equs.(43), (58) for T=80K. (a) 3-dimensional plot; (b) lines of equal height in Im$-1/\varepsilon(\Omega,K)$. Thick solid line: pronounced maxima in Im$-1/\varepsilon$. Dashed line: calculation with Re$\varepsilon(\Omega,K)=0$. Shades area: SPE region for T=0.

## 6. CONCLUSIONS

We have investigated experimentally and theoretically the spatial dispersion of coupled plasmon-LO-phonon modes. The problem is stated in Fig.1 to which the answer is given in Fig.18. The comparison between experimental data and calculations gave us the possibility to judge on different models for the Ω and K dependent free carrier electric susceptibility. This quantity is important in all problems where the restriction K=0 has not to be obeyed (screening,transport). In those cases, where the models are able to describe the data, we are also now in the position, to determine very accurate free carrier parameters. That is, for example, the carrier concentration, but even more important the effective mass.

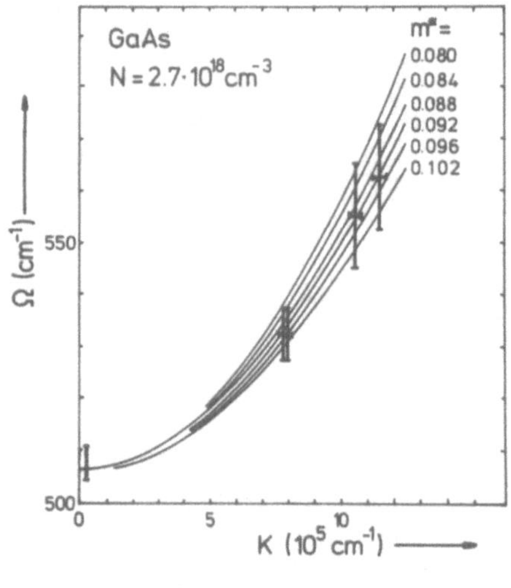


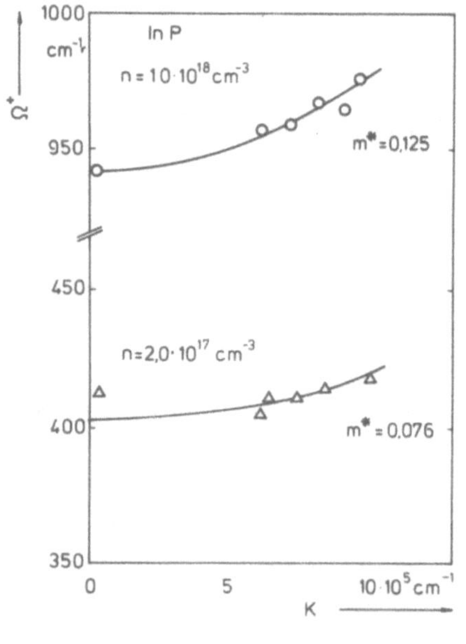


Fig. 19. (Left) Spatial dispersion of the $\Omega^+$-mode in GaAs calculated for different effective masses $m^*$, but with a fixed ratio $N/m^*$, compared to a set of experimental data.

Fig. 20. (Right) Spatial dispersion of the $\Omega^+$-mode for two different carrier concentrations in InP. The effective masses, needed to describe the slope, are indicated.

Fig. 19 gives a demonstration of the sensitivity, obtained in adjusting the dispersion. Two results are given for InP in Fig.20, where the effective mass variation with carrier concentration (non-parabolic bands) can be seen. Finally, in Fig. 21, we give experimental results on the free carrier relaxation time obtained via the PLP-mode linewidth. These results may be directly reinterpreted into a mobility. This is, of course, not a DC mobility, but a mobility measured at the frequencies $\Omega=\Omega^-$ or $\Omega=\Omega^+$. The temperature independence, as seen in Fig. 21, follows from the fact that $\hbar\Omega^+ \gg k_B \cdot T$ in this case.

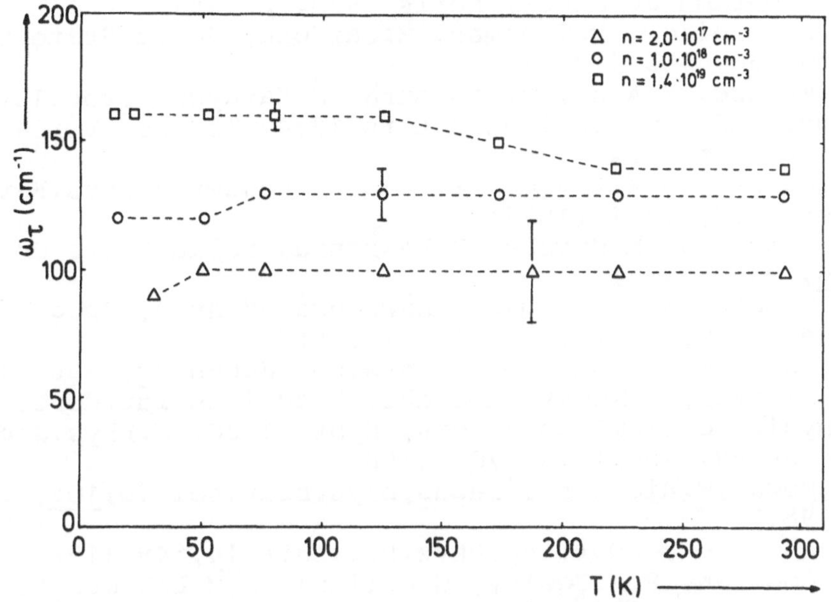


Fig.21. Relaxation frequency $\Omega_{\tau e}$ (here denoted $\omega_\tau$) as a function of temperature. Obtained from a fit with (43), (54) and (75) to the $\Omega^+$-mode linewidth in three differently doped samples.

Many other examples could be given. They all gain additional advantage from the fact that only contactless optical measurements are involved, and, because of laser illumination, very small volumina can be probed.

ACKNOWLEDGEMENTS

I am grateful to R.Geerdt and U.Nowak for providing some of their unpublished data (Figs.11,13,20,21).

Thanks are also due to B. Harbecke for several discussions concerning the dissipation-fluctuation theorem. Finally apparative and financial support by the Deutsche Forschungsgemeinschaft is acknowledged.

REFERENCES

/ 1/ B.B. Varga, Phys.Rev. A137, 1896 (1965)
/ 2/ A.Pinczuk, G.Abstreiter, R.Trommer, M.Cardona, Solid State Commun. 21, 959 (1977)
/ 3/ G.Abstreiter, A.Pinczuk, R.Trommer, M.Cardona in: Proc.Int.Conf. Lattice Dynamics, ed.by M.Balkanski, Flammarion Sciences, Paris 1978, p. 191
/ 4/ G.Abstreiter, R.Trommer, M.Cardona, Solid State Commun. 30, 703 (1979)
/ 5/ K.Murase, Y.Ando, H.Kawamura, S.Katayama,Proc.12th Int.Conf.Phys.Semiconductors 1974, Teubner Verlag Stuttgart, p.458
/ 6/ K.Murase, S.Katayama, Y.Ando, H.Kawamura,Phys.Rev. Lett. 33, 1481 (1974)
/ 7/ S.Katayama, M.Murase, H.Kawamura, Solid State Comm. 16, 945 (1975)
/ 8/ K.Murase, S.Katayama, H.Kawamura, Y.Ando, Prog. Theor.Phys.Suppl. 57, 115 (1975)
/ 9/ S.Katayama, K.Murase, J.Phys.Soc.Japan 42, 886 (1977)
/10/ W.Richter, U.Nowak, A.Stahl, Proc.15th Int.Conf. Physics of Semiconductors, Kyoto 1980, J.Phys.Soc. Japan 49, Suppl. A, 703 (1980)
/11/ U.Nowak,W.Richter,G.Sachs,phys.stat.sol.(b)108,131 (1981)
/12/ R.A.Cowley,G.Dolling,Phys.Rev.Lett.14,549 (1965(
/13/ W.Cochran, R.A.Cowley, G.Dolling, M.M.Elcombe,Proc. Royal Soc.A 293, 433 (1966)
/14/ L.F.Lemmens, J.T.Devreese,Solid State Comm. 14,1339 (1974)
/15/ I.G.Lang, U.S.Pashabekova, Soviet Physics-Solid State 7, 2829 (1966)
/16/ K.S.Singwi, M.P.Tosi,Phys.Rev. 147, 658 (1966)
/17/ L.D.Landau, E.M.Lifschitz, Course of Theoretical Physics, Vol.8, Pergamon Press, Oxford 1960(Chapter 8,§ 88)
/18/ W.Hayes, R.Loudon, Scattering of Light by crystals, J.Wiley & Sons, New York (1978)
/19/ J.Lindhard, Mat.Fys. Medd.28,881 (1954)
/20/ N.D.Mermin, Phys.Rev.B1, 2362 (1970)
/21/ e.g.D.Pines, Elementary Excitation in Solids,Benjamin, New York (1964)
/22/ K.Sturm, Solid State Commun. 27, 645 (1978)
/23/ A.Stahl, W.Richter, to be published

/24/ N.G. Van Kampen, Phys.Rep. 24, 171 (1976)
/25/ F.Forstmann, Z.Phys. B32, 385 (1972)
/26/ H.R.Philipp, H.Ehrenreich, J.Appl.Phys.35,1416(1964)
/27/ R.J.Bell, Introductory Fourier Transform Spectroscopy,Academic Press, N.Y.,1972
/28/ C.E.Hathaway: in The Raman Effect, Vol.1, ed. by A.Anderson (Marcel Dekker Inc., N.Y.,1971)
/29/ R.Claus, L.Merten, J.Brandmüller, Springer Tracts in Modern Physics, Vol. 75, ed.by G.Höhler(Springer Verlag Berlin 1975)
/30/ E.Burstein, A.Pinczuk, S.Iwasa, Phys.Rev.157,611 (1967)
/31/ S.Katayama, M.Murase,H.Kawamura, Solid State Commun. 16, 945 (1975)
/32/ G.Abstreiter, R.Trommer, M.Cardona, Solid State Commun. 30, 703 (1979)
/33/ W.Cochran, R.A.Cowley,G.Dolling, M.M.Elcombe, Proc.Roy.Soc. A293, 433 (1966)
/34/ R.Cowley, G.Dolling, Phys.Rev.Lett. 14, 549 (1965)
/35/ L.F.Lemmens, F.Brosens, J.T.Devreese, Solid State Commun. 17, 337 (1975)
/36/ L.F.Lemmens, J.T.Devreese, Solid State Commun. 14, 1339 (1974)

# GENERAL ASPECTS OF THE FUNCTIONAL-INTEGRAL APPROACH TO THE POLARON AND RELATED SYSTEMS*)

Janusz Adamowski[1)], Bernd Gerlach[2)], and Hajo Leschke[3)]

[1)]Solid State Physics Department, Academy of Mining and Metallurgy-AGH, 30-059 Kraków, Poland

[2)]Institut für Physik, Universität Dortmund
4600 Dortmund 50, Federal Republic of Germany

[3)]Institut für Theoretische Physik **)
Universität Düsseldorf, 4000 Düsseldorf 1
Federal Republic of Germany
and
Fachbereich Physik, Universität Essen-GHS
4300 Essen 1, Federal Republic of Germany

"Functional integrals are useful in many branches of theoretical physics, and we may regard them as an 'integral calculus' for modern physics"

V.N. Popov[1]

ABSTRACT

The aim of these notes is to give a tutorial but concise introduction to the functional-integral method as being relevant to the (understanding of) applications to the polaron and related systems. For some of the more recent applications to the statics of these systems the reader is referred to the quoted literature.

*) Notes accompanying the lectures given by one of us (H.L.) at the NATO Advanced Study Institute on "Physics of Polarons and Excitons in Polar Semiconductors and Ionic Crystals", Antwerpen, Belgium, July 26 - August 5, 1982

**) permanent address

## INTRODUCTION

After more than a quarter of a century, when Feynman[2] invented the functional-integral approach to the polaron problem, one may feel somewhat surprised in finding the following lines in Mahan's[3] recent and rather comprehensive textbook on advanced solid-state theory (see p. 488):

> "We shall not present [Feynman's] discussion, since his theory is very lengthy, and the variational methods have not been applied to other problems. The Feynman results will, however, be used as the standard to which other theories are compared."

While we agree with Mahan in judging the quality of the Feynman theory, we do not agree with the objections brought up to present the underlying method. In fact, as most concisely stated by Popov in our heading quotation, functional or path integration is being considered by many physicists as a convenient and effective language to formulate dynamical theories, which are affected by stochastic and/or quantum fluctuations. There are several reasons for the increasing popularity enjoyed by this language. Roughly, one may distinguish an heuristic and a computational value. The former is due, e.g., to the fact that functional integral formulations are serving as a unifying bridge between seemingly disconnected disciplines, and that they (thereby) provide a fresh way of looking at a particular problem, which almost always gives new qualitative insights and often stimulates novel quantitative results. Typically, these results rest on techniques especially designed for functional integrals. Therefore it is usually difficult or at least much less economical to obtain the same results within the more traditional approaches. Sometimes it is even unknown, how to obtain them without functional integrals. The polaron problem is an early paradigm of that computational value. A quotation from the preface of Schulman's[4] recent book may underline this:

> "What makes the polaron special from the standpoint of selling path integrals is that it is one of the few places where the path integral not only helps you discover an answer, but also remains the best way to calculate the answer even after you know it."

Summarizing these remarks, one may say that there has emerged during the last some ten years a consensus on the usefulness of functional integration in various fields of theoretical physics and especially in polaron theory.

In view of this, what can be said to Mahan's objections to present the Feynman theory? Is it really "very lengthy"? Is it true that the associated variational method has "not been applied to other problems"?

We want to offer the following answers to these questions. The lengthiness of any theory depends on what the reader is supposed to know a priori. Given the language of functional integration, the Feynman theory of the polaron is (in our opinion) no more difficult or tedious than any other theory formulated in a more conventional (and therefore presupposed) language. Simply because functional integration is even today not yet conventional, i.e. does not yet belong to the standard repertoire of standard theoretical physicists, there is usually felt a need to include in a textbook some sort of expository material. Such an inclusion might lengthen the book but not the theory. Similarly a textbook on classical mechanics is extended by the addition of a chapter on calculus.

As to Mahan's second objection we want to point out that the variational method has been applied, e.g., to the problems of calculating the density of electronic states in disordered systems[5,6], the polymer statistics[7] and the velocity correlations in a randomly stirred fluid[8].

Despite its successes in theoretical physics the place of functional integration is not as a substitute but rather as an alternative to other methods, e.g., operator methods. Like those it has its specific advantages and disadvantages making it more or less suitable for the solution of a particular problem. In short, adopting right words of Simon[10], functional integration should be considered neither as a "secret weapon" nor as a "panacea".

Our main goal in these lecture notes is the following. We want to give a concise but (hopefully) reasonably self-contained introduction to functional integration and present the basic general steps which have been used in most applications to the polaron and related systems. The purpose of these notes would be fulfilled, if they enable some readers, previously unfamiliar with functional integrals, to follow up articles on those applications.

There is considerable overlap with the material presented in Schultz's[11] excellent lectures twenty years ago. Nevertheless we hope that these notes may be viewed in part as a supplement and to some extent as a modern alternative to his lectures.

The end of this introduction might be the appropriate place to mention some expository literature on functional integration. Besides a "classic"[13] there are now four further textbooks[4,10,14,15] available which are explicitly devoted to this theme and which also apply to (theoretical) physicists. Schulman's book[4] is similar in purpose and in style to the book of Feynman and Hibbs[13] with emphasis on the more recent techniques and applications. The arguments presented by Schulman are well balanced between the demands on physical conceivability and mathematical rigor. The book of Langouche, Roekaerts, and Tirapegui[15] constitutes a fairly complete presentation of the discretization approach to functional integrals and their asymptotic evaluation by perturbation and WKB-type expansions. In contrast to these three books the books of Simon[10] and of Glimm and Jaffe[14] are addressed to the mathematically well trained reader, in part, because they presuppose some working knowledge of functional analysis. On the other hand these two latter books are likely to become standard sources on (some) rigorous applications of functional integration to physical problems.

Concerning earlier introductory and/or review type literature on the functional-integral approach to the polaron and related problems we refer the reader to the corresponding chapters in the books of Feynman and Hibbs and of Schulman, to the aforementioned lectures of Schultz[11], and to the more recent lectures of Devreese[16] which also contain a thorough discussion of dynamical aspects.

## FUNCTIONAL INTEGRALS RELATED TO THE DENSITY MATRIX

The purpose of this section is to give a concise introduction to functional integrals related to the thermal equilibrium state of a quantum mechanical system. Correspondingly, we "derive" these integrals as a possible tool or language to represent a canonical density matrix and some related averages.

The general idea of functional integration is most clearly illustrated for solid-state physicists by the example of a non-relativistic quantum mechanical system[*)] consisting of a single (Cartesian) degree of freedom with position operator Q and momentum operator P, which obey the Bose-type commutation relation

$$QP - PQ = i\hbar \quad . \tag{1}$$

The extension to systems with several degrees of freedom is for all problems which we will consider here essentially a matter of notation.

Let us assume that the dynamics of the system is generated by a Hamiltonian of the type

$$H = \frac{1}{2m} P^2 + v(Q) \tag{2}$$

*) A modern presentation of the operator formulation of quantum mechanics can be found, e.g., in Ref. 17.

where, as usual, the first term represents kinetic energy and the second potential energy corresponding to a real-valued function v(x) of the classical coordinate x.

In the sequel we will need the (generalized) eigenbasis $\{|x\rangle\}$ of the position operator. It is characterized by the relations

$$Q|x\rangle = x|x\rangle , \qquad \langle x|x'\rangle = \delta(x-x') \qquad \int dx\ |x\rangle\langle x| = 1 \ . \tag{3}$$

Here and in the following position-space integrals extend over all of space, if not stated otherwise.

The thermal equilibrium properties of our system follows from the (canonical) statistical operator $\exp(-\beta H)$, where $\beta>0$ has the meaning of an inverse temperature (times Boltzmann's constant). Clearly, from a formal point of view the physical interpretation of $\beta$ is not important. What is important for our purposes is that (the spectral properties of) H can be recovered from (those of) $\exp(-\beta H)$. And, as a matter of fact, in order to get information about H it is often more convenient to study $\exp(-\beta H)$, even for systems (like the ones with a single degree of freedom) for which the concept of thermal equilibrium is problematical.

The kernel or the set of matrix elements

$$\langle x|\ e^{-\beta H}\ |x_o\rangle \tag{4}$$

of the statistical operator is equivalent to the operator itself and is usually called the (canonical) density matrix (with respect to the position basis). It is the basic object of these lectures.

In general it is difficult to get a fairly explicit expression for the density matrix. However, in the simple case of a free particle (i.e. v=0) one easily computes

$$\langle x|\ e^{-\beta P^2/2m}|x_o\rangle = \frac{1}{\lambda_1}\ e^{-(x-x_o)^2\pi/\lambda_1^2} \ . \tag{5}$$

Here $\lambda_1$ is the thermal de-Broglie wavelength. Its definition follows from the more general definition

$$\lambda_N := \left(\frac{2\pi\hbar^2\beta}{mN}\right)^{1/2} \tag{6}$$

by setting N=1. Evidently the expression (5) is non-negative and gives 1 upon integration over x for each fixed $x_o$. It may therefore be interpreted for each $x_o$ as a (Gaussian) probability density. More specifically, it can be interpreted as the transition (or conditional) probability density to find a particle - moving in a medium with diffusion constant $\hbar/2m$ - near x at time $\beta\hbar>0$, if it has been observed at $x_0$ at time 0. In other words, (5) can be used to characterize the usual mathematical model of Brownian motion (in one dimension).

Due to the non-commutativity of kinetic and potential energy the statistical operator $\exp(-\beta H)$ does not factorize into a product of two statistical operators, one corresponding to the kinetic energy alone and one to the potential energy alone. This is the reason, why it is in general difficult or even impossible to compute the density matrix for a particle which is not free.

To proceed further, let us consider two general operators A and B, both of which self-adjoint and bounded from below. While for $AB \neq BA$

$$e^{-\beta(A+B)} \neq e^{-\beta A}\, e^{-\beta B} \quad , \tag{7}$$

a formal expansion of both sides around $\beta = 0$ and dropping terms of second and higher order in $\beta$ suggest that there might hold an approximate equality for sufficiently small $\beta > 0$. In fact, one can prove[10,18] for all $\beta > 0$

$$e^{-\beta(A+B)} = \lim_{N\to\infty} \left(e^{-\frac{\beta}{N}A}\, e^{-\frac{\beta}{N}B}\right)^N \tag{8}$$

This relation, known as the <u>(Lie-) Trotter product formula</u>, is almost equivalent to the functional-integral representation of the density matrix we are after.

Choosing $A = P^2/2m$, $B = v(Q)$ and rewriting (8) in the position basis we get with the help of (3), (5) and (6)

$$\langle x|e^{-\beta H}|x_0\rangle = \lim_{N\to\infty} I_N(x,x_0) \tag{9}$$

$$I_N(x,x_0) := \int \frac{dx_N}{\lambda_N} \cdots \int \frac{dx_1}{\lambda_N}\, \delta(x-x_N)\, e^{-S_N(x_N,\ldots,x_1,x_0)} \tag{10}$$

$$S_N(x_N,\ldots,x_1,x_0) := \sum_{\nu=1}^{N} \frac{\beta}{N}\left\{\frac{m}{2\hbar^2}\left(\frac{x_\nu - x_{\nu-1}}{\beta/N}\right)^2 + v(x_{\nu-1})\right\}. \tag{11}$$

Hence, we conclude that the density matrix can be represented as the limit $I_\infty$ of the sequence $(I_N)$, where $I_N$ is a N-dimensional integral. Since this limit is, in a sense, an $\infty$-dimensional or continual integral, it is customary and mnemonicly convenient to use a continuum notation for it. Let us first introduce the notation used by most physicists. It is suggested by the form of $S_N$. Although it does not make sense to take the limit in (9) separately for $S_N$ and the N integrals, the sum $S_N$ may be viewed as an approximation to the ("imaginary time" or "Euclidean") <u>action functional</u>

$$S[q] := \int_0^\beta d\tau \left\{ \frac{m}{2\hbar^2} \left(\frac{dq}{d\tau}\right)^2 + v(q(\tau)) \right\} \tag{12}$$

evaluated at a differentiable path $q:[0,\beta] \to \mathbb{R}$, $\tau \mapsto q(\tau)$ which goes through $x_\nu$ at $\tau^{(\nu)} := \nu\beta/N$; i.e. $q(\tau^{(\nu)}) = x_\nu$. Since this approximation which basically stems from replacing the actual path by an associated polygonal one, is getting better with increasing N, the following suggestive notation for $I_\infty$ presents itself

$$\int_{q(o)=x_o} \delta q \; \delta(x-q(\beta)) \; e^{-S[q]} := I_\infty(x,x_o) \quad . \tag{13}$$

This is (similar to) the usual physicists notation for a particular functional integral or path integral. The formal (dimensionless) integration symbol $\int_{q(o)=x_o} \delta q$ is meant to indicate "summation" over all functions or paths q starting at $\tau = 0$ from $x_o$.

Fortunately, with a slight modification this interpretation and the underlying notation may be rigorously justified. While both

$$\int_{q(o)=x_o} \delta q \qquad \text{and} \qquad \exp\left\{ -\frac{m}{2\hbar^2} \int_0^\beta d\tau \left(\frac{dq}{d\tau}\right)^2 \right\} \tag{14}$$

have only formal meanings, the combined expression

$$\langle \cdot \rangle_{x_o} := \int_{q(o)=x_o} \delta q \; \exp\left\{ -\frac{m}{2\hbar^2} \int_0^\beta d\tau \left(\frac{dq}{d\tau}\right)^2 \right\} (\;\cdot\;) \tag{15}$$

can be re-interpreted[10] as a mathematically well-defined integration or average over (the continuous but nowhere differentiable) paths of one-dimensional Brownian motion with diffusion constant $\hbar/2m$ starting at time 0 from $x_o$ and ending at time $\beta\hbar > 0$. Since the existence of the underlying probability distribution was shown for the first time by N. Wiener, the functional integral defined in (13) is called in the mathematical literature a <u>Wiener integral.</u>

Eqs. (10),(11), and (13) give its so-called sequential-limit definition.

Summarizing the above, we have arrived at the desired functional-integral representation of the density matrix. It can be written either as

$$<x|e^{-\beta H}|x_o> = \int_{q(o)=x_0} \delta q \; \delta(x-q(\beta)) \; e^{-S[q]} \tag{16}$$

or as

$$<x|e^{-\beta H}|x_o> = \left\langle \delta(x-q(\beta)) \; e^{-\int_o^\beta d\tau \; v(q(\tau))} \right\rangle_{x_0} . \tag{17}$$

This relation is usually called the Feynman-Kac(-Wiener) formula. It holds[10] under mild restrictions on v. The appearance of the Dirac delta function shows that only those paths contribute which end precisely at x.

It is interesting to note that one can get rid of the position basis and derive, at least formally, an operator version of the Feynman-Kac formula

$$e^{-\beta H} = \left\langle e^{-\frac{i}{\hbar} q(\beta)P} \; e^{-\int_o^\beta d\tau \; v(Q+q(\tau))} \right\rangle_o . \tag{18}$$

In fact, because of

$$<x|e^{-\frac{i}{\hbar} q(\beta)P}|x_o> = \delta(x-x_o-q(\beta)) \tag{19}$$

the $<x|\cdot|x_o>$ -matrix element of the r.h.s. of (18) is seen to be identical with the r.h.s. of (16) or (17) when integrating in (18) over $q(\tau)+x_0$ instead of $q(\tau)$. Eq. (18) expresses the close relation of the Feynman-Kac formula to Feynman's so-called disentangling formalism[19] and/or to the so-called Hubbard-Stratonovich-trick[20]. The formalism is essentially time-dependent perturbation theory and the trick is "uncompleting the square" in a Gaussian integral[4].

The Feynman-Kac formula can be easily extended to potentials v with an explicit (continuous) $\tau$ -dependence. The corresponding operator version reads as follows

$$T\left\{e^{-\int_0^\beta d\tau\, H(\tau)}\right\} = \left\langle e^{-\frac{i}{\hbar} q(\beta)P}\, e^{-\int_0^\beta d\tau\, v(Q+q(\tau),\tau)}\right\rangle_0 \quad . \tag{20}$$

Here the $\tau$-ordering T arranges a product of operators in such a way that operators belonging to larger $\tau$-values stand further to the left. The $\tau$-ordered exponential (or product integral[22]) on the l.h.s. of (20) generalizes the (canonical) statistical operator. It is the solution of the Bloch equation

$$\frac{\partial}{\partial\beta} W_\beta = - H(\beta)\, W_\beta \quad , \quad W_0 = 1 \tag{21}$$

associated with the Hamiltonian

$$H(\tau) := \frac{1}{2m} P^2 + v(Q,\tau) \quad . \tag{22}$$

The proof of (20) or, equivalently, of its matrix version with respect to the position basis, parallels the one for $\tau$-independent v, except that it starts with a discretization of the integral on the l.h.s. of (20).

Let us now apply (20) to a potential of the form

$$v(x,\tau) = v(x) - \eta(\tau)x \tag{23}$$

where v(x) is the potential entering our original Hamiltonian H of (2) and $\eta$ is an unspecified real-valued function of $\tau$. Defining

$$Q_\tau := e^{\tau H}\, Q\, e^{-\tau H} \quad , \tag{24}$$

we can write the l.h.s. of (20) as

$$T\left\{e^{-\int_0^\beta d\tau\, H(\tau)}\right\} = e^{-\beta H}\, T\left\{e^{\int_0^\beta d\tau\, \eta(\tau) Q_\tau}\right\} \quad . \tag{25}$$

This is true, because the r.h.s. of (25) solves (21) due to $H(\tau) = H-\eta(\tau)Q$. Now let us combine (25) with (20) and perform a mixed functional differentiation $\delta^n/\delta\eta(\tau_1)\ldots\delta\eta(\tau_n)$ of order n at $\eta = 0$ on both sides of the resulting equation. Then we get

$$e^{-\beta H}\, T\{Q_{\tau_1}\ldots Q_{\tau_n}\} = \langle \Omega\cdot(Q+q(\tau_1))\ldots(Q+q(\tau_n))\rangle_0 \tag{26}$$

where we have introduced the operator-valued functional of Brownian paths

$$\Omega[q] := e^{-\frac{i}{\hbar} q(\beta)P} \, e^{-\int_0^\beta d\tau \, v(Q+q(\tau))} \tag{27}$$

as an abbreviation. From (26) follows the more general relation

$$e^{-\beta H} T\{f_1(Q_{\tau_1})\ldots f_n(Q_{\tau_n})\} = \langle \Omega f_1(Q+q(\tau_1))\ldots f_n(Q+q(\tau_n))\rangle_0 \tag{28}$$

for complex-valued functions $x \mapsto f_j(x)$ $(j = 1,\ldots,n)$. As it should be, both sides of (28) are invariant under a permutation of the index set $\{1,\ldots,n\}$.

Clearly, (28) is a generalization of the operator version (18) of the Feynman-Kac formula. The equivalent generalization of the more usual matrix version (16) follows by taking the appropriate matrix element of (28). The result can be written as

$$\boxed{\langle x|e^{-\beta H} T\{\prod_{j=1}^{n} f_j(Q_{\tau_j})\}|x_0\rangle = \int_{q(o)=x_0} \delta q \, \delta(x-q(\beta)) e^{-S[q]} \prod_{j=1}^{n} f_j(q(\tau_j))} \; . \tag{29}$$

Eq. (29) or (28) constitutes a sufficiently broad basis for the functional-integral approach to the polaron and related problems. It serves as sort of a dictionary to translate expressions and relations from operator language into functional-integral language and vice versa. In particular, eq. (29) enables one to translate some thermal expectation values and certain perturbation series.

We close this section by mentioning two generalizations of (29) into different directions. One is simple and is needed in polaron theory, the other one is non-trivial in general but will not be used in these lectures. The first one is the extension to several degrees of freedom, e.g. to three, corresponding to dimension of space. In the above formulas one simply has to re-interpret the symbols Q,P,x,q etc. as multi-component quantities and products of two such symbols as a scalar product *). The second direction is concerned with the generalization of (29) for $f_j$-operators and/or Hamiltonians H which depend on P (more generally than in (2)). In such situations, being typical for some "non-linear" dynamical (field-) theories, functional-integral formulations are plagued by factor-ordering questions. For a recent discussion of this point we refer the reader to Ref.23 and especially to the book of

*) compare (46) below

Langouche, Roekaerts, and Tirapegui[15].

## MAIN STEPS IN THE APPLICATIONS TO THE POLARON AND RELATED SYSTEMS

The purpose of this section is to present and discuss the main general steps and arguments which have been used in most applications of the functional integral formalism to the polaron and related systems.

### Elimination of Harmonic Oscillator Variables

One reason for the power of the functional-integral approach to the polaron and related systems is due to the fact that the functional-integral representation of the density matrix allows one to eliminate or "integrate out" the phonon variables leading to an "effective" formulation of the (sub-) system of the remaining variables.

The procedure of this elimination does work, more generally, for all systems, in which a set of harmonic oscillators or, equivalently, a set of non-interacting boson modes is linearly coupled to a system of distinguishable particles or bosons. Again, following an early suggestion of Haken[24], the general aspects and the result of the method can be most clearly illustrated by working with a simplified Hamiltonian which, nevertheless, contains the essential features

$$\boxed{H = \frac{1}{2m} P^2 + v(Q) + \hbar\omega b^+ b + g(Q) b^+ + g^*(Q) b} \quad . \tag{30}$$

While the pair P,Q represents the single degree of freedom considered in the last section, called particle from now on, the pair $b, b^+$ represents the annihilation and creation operator of a single boson mode of energy $\hbar\omega > 0$ which the usual commutation relation

$$b\, b^+ - b^+ b = 1 \quad . \tag{31}$$

Clearly, b and $b^+$ commute with P and Q. The coupling of the particle to the boson mode is given in terms of a complex-valued function $x \mapsto g(x)$.

According to $\tau$-dependent perturbation theory similar to (25) the statistical operator may be written as

$$e^{-\beta H} = e^{-\beta(\frac{1}{2m} P^2 + v(Q))} T\left\{ e^{-\int_0^\beta d\tau \, H_o(Q_\tau)} \right\} \tag{32}$$

with a "displaced" harmonic oscillator Hamiltonian

$$H_o(x) := \hbar\omega b^+ b + g(x) b^+ + g^*(x) b \quad . \tag{33}$$

By expanding the τ-ordered exponential in (32), using a slight extension of (28), necessary because $H_o(x)$ and $H_o(x')$ do not commute for $x \neq x'$, and a final re-exponentiation we get

$$e^{-\beta H} = \left\langle \Omega T \left\{ e^{-\int_0^\beta d\tau \, H_o(Q+q(\tau))} \right\} \right\rangle_o \quad . \tag{34}$$

In this form it is easy to eliminate the boson mode, i.e. to perform the corresponding (partial) trace $tr_{bo}$. This is because the boson mode only appears in the τ-ordered exponential. From the dynamics of a classically driven quantum mechanical harmonic oscillator (for "imaginary times") one thus finds for the resulting reduced statistical operator of the particle

$$tr_{bo} e^{-\beta H} = Z_{bo}^o \left\langle \Omega \exp\{\int_0^\beta d\tau \int_0^\beta d\tau' G_{\hbar\omega}(\tau-\tau') g^*(Q+q(\tau)) g(Q+q(\tau'))\} \right\rangle_o \quad . \tag{35}$$

Here

$$Z_{bo}^o := tr_{bo} \, e^{-\beta\hbar\omega b^+ b} = (1 - e^{-\beta\hbar\omega})^{-1} \tag{36}$$

is the partition function and

$$G_{\hbar\omega}(\tau-\tau') := (2Z_{bo}^o)^{-1} tr_{bo} (e^{-\beta\hbar\omega b^+ b} T\{b_\tau b_{\tau'}^+ + b_{\tau'} b_\tau^+\})$$

$$= \frac{\cosh[(\frac{\beta}{2} - |\tau-\tau'|)\hbar\omega]}{2 \sinh \frac{\beta\hbar\omega}{2}} = \frac{\hbar\omega}{\beta} \sum_{\ell=-\infty}^{\infty} \frac{e^{i(\tau-\tau')\nu_\ell}}{(\hbar\omega)^2 + \nu_\ell^2} \tag{37}$$

the temperature Green function of the free harmonic oscillator, where we have used the notations $\nu_\ell := 2\pi\ell/\beta$ and

$$b_\tau := e^{\tau\hbar\omega b^+ b}\, b\, e^{-\tau\hbar\omega b^+ b} = b\, e^{-\tau\hbar\omega}$$
$$b_\tau^+ := e^{\tau\hbar\omega b^+ b}\, b^+ e^{-\tau\hbar\omega b^+ b} = b^+ e^{\tau\hbar\omega} \; . \tag{38}$$

The matrix version of (35) gives the representation of the reduced density matrix as a Wiener integral

$$\langle x|\, \mathrm{tr}_{bo} e^{-\beta H} |x_o\rangle = Z^o_{bo} \int_{q(o)=x_o} \delta q \;\, \delta(x-q(\beta))\; e^{-S[q]} \tag{39}$$

with the reduced or effective action

$$S[q] := \int_0^\beta d\tau \{\frac{m}{2\hbar^2}\left(\frac{dq}{d\tau}\right)^2 + v(q(\tau))\} - \int_0^\beta d\tau \int_0^\beta d\tau' G_{\hbar\omega}(\tau-\tau') g^*(q(\tau)) g(q(\tau')) \tag{40}$$

This result is central to the functional-integral approach to the polaron and related problems. Most articles on these problems which make use of functional integrals, start with relations directly obtainable from (39) and (40).

Let us list some general remarks:
i) The exact elimination of the boson mode leads to the appearance of the double-integral term

$$U[q] := \int_0^\beta d\tau \int_0^\beta d\tau' \; G_{\hbar\omega}(\tau-\tau') g^*(q(\tau)) g(q(\tau')) \tag{41}$$

in the effective action representing physically a non-instantaneous (self-) interaction of the effective particle system due to boson exchange in the original one. As a consequence, there is no effective potential in the strict sense. In more mathematical terms, there is no Feynman-Kac formula for double-integral actions. Still another, instructive way of saying the same thing goes as follows. Interpret exp (-S), with S given by (40), (formally) as the (unnormalized) "probability weight" of a stochastic process, i.e. of an ensemble {q} of (continuous) functions q considered as random. Then Brownian motion is the stochastic process corresponding to

$v = g = 0$, i.e. to a free particle. For $v \neq 0$ but $g = 0$ we get a more general stochastic process which, however, shares with Brownian motion the property of being Markovian. "Colloquially, this says that the future depends an the past only through the present" (Ref.10). For $g \neq 0$ this property is lost. Accordingly, the effective particle system defined by (40) corresponds, in general, to a non-Markovian stochastic process. On the other hand, an effective Hamiltonian $H_{eff}$ defined by

$$e^{-\beta H_{eff}} := \mathrm{tr}_{bo}\, e^{-\beta H} \tag{42}$$

not only is a complicated (unknown) function of Q and (!) P, corresponding to a non-local potential, but also depends on $\beta$.
ii) Using the Fourier-representation of $G_{\hbar\omega}(\tau-\tau')$ one finds

$$U[q] = \frac{1}{\beta}\sum_{\ell} \frac{\hbar\omega}{(\hbar\omega)^2+\nu_\ell^2}\left|\int_0^\beta d\tau\, e^{i\tau\nu_\ell} g^*(q(\tau))\right|^2 \geqslant 0 \tag{43}$$

Hence, there occurs for any $g \neq 0$ a <u>lowering of the free energy</u>

$$F(\beta) := -\beta^{-1}\ell n \int dx\, \langle x|\mathrm{tr}_{bo} e^{-\beta H}|x\rangle \tag{44}$$

and the ground-state energy

$$E_0 = \lim_{\beta\to\infty} F(\beta) \tag{45}$$

in comparison to the uncoupled case $g = 0$. In this sense the particle-boson interaction is always attractive.
iii) The generalization of (39) and (40) to <u>several particle degrees of freedom</u> requires, as has been said above, the re-interpretation of x and q as quantities with sufficiently many components. For example, in the case of n distinguishable particles in three dimensions this re-interpretation can be symbolized as follows:

$$\begin{aligned} x &\longrightarrow (\vec{x}_1,\dots,\vec{x}_n) \\ dx &\longrightarrow \prod_{j=1}^{n} d^3x_j \\ q(\tau) &\longrightarrow (\vec{q}_1(\tau),\dots,\vec{q}_n(\tau)) \\ m\left(\frac{dq}{d\tau}\right)^2 &\longrightarrow \sum_{j=1}^{n} m_j \left(\frac{d\vec{q}_j}{d\tau}\right)^2 \end{aligned} \tag{46}$$

$$\delta q \longrightarrow \prod_{j=1}^{n} \delta^3 q_j \quad . \tag{46}$$

iv) The generalization to several boson modes enumerated, e.g., by a wave-vektor $\vec{k}$ has to take into account that, in general, the mode energy $h\omega$ and the coupling function g will depend on $\vec{k}$. The resulting sum over $\vec{k}$ in the Hamiltonian then leads to the replacement

$$G_{\hbar\omega}(\tau-\tau')g^*(q(\tau))g(q(\tau')) \longrightarrow \sum_{\vec{k}} G_{\hbar\omega_{\vec{k}}}(\tau-\tau')g^*_{\vec{k}}(q(\tau))g_{\vec{k}}(q(\tau')) \tag{47}$$

in the action (40). Correspondingly, there is an additional sum over $\vec{k}$ in the representation (43). Moreover, $Z^o_{bo}$ has to be replaced by a product over $\vec{k}$.

It should be clear that the two generalizations given in iii) and iv) do not destroy the truths of the statements made under i) and ii). An example where these statements have been used with advantage in the investigation of (the thermodynamic limit of) a many-particle system coupled to bosons, is due to Gallavotti, Ginibre, and Velo[25].

## Jensen-Feynman-Inequality and the Quadratic Simulation of the Non-Instantaneous Interaction

In physically interesting cases the coupling functions g(x) does not vary linearly with x but rather as $\sim\exp(-ikx)$ due to translation invariance. Consequently, the effective action (40) is in most cases (even for v=0) not a quadratic functional of the path which implies that the functional integral in (39) is not Gaussian. This, in turn, seems to be almost equivalent to the statement that it cannot be evaluated exactly.

The question arises, if one can proceed, at least approximately, via Gaussian integrals. An affirmative answer, including to the sign of the error made, has been given by Feynman in his ingenious polaron paper[2].

His method rests on an inequality which we want to describe now. Besides the given action S consider another action $\tilde{S}$ which need not yet be specified except that it should be real-valued and should contain, like S, an additive (formal) term

$$\int_0^\beta d\tau \, \frac{\tilde{m}}{2\hbar^2} \left( \frac{dq}{d\tau} \right)^2 \tag{48}$$

(possibly with $\tilde{m} \neq m$) in order to give rise to a well-defined Wiener integral. Now define an associated free energy

$$F_{\tilde{S}} := -\beta^{-1} \ln \oint \delta q \; e^{-\tilde{S}[q]} \tag{49}$$

and an associated average or expectation value

$$\langle \cdot \rangle_{\tilde{S}} := e^{\beta F_{\tilde{S}}} \oint \delta q \; (\cdot) \; e^{-\tilde{S}[q]} \tag{50}$$

with respect to the "probability weight" $\exp(-\tilde{S})$ . Here we have used the notation

$$\oint \delta q \; (\cdot) := \int dx \int_{q(o)=x} \delta q \; \delta(x-q(\beta)) \; (\cdot) \tag{51}$$

to indicate (Wiener) integration over all closed paths. Then the following inequality

$$\boxed{F_S \leqslant F_{\tilde{S}} + \beta^{-1} \langle S - \tilde{S} \rangle_{\tilde{S}}} \tag{52}$$

holds. We call it the Jensen-Feynman inequality, because it is a special case of Jensen's inequality in probability theory[26,27] belonging to $\langle \exp(\cdot) \rangle \geqslant \exp(\langle \cdot \rangle)$.

By minimizing the r.h.s. of (52) with respect to an appropriate class of "trial" actions $\tilde{S}$ one gets a variational upper bound on $F_S$ and therefore, according to (39) and (44), on the (true) free energy

$$F = F^{o}_{bo} + F_S \tag{53}$$

and, letting $\beta \to \infty$ , on the ground-state energy of the Hamiltonian (30).

The actual choice of an optimal $\tilde{S}$ is of course restricted by one's ability to compute the corresponding functional averages. For this reason all trial actions which have been used so far in the polaron and in related systems (in the absence of magnetic fields), are of the type*)

$$\boxed{\tilde{S}[q] = \int_0^\beta d\tau \{ \frac{m}{2\hbar^2} (\frac{dq}{d\tau})^2 + \tilde{v}(q(\tau)) \} + \int_0^\beta d\tau \int_0^\beta d\tau' \tilde{f}(\tau-\tau') q(\tau) q(\tau')} \tag{54}$$

.

*) with the possible exception of a different mass $\tilde{m}$

While the double-integral term represents a quadratic simulation of the non-instantaneous interaction (41), the single-integral term may serve to correct for some faults implied by this approximation and/or to approximate the potential term in the true action (40). In principle, the potential $\tilde{v}$ and the "memory kernel" $\tilde{f}$ can be considered as variational functions. But, of course, in most applications these functions must be restricted further in order to get explicit results. To understand the practical necessity for this restrictions, let us first discuss two extreme choises

i) $\tilde{v} = 0$ , $\tilde{f} \neq 0$

clearly, in this case only averages with respect to a (non-Markovian) Gaussian stochastic process have to be computed. These are uniquely determined by the covariance $\langle q(\tau)q(\tau')\rangle_{\tilde{S}}$, because the mean $\langle q(\tau)\rangle_{\tilde{S}}$ vanishes for symmetry reasons. Moreover, as is suggested from (formal) analogy to finite dimensional Gaussian integrals, the covariance is the inverse kernel of the action with the appropriate boundary conditions. This can be computed explicitly by Fourier analysis, if (and only if) $\tilde{f}$ is supposed to fulfill

$$\tilde{f}(\tau-\beta) = \tilde{f}(\tau) = \tilde{f}(-\tau) \, , \quad 0 \leqslant \tau \leqslant \beta \quad . \tag{55}$$

The result is[28]

$$\langle q(\tau)q(\tau')\rangle_{\tilde{S}} = \frac{1}{\beta} \sum_{\ell}{}' \frac{e^{i(\tau-\tau')\nu_\ell}}{\beta\tilde{f}_\ell + (m\nu_\ell^2/\hbar^2)} \tag{56}$$

where

$$\tilde{f}_\ell := \frac{1}{\beta} \int_0^\beta d\tau \; \tilde{f}(\tau) \; e^{-i\tau\nu_\ell} \; . \tag{57}$$

The prime in (56) is to indicate that the term corresponding to $\ell = 0$ is to omitted from the sum if $\tilde{f}_0 = 0$, i.e. if $\tilde{S}$ is invariant with respect to constant translations

$$q(\tau) \longrightarrow q(\tau) + \text{const} \quad . \tag{58}$$

For details of the derivation of (56) and related Gaussian averages relevant to the polaron and similar systems we refer the reader to Ref. 29.

ii) $\tilde{f} = 0$, $\tilde{v} \neq 0$

In order to understand the features of this choice, it is helpful to write down the corresponding Jensen-Feynman inequality more explicitly

$$F \leqslant F^{o}_{bo} + F_{\tilde{S}} + \beta^{-1}\int_{o}^{\beta} d\tau\,(\langle v(q(\tau))\rangle_{\tilde{S}} - \langle\tilde{v}(q(\tau))\rangle_{\tilde{S}})$$

$$- \beta^{-1}\int_{o}^{\beta} d\tau \int_{o}^{\beta} d\tau'\; G_{\hbar\omega}(\tau-\tau')\langle g^{*}(q(\tau))g(q(\tau'))\rangle_{\tilde{S}} \quad . \qquad (59)$$

Now remember that $\tilde{S}$ is a completely instantaneous action. The only difference between $\tilde{S}$ and the action (12) is that v is replaced by $\tilde{v}$. Therefore we may re-translate the r.h.s. of (59) into operator language by using a "dictionary" which results from (29) for $x = x_o$ and upon integrating over x. In combination with the cyclic invariance of the particle trace

$$tr_{par}(\cdot) := \int dx \langle x|\cdot|x\rangle \qquad (60)$$

this gives

$$F \leqslant F^{o}_{bo} + F_{\tilde{H}} + \langle v(Q) - \tilde{v}(Q)\rangle_{\tilde{H}}$$

$$- \beta^{-1}\int_{o}^{\beta} d\tau \int_{o}^{\beta} d\tau'\; G_{\hbar\omega}(\tau-\tau')\langle T\{g^{*}(\tilde{Q}_{\tau})g(\tilde{Q}_{\tau'})\}\rangle_{\tilde{H}} \quad . \qquad (61)$$

Here

$$F_{\tilde{H}} := -\beta^{-1} \ln tr_{par}\, e^{-\beta\tilde{H}} \qquad (62)$$

denotes the free energy of the particle Hamiltonian

$$\tilde{H} := \frac{1}{2m} P^2 + \tilde{v}(Q) \qquad (63)$$

and

$$\langle\cdot\rangle_{\tilde{H}} := e^{\beta F_{\tilde{H}}}\, tr_{par}(e^{-\beta\tilde{H}}(\cdot)) \qquad (64)$$

the corresponding thermal average. $\tilde{H}$ also governs the $\tau$-dependence of the position operator, i.e.

$$\tilde{Q}_{\tau} := e^{\tau\tilde{H}}\, Q\, e^{-\tau\tilde{H}} \quad . \qquad (65)$$

Let us make two remarks on (61). First, (61) clearly shows that any actual choice of $\tilde{v}$ is restricted by the requirement that the spectral properties of $\tilde{H}$ must be more or less explicitly known.

This, typically, restricts $\tilde{v}$ to potentials of the harmonic or Coulombic type. Second, we may identify all terms on the r.h.s. of (61)as occurring in a (thermodynamic) perturbation expansion of our starting Hamiltonian (30), if we consider $\tilde{H} + \hbar\omega b^{+}b$ as the unperturbed part and $v - \tilde{v} + gb^{+} + g^{*}b$ as the perturbation. With this decomposition*) the first and the second term in (61) make up the 0-th order contribution, the third term is the 1-st order contribution and the fourth (negative) term**) is the 2-nd order contribution originating from the particle-boson coupling. The 2-nd order contribution coming from $v - \tilde{v}$ does not occur in (61). Due to this identification one may view the r.h.s. of (61) as a sharpening of the upper bound which (universally) arises from 1-st order perturbation theory, i.e. from a corresponding version of the Peierls-Bogolyubov inequaliy[30]. In particular, for $\tilde{v} = v$ which implies that the 1-st order contribution vanishes, the assertion is that <u>2-nd</u> order perturbation theory gives an upper bound.

So far we have only discussed in some detail the extreme choises for $\tilde{v}$ and $\tilde{f}$, namely that either $\tilde{v}$ or $\tilde{f}$ vanishes (identically). In particular, we have found practically necessary restrictions on $\tilde{v}$ and $\tilde{f}$ in order to arrive at explicitly computable upper bounds. Now it is natural to ask the following question. Given these restrictions, is it possible to get fairly explicit results for the more general choice of a $\tilde{v}$ and a $\tilde{f}$, both being different from zero? Not surprisingly, the answer to this question is, in general, negative. There are essentially two exceptions to this rule. One occurs when $\tilde{v}$ is of the harmonic-oscillator type. In this case $\tilde{v}$ may be incorporated into $\tilde{f}$ as a delta function. The other exception occurs when there is more than one particle degree of freedom and (the matrix-kernel) $\tilde{f}$ couples other degrees of freedom than those appearing in $\tilde{v}$. Both exceptions, especially the latter one, are of particular importance in applications to the exciton-phonon system[31,32].

We close this section with the remark that the Jensen-Feynman inequality may be viewed as the first member in a chain of - in principle - successively sharper inequalities[16,33].

## APPENDIX: APPLICATIONS TO THE POLARON AND RELATED SYSTEMS

In order to make these notes reasonably self-contained, the main issue of this appendix is to derive the effective action for Fröhlich-type electron-phonon systems from the general formulas given in the last section. Moreover, we list some references on recent applications to the polaron and the exciton-phonon system.

---

*) which is different from the one in (32)

**) compare the definition (37) of $G_{\hbar\omega}$

## Effective Action for Fröhlich-Type Electron-Phonon Systems

With the prescriptions (46) and (47) in mind let us re-interpret the Hamiltonian (30) as representing a system of n distinguishable (spinless) particles in three dimensional space interacting with a set of boson modes. Then the effective action (40) takes the form

$$S = \int_0^\beta d\tau \left\{ \frac{1}{2\hbar^2} \sum_{j=1}^{n} m_j \left( \frac{d\vec{q}_j}{d\tau} \right)^2 + v(\vec{q}_1(\tau),\ldots,\vec{q}_n(\tau)) \right\} - U$$

$$U = \int_0^\beta d\tau \int_0^\beta d\tau' \sum_{\vec{k}} G_{\hbar\omega_{\vec{k}}}(\tau-\tau') g_{\vec{k}}^*(\vec{q}_1(\tau),\ldots,\vec{q}_n(\tau)) g_{\vec{k}}(\vec{q}_1(\tau'),\ldots,\vec{q}_n(\tau')) \geq 0 . \tag{66}$$

Now let us interpret the particles and bosons, respectively, as (defect-) electrons and phonons of a (dispersionless) branch of LO-lattice vibrations in an ionic crystal or polar semiconductor. Then, in order to account for the corresponding Fröhlich-type (defect-) electron-phonon interaction, we must choose

$$g_{\vec{k}}(\vec{x}_1,\ldots,\vec{x}_n) = -\frac{1}{\sqrt{V}} \frac{g}{|\vec{k}|} \sum_{j=1}^{n} \delta_j \, e^{-i\vec{k}\cdot\vec{x}_j}$$

$$g := -i\hbar\omega \left(\frac{\hbar}{2m_j\omega}\right)^{1/4} (4\pi\alpha_j)^{1/2} \tag{67}$$

$$\alpha_j := \frac{e^2}{2\hbar\omega} \left(\frac{2m_j\omega}{\hbar}\right)^{1/2} \left(\frac{1}{\varepsilon_\infty} - \frac{1}{\varepsilon_0}\right) .$$

Here V is the quantization volume and $\delta_j$ is a charge sign defined as 1 or −1, if j corresponds to an electron or a defect-electron (= hole), respectively. All other symbols have their usual meanings.

From (66) and (67) we get due to $\omega_{\vec{k}}$ = const

$$U = |g|^2 \int_0^\beta d\tau \int_0^\beta d\tau' G_{\hbar\omega}(\tau-\tau') \sum_{j=1}^{n} \delta_j \sum_{j'=1}^{n} \delta_{j'} \frac{1}{V} \sum_{\vec{k}} \frac{1}{k^2} e^{i\vec{k}\cdot(\vec{q}_j(\tau)-\vec{q}_{j'}(\tau'))}$$

$$\overset{V\to\infty}{=} \frac{|g|^2}{4\pi} \sum_{j=1}^{n} \delta_j \sum_{j'=1}^{n} \delta_{j'} \int_0^\beta d\tau \int_0^\beta d\tau' \frac{G_{\hbar\omega}(\tau-\tau')}{|\vec{q}_j(\tau)-\vec{q}_{j'}(\tau')|} \geq 0 . \tag{68}$$

This expression is characteristic for all Fröhlich-type (defect-) electron-phonon systems. It consists on the one hand of non-instantaneous interaction terms ($j \neq j'$) and on the other hand of non-instantaneous self-interaction terms ($j = j'$). The overall sign is positive, although the electron-hole terms are negative.

Let us further specialize to the usual polaron system and the exciton-phonon system in the absence of external fields. The polaron system is characterized by $n = 1$ and $v = 0$. The exciton-phonon system or the exciphon system, as we would like to call the corresponding quasi particle, is characterized by $n = 2$, $\delta_1 = -\delta_2$ and the Coulombic potential

$$v(\vec{x}_1, \vec{x}_2) = -\frac{e^2}{\varepsilon_\infty |\vec{x}_1 - \vec{x}_2|} \quad . \tag{69}$$

Some References on Recent Applications

For a general background on the physics of polarons and excitons we refer the reader to the two precursers[12,34] of this conference, to the appropriate chapters in Mahan's book[3], and to a recent review of Büttner and Pollmann[35].

Let us now briefly mention some recent works on applications of the functional-integral formalism to the polaron and related systems. In keeping with the restricted formalism we have presented in these notes, we will confine ourselves to works devoted to equilibrium properties*). Nevertheless, this short compilation is far from being complete but, hopefully, in some sense representative. We apologize for omissions.

Approaches to the polaron problem using general quadratic trial actions are due to Saitoh[36] and ourselves[28]. A variant of these more standard approaches has been shown by Luttinger and Lu[37] to account for the correct strong-coupling limit of polaron energy. In fact, this limit can be exactly extracted from the polaron functional integral, even for non-zero temperatures. This has been proved by Donsker and Varadhan[38] and ourselves[39] by using a new method for the asymptotic evaluation of Wiener integrals due to Donsker and Varadhan[40] and Friedberg and Luttinger (see Ref.41). New insights in the (derivation of the) polaron mass and its temperature dependence have

*) For applications to non-equilibrium problems and external-field effects the reader is referred to the lecture notes of Prof. Devreese and Prof. Evrard.

been gained by Sa-yakanit[42] and by Kochetov and Smondyrev[43]. A clary-fying discussion on the possible existence of a so-called phase transition ascribed to the polaron subjected to a magnetic field or not is due to Peeters and Devreese[44].

Recent applications to the exciton-phonon problem seem, to the best of our knowledge, only be made by ourselves[31,32].

It is interesting to speculate, if further progress will also come from a Monte-Carlo evaluation of Wiener integrals. A recent application of this method to the polaron is due to Sabel'fel'd[45].

REFERENCES

1. V.N. Popov, Functional Integrals in Quantum Field Theory, Ref. TH 2424-CERN (1977)
2. R.P. Feynman, Slow Electrons in a Polar Crystal, Phys. Rev. 97:660 (1955)
3. G.D. Mahan, "Many-Particle Physics", Plenum, New York (1981)
4. L.S. Schulman,"Techniques and Applications of Path Integration", Wiley-Interscience, New York (1981)
5. E.P. Gross, Partition Function for an Electron in a Random Potential, J. Stat. Phys. 17:265 (1977)
6. J.M. Luttinger, Useful Bounds on Interesting Quantities by Path Integrals, in : see Ref.9
7. K.F. Freed, Functional Integrals and Polymer Statistics, Adv. Chem. Phys. 22:1 (1972)
8. M. Lücke, Velocity Correlations in a Randomly Stirred Fluid: A Variational Principle for Path-Integral Functionals, Phys. Rev.A 18:282 (1978)
9. G.J. Papadopoulos, J.T. Devreese, eds., "Path Integrals and their Applications in Quantum, Statistical, and Solid State Physics", Plenum, New York (1978)
10. B. Simon, "Functional Integration and Quantum Physics", Academic, New York (1979)
11. T.D. Schultz, Feynman's Path-Integral Method Applied to the Equilibrium Properties of Polarons and Related Problems, in: see Ref.12
12. C.G. Kuper and G.D. Whitfield, eds., "Polarons and Excitons", Oliver and Boyd, Edinburgh (1963)
13. R.P. Feynman, A.R. Hibbs, "Quantum Mechanics and Path Integrals", McGraw-Hill, New York (1965)

14. J. Glimm, A. Jaffe, "Quantum Physics, a Functional Integral Point of View", Springer, New York (1981)

15. F. Langouche, D. Roekaerts, E. Tirapegui, "Functional Integration and Semiclassical Expansions", Reidel, Dordrecht (1982)

16. J.T. Devreese, Path Integrals and Continuum Fröhlich Polarons, in : see Ref.9

17. A. Böhm, "Quantum Mechanics", Springer, New York (1979)

18. E.B. Davies, "One-Parameter Semigroups", Academic, London (1980)

19. R.P. Feynman, An Operator Calculus Having Applications in Quantum Electrodynamics, Phys. Rev. 84 : 108 (1951)

20. B. Mühlschlegel, Functional Integrals and Local Many-Body Problems : Localized Moments and Small Particles, in: see Ref.21

21. A.M. Arthurs, ed., "Functional Integration and its Applications", Clarendon, Oxford (1975)

22. J.D. Dollard, C.N. Friedman, "Product Integration with Applications to Differential Equations", Addison-Wesley, Reading (1979)

23. H. Leschke, Path Integral Approach to Fluctuations in Dynamic Processes, in : "Chaos and Order in Nature", H.Haken, ed., Springer, Berlin (1981)

24. H. Haken, Berechnung der Energie des Exzitonen-Grundzustandes im polaren Kristall nach einem neuen Variationsverfahren von Feynman. I, Z. Physik 147:323 (1957)

25. G. Gallavotti, J. Ginibre, G. Velo, Statistical Mechanics of the Electron-Phonon System, Lett. Nuov. Cim. 4:1293 (1970)

26. R.G. Laha, V.K. Rohatgi, "Probability Theory", Wiley, New York (1979)

27. A.W. Marshall, I. Olkin, "Inequalities: Theory of Majorization and its Applications", Academic, New York (1979)

28. J. Adamowski, B. Gerlach, H. Leschke, Feynman's Approach to the Polaron Problem Generalized to Arbitrary Quadratic Actions, in : "Functional Integration, Theory and Applications", J.-P. Antoine, E. Tirapegui, eds., Plenum, New York (1980)

29. J. Adamowski, B. Gerlach, H. Leschke, Explicit Evaluation of Certain Gaussian Functional Integrals Arising in Problems of Statistical Physics, J. Math. Phys. 23: 243 (1982)

30. A. Huber, Variational Principles in Quantum Statistical Mechanics, in : "Mathematical Methods in Solid State and Superfluid Theory", R.C. Clark, G.H. Derrick, eds., Oliver and Boyd, Edinburgh (1969)

31. J. Adamowski, B. Gerlach, H. Leschke, Treatment of the Exciton-Phonon Interaction via Functional Integration. I. Harmonic Trial Actions, Phys. Rev.B 23: 2943 (1981)

32. J. Adamowski, B. Gerlach, H. Leschke, Polaronic versus Bare Excitons: A Functional-Integration Approach to the Exciton-Phonon Problem, Proceedings of 16-th Int. Conf. on the Physics of Semiconductors, Montpellier, France, Sept. 6-9, 1982; to appear in Physica B

33. K. Zeile, A Generalization of Feynman's Variational Principle for Real Path Integrals, Phys. Lett. 67A: 322 (1978)

34. J.T. Devreese, ed., "Polarons in Ionic Crystals and Polar Semiconductors", North-Holland, Amsterdam (1972)

35. H. Büttner, J. Pollmann, Excitons in Polar Semiconductors, Proceedings of 16-th Int. Conf. on the Physics of Semiconductors, Montpellier, France, Sept. 6-9, 1982; to appear in Physica B

36. M. Saitoh, Theory of a Polaron at Finite Temperatures, J.Phys. Soc. Jpn. 49: 878 (1980)

37. J.M. Luttinger, C.-Y. Lu, Generalized Path-Integral Formalism of the Polaron Problem and its Second-Order Semi-Invariant Correction to the Ground-State Energy, Phys. Rev.B 21: 4251 (1980)

38. M.D. Donsker, S.R.S. Varadhan, The Polaron Problem and Large Deviations, Phys. Rep. 77: 235 (1981)

39. J. Adamowski, B. Gerlach, H. Leschke, Strong-Coupling Limit of Polaron Energy, Revisited, Phys. Lett. 79A: 249 (1980)

40. M.D. Donsker, S.R.S. Varadhan, Asymptotic Evaluation of Certain Wiener Integrals for Large Time, in: see Ref.21

41. J.M. Luttinger, A New Method for the Evaluation of a Class of Path Integrals, J. Math. Phys. 23: 1011 (1982)

42. V. Sa-yakanit, The Feynman Effective Mass of the Polaron, Phys. Rev.B 19: 2377 (1979)

43. E.A. Kochetov, M.A. Smondyrev, Thermal Effects in the Polaron Model, Theor. Math. Phys. 47: 524 (1981)

44. F.M. Peeters, J.T. Devreese, On the Existence of a Phase Transition for the Fröhlich Polaron, phys. stat. sol.(b) 112: 219 (1982)

45. K.K. Sabel'fel'd, Approximate Evaluation of Wiener Continual Integrals by the Monte Carlo Method, U.S.S.R. Comput. Maths. Math. Phys. 19: 27 (1979)

# EXCITONS AND ELECTRON-HOLE CONDENSATION

## INTRODUCTION TO THE THEORY OF EXCITONS

R.J. Elliott

Department of Theoretical Physics
1 Keble Rd
Oxford, OX1 3NP, England

### INTRODUCTION

Excitons are the lowest lying excited states of semiconducting and insulating crystals. Such crystals have a single ground state and all excited states are separated from it by a finite (and often large) energy gap, $E_G$. Except in a very few special cases this gap is much larger than kT so that the thermal occupation of these states is vanishingly small. They are therefore normally studied by optical techniques, absorption and fluourescence, and excitons are responsible for the rich structure which is often found near the absorption edge. In recent years, it has been possible to reach highly excited states containing many excitons in quasi-equilibrium, using lasers. Crystals in these states have very interesting properties and raise some fascinating theoretical problems. These lectures will present a summary of the theory of single excitons and their optical properties, using band theory as a starting point so that it is mainly applicable to semiconductors. Other aspects of this theory will be discussed in Dr. Baldereschi's lectures. In the second part of the lectures the properties of high densities of excitons will be discussed, although a fuller discussion of the electron-hole plasma which occurs at the highest density will be given by Dr. Reinecke.

### BAND THEORY

The single particle states are taken to have the Bloch form

$$\psi_{n\underline{k}}(r) = \frac{1}{\sqrt{N\Omega}} u_{n\underline{k}}(\underline{r}) e^{i\underline{k}.\underline{r}} \qquad (2.1)$$

for a crystal of N cells of volume $\Omega$. The function u has the periodicity of the lattice and is the same in each cell. Near to the atomic cores it is a rapidly varying function, similar to an atomic wave function. The possible values of $\underline{k}$ form a quasi-continuous set covering the first Brillouin Zone. The energy $E_n(\underline{k})$ is continuous over the zone and forms a band of N states for each value of n. We shall assume that particles are created and destroyed in these states by fermion operators $a_n^+(\underline{k})$, $a_n(\underline{k})$.

The ground state wave function is an antisymmetric product of the $\psi_{nk}$ for occupied states which cover an integral number of bands. Thus the total momentum

$$\underline{K} = \Sigma \underline{k} = 0 \tag{2.2}$$

The energy of the state includes a contribution from the electron-electron coulomb interactions. The first excited states correspond to the promotion of a single electron across the gap into the conduction band c, leaving behind a hole in the valence band, v. Such a state is obtained by the use of the operator

$$|p,\underline{K}\rangle = a_c^+(\underline{k})a_v(\underline{k}')|0\rangle \tag{2.3}$$

$$\underline{K} = \underline{k}-\underline{k}' \tag{2.4}$$

is the total momentum of the pair and it has energy

$$E = E_c(\underline{k}) - E_v(\underline{k}') \tag{2.5}$$

It is sometimes convenient to think in terms of the hole as an actual particle created by $\alpha_v^+(-\underline{k}') = a_v(\underline{k}')$ with momentum $\underline{k}_h = -\underline{k}'$ and energy $E_h(\underline{k}_h) = -E_v(\underline{k}')$.

The lowest energy pair states will occur when both the electron and hole are close to the band edges on either side of the gap. Such edges occur near extrema where $E_n(\underline{k})$ may be expanded in powers of $(\underline{k}-\underline{k}_o)$. The first term is quadratic and taking $\underline{k}_o = 0$ it may be written in the form

$$E_n(\underline{k}) - E_n(0) = \sum_{\alpha\beta} \frac{\hbar^2 k_\alpha k_\beta}{2m_{\alpha\beta}} \tag{2.8}$$

where $m_{\alpha\beta}$ is the effective mass tensor. For small $\underline{k}$, $E_n(\underline{k})$ and $u_{nk}$ may be obtained by perturbation theory in the so-called effective mass approximation.

$$u_{nk} = u_{no} + \sum_{n'\neq n} u_{n'o} \frac{\hbar \underline{k}}{m} \cdot \langle n'|\underline{p}|n\rangle \tag{2.9}$$

$$\frac{1}{2m_{\alpha\beta}} = \frac{1}{2m}\left[\delta_{\alpha\beta} + \sum_{n'\neq n} \frac{<n|p_\alpha|n'><n'|p_\beta|n>}{m(E_n(0) - E_{n'}(0))}\right] \tag{2.10}$$

where $<n'|\underline{p}|n> = \frac{1}{\Omega} \int_{cell} u^*_{n'0}(\underline{r})\ \underline{p}\ u_{n0}(\underline{r})\ d\underline{r}$ (2.11)

OPTICAL PROPERTIES

The interaction between the electrons and the radiation field is $\frac{1}{c}\ \underline{j}.\underline{A}$ depending on the current operator, $\underline{j} = \frac{e}{m} \sum_i \underline{p}_i$ and the vector potential for the radiation field which for photons with wave vector $\underline{q}$ and polarisation in a medium of refractive index $\mu$ has the form

$$A_\lambda = \underline{\varepsilon}_\lambda \frac{c}{\eta}\left(\frac{\hbar}{V\nu}\right)^{\frac{1}{2}}[\gamma_\lambda(\underline{q})e^{i(\eta\underline{q}.\underline{r} - 2\pi\nu t)} + c.c] \tag{3.1}$$

Compared to the dimension of the Brillouin zone and the typical values of the electron $\underline{k}$ involved in the process, $\underline{q}$ may be taken to be zero except in unusual cases. Then the optical transitions involve the matrix elements of

$$\underline{\varepsilon}_\lambda.\underline{j} = \frac{e}{m} \sum_{nn'} <n\ \underline{k}|\underline{\varepsilon}_\lambda.\underline{p}|n'\underline{k}>\ a^+_{n'}(\underline{k})\ a_n(\underline{k}) \tag{3.2}$$

The matrix element is zero unless the two $\underline{k}$-values are the same. The absorption coefficient for photons

$$K(\nu_\lambda) = \frac{2\pi e^2}{m^2 c\eta\nu} |<o|\underline{\varepsilon}_\lambda.\underline{p}|f>|^2\ S(h\nu) \tag{3.3}$$

where $S(h\nu)$ is the density of final states with total $K = 0$ at an energy E equal to the photon energy $h\nu$ above the ground state. If the final state is discrete the transition may be characterised by an f-value

$$f = 2|<0|\underline{\varepsilon}_\lambda.\underline{p}|f>|^2/Nmh\nu \tag{3.4}$$

These formulae may be readily applied to the independent particle approximation for band to band transitions. The density of electron states near an extremum has the form

$$\frac{1}{2\pi^2}\left(\frac{2m_c}{\hbar^2}\right)^{3/2} (E-E_o)^{\frac{1}{2}} \tag{3.5}$$

where $m_c$ is some appropriate angular average of the effective mass tensor. The density of pair states with $\underline{K} = 0$ is

$$\frac{1}{(2\pi)^2}\left(\frac{2\mu}{\hbar^2}\right)^{3/2} (h\nu - E_G)^{\frac{1}{2}} \tag{3.6}$$

where $\mu$ is the reduced mass

$$\frac{1}{\mu} = \frac{1}{m_c} + \frac{1}{m_v} \tag{3.7}$$

The matrix element of $\underline{p}$ in (3.3) depends in general on $\underline{k}$. Near the band edge where $\underline{k}-\underline{k}_o$ is small we can expand

$$\frac{1}{\Omega} \int_{cell} u^*_{ck}(\underline{r})\underline{\varepsilon}.\underline{p}\, u_{vk}(\underline{r})\, d\underline{r} = \langle c|\underline{\varepsilon}.\underline{p}|v\rangle + \frac{\hbar\underline{k}}{m}.\langle c|\underline{M}|v\rangle \tag{3.8}$$

$$\text{where } \underline{M} = \sum_i \frac{\langle c|\underline{p}|i\rangle\langle i|\underline{\varepsilon}.\underline{p}|v\rangle}{E_c - E_i} + \frac{\langle c|\underline{\varepsilon}.\underline{p}|i\rangle\langle i|\underline{p}|v\rangle}{E_v - E_i} \tag{3.9}$$

If the first term in (3.8) is finite it dominates at the band edge. This is called an allowed transition

$$K_A = \frac{e^2}{m^2 c\pi\eta\nu}\left(\frac{2\mu}{\hbar^2}\right)^{3/2} (h\nu - E_G)^{\frac{1}{2}}\ |\langle c,\underline{k}_o|\underline{p}|v,\underline{k}_o\rangle|^2 \tag{3.10}$$

If the first term is zero it is called a forbidden transition and

$$K_F = \frac{e^2\hbar^2}{m^4 c^3 \pi\eta\nu}\left(\frac{2\mu}{\hbar^2}\right)^{3/2} |\langle c|\underline{M}|v\rangle|^2\ (h\nu - E_G)^{3/2} \tag{3.11}$$

## EFFECTIVE MASS THEORY FOR EXCITONS [1]

In the pair states discussed in the last section the interaction between the particles was ignored. The coulomb repulsion energy for a full band of electrons may be taken as being included in ground state energy. Then the energy of the pair state relative to $E_o$ should include the effect of the residual interaction between the hole and the electron. The polarisation of the background both by distortion of the electron clouds[2], and by ion displacements[3] will reduce the bare coulomb interaction. To a good approximation these effects lead to the introduction of a term

$$U = -\frac{e^2}{\varepsilon|\underline{r}_e - \underline{r}_h|} \tag{4.1}$$

where $\varepsilon$ is the dielectric constant. In general $\varepsilon$ depends on frequency because the dielectric screening depends on the speed of the relative electron-hole motions. The electronic part only varies significantly at frequencies comparable with the band gap and for the relative motions of particles near the band edges the static electronic part of $\varepsilon$ should be taken in (4.1). For ionic motion however the characteristic frequency at which $\varepsilon$ varies is the optical phonon $\omega_L$. For relative motion below this frequency we must use the static dielectric constant $\varepsilon_o$, and the polaron mass for $m^*$. At higher frequencies however both $m^*$ and $\varepsilon$ will change in a complicated way. The latter may also be seen, phenomenolgically, as a change in (4.1) with distance when the polarisation clouds of the hole and the electron overlap. To allow for this Haken[3] suggests an approximate form of the electron hole interaction as

$$-\frac{e^2}{r}\left[\frac{1}{\varepsilon_o}+\frac{1}{2}\left(\frac{1}{\varepsilon_\infty}-\frac{1}{\varepsilon_o}\right)\left(e^{-x_1 r}+e^{-x_2 r}\right)\right] \tag{4.2}$$

where $x_1 = (2m_1\omega_L/\hbar)^{\frac{1}{2}}$

In the presence of the perturbations (4.1), the exciton states may be found as combinations of the pair states.

$$\Psi(\underline{r}_e,\underline{r}_h) = \sum_{k_e k_h} \Phi(\underline{k}_e,\underline{k}_h)\ \psi_{ck_e}(\underline{r}_e)\psi_{vk_h}(\underline{r}_h)$$

so that an exciton is created by the operator.

$$\sum_{kk'} \Phi(\underline{k},\underline{k}')\ a_c^+(\underline{k}')a_v(-\underline{k}) = b^+(\underline{K}) \tag{4.3}$$

The matrix elements of (4.1) which depends only on the relative position

$$\underline{r} = \underline{r}_e - \underline{r}_h \tag{4.5}$$

and not on the mean position

$$\underline{R} = \tfrac{1}{2}(\underline{r}_e + \underline{r}_h) \tag{4.6}$$

when taken between the pair states depends only on the relative momentum

$$\underline{\kappa} = \underline{k}_e - \underline{k}_h \tag{4.7}$$

and not on the total momentum

$$\underline{K} = \underline{k}_e + \underline{k}_h \tag{4.8}$$

Thus the exciton has a uniquely defined $\underline{K}$ but is composed of a combination of pairs with different $\underline{\kappa}$. The equation for $\Phi$ takes the form

$$(E_c(\underline{k}_e) - E_v(\underline{k}_h) - E)\Phi(\underline{k}_e,\underline{k}_h) + \sum_{k_e'k_h'} U(\underline{\kappa},\underline{\kappa}')\Phi(\underline{k}_e',\underline{k}_h') = 0 \tag{4.9}$$

where

$$U(\underline{\kappa},\underline{\kappa}') = \frac{1}{N^2\Omega^2}\int \frac{e^2}{\varepsilon_o r} e^{i(\underline{\kappa}-\underline{\kappa}').\underline{r}} u^*_{ck_e}(\underline{r}_e)u_{vk_h'}(\underline{r}_h)u_{ck_e}(\underline{r}_e)u_{vk_h}(\underline{r}_h) \, d\underline{r}_e \, d\underline{r}_h \tag{4.10}$$

If $\varepsilon_o$ is large so that $U(\underline{r})$ is slowly varying on an atomic scale, (4.10) is large only when $\underline{\kappa}-\underline{\kappa}'$ is small, and since $\underline{K}$ is fixed this implies $\underline{k}_e \sim \underline{k}_e'$ and $\underline{k}_h \sim \underline{k}_h'$. Then neglecting the variation of u with $\underline{k}$, (4.9) takes the simpler form

$$\frac{1}{N\Omega}\int \frac{e^2}{\varepsilon_o r} e^{i(\underline{\kappa}-\underline{\kappa}').\underline{r}} \, d\underline{r} = \frac{4\pi e^2}{\varepsilon_o |\underline{\kappa}-\underline{\kappa}'|^2} \tag{4.11}$$

In this approximation

$$\Psi = \Phi(\underline{r}_e,\underline{r}_h) \, u_c(\underline{r}_e) \, u_v(\underline{r}_h) \tag{4.12}$$

where $$\Phi(\underline{r}_e,\underline{r}_h) = \frac{1}{N\Omega}\int \Phi(\underline{k}_e,\underline{k}_h) \, e^{i\underline{k}_e\cdot\underline{r}_e} \, e^{i\underline{k}_h\cdot\underline{r}_h} \tag{4.13}$$

is the envelope function of relative hole electron motion which is modulated by the $u(\underline{r}_e)$ $u(\underline{r}_h)$ on an atomic scale. This envelope function is the solution of the differential equation, obtained by fourier transforming the integral equation (4.9)

$$(E_c(-i\nabla_e) - E_V(-i\nabla_h) - \frac{e^2}{\varepsilon_o r}) \, \Phi = E\Phi \tag{4.14}$$

For simple spherical bands it is exactly like the hydrogen atom equation with solution

$$\Phi_{i\underline{K}} = \frac{1}{\sqrt{N\Omega}} \, \phi_i(\underline{r}) \, e^{i\underline{K}.\underline{\rho}} \tag{4.15}$$

where $\underline{\rho}$ is the centre of mass co-ordinate, $\phi_i$ is the normalised hydrogen atom wave function with reduced mass $\mu$, and i represents the usual set of three quantum numbers. The energy of the exciton is

$$E = E_G - R/n^2 + \hbar^2K^2/2 \qquad (4.16)$$

where $M = m_e + m_h$, giving a band of exciton states for each relative motion. A sketch of the energy spectrum of single particle and exciton (pair) states is shown in fig. 1.

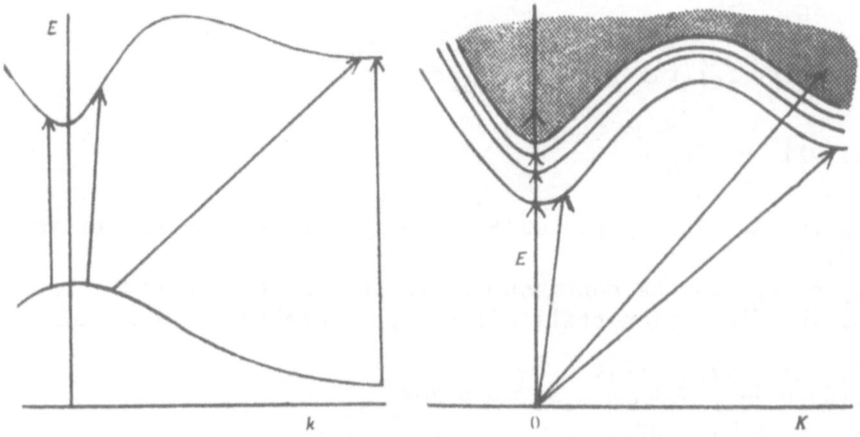


FIG. 1. A. Energy band plot of one particle states.
B. Exciton bands (lines) and pair states (shaded). The ground state is at the origin. Direct transitions are vertical, indirect are diagonal.

TRANSITIONS TO EXCITON STATES[4]

Direct Transitions

The probability of optical transitions from the ground state to the exciton states given as combinations (4.2) is readily obtained from the matrix element (3.2) for pair creation. Substituting $\underline{k}' = \underline{k}_e$, $\underline{k} = -\underline{k}_h$ since $\underline{K} = 0$ we find

$$\int \Psi^*_{ex}\ \underline{\varepsilon}.\underline{p}\ \Psi_o = \sum_{k_e k_h} \Phi(\underline{k}_e,\underline{k}_h)\langle c,\underline{k}_e|\underline{\varepsilon}.\underline{p}|v,-\underline{k}_h\rangle \qquad (5.1)$$

If the range of k which are significant is small and the transition is allowed so that the first term in (3.8) is finite, (5.1) becomes

$$\langle c|\underline{\varepsilon}.\underline{p}|v\rangle \sum_{k_e k_h} \Phi(\underline{k}_e,\underline{k}_h) = \langle \underline{c}|\underline{\varepsilon}.\underline{p}|v\rangle\ \phi_i(0)/(N\Omega)^{\frac{1}{2}} \qquad (5.2)$$

Because of the selection rule $\underline{K} = 0$ the exciton spectrum is a series of lines with f values

$$f = \frac{2\Omega}{mh\nu} |\langle c|\underline{\varepsilon}.\underline{p}|v\rangle|^2 \; |\phi_i(0)|^2 \tag{5.3}$$

which is like an atomic transition multiplied by the probability that the electron and hole are on the same atom. For the simple case of spherical bands at $\underline{K} = 0$ we have a hydrogenic spectrum at

$$E = E_G - R/n^2 \tag{5.4}$$

with intensity falling like $n^{-3}$ since

$$|\phi_1(0)|^2 = (\pi a_o^3 n^3)^{-1} \tag{5.5}$$

where $a_o$ is the exciton radius $\hbar^2\varepsilon_o/\mu e^2$. Only s-like states are observed.

Beyond $E_G$, in the continuum corresponding to ionised pair states the Coulomb interaction still affects the relative motions and

$$\phi_i(0) = \frac{1}{N\Omega}\left(\frac{\pi\alpha\, e^{\pi\alpha}}{\sinh\pi\alpha}\right)^{\frac{1}{2}} , \quad \alpha^2 = R/E\text{-}E_G \tag{5.6}$$

The absorption coefficient form (3.3) can be written

$$\kappa = \kappa_A \; \pi\alpha \; e^{\pi\alpha}/\sinh \pi\alpha \tag{5.7}$$

ie. the absorption $\kappa_A$ due to free pairs (3.10) multiplied by the Sommerfeld factor representing the final state interaction. This greatly increases the absorption close to the edge.

If transitions are forbidden between band states at the extremum so that only the second term in (3.8) is present

$$\int\Psi^*_{ex}\; \underline{\varepsilon}.\underline{p}\; \Psi_o = \sum_{k_e k_h} \hbar\; \Phi(\underline{k}_e,\underline{k}_h)\underline{k}_h.\langle c|\underline{M}|v\rangle/m \tag{5.8}$$

with $\underline{k_e} = -\underline{k_h}$ and $\underline{K} = 0$. Then the f value becomes

$$f_1 = \frac{2\Omega\hbar^2}{3m^2h\nu} |\langle c|\underline{M}|v\rangle|^2 \; |\nabla\phi_i(0)|^2 \tag{5.9}$$

In the hydrogenic case

$$|\nabla\phi_i(0)|^2 = (n^2-1)/\pi a_o^5 n^5 \tag{5.10}$$

for p relative motion only. The excitons is again a hydrogenic series but the n=1 line is missing. Beyond the continuum limit the absorption is increased over the free pair result (3.11).

$$\kappa = \kappa_F \pi\alpha(1+\alpha^2)e^{\pi\alpha}/\sinh \pi\alpha \tag{5.11}$$

Sketches of the absorption near the band edge are shown in figure 2

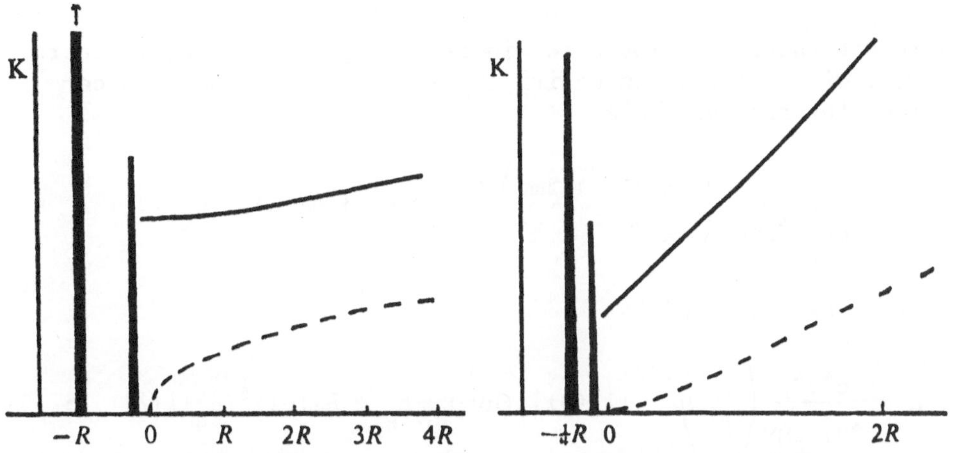


FIG. 2. Absorption near edge for (A) Direct allowed and (B) Indirect allowed cases. Only first two lines are shown. Free pair absoprtion is dashed curves.

Indirect Transitions

In the preceeding section we saw that the only allowed optical transitions were to excitons with $\underline{K} = 0$. In some important semi-conductors the extrema of the valence and conduction bands are at different $\underline{k}_o$ so that the gap is indirect. The lowest energy excitons will leave $\underline{K} \sim \underline{k}_{co}-\underline{k}_{vo}$. In fact weak transitions to such states can take place if the momentum is conserved by the creation or destruction of a phonon with $\pm\underline{K}$. The hamiltonian for electron phonon is

$$H = \sum_{kq} V(\underline{k},\underline{q})[c(\underline{q})+c^+(-\underline{q})]a^+_{n'}(\underline{k}+\underline{q})a_n(\underline{k}) \tag{5.12}$$

Using perturbation theory there is a process in which the photon creates a virtual pair at $\underline{K} = 0$ and the electron or the hole are scattered by (5.12) to reach a final pair state with $\underline{K} = \underline{q}$. The details of the matrix element are complicated but it will be approximately constant over the small regions of $\underline{k}$ and $\underline{k}'$ which are important. The absorption profile is therefore controlled by the density of final states. For free pairs with parabolic bands, the joint density of states is proportional to $(E-E_G)^2$, while for a single exciton band it is $(E-E_G-E_{ex})^{\frac{1}{2}}$. In all processes we may create an exciton with probability $(n_q+1)$ or destroy one with probability $n_q$ where

$$-n_q = [1 - \exp(\hbar\omega(\underline{q})/kT)]^{-1} \tag{5.13}$$

As a result there are always two indirect edges seperated in energy by $2\hbar\omega(\underline{q})$ with relative intensity $(n_q+1)/n_q$. The absorption coefficient for free pairs is

$$\kappa = \frac{e^2V^2}{16\pi^2m^2c\eta\nu}\left(\frac{2m_c}{\hbar^2}\right)^{3/2}\left(\frac{2mv}{\hbar^2}\right)^{3/2}(\hbar\nu-E_G \mp \hbar\omega(q))^2(n_q+\tfrac{1}{2}\pm\tfrac{1}{2}) \tag{5.14}$$

and for an exciton band is

$$\kappa = \frac{e^2V^2}{2\pi m^2c\eta\nu}\left(\frac{2M}{\hbar^2}\right)^{3/2}|\phi_1(0)|^2(h\nu-E_G-E_{ex}\mp \hbar\omega(q))^{\frac{1}{2}}(n_q+\tfrac{1}{2}\pm\tfrac{1}{2}) \tag{5.15}$$

## EXAMPLES OF EXCITON SPECTRA

Exciton spectra may be broadly classed either as direct with a series of sharp lines leading to a continuum, or indirect beginning with a series of relatively weak steps. Many semiconductors, particularly Si, Ge, III-V and II-VI compounds have been studied and examples of both types are found. The elements Si and Ge[(5)] have indirect gaps as do AgBr and PbTe. A sketch of the absorption spectra of Ge is shown in Fig. 3.

In addition to the indirect edge, a direct exciton has been observed together with other features at higher energy. These may be interpreted in terms of the known band structure, sketched in fig. 4.

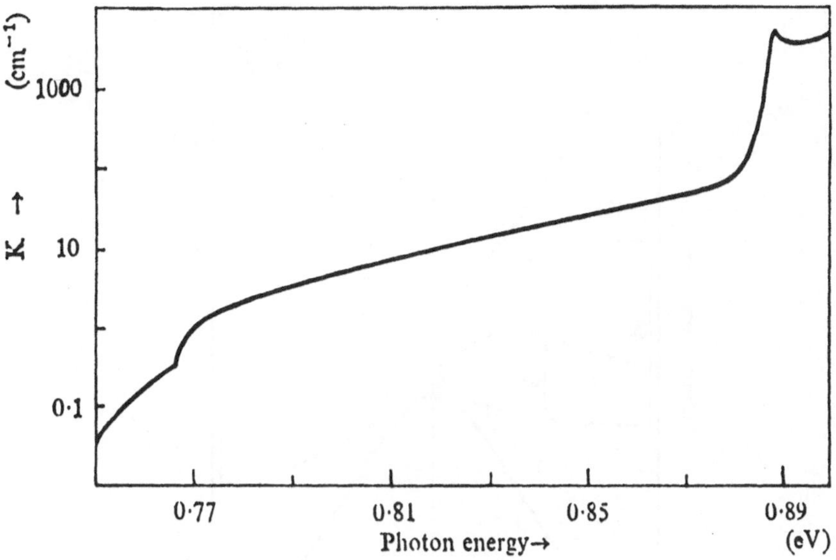


FIG. 3.A. The weak indirect edge in Ge (after ref. 5).

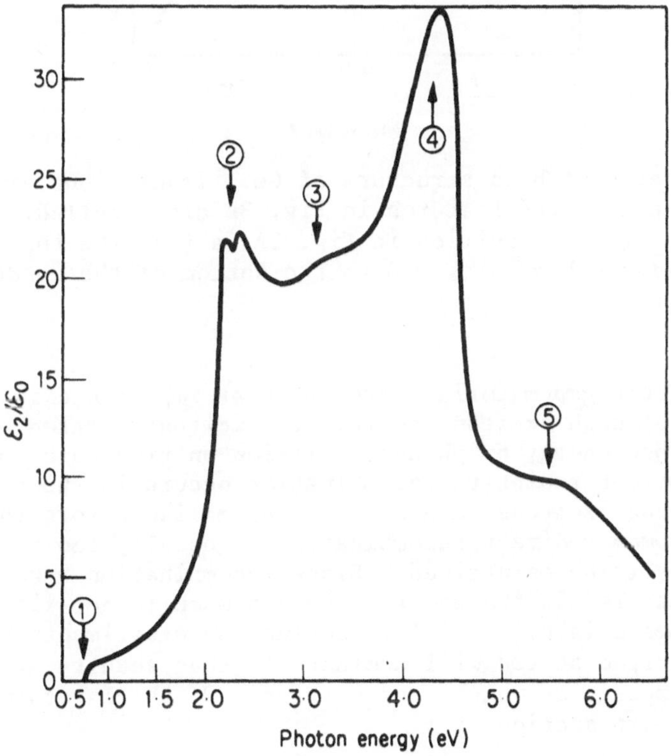


FIG. 3.B. Absorption (imaginary part of dielectric constant) of Ge deduced from reflectivity data (after ref. 6).

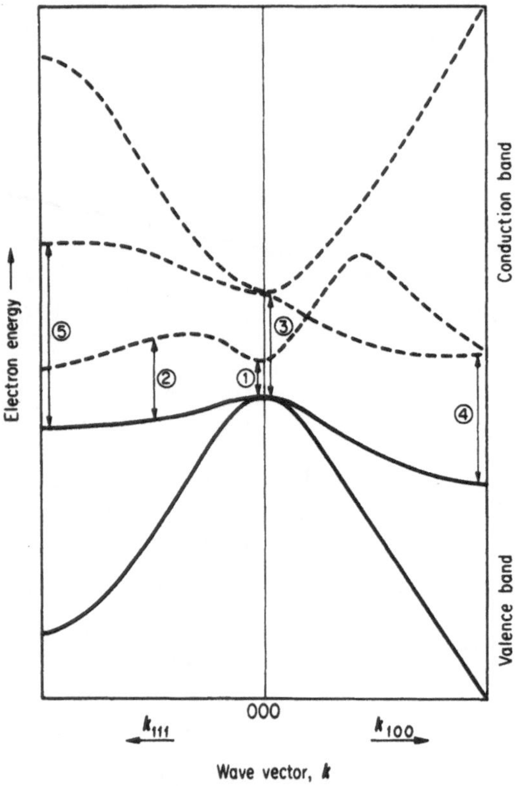


FIG. 4. Sketch of band structure of Ge. Transitions giving the numbered features in Fig. 3B are labelled. The indirect transition in Fig. 3A is from the top of the valence band at $\underline{k} = 0$ to the bottom of the conduction band at the left hand edge.

The indirect gap materials are of great importance in the modern study of high exciton densities. Excitons created with high energy will lose energy by phonon collision on timescales of order $10^{-12}$ sec. Direct radiative recombination occurs in about $10^{-8}$ sec. so that the excitons have time to thermalise before they are destroyed. However direct recombination is usually too rapid for a high density to be maintained. Since recombination takes place from the lowest levels the excitons have a much longer life in indirect gap materials. In all recombination experiments however the lowest excited states will dominate so that results are often affected by impurities which create traps. We shall return to this point in the next section.

The most detailed investigations of direct exciton spectra have been made in $Cu_2O$(7). This is one of the few known forbidden spectra ($SnO_2$ is another) and has a particularly rich structure. Basically it shows two well defined hydrogenic series in the yellow and the green which arise from a spin-orbit split valence band[8]. The highest hole band is single of $\Gamma_7$ symmetry

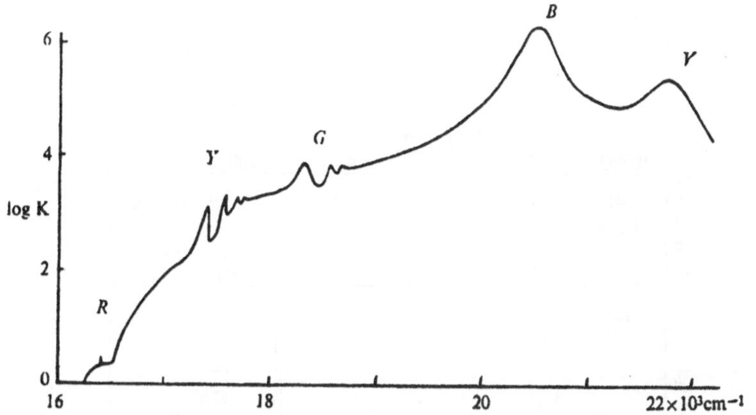


FIG. 5. Sketch of absorption in $Cu_2O$. The absorption edges at R are due to indirect transitions to 1s excitons. The weak 1s line is halfway between. Y and G are the yellow and green exciton series. The blue (B) and violet (V) features are due to higher conduction bands.

and the lower (green) band is double $\Gamma_8$. It is therefore slightly surprising that the hydrogenic series is not more perturbed. Detailed calculations for such a band structure have been made by Baldereschi and Lipari[9]. The degeneracy of the bands and the fluting which causes deviations from spherical symmetry give important effects. The only other exciton which is easily observable in absorption is the lowest 1s yellow. Although forbidden to dipole radiation it is allowed weakly by quadrupole, and appears as a weak sharp line. It also displays indirect edges by the simultaneous creation of optic phonons. The position of this line is not given satisfactorily by the hydrogenic series formula. This is due in part to its relatively small radius. The approximate interaction (4.1) is not valid at small r, partly because of the polaron effect (4.2) and partly because of the atomic structure. It is usual to include a central cell correction to represent the electron-hole interaction when they are on the same atom. Further complications arise from the electron spin. All the yellow excitons in the model are four-fold degenerate while the green excitons have eight fold degeneracy. There is also an effective exchange interaction when the electron and hole are on the central atom which splits the s-excitons into triplet ortho excitons and singlet para excitons. The observed 1s line is ortho, the para component is at lower energies. A much more detailed understanding of the $Cu_2O$ spectrum has resulted from the work of Frohlich et al. (11) who measured the two-photon absorption cross-section (see fig. 6) which involves the even parity s and d excitons. They are able to give a satisfactory interpretation of the observed s p and d excitons of the yellow and green series using six valence and

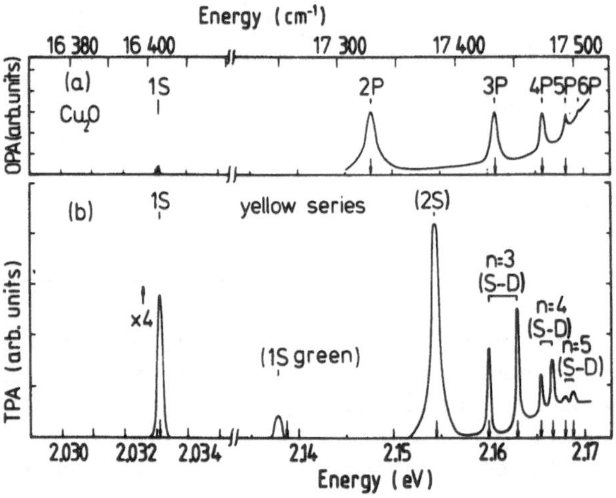


FIG. 6. One and two photo spectra of $Cu_2O$ (after ref. 11)

two conduction bands. The valence bands are labelled in terms of a pseudo angular momentum I = 1 and a spin $S_h$. The hamiltonian of relative motion is then the sum of the usual term

$$H_o = \frac{\hbar^2}{2\mu}\nabla^2 - \frac{e^2}{\varepsilon_o r} \tag{6.1}$$

the central cell correction and exchange terms

$$H_1 = [V_o - J_o(\underline{S}_e \cdot \underline{S}_h)]\delta(\underline{r}) \tag{6.2}$$

the spin orbit coupling

$$H_2 = 2/3\ \Delta(\underline{I}.\underline{S}_h) \tag{6.3}$$

and the term which describes the valence band structure

$$H_3 = \frac{\hbar^2}{3\gamma}(P^{(2)}.I^{(2)}) \tag{6.4}$$

where $P^{(2)}$, $I^{(2)}$ are the second order harmonics of the operators $\nabla$ and $\underline{I}$.

EXCITON - IMPURITY COMPLEX

In the effective mass approximation, with an impurity at the origin, the hamiltonian is

$$H = -\frac{\hbar^2\nabla_e^2}{2m_e} - \frac{\hbar^2\nabla_h^2}{2m_h} - \frac{Ze^2}{\varepsilon r_e} + \frac{Ze^2}{\varepsilon r_h} - \frac{e^2}{\varepsilon|r_e - r_h|} \tag{7.1}$$

The resulting equation for $\Phi(\underline{r}_e,\underline{r}_h)$ is too difficult to solve analytically. For a donor impurity $Z > 0$, we can get a feeling for the result in various limits. For example if $m_h \gg m_e$ the system is analagous to a hydrogen molecule ion. The binding energy $E_c$ is then a considerable fraction of the exciton Rydberg. On the other hand if $m_h$ is light ($\sim m_e$) it is a reasonable approximation to consider the exciton as an entity weakly bound to the impurity. Then

$$\Phi(\underline{r}_e,\underline{r}_h) = \chi(\underline{R})\phi(\underline{r}) \tag{7.2}$$

where $\phi$ is the usual exciton function of relative motion and $\chi(R)$ describes the motion of the exciton mid-point around the origin.

In analogy with equation (5.3) the f-value of the transition

$$f = \frac{2}{mh\nu}\,|\langle c|\underline{\varepsilon}.\underline{p}|v\rangle|^2\,|\int\Phi(\underline{r}_e,\underline{r}_c)\,d\underline{r}_e|^2\,N_I \tag{7.3}$$

The electron-hole pair creation again takes place only when $\underline{r}_e = \underline{r}_h$ The integral over $\Phi$ gives roughly the volume of the bound state modulated by the probability that the electron and hole will be together. For the approximation (7.2)

$$\int\Phi(\underline{r}_e,\underline{r}_e)d\underline{r}_e = \int\chi(\underline{R})d\underline{R}\;\phi(0) \tag{7.4}$$

so the second factor is the same as in the free exciton. In this case

$$\frac{f_I}{f_{ex}} = \frac{N_I}{N}\,\frac{V}{\Omega} \tag{7.5}$$

ie. it contains the ratio of V, the volume of the impurity function to $\Omega$ the atomic volume. This can be very large giving the so-called Giant Effects[12].

However one must not expect the impurity line ever to exceed the exciton intensity. If $N_I V$ becomes comparable to $N\Omega$ the assumptions break down. The impurity is only attracting intensity from the exciton line.

The effects are however very important in fluorescense. Since the impurity level is usually at a lower energy it can trap the excitation energy and flourescence takes place preferentially from the sharp impurity lines.

Other interesting effects are expected to occur in this energy region, unfortunately they are difficult to distinguish from the impurity lines.

BIEXCITONS

Excitons may also be bound to each other to form biexcitons. Again we can make an analogy with hydrogen molecules in the limit when $m_h >> m_e$. The binding energy decreases when $m_e \sim m_h$ - which is analagous to the case of positronium molecules. It now costs energy to localise the holes as well as the electrons. The best variational calculations give

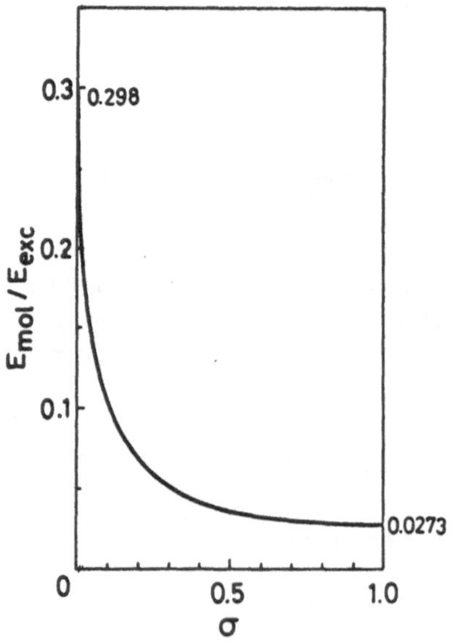


FIG. 7. The binding energy of the biexciton as a function of $\sigma = m_e/m_h$.

Near $m_e \sim m_h$ the binding energy increases somewhat with mass anistropy.

As for $H_2$ the best binding is for S = 0 states when effects of the exclusion principle can be minimised. For this reason in many-valley semiconductors large molecules, 8 excitons in Ge and 12 in Si may be stable.

Calculated binding energies in meV

| | $E_{ex}$ | $E_B$ |
|---|---|---|
| Cu Cl | 190 | 44 |
| $Cu_2O$ (yellow) | 97 | 3.3 |

Optical absorption by biexcitons is difficult to find experimentally since it presupposes a high density present. Flourescence is however likely since the biexcitons are lower in energy.

The transition probability where a pair of momentum $\underline{K}$ transform to a single exciton of $\underline{K}$ is

$$\int \Phi_B(\underline{r}_e, \underline{r}_e', \underline{r}_e, \underline{r}_h') \, \Psi_{ex}(\underline{r}_e' \underline{r}_h') \, d\underline{r}_e \, d\underline{r}_e' \, d\underline{r}_h \tag{8.1}$$

since, as usual, the transition requires $\underline{r}_h = \underline{r}_e$. If we regard the biexciton as a weakly bound pair of excitons

$$\Phi_B = [\Phi_{ex}(\underline{r}_e - \underline{r}_h) \, \Phi_{ex}(\underline{r}_e' - \underline{r}_h') - \Phi_{ex}(\underline{r}_e - \underline{r}_h') \, \Phi_{ex}(\underline{r}_e' - \underline{r}_h)] \times$$

$$\times \, \chi(\underline{R} - \underline{R}') \exp i\tfrac{1}{2}\underline{K}.(\underline{R} + \underline{R}') \tag{8.2}$$

and the result is approximately

$$\frac{f_B}{f_{ex}} = 2 \left| \int \chi(\underline{R} - \underline{R}') \exp i\tfrac{1}{2}\underline{K}(\underline{R} - \underline{R}') \, d(\underline{R} - \underline{R}') \right|^2 \tag{8.3}$$

which may again be large at small K, as in the impurity case and falls off as $(K^2 R_B^2) \sim 1$. Thus the emission should be broad and similar in shape to that observed from indirect excitons. The absorption is roughly of the form $\Delta E^{\frac{1}{2}}[1-(\Delta E/E_B]^2$ while the emission has an additional thermal factor $\exp(-\Delta E/k_T)$. The T dependence of the width should seperate biexcitons form impurity bound excitons.

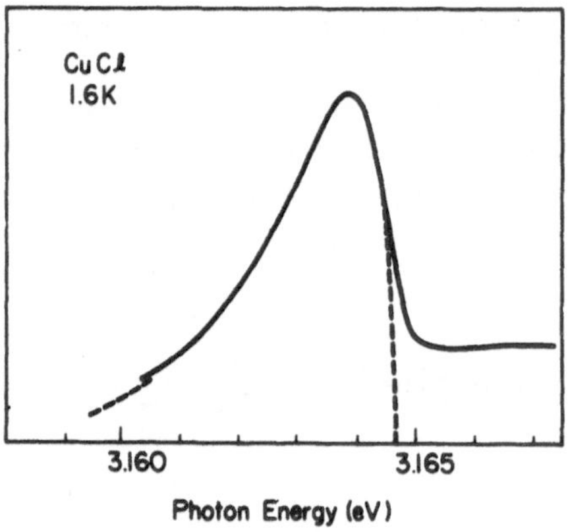


FIG. 8. Emission spectra of biexcitons in CuCl (after ref. 15). The dashed curve is the theoretical prediction for $T_{ex}$ = 26K.

## EFFECTS AT HIGH EXCITON DENSITY - BOSE CONDENSATION

If the exciton density is increased, usually by laser excitation in indirect gap materials where recombination is relatively slow, the possibility exists of a number of interesting phases. This can be seen by continuing the analogy with hydrogen. We know that this forms a gas of molecules which at low T condenses into a solid, because of the intermolecular forces. If these were smaller, Bose-Einstein condensation is expected, a form of which may be said to occur in He. On the other hand at higher densities as overlap increases and ionisation of the constituents is made easier by screening, hydrogen is expected to be a metal. These phenomena are changed quantitatively in excitons by the easier delocalisation of the lighter hole. We have already seen that the relative binding energy of biexcitons decreases as the hole-electron mass ratio decreases. We now consider the possible form and manifestation of these phases in excitonic matter. In so doing we shall normally assume simple isotropic bands and spinless particles, more complicated effects may exist when there are many degenerate bands, and for orthoexcitons.

The possibility of Bose condensation of excitons was first raised by Keldysh[16] and has been pursued by many authors [17]. In principle this is expected for a gas of excitons or of biexcitons since both are bosons. A perfect non-interacting gas of such particles condenses at

$$kT_c \sim \frac{2\pi\hbar^2}{M}\left(\frac{N}{V}\right)^{2/3} 0.575 \qquad (9.1)$$

which is several $^oK$ for densities $\sim 10^{17}/cm^3$. Below $T_c$ a finite fraction of the excitons condense into the $\underline{K} = 0$ state; this fraction varies with T as

$$\frac{No}{N} = 1 - \left(\frac{T}{T_c}\right)^{3/2} \tag{9.2}$$

Direct observation of this effect should be visible in emission from indirect excitons. The profile shape of the continuum will be like that in figure 8 , and the condensate will give a sharp line below this, but unfortunately this is easily confused with bound exciton emission. For biexcitons a similar result is expected (17) An alternative possibility is the creation of biexcitons by two-photon absorption in direct gap materials.

This picture of non-interacting bosons is oversimplified in two ways. For one thing the particles are not true bosons since they are built of fermions, and the operators (4.3) only have approximate Bose commutation rules. This problem has recently been considered again in detail by Nozieres and Comte (18). The ground state of the system with $N_0$ particles is approximately $(b_0)^{N0}|0>$ that the operators $b_0, b_0^+$ act like c-numbers with value $N_0^{\frac{1}{2}}$. A better ground state function which allows for fluctuations in $N_0$ is $e^{\lambda b0}|0>$ where $\lambda$ fixes the number of particles. Using (4.3) this may be written

$$\prod_k \exp(\lambda\Phi(\underline{k},\underline{k})a_c^+(\underline{k})a_v(\underline{k}))|0> = \prod_k [x_k + y_k\, a_c^+(\underline{k})a_v(\underline{k})]|0> \tag{9.3}$$

where $y_k/x_k = \lambda\Phi(\underline{k},\underline{k})$ and $x_k^2 + y_k^2 = 1$. In fact this state is the dilute Bose condensed state; $y_k$ gives the shape of the fermion occupation due to the excitons and $\lambda = N_0^{\frac{1}{2}}$ fixes the density. However as $N_0$ increase the normalisation condition stops the growth of $y_k$ which must remain < 1. The $N_0$ excitons exhaust the underlying stock of exciton states. In order to increase the number $y_k$ must spread farther in $\underline{k}$-space. For the 1s exciton state

$$\Phi(\underline{k},\underline{k}) = \frac{(4\pi a_0^3)^{\frac{1}{2}}}{(1+k^2a_0^2)^2} \tag{9.4}$$

so that this occurs when $N_0 a_0^3 \sim 1$, ie. when the excitons begin to overlap.

At high density $N_0 a_0^3 >> 1$, the state describes the degenerate electron-hole plasma with

$$y_K^{\ 2} = 1,\ k<k_F; \quad \text{and} \quad y_K^{\ 2} = 0 \quad k<k_F \tag{9.5}$$

with $k_F = (6\pi^2 N)^{1/3}$ as the fermi wave vector. The variation of $y_K^{\ 2}$ with density is shown in figure 9.

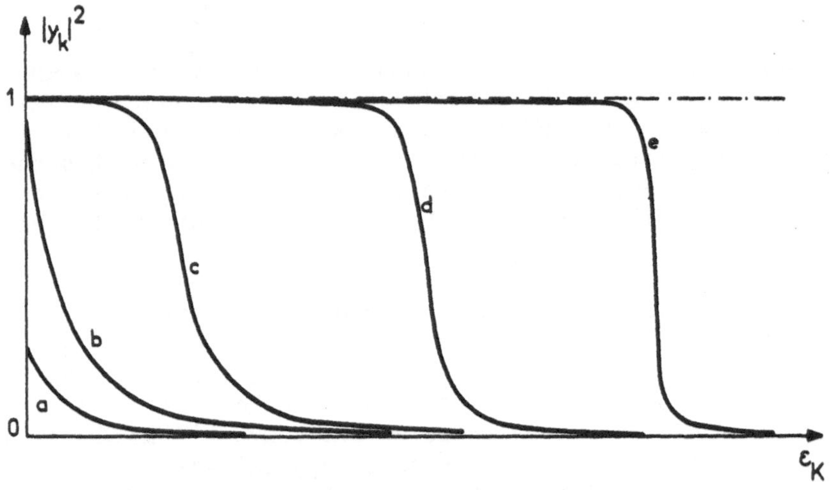


FIG. 9. The ground state fermion distribution $y_K^{\ 2}$ as a function of pair energy $\varepsilon(k)$ for various densities. Curves (a) and (b) correspond to low density, (c) to intermediate density, (d) and (e) to high density.

Thus for non-interacting excitons this theory predicts a continuous transition from a Bose condensate to an electron-hole plasma with increasing density. However the residual interaction between the bosons will affect this result. Experience of normal matter suggest that this will cause a discontinuous transition from the plasma to the exciton gas and that the latter will become Bose condensed.

A residual interaction between bosons modifies the hamiltonian

$$H = \sum_k \varepsilon_k\, b_k^+ b_k + \sum_{kk'q} v(q)\, b_{k+q}^+\, b_{k'-q'}^+\, b_k b_{k'} \tag{9.6}$$

We can get a crude result by treating $b_0\ b_0^+$ as number equal to $(N_0)^{\frac{1}{2}}$ where

$$N_0 = N - \sum_k \langle b_k^+\ b_k \rangle \tag{9.7}$$

Then

$$H_{eff} = \tfrac{1}{2}N\rho v(0) + \sum_k \varepsilon_k b_k^+ b_k + \tfrac{1}{2}\rho \sum_k v(k)(2b_k^+ b_k + b_k^+ b_{-k}^+ + b_k b_{-k}) \tag{9.8}$$

this can be diagonalised by a Bogoliubov transformation

$$b_k = \frac{1}{(1-u_k^2)^{\frac{1}{2}}} (B_k + u_k B_{-k}^+) \tag{9.9}$$

when

$$H = \sum_k E_k B_k^+ B_k \tag{9.10}$$

and

$$E_k^2 = [\varepsilon_k^2 + 2\varepsilon_k \rho v(k)]$$

$$\rightarrow \hbar s.k \qquad k \rightarrow 0 \quad \text{(sound waves)}$$

$$\rightarrow \varepsilon_k \qquad k \rightarrow \infty \quad \text{(single particle)} \tag{9.11}$$

From this we find

$$n_k = \langle b_k^+ b_k \rangle = \frac{(\varepsilon_k + \rho v(k))}{E_k} (\nu(k) + \tfrac{1}{2}) - \tfrac{1}{2} \tag{9.12}$$

where $\nu(k)$ is the number of quasi-particles $[\exp(\beta E_k)-1]^{-1}$ $n_k$ is increased and $N_0$ decreased by the interaction. The shape of the emission will also be changed. These effects have not yet been observed in detail. The plasma state has however been investigated in detail and will be the subject of Dr. Reinecke's lectures. The possibility of an exciton solid has been discussed by Nikitine[(19)].

REFERENCES


1. The original ideas are due to G. Wannier, Phys. Rev. 52, 191, (1937) and have been extended by many authors, e.g. G. Dresselhaus, J. Phys. Chem. Sol. 1, 14, (1955).
2. L. Roth and G. Platt, J. Phys. Chem. Sol. 8, 47, (1957).
3. H. Haken in "Polarons and Excitons", Kuper and Whitfield ed; p.294 (Oliver and Boyd, Edinburgh 1963).
4. R.J. Elliott, Phys. Rev. 108, 1384, (1957).
5. T.P. McLean, Prog. in Semiconductors, 5, 221, (1960).
6. D. Brust, J.C. Phillips and G.F. Bassani, Phys. Rev. Lett, 9 94, (1962).

G. Harbeke, Z.F. Naturforsch, 19a, 548, (1964).
7. S. Nikitine, Prog. in Semiconductors, 6, 235, (1962).
8. L. Kleinman and K. Mednick, Phys. Rev. B21, 1549, (1979).
9. A. Baldereschi and N.O. Lipari, Phys. Rev. B8, 2697, (1973).
10. R.J. Elliott, Phys. Rev. 124, 340, (1961).
11. D. Frohlich, C.H. Uihlein and R. Kenklies, Phys. Rev. B23, 2731, (1981).
12. E.I. Rashba, Springer Tracts Mod. Phys. 73, 150, (1975).
13. S. Nikitine, Springer Tracts Mod. Phys. 73, 18, (1975).
14. O. Akimoto and H. Hanamura, Solid State. Comm. 12, 227, (1969)
15. N. Nagusawa, N. Nakata, Y. Doi and M. Ueta, J. Phys. Soc. Japan, 38, 593; 39, 987, (1975).
16. L.V. Keldysh et al., Sov. Phys; Solid State 6, 2219, (1965), JETP, 27, 521, (1968).
17. For an extensive review see E. Hanamura and H. Haug, Phys. Rev. Rep. 33, 210, (1977).
18. P. Nozières and C. Comte, J. de Phys. (1982 in press).
19. S. Nikitine, Optics Comm. 35, 377 (1980).

# EXCITONS AND EXCITON RELAXATION IN SILVER HALIDES

W. von der Osten

Fachbereich 6 - Naturwissenschaften I
Universität - Gesamthochschule
4790 Paderborn, West Germany

## 1. INTRODUCTION

The systematic study of excitons in silver halides goes back to the early Thirties when Fesefeldt (1930) in Göttingen started to measure thin film absorption of these materials. Since that time a huge amount of experimental and theoretical results concerning their optical as well as many other properties was reported and, at the times, reviewed in several articles (for more recent ones see Brown,1973; Kanzaki, 1980; von der Osten, 1982). It is not intended in this lecture to go over the grounds covered by these. Rather its purpose is to describe and interpret more recent results concerning properties of excitons in these materials as revealed by absorption and luminescence thereby stressing one specific aspect, namely exciton relaxation. A central question that we will try to answer is, how an exciton created by optical excitation in some higher excited state will lose its energy during relaxation going into a state of lower energy from where it finally may radiatively return into the crystal ground state.

The content of this lecture will be organized in the following way: in the next section the necessary background to the problem will briefly be reviewed by introducing the electronic band structure and the phonon dispersion of silver halides. In section 3, absorption and luminescence due to phonon-assisted "indirect" exciton transitions will be treated with special emphasis to lineshapes and exciton binding energies. In this chapter the relaxed exciton state will be discussed too, which in silver halides may either be a free exciton (FE) or a self-trapped exciton (STE) depending on the substance. Finally section 4 deals with resonant Raman scattering (RRS), a modern technique in which tunable lasers are employed as intense and monochromatic light sources. Starting out from the discussion of nonresonant second-order Raman scattering (RS) expressions for the RRS cross-section are derived. It will then be demonstrated that RRS is capable to reveal hitherto unknown exciton properties, including spin properties, and provides new information on exciton dispersion. In particular, the interaction of the exciton with various lattice phonons may be investigated, the results alltogether giving a consistent picture of exciton relaxation in silver halides.

The discussion exclusively will refer to AgCl and AgBr and, to certain extent, to mixed crystals of $AgBr_{1-x}Cl_x$. The results, however, in many respects are of more general relevance applying to other polar semiconductors too. In fact, silver halides occupy an intermediate position between semiconducting materials and ionic crystals. A measure for the ionic contribution to the binding is Phillips' ionicity (Phillips, 1970) that for AgCl and AgBr is just large enough to favour the rocksalt structure. Although the silver halides exhibit many properties typical for semiconductors it is this simple crystal structure which is of specific advantage in the study of excitons resulting in simple selection rules and therefore incomplicated spectra. In contrast, due to its smaller ionicity AgI already is tetrahedrally coordinated showing quite different properties and thus not being considered here.

## 2. ELECTRONS AND PHONONS IN SILVER HALIDES

### 2.1 Electronic Band Structure

The basis for the discussion of excitons and exciton properties is the electronic band structure. For the two silver halides AgBr and AgCl various methods of computation were applied. Among these are the early calculations by Bassani et al. (1965), who applied the tight-binding approach introducing spin-orbit effects only in a semiempirical way. Although highly speculative at the time it was proposed, since then these calculations were confirmed in their essential features by optical and transport experiments. A set of very recent KKR-computations of band structure for AgCl, AgBr and also f.c.c. AgI including spin-orbit interaction was recently performed by Overhof (1981; see von der Osten, 1982). These nicely demonstrate the systematic changes that occur for these materials in varying the anion (Fig. 1 for AgBr).

From these calculations the lowest conduction band is simple, isotropic and s-like with the minimum at $\Gamma$ ($\Gamma_1$ or, including spin, $\Gamma_6^+$). It essentially compares to the case of the alkali halides. In contrast to these, however, the valence band structure of the silver halides is more complex and rather compares to that of

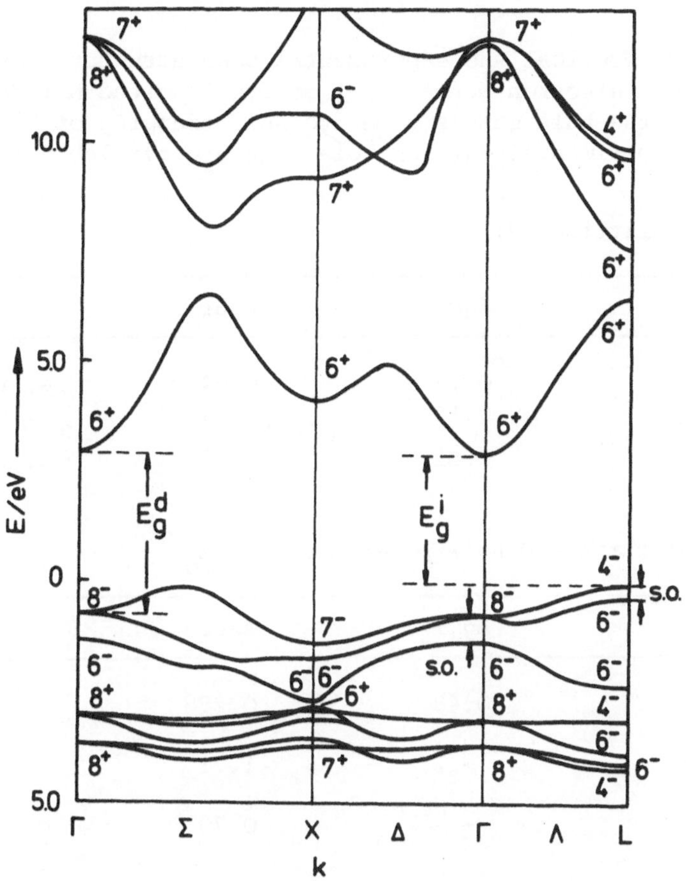


Fig. 1. Electronic band structure of AgBr calculated by means of the relativistic KKR method (Overhof, 1981). $E_g^d$ and $E_g^i$ denote energies of the lowest direct and indirect gap, s.o. marks spin-orbit split valence band states at Γ and L.

many semiconducting materials. It reflects a considerable degree of covalency and is largely due to the close proximity of the silver 4d- and halogen p-states that lie within about 1 eV to each other. Due to the inversion symmetry no mixing of these states occurs at Γ ($\vec{k} = 0$). There the valence band is either pure p of the halogen ($\Gamma_{15}$) or d of the metal ($\Gamma_{12}$, $\Gamma_{25'}$). Because of strong mixing for $\vec{k} \neq 0$ the valence band states repel each other, resulting in the uppermost maximum at L and giving rise to the indirect character. The valence band maximum at L is nearly degenerate with another maximum at Σ, only slightly lower in energy. Considering the effect of spin-orbit interaction, $\Gamma_{15}$ splits into $\Gamma_8^-$ and $\Gamma_6^-$ while $L_3$ splits into $L_4^-$, $L_5^-$ (degenerate) and $L_6^-$, with $L_6^-$ lower in energy (Fig. 1). Representative electronic band structure parameters in comparison with various experimental data are compiled in Table 1.

Table 1. Theoretical and experimental band structure parameters. $m_o$: electron mass; $\parallel$, $\perp$ denote longitudinal and transverse hole masses. Due to self-trapping of holes no experimental data on hole mass are available for AgCl.

Spin-orbit splitting/eV:

| | AgCl | AgBr | Ref. |
|---|---|---|---|
| $\Gamma_8^- - \Gamma_6^-$ | 0.16 | 0.64 | a, theor. |
| | 0.14 | 0.54 | b, exp. |
| $L_{4,5}^- - L_6^-$ | 0.02 | 0.28 | a, theor. |

Effective electron and hole polaron masses/$m_o$:

| | AgCl | AgBr | Ref. |
|---|---|---|---|
| $m_e^{**}(\Gamma)$ | 0.43 | 0.288 | c,d, exp. |
| $m_{h,\parallel}^{**}$ (L) | - | 1.71 | c, exp. |
| $m_{h,\perp}^{**}$ (L) | - | 0.79 | c, exp. |

Effective electron and hole band masses/$m_o$:

| | AgCl | AgBr | Ref. |
|---|---|---|---|
| $m_e^{*}(\Gamma)$ | 0.302 | 0.215 | e, exp. |
| | 0.40 | 0.366 | a, theor. |
| $m_{h,\parallel}^{*}$ (L) | - | 1.25 | e, exp. |
| | 1.563 | 2.069 | a, theor. |
| $m_{h,\perp}^{*}$ (L) | - | 0.52 | e, exp. |

a from KKR calculation (Overhof, 1981)
b from uv transmission (Carrera and Brown, 1971)
c from cyclotron resonance (Tamura and Masumi, 1973)
d from cyclotron resonance (Hodby, 1971)
e from cyclotron resonance data using various coupling calculations

As will be discussed later, the lowest exciton state being the main topic in this lecture is constructed out of a hole and an electron with $L_{4,5}^-$ and $\Gamma_6^+$ symmetry, respectively, resulting in an exciton with finite wavevector ($\vec{k}_L$).

## 2.2 Phonon Dispersion

Compared to alkali halides, the presence of the silver d-electrons gives also rise to qualitative differences in the lattice properties. One example of this kind is the invalidity of Cauchy's relation that holds for most alkali halides relating to each other the elastic constants $c_{12} = c_{44}$, the violation for silver halides indicating that notcentral forces play a major role in binding. A particularly clear picture of the differences in lattice dynamics is contained in the extended shell model by Fischer et al. (1972) which was successfully used to describe the measured phonon dispersion of both AgCl (Vijayaraghavan et al., 1970) and AgBr (Dorner et al., 1976). Without going into detail the most striking consequence of this model, as far as the following discussion of the indirect exciton is concerned, is the inversion of the transverse modes with regard to their eigenvectors near the L point in AgBr (von der Osten and Dorner, 1975). This effect is not observed for AgCl leading to the different notations of the momentum-conserving phonons for both materials in section 3.3. Since phonons are important in context with the indirect exciton and exciton relaxation, in Fig. 2 the phonon dispersion of AgBr is representatively reproduced.

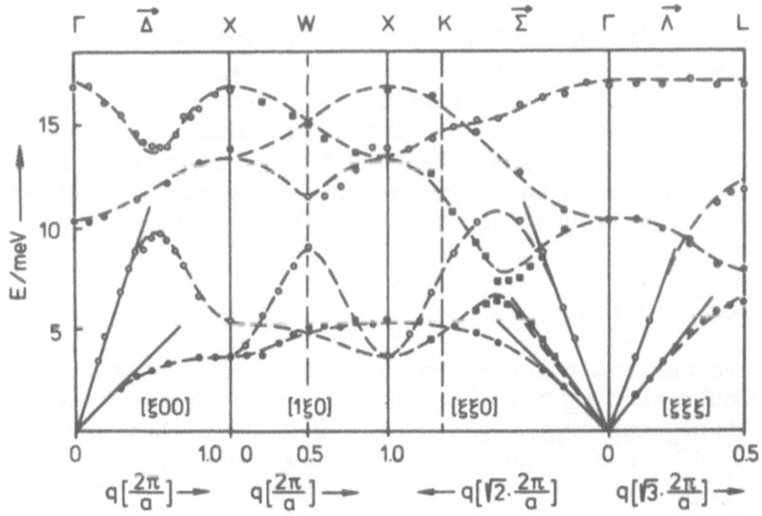


Fig. 2. Phonon dispersion in AgBr at 85K as obtained from inelastic neutron scattering. Broken curves are a fit of an extended shell model (Dorner et al., 1976).

## 3. THE INDIRECT EXCITON

The indirect character of the band structure of silver halides is confirmed by the long wavelength tail of absorption that extends below the direct exciton peaks down to about 1.5 to 2.0 eV. It is associated with the phonon-assisted "indirect" exciton formation. At room temperature, this weak absorption tail is the origin for the light sensitivity of these materials that causes a photolytic darkening of the crystals even without any sensitizer used in modern photographic silver halide emulsions. At sufficiently low temperatures, however, the formation of photolytic silver is suppressed due to the frozen-in ionic mobility. Then the investigation of the exciton edge by means of various spectroscopic techniques, including conventional absorption, stress- and wavelength-modulation, luminescence and resonant light scattering spectroscopy, reveals details of the excitonic structure and exciton-phonon interaction.

### 3.1 Selection Rules

For transitions between a valence band maximum and a conduction band minimum with wavevectors $\vec{k}_v$ and $\vec{k}_c$ in the Brillouin zone involving only a photon, wavevector conservation requires that

$$\vec{k}_v - \vec{k}_c = \vec{k}_{photon} \qquad (1).$$

Compared to the extension of the Brillouin zone ($\approx 10^8$ $cm^{-1}$) and in the energy range of interest ($\approx$ 2 eV) $\vec{k}_{photon} \approx 0$. Eq.1 therefore allows only for "direct" transitions with $\vec{k}_v \approx \vec{k}_c$. If the wavevectors of the valence band maximum and conduction band minimum differ, a transition between these may become allowed in a second-order process involving an additional phonon with wavevector $\vec{q}$. The wavevector requirement in this case is given by

$$\vec{k}_v - \vec{k}_c = \vec{k}_{photon} + \vec{q} \qquad (2)$$

where we have omitted a reciprocal lattice vector at the right-hand side that would account for the translational invariance of the crystal. Since again $\vec{k}_{photon} \approx 0$ eq. 2 implies that any difference in $\vec{k}_v$ and $\vec{k}_c$ has to be compensated by a phonon of appropriate wavevector $\vec{q}$.

For this photon-phonon process the probability for a transition from the valence band (energy $E_v$, wavevector $\vec{k}$) to the conduction band ($E_c$, $\vec{k}'$) is obtained from applying second-order perturbation theory (Elliott, 1957). It is proportional to

$$W_{v,\vec{k} \to c,\vec{k}'} \sim \left| \sum_{\lambda} \frac{<c,\vec{k}',n_q|H_{ER}|\lambda,\vec{k}',n_q><\lambda,\vec{k}',n_q|H_{EP}|v,\vec{k},n_q\pm 1>}{E_v(\vec{k}) - E_\lambda(\vec{k}') \pm \hbar\omega(\vec{q})} \right.$$

$$\left. + \frac{<c,\vec{k}',n_q|H_{EP}|\lambda,\vec{k},n_q\pm 1> <\lambda,\vec{k},n_q\pm 1|H_{ER}|v,\vec{k},n_q\pm 1>}{E_c(\vec{k}') - E_\lambda(\vec{k}) \pm \hbar\omega(\vec{q})} \right|^2$$

$$\cdot \ \delta(E_c - E_v \pm \hbar\omega - E_{photon}) \qquad (3)$$

where the sum is over all intermediate states $\lambda$. For the reversed process that would correspond to emission,v and c have to be exchanged. $H_{ER}$ and $H_{EP}$ is the electron-radiation and electron-phonon interaction operator, respectively, $H_{EP}$ changing the phonon population number $n_q$ by one. The plus and minus signs refer to creation and destruction of a phonon of energy $\hbar\omega$.

Each term in eq. 3 consists of two virtual processes, a direct photon transition and a phonon scattering process involving the intermediate state $\lambda$. Obviously those processes will be most probable for which the energy denominator, i.e. the energy difference between initial and intermediate state will be smallest. Inspection of the band structure for AgBr in Fig. 1 and similarly for AgCl shows that in silver halides the $\Gamma_8^-$ valence band minimum will be the energetically favoured intermediate state for both absorption and emission processes so that the second term in eq. 3 may be neglected. Since the direct transition $\Gamma_8^- \leftrightarrow \Gamma_6^+$ is dipole-allowed resulting in a non-vanishing matrix element $< \Gamma_8^- |H_{ER}| \Gamma_6^+ >$ in eq. 3 it will be the favoured virtual photon process.

The other process then will be associated with scattering of a hole within the valence band between $\Gamma_8^-$ and $L_4^-$ by means of a phonon. By applying group theory and working out the selection rules (Weber, 1975), phonons of appropriate symmetry for this process are the TA(L) and LA(L) phonons that are consequently expected to be associated with the indirect absorption or emission processes in silver halides. Processes involving TO(L) and LO(L) phonons are phonon-symmetry forbidden and therefore, if at all, would lead to transitions weaker by orders of magnitude*.

If we consider allowed phonon scattering assuming that the momentum-conserving phonons have negligible dispersion, which is approximately true for the L-point phonons in silver halides (see Fig. 2), the matrix elements in eq. 3 become independent on $\vec{k}$ so that the transition probability for absorption may be written

---

*See, however, footnote in section 3.3 about notation of phonons in AgBr.

$$W^{+}_{v,\vec{k}\to c,\vec{k}'} \sim |M_{abs}|^2 \cdot (n_q+1) \quad \text{for phonon creation} \quad (4a),$$

and
$$W^{-}_{v,\vec{k}\to c,\vec{k}'} \sim |M_{abs}|^2 \cdot n_q \quad \text{for phonon destruction} \quad (4b).$$

Here $|M_{abs}|^2$ denotes the constant electron-phonon matrix element for absorption while $(n_q+1)$ and $n_q$ pay regard of the phonon population and lead to the temperature dependence of the transition probability acc. to

$$n_q = [\exp(\hbar\omega/kT) - 1]^{-1} \quad (5).$$

In case of a phonon-symmetry <u>forbidden</u> transition, the matrix element in eq. 3 has to be expanded in powers of $\vec{k} = (\vec{k} - \vec{k}') - \vec{k}_o$ with $\vec{k}_o = \vec{k}_v - \vec{k}_c$ (Fig. 3a) so that, neglecting quadratic and higher order terms

$$M_{abs} \sim M^{o}_{abs} + M^{1}_{abs} \cdot k + \ldots \sim M^{1}_{abs} \cdot k \quad (6)$$

since the transition is forbidden in zeroth order.

### 3.2 <u>The Wannier Exciton Model</u>

The band structure discussed so far is based on a single-particle model that describes the energy states of an electron e.g. in the conduction band with any interaction with the electrons in the valence band neglected (Fig. 3a). Actually, in exciting the electron to the empty conduction band, one electron is missing in the valence band resulting effectively in a positive charge. Taking into account the Coulomb interaction between the electron and the positive hole in the conduction and valence bands, respectively, a bound electron-hole pair i.e. an exciton is formed.

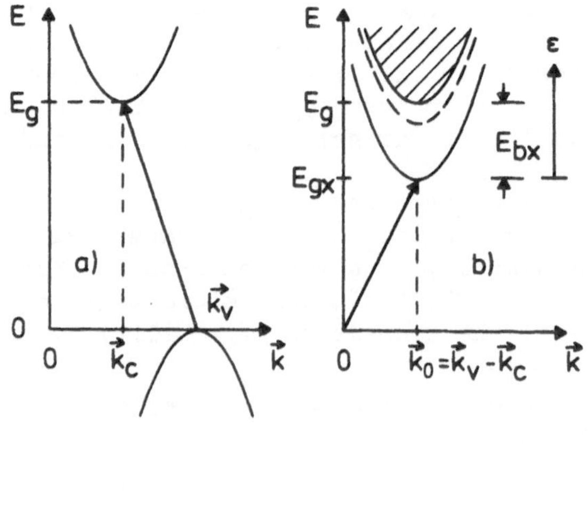


Fig. 3.

Schematic representation of a transition into (a) single-particle states and (b) exciton states.
$E_g$: band gap energy,
$E_{gx}$: exciton gap energy,
$E_{bx}$: exciton binding energy,
$\varepsilon$ : kinetic energy of the exciton

As in the case of the hydrogen atom, the formation of an exciton leads to a series of bound states that are lowered in energy with regard to the continuum of free electron-hole pair states as given by band theory (Fig. 3b). In crystals with a relatively high dielectric constant, i.e. in most semiconductors including the silver halides, the Coulomb interaction is weak. Therefore the exciton is of the Wannier-Mott type with the wave function, describing the electron-hole pair separation, extending over many unit cells of the crystal structure. The binding energy, $E_{bx}$, in this case is small (of the order of 10 ∿ 100 meV) and may be well treated in effective mass-approximation. Assuming that the valence and conduction bands are parabolic and isotropic in $\vec{k}$-space adjacent to the relevant band extrema and neglecting spin, the energy states of the exciton can be written as (see e.g. Knox, 1963)

$$E_n(\vec{k}) = E_g - \frac{1}{n^2} \cdot \frac{\mu e^4}{2\hbar^2 \varepsilon_{eff}^2} + \frac{\hbar^2 (\vec{k} - \vec{k}_o)^2}{2M^*} \qquad (7).$$

$E_g$ is the gap energy between the valence band maximum and the conduction band minimum. The second and third term of the right-hand side of eq. 7 describe the motion of the electron and hole relative to each other and the translational motion of the exciton as a whole, respectively. In this elementary stage of approximation these motions are assumed to be decoupled so that n is the hydrogen-like main quantum number. The exciton binding energy then is given by

$$E_{bx} = \frac{\mu e^4}{2\hbar^2 \varepsilon_{eff}^2} = R_H \cdot \frac{\mu}{\varepsilon_{eff}^2 m_o} \qquad (8),$$

where the Rydberg energy $R_H$ of the hydrogen atom is introduced. In eqs. 7 and 8 $\varepsilon_{eff}$ is an appropriate dielectric constant of the crystal and $m_o$ is the electron mass. $M^*$ and $\mu$ are the effective translatorial mass and the effective reduced mass, respectively, given by $M^* = m_e^* + m_h^*$ and $\mu^{-1} = (m_e^*)^{-1} + (m_h^*)^{-1}$ with $m_e^*$ and $m_h^*$ being the effective electron and hole masses. $M^*$, $\mu$ and $\varepsilon_{eff}$ contain any polaron effect. The kinetic term in the total energy of the exciton implies that the electron and the hole and consequently the exciton is freely mobile in the crystal. The term $\vec{k}_o = \vec{k}_v - \vec{k}_c$ takes into account the possibly different positions of the band extrema in the Brillouin zone (Fig. 3b).

Referring to the indirect exciton (upper index i) in silver halides, the electron is at Γ and the hole at L giving $\vec{k}_o = \vec{k}_L$. The energy of the lowest exciton state (n = 1) acc. to eqs. 7 and 8 may then be written as

$$E^{1s}(k) = E_g^i - E_{bx}^i + \frac{\hbar^2(\vec{k} - \vec{k}_L)^2}{2M^*} = E_{gx}^i + \varepsilon(|\vec{k} - \vec{k}_L|) \qquad (9).$$

Here we have introduced the indirect exciton gap energy, $E_{gx}^i$, and the kinetic part of the exciton energy, $\varepsilon$, that acc. to Fig. 3b is measured from the bottom of the lowest exciton band. With regard to the indirect gap energy $E_g^i$ the 1s exciton state is lowered in energy by the exciton binding energy $E_{bx}^i$.

It has to be kept in mind that in deriving eqs. 7 and 8 the band masses $m_e^*$ and $m_h^*$ are assumed to be isotropic. This condition is generally invalid for band extrema off the Brillouin zone center. Still the formalism remains qualitatively applicable to indirect excitons as in silver halides. Also for reasons of symmetry in these cases there exist several equivalent positions in the Brillouin zone. In cubic crystals there in all exist four L-points. An exciton constructed of an electron at $\Gamma$ and a hole at L therefore is fourfold k-degenerate. This "k-star"-degeneracy is also neglected.

### 3.3 Exciton Absorption Processes

Eq. 7 implies that transitions to exciton states extend below the band-to-band absorption at $E_g$. In case of direct transitions where $\vec{k}_v = \vec{k}_c$ and $\vec{k}_o = 0$, only excitons of vanishing kinetic energy may be excited in the minimum of the exciton band (Fig. 3b). In principle this results in a series of discrete exciton absorption lines. In contrast, indirect transitions to exciton states give rise to a continuous absorption. The phonons needed in this case for reasons of momentum conservation in eq. 2 form a "rêservoir" of momentum so that transitions to all states $(\vec{k} - \vec{k}_o)$ of the exciton band may occur and excitons of different kinetic energies are created. The lineshape of the indirect absorption reflects the density of final states, i.e. the density of states Z of excitons which for a single parabolic band is given by

$$Z(\varepsilon) = (1/2\pi^2)\ (2M^*/\hbar^2)^{3/2} \cdot \varepsilon^{1/2} \qquad (10).$$

Combining eqs. 4a,b and 10, the resulting absorption coefficient for the case of an indirect allowed exciton with main quantum number n and momentum-conserving phonon of energy $\hbar\omega$ will become (Elliott, 1957)

$$\alpha_n = (n_q + \tfrac{1}{2} \pm \tfrac{1}{2})\ (A/\nu)\,|M_{abs}|^2|\phi(0)|^2 \cdot (2M^*/\hbar^2)^{3/2} \cdot (E_{photon} - E_g^i + E_{bx}^i/n^2 \mp \hbar\omega)^{1/2} \qquad (11).$$

Here the kinetic energy ε is expressed in terms of the incident photon, band gap, exciton binding and phonon energies, while A comprises various constants. The threshold of absorption is determined by transitions into the n=1 state and occurs for

$$E_{photon} \geq E_g^i - E_{bx}^i \pm \hbar\omega \qquad (12).$$

As schematically illustrated in Fig. 4a, for n=1 two contributions to the absorption will occur. Neglecting the weak $\nu^{-1}$-dependence in eq. 11 each of these will show a characteristic square-root dependence on energy. The contributions are due to phonon creation ($n_q$+1) and destruction ($n_q$), respectively, and are shifted to each other by $2\hbar\omega$. According to the thermal population of phonon states given by eq. 5, the components have different temperature dependence. At low temperatures only the high energy component due to emission of phonons into the lattice will be present, while at sufficiently high temperatures $kT \geq \hbar\omega$ additional exciton transitions with phonon absorption will become observable. As in silver halides, different phonons may be associated with exciton formation giving rise to contributions with different intensities, determined by the exciton-phonon matrix elements in eqs. 3 and 4a,b, and with different absorption thresholds.

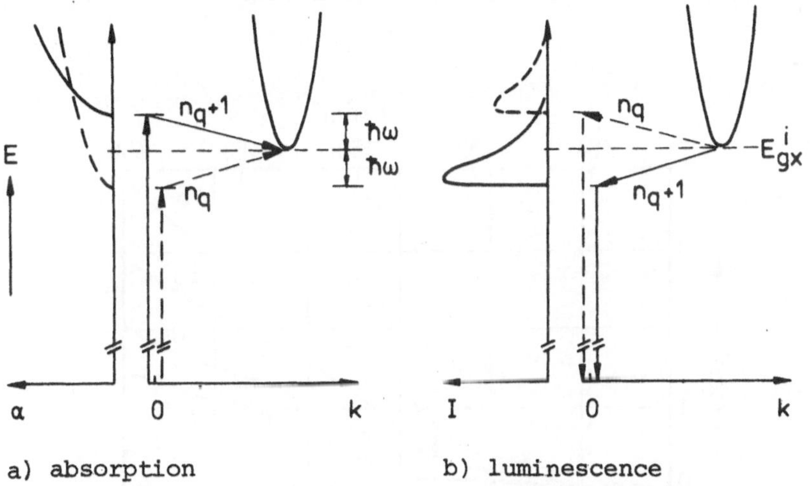


Fig. 4. Schematic representation of absorption (a) and emission (b) processes of the indirect exciton and resulting lineshapes. Full and dashed lines correspond to phonon creation and destruction, respectively. Thresholds are shifted with regard to $E_{gx}^i$ by plus or minus one phonon energy $\hbar\omega$ (see text).

$\phi(0)$ is the wave function describing the relative electron-hole motion, so that $|\phi(0)|^2$ in eq. 11 stands for the probability to find the electron and the hole at the same atom. Since $|\phi(0)|^2 \sim n^{-3}$, higher exciton bands are expected to contribute much less to the total absorption. For continuum states the absorption can be shown to be (Elliott, 1957)

$$\alpha \sim \sum_i (E_{photon} - E_g^i \pm \hbar\omega_i)^2 \qquad (13),$$

where i enumerates the various possible phonons. Thus the square-root dependence for each phonon contribution near the exciton absorption edge will change into a quadratic dependence for energies considerably higher than $E_g^i$ with a more complicated dependence in the intermediate energy range. At still larger energies, however, the result might be affected by higher-order effects neglected in the considerations above.

Experimentally, the indirect character of the absorption edge was first strikingly revealed in the temperature dependence of absorption (Joesten and Brown, 1966). As illustrated in Fig. 5 and as predicted above, with increasing temperature the weak absorption component due to phonon-destruction becomes observable that extends below the threshold at low temperature, the difference in threshold energies (arrows in Fig. 5) giving twice the phonon energy. Later and more detailed studies of the band edge in AgBr and AgCl in accordance with the selection rules proved that

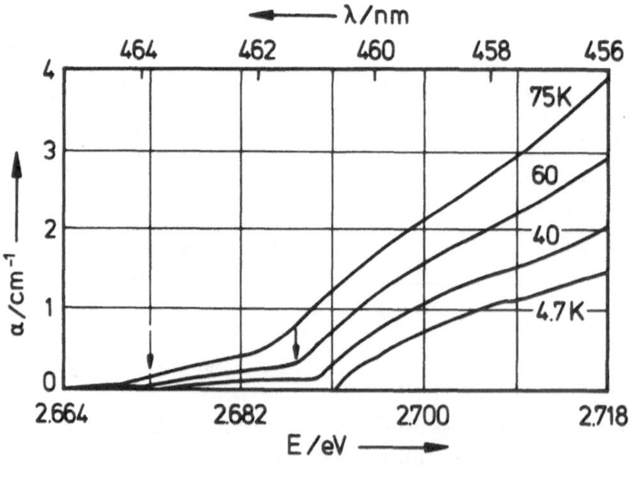


Fig. 5. Temperature dependence of exciton absorption edge in AgBr. Arrows indicate absorption thresholds for phonon creation and destruction, respectively, as obtained from analyzing the edge at 60K (Joesten and Brown, 1966).

momentum-conservation in the optical transition is accomplished by a transverse and a longitudinal acoustic* L-point phonon, respectively. These studies included conventional absorption measurements, stress- and wavelength modulation as well as luminescence and resonant Raman spectroscopy (for references see von der Osten, 1982). While from transport studies the conduction band minimum at Γ was well established, piezo-optical and cyclotron resonance experiments were only able to place the uppermost valence band maximum somewhere along the <111> direction in the Brillouin zone. Thus conclusive evidence for the position of the latter exactly at the L-point came only from comparison with the phonon dispersion data when these were available (Vijayaraghavan, 1970; von der Osten and Dorner, 1975; Dorner et al., 1976). As a result of a recent investigation Fig. 6 represents the low temperature exciton absorption edge of AgBr measured with high resolution together with the wavelength-modulated signal. This technique enhances weak structure contained in the absorption demonstrating two different phonons to be associated. Even though some finer details in the absorption edge of both halides still have to await interpretation, the absorption coefficient for each of the two

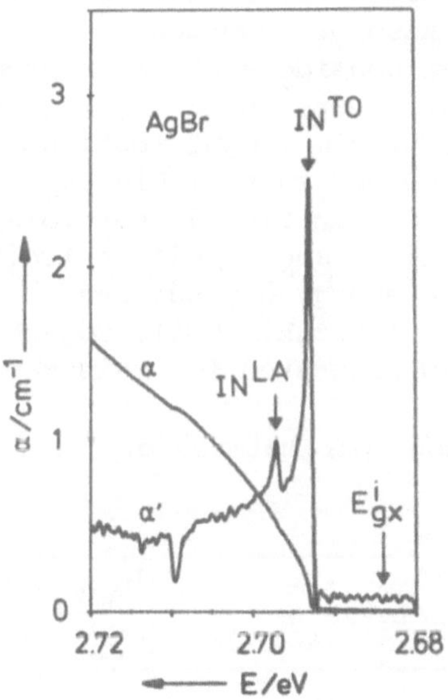


Fig. 6.

Indirect exciton absorption edge of AgBr at 1.5 K. Absorption coefficient (α) shown together with its energy-derivative (α') in arb. units. $IN^{TO}$, $IN^{LA}$ denote threshold for TO(L) and LA(L) assisted transitions, $E^{i}_{gx}$ indirect exciton gap energy (Śliwczuk, 1981).

*For AgCl these are denoted TA(L) and LA(L). Due to an inversion of the transverse phonon branches with regard to their eigenvectors near L, the transverse momentum-conserving phonon in AgBr belongs to the optical branch and accordingly is denoted by TO(L) (von der Osten and Dorner, 1975).

phonon components (lower index i) at low temperature follows grossly the energy dependence

$$\alpha_i = \alpha^o |M_{abs,i}|^2 (E_{photon} - E^i_{gx} - \hbar\omega_i)^{1/2} = \alpha^o |M_{abs,i}|^2 \varepsilon_i^{1/2} \quad (14)$$

suggested by eq. 11 for n = 1 with the parameters listed in Table 2. The constant $\alpha^o$ summarizes all quantities except the exciton-phonon matrix element.

Basically, absorption involving symmetry-forbidden phonons (TO(L) and LO(L) in AgCl, TA(L) and LO(L) in AgBr) is also expected to occur. The energy dependence of the absorption coefficient for n=1 in this case is obtained from eq. 6 (see Nakahara et al., 1974 for a detailed treatment) giving, with $\varepsilon \sim k^2$ acc. to eq. 9,

$$\alpha \sim |M^1_{abs}|^2 \cdot \varepsilon^{3/2} \quad (15).$$

Due to their energy dependence and to small electron-phonon matrix elements these processes in silver halides so far have not been detected in any absorption measurement. As will be pointed out in section 4.5, however, in AgBr RRS is observed that involves the TA(L) phonon indicating that in this substance transitions associated with symmetry-forbidden phonons may also be important. The detailed analysis in terms of a model that considers RRS as a process of absorption followed by emission suggests that, in describing the energy dependence of the total exciton absorption, contributions from these may not be neglected but will become appreciable especially at higher energies. For the fit shown in Fig. 7 therefore contributions from both allowed (TO(L), LA(L)) and forbidden (TA(L)) absorption processes with their respective energy dependences (eqs. 14 and 15) were taken into account (Sliwczuk, 1981), together with a weak zero-phonon component* as also suggested for this sample

Table 2. Indirect absorption edge parameters for allowed transitions

| | $E^i_{gx}$/eV | $\hbar\omega_1$/meV | $\hbar\omega_2$/meV | $\left\|\frac{M_{abs,2}}{M_{abs,1}}\right\|^2$ |
|---|---|---|---|---|
| AgCl | 3.248 | 8.2 (TA(L)) | 12.9 (LA(L)) | - |
| AgBr | 2.684 | 8.3 (TO(L)) | 11.8 (LA(L)) | 0.16 |

*Zero-phonon transitions are forbidden due to the translational invariance of an ideal crystal. Due to small amounts of defects and impurities, however, $\vec{k}$-selection may be relaxed allowing for weak absorption or emission zero-phonon components.

from RRS. The components are properly shifted with regard to the indirect exciton gap energy $E^i_{gx}$, their relative strength obtained to get the best fit giving the relative matrix elements listed in Table 2. To take into account the forbidden process, a relative matrix element $|M^1_{abs,TA(L)}|^2/|M^o_{abs,TO(L)}|^2 = 3.6 \cdot 10^{-3}$ meV$^{-1}$ was used as implied by corresponding RRS results. The overall fit ($\alpha_c$) to the experimental data ($\alpha$) is satisfactory except some rest of less than 4% to the measured absorption. The misfit occurs for both the TO(L) and LA(L) components suggesting that it is due to a pecularity in the exciton band itself (see also section 4.3).

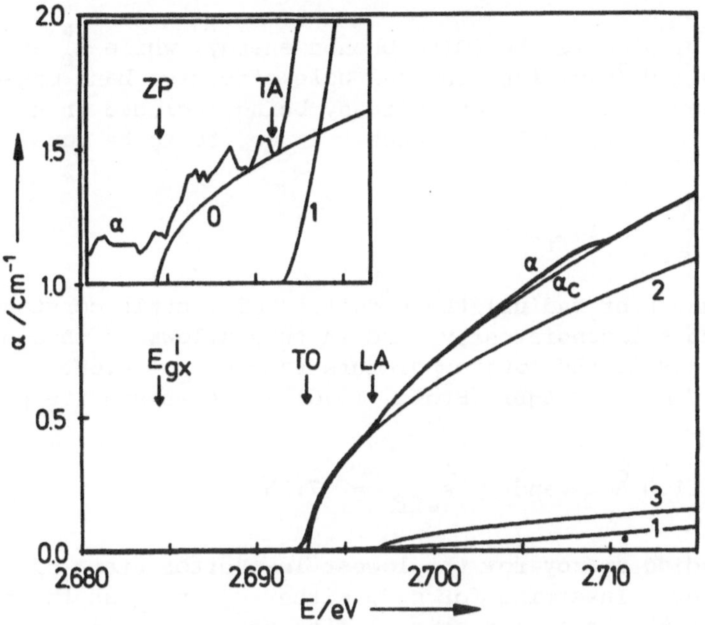


Fig. 7. Exciton absorption edge of AgBr at 1.8K. $\alpha$: experimental data; $\alpha_c$: computed from various contributions 0 to 3 as described in the text. The components shown are due to the following processes: 0 = zero-phonon (ZP), 1 = forbidden TA(L), 2 = TO(L), 3 = LA(L). Corresponding thresholds are marked. $E^i_{gx}$: exciton gap energy. Ordinate of insert enlarged by a factor of 300 (Sliwczuk, 1981).

## 3.4 Binding Energy of the Indirect Exciton

Although efforts were undertaken to determine the binding energy $E^i_{bx}$ of the indirect exciton, there are hardly reliable values for this important quantity for both AgCl and AgBr found in the literature. The binding energy may be estimated within the hydrogen model of the exciton introducing an effective dielectric constant $\varepsilon_{eff}$. According to Haken (1958) this depends on the electron-hole distance r and may be written

$$\frac{1}{\varepsilon_{eff}(r)} = \frac{1}{\varepsilon_\infty} - \left( \frac{1}{\varepsilon_\infty} - \frac{1}{\varepsilon_o} \right) \left(1 - \frac{1}{2} \left(e^{-r/\rho_e} + e^{-r/\rho_h}\right)\right) \quad (16)$$

Here $\varepsilon_\infty$ and $\varepsilon_o$ are the high-frequency and static dielectric constants, $\rho_e$ and $\rho_h$ are the polaron radii defined as

$$\rho_e = (\hbar^2/2m_e^*\hbar\omega_{LO})^{1/2} \quad \text{and} \quad (17a)$$

$$\rho_h = (\hbar^2/2m_h^*\hbar\omega_{LO})^{1/2} \quad (17b).$$

In eqs. 17a,b, $\hbar\omega_{LO}$ is the LO(Γ) phonon energy, while $m_e^*$ and $m_h^*$ are the electron and (say) longitudinal hole effective band masses (Table 1), any polaron effects already being included in $\varepsilon_{eff}$. By choosing the electron-hole distance r in eq. 16 to be equal to the exciton Bohr radius

$$a_o = \frac{m_o}{\mu} \cdot \varepsilon_{eff} \cdot a_H \quad (18)$$

($a_H$ = hydrogen Bohr radius) the effective dielectric constant may be determined selfconsistently, and in turn allows to obtain $E^i_{bx}$ from eq. 8. Using the low temperature values $\varepsilon_\infty = 4.68$, $\varepsilon_o = 10.6$ and $\hbar\omega_{LO}$ = 17 meV for AgBr (Stolz, 1976) a representative pair of values will be

$$a_o = 21.0\ \text{Å} \quad \text{and} \quad \varepsilon_{eff} = 7.25$$

giving a binding energy for the lowest 1s-exciton state of $E^i_{bx}$ = 47.3 meV. Inserting for $\varepsilon_{eff}$ either $\varepsilon_\infty$ or $\varepsilon_o$ as the two extreme cases the lower and upper limits for $a_o$ and $E^i_{bx}$ will be

$$13.6\ \text{Å} < a_o < 30.7\ \text{Å} \quad \text{and} \quad 114\ \text{meV} > E^i_{bx} > 22.1\ \text{meV}.$$

Since the contributions to the absorption from higher exciton states have relative strengths given by $|\phi(0)|^2 \sim n^{-3}$ in eq. 11, structure that unambigeously may be interpreted in terms of excited exciton states is not observed. Any experimental determination of $E^i_{bx}$ by using eq. 11 therefore seems to be questionable. Values of 23 meV for AgCl and 16 to 20 meV for AgBr are quoted in the literature Stulen and Ascarelli,1976;Kanzaki and Sakuragi, 1969; Ascarelli and Baxter, 1972), but although the order of magnitude appears reasonable, the interpretation of the experimental data on which these values are based on is not beyond any doubt. Extending the fit of Fig. 7 for AgBr to energies higher than shown there, strong deviations occur unless an additional contribution is taken into account that exhibits a quadratic energy dependence. This is believed to be due to absorption into the continuum of free electron-hole pair states. From the fit shown in Fig. 8 the free electron-

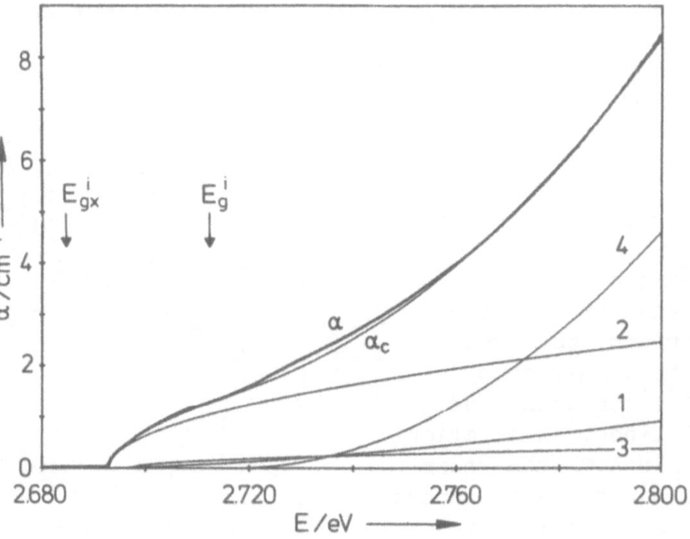


Fig. 8. Same as Fig. 7, but with an additional contribution (4) due to absorption by free electron-hole pairs fitted at higher energies to the measured absorption. $E^i_g$: band gap energy obtained from fit (Sliwzuk, 1981).

hole pair gap energy $E^i_g$ is approximately obtained at 2.713 eV, from which by applying eq.9 an exciton binding energy $E^i_{bx}$ = (28±5) meV is derived. Again some rest to the measured absorption remains unexplained ($\alpha - \alpha_c$ in Fig. 8), the difference being partly due to the contributions from excited states of the exciton.

3.5 Exciton Luminescence

While the electronic band structure and the indirect absorption edge like most other properties are qualitatively similar for AgCl and AgBr, their intrinsic luminescence spectra show remarkable differences (Kanzaki et al., 1971). These obviously for the two materials reflect the contrasting nature of the relaxed exciton state probed by luminescence. As shown in Fig. 9 (top) for AgBr the most prominent transitions for excitation in the exciton absorption at low temperatures are two narrow luminescence lines, $IN^{TO(L)}$ and $IN^{LA(L)}$. They roughly exhibit lineshapes acc. to a Maxwell-Boltzmann law with halfwidths of about 1.5 meV (von der Osten and Weber, 1974) and, as will be discussed below, correspond to the radiative recombination of the indirect free exciton (FE) involving the same momentum-conserving phonons like in absorption. In contrast, for AgCl a much broader luminescence band (halfwidth about 0.3 eV) of nearly Gaussian lineshape is observed (Fig. 9, bottom). The width of the band and its large Stokes shift of about 0.7 eV with regard to the indirect exciton

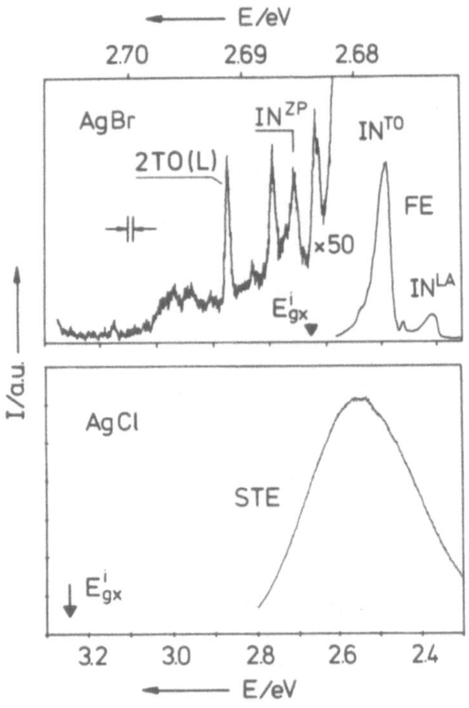


Fig. 9.

Exciton recombination luminescence in AgBr and AgCl at 1.8K. In AgBr $IN^{TO}$ and $IN^{LA}$ are phonon-assisted free exciton (FE) transitions, observed together with a weak zero-phonon ($IN^{ZP}$) component and several RRS lines. Excitation at 4579 Å. In AgCl the recombination is due to self-trapped excitons (STE). $E^i_{gx}$: indirect gap energies.

gap energy $E^i_{gx}$ suggest that in AgCl the exciton is self-trapped during relaxation and strongly interacts with the surrounding lattice.

The occurrence of two basically different types of relaxed exciton states as dramatically reflected in the luminescence spectra of the two silver halides may be understood in terms of different strengths in exciton-phonon interaction (see e.g. Toyozawa, 1974; Toyozawa, 1980). As schematically summarized in Fig. 10, an exciton in the deformable lattice is subject to the competition of two opposing interactions leading to delocalization and localization, respectively. The delocalization originates from the transfer of excitation energy between neighbouring lattice sites. In the presence of transfer and neglecting any lattice distortion, the exciton moves freely through the crystal as represented by the exciton band in Fig. 10c, the bandwidth 2B in this case being determined by the rate of transfer. On the other hand, in the absence of transfer, the exciton would be localized at a single lattice site leading to the well-known configuration coordinate diagram often used in context with localized defect centers e.g. in ionic crystals (Fig. 10a). In this case, the lattice relaxation energy $E_{LR}$ with regard to the perfectly localized state at $E = E_a$ determines the

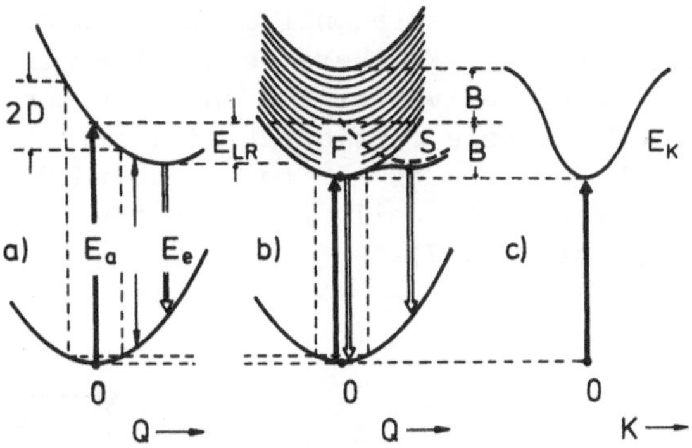


Fig. 10. Configurational coordinate models for a localized excitation (a) and an exciton (b), and excitonic band in the rigid lattice (c). Energy shown vs. configurational coordinate Q and exciton wavevector K, respectively (Toyozawa, 1974).

position of the relaxed exciton state and the energy gain by localization due to the strong interaction with the surrounding lattice distortions is greater than that by purely electronic band motion.

Combining the two considered limiting cases as illustrated in Fig. 10b, two distinctly different types of relaxed exciton states are obtained that may be interpreted as free (F) and self-trapped (S) exciton states. In the detailed theoretical treatment of self-trapping the spatial range of exciton-phonon interaction turns out to be crucial (Toyozawa, 1961). In particular, the deformation potential interaction being of short range gives a sharp criterion for the occurrence of the self-trapped state. It allows the coexistence of free and self-trapped states and, depending on the coupling strength $g = E_{LR}/B$, either the free or the self-trapped state is stable, while the other is metastable in each case.

Inspection of Fig. 10 readily shows that for a strongly coupled exciton-phonon system a broad and strongly Stokes-shifted luminescence band is to be expected from the configurational coordinate model. Obviously, AgCl where the exciton becomes self-trapped is of this type. The phonon cloud associated with the exciton implies a practically infinite translational mass making the dressed particle immobile, as is the hole (compare Table 1). As to the microscopic structure of the self-trapped exciton (STE) it is not yet undoubtedly established. From the early EPR measurements of Höhne and Stasiw (1968) and more recent studies of other groups (for references see von der Osten, 1982) the self-trapped hole in AgCl is known to be localized at the silver ion forming a $Ag^{2+}$ center with a lattice distortion of tetragonal symmetry around as pictured in Fig. 11. In case of the STE, there is still some discussion if the localization is at the silver or the chlorine ion. Although basically a change of the hole configuration in the STE from that in the self-trapped hole alone is not to be excluded, recent ODMR measurements seem to favour the silver ion as the trapping site for the exciton (Hayes et al., 1977).

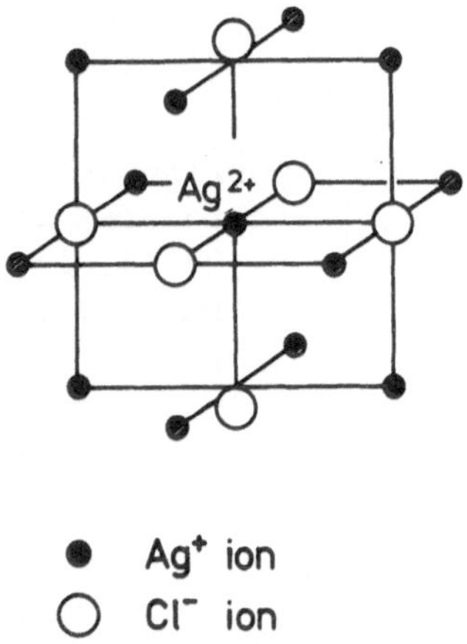


Fig. 11.

Microscopic structure of the self-trapped hole in AgCl showing tetragonal distortion of the surrounding lattice.

The other extreme, i.e. weak exciton-phonon interaction, occurs in AgBr. Here, down to the lowest temperatures studied so far (1.6K), the exciton, like the hole remains free during relaxation. As revealed in detail from resonant Raman scattering (section 4), the exciton excited high in the band relaxes down to lower exciton states by interaction with various well-defined phonons. It still remains mobile since no sufficiently strong lattice distortion occurs that would force the exciton to be self-trapped. In this case, the perturbational picture may be used to describe exciton-phonon interaction starting from free excitons and phonons.

The luminescence in this case will clearly be of the resonance type (phonon-assistance neglected), as illustrated in Fig. 10c. To derive the observed Maxwell-Boltzmann lineshape, the exciton total lifetime is assumed to be sufficiently large so that thermal equilibrium may be established between the exciton system and the lattice. Then the distribution of excitons among the various energy states (eq. 7) will be given by the density of exciton states (eq.10) weighted by the thermal distribution function

$$f(\varepsilon) \sim \left( \frac{2M^*}{\hbar^2} \right)^{3/2} \cdot \varepsilon^{1/2} \cdot \frac{1}{e^{(\varepsilon-\mu)/kT}-1} \sim \varepsilon^{1/2} \cdot e^{-\varepsilon/kT} \qquad (19)$$

for not too high densities of excitons ($\mu$ = chemical potential).

Here, $\varepsilon$ again is the kinetic part of the exciton energy as introduced in eq. 9. In complete analogy to absorption, in a direct emission process of the excitons only those near the band minimum may recombine radiatively resulting in a discrete and narrow luminescence line. Indirect emission on the other hand may occur for each exciton independent on its momentum $\vec{k}$. Assuming again dispersion-free phonons, the intensity distribution of the recombination luminescence can be calculated from eqs. 3 and 19 to give

$$I(E_{photon}) = (n_q + \frac{1}{2} \pm \frac{1}{2}) \cdot B \cdot |M_{em}|^2 \cdot \varepsilon^{1/2} \cdot e^{-\varepsilon/kT} \tag{20}$$

$$\text{with} \quad \varepsilon = E_{photon} - E^i_{gx} \pm \hbar\omega \geq 0,$$

reflecting the population of exciton states in thermal equilibrium with the lattice. Corresponding to the binding energies in question for the higher ($n > 1$) exciton states (compare 3.4), in AgBr these are not thermally populated at the low temperatures of the experiments. Thus in deriving $I(E_{photon})$ emission from the lowest ($n = 1$) exciton state only had to be taken into account. $|M_{em}|^2$ is the exciton-phonon matrix element of eqs. 3 and 4 with c and v interchanged corresponding to the process of emission and equal to $|M_{abs}|^2$. B comprises the remaining constants. Like for absorption and as shown schematically in Fig. 4b, for $T > 0K$ two luminescence components exist shifted by twice the energy of the phonon that assists the indirect emission process. From the thermal phonon population factors their total intensities depend on temperature, the higher energy component becoming observable at high temperatures only. Since in the exciton transitions described by eqs. 11 and 20 the same matrix elements are involved as for the band-to-band transitions treated in section 3.1 (eq. 3), the selection rules derived there remain unaltered.

Regarding the experimental situation,the free exciton recombination in AgBr, observed at very low temperatures and for broad band excitation at high energies, grossly exhibit the lineshape expected from the considerations above. The detailed analysis in terms of exciton temperature, $T_{exc}$, used as the fitting parameter shows, however, that at low temperatures $T_{exc} >> T_L$, where $T_L$ is the crystal lattice temperature as determined from the bath temperature of the cooling medium (von der Osten and Weber, 1974). This implies that within the exciton lifetime no thermal equilibrium is established between excitons and phonons. Since for indirect exciton transitions being second-order processes the radiative lifetime, $\tau_{rad}$, is expected to be long and for AgBr may be estimated to be about 100 µs, this is surprising. In fact, the actual (total) lifetime is measured to be $\tau_{tot} \approx 60$ ns (Stolz et al., 1976) and sample dependent, suggesting that it is determined by very effective trapping processes in the crystal. As will be shown in section 4.3, the luminescence lineshape at low temperatures is indeed determined by exciton-phonon scattering that is revealed by using RRS. Phonon emission associated with exciton relaxation will typically lead to a non-thermal population of exciton states that is observed in luminescence.

## 4. RESONANT RAMAN SCATTERING

The results discussed so far are obtained with standard techniques of exciton spectroscopy that make use of one-photon processes like absorption and luminescence. With the advent of powerful laser light sources and in particular tunable lasers the application of novel methods became possible involving non-linear processes, too. In studying excitons in semiconductors, RRS proved to be an extremely effective method. Conventional Raman scattering in crystals is used since long to investigate various kinds of elementary excitations of low energies like phonons or plasmons. By taking advantage of resonances in scattering cross-section this method since recently is exploited to also investigate high energetic excitations like excitons providing information on their dispersion and their relaxation behaviour. Depending on the particular case, the exciton interaction with the surrounding lattice may be revealed in great detail which in silver halides may either be coupling to acoustic and non-polar optical phonons via the deformation potential interaction or with longitudinal optical phonons via the Fröhlich interaction.

A great deal of work published in the literature refers to RRS at direct gaps in covalent or weakly polar semiconductors where excitonic effects, however, play a minor role (for references see e.g. Cardona, 1975; Richter, 1976). Much less work has been done in strongly polar materials like e.g. in thallous halides (see e.g. Stolz and von der Osten, 1980), although these would be favourable candidates due to their discrete exciton states. One reason for this is that the increasingly larger gap energies make it more difficult to find suitable lasers that will match in photon energies and are tunable.

AgCl and AgBr are two of a very few systems where Raman scattering is observed in resonance with an indirect gap, other examples being Si, GaP (Klein et al., 1974) and $BiI_3$ (Komatsu et al., 1981). In fact, AgBr was the first material at all to show RRS at an indirect exciton (von der Osten et al., 1974). As obvious from Fig. 9 (top), for excitation above the indirect absorption edge with a monochromatic laser beam in addition to the FE luminescence at low temperature a number of narrow, resonantly enhanced Raman lines are observed. They can be discriminated against the luminescence lines by varying the excitation photon energy. Since in a Raman-type process the correlation between the scattered photon (energy $E_s$) and the incident photon ($E_i$) is preserved, the Raman energy shift $E_i-E_s$ remains constant, i.e. the scattering line in a Raman process "follows" the incident laser line. On the other hand, photoluminescence occurs after absorption of the incident photon followed by relaxation within the excited state. In general therefore luminescence probes the relaxed excited state resulting in transitions with energies independent on $E_i$.

By comparing the Raman energy shift in AgBr, the dominant scattering process is found to involve a pair of TO(L) phonons. It is important to realize that this phonon is the one to participate in the indirect optical exciton transitions discussed in section 3. Also a weaker 2LA(L) process is observed as well as TO(L)+LA(L) scattering. The two-phonon scattering lines are initiated at the indirect exciton absorption threshold implying a resonantly enhanced second-order Raman process.

## 4.1 Raman Scattering: Relevant Background

In Raman scattering (RS) in crystals a photon is inelastically scattered with absorption and/or emission of one or more phonons, depending on the details of electron - phonon interaction. Due to

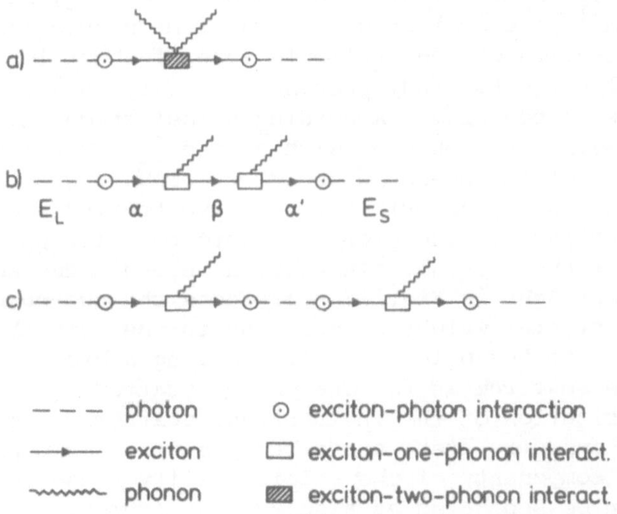


Fig. 12. Diagrammatic representation of possible two-phonon Raman processes. Process b) corresponds to RRS at the indirect exciton in silver halides with α, α' and β being direct and indirect exciton states, respectively. The phonons associated in this case are momentum-conserving L-point phonons.

the inversion symmetry of the silver halides, first-order RS is forbidden in these materials so that only second-order processes may be observed (von der Osten, 1974). As illustrated in the diagrammatic representation of Fig. 12, in second-order RS principally three different types of processes have to be considered (see e.g. Loudon, 1964). The kinetics of these are governed by the conservation laws for energy and momentum so that

$$E_i = E_s \pm \hbar\omega_1 \pm \hbar\omega_2 \tag{21a}$$

$$\hbar\vec{k}_i = \hbar\vec{k}_s \pm \hbar\vec{q}_1 \pm \hbar\vec{q}_2 \tag{21b}$$

Here $\hbar\omega_1$, $\hbar\omega_2$ and $\hbar q_1$, $\hbar q_2$ are the energies and momenta, respectively, of the two phonons involved, while $\hbar k_i$ and $\hbar k_s$ are the incident (laser) and scattered photon momenta. The positive sign refers to phonon emission into the lattice, a process that gives rise to Stokes scattering. The negative sign corresponds to phonon absorption and anti-Stokes scattering. Difference processes are also possible.

Experimentally the Raman energy shift, $E_i$-$E_s$ (denoted by $E_L$-$E_S$ in the following), is measured that contains the information on the energies of the phonons involved. Since $\vec{k}_i$ and $\vec{k}_s$ are only slightly different and therefore $\vec{k}_i-\vec{k}_s \simeq 0$, it is obvious from eq. 21b that in general $\vec{q}_1 = -\vec{q}_2$, i.e. the two phonons involved are expected to have equal but opposite momenta. Although $k_i$ and $k_s$, being of the order of $10^5$ cm$^{-1}$, are smaller by orders of magnitude compared to the maximum extension of the Brillouin zone of about $10^8$ cm$^{-1}$, this condition enables one to study phonons that vary in $\vec{q}$ from the zone center to the zone boundary. According to selection rules imposed by crystal symmetry, the phonons with $\vec{q}_1$ and $\vec{q}_2$ may belong to various branches of the phonon dispersion. Thus a great number of phonon combinations may contribute to the scattering spectrum that principally contains all energies from zero to twice the maximum phonon energy of the crystal. In fact, a second-order Raman spectrum as shown for AgBr in Fig. 13 represents the two-phonon density of states of a crystal weighted according to the contribution of each phonon pair to the polarizability, giving a broad rather than a discrete line spectrum as for one-phonon scattering. By proper choice of polarization of the incident and scattered light relative to the crystal symmetry axes, contributions of the two-phonon combinations to the components of the polarizability tensor of the crystal $\alpha_{ij}$ can be separated as also evident from Fig. 13.

In a microscopic theory expressions for the Raman cross-section for the processes in Fig. 12 are obtained applying time-dependent perturbation theory (see e.g. Loudon, 1963). For example we consider process b) which for reasons of momentum-conservation turns out to be relevant for the situation in silver halides, since only this process contains the indirect exciton as resonant intermediate

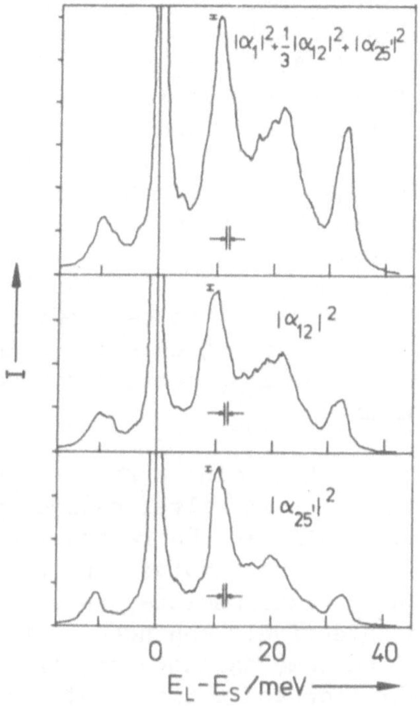


Fig. 13.

Stokes and anti-Stokes Raman spectra of AgBr at 77K for excitation at 4880 Å. Spectra (from top) are measured in Z(Y,X)Y, Z(X,Z)Y, and Z(Y,X)Y configurations (von der Osten, 1974).

state. If we restrict ourselves to phonon emission, in this process the incident laser photon (energy $E_i = E_L$) via the exciton-photon interaction creates an exciton in state α, followed by the emission of two phonons, each via the exciton-one-phonon interaction, and scattering of the exciton successively through two different states β and α'. Finally the scattered photon (energy $E_S$) is emitted annihilating the exciton.

The Raman cross-section for this process that is proportional to the scattered intensity I has the form (Ganguly and Birman, 1967)

$$R(E_L,E_S) \sim \sum_{\vec{q}} \left| \sum_{\alpha\beta\alpha'} \frac{\langle O|H_{ER}(E_S)|\alpha'\rangle\langle\alpha'|H_{EP}^{(2)}|\beta\rangle\langle\beta|H_{EP}^{(1)}|\alpha\rangle\langle\alpha|H_{ER}(E_L)|O\rangle}{(E_{\alpha'}-E_S)\,(E_\beta+\hbar\omega_1(\vec{q})-E_L)\,(E_\alpha-E_L)} + 23 \text{ other terms} \right|^2 \delta[E_L - E_S - \hbar\omega_1 - \hbar\omega_2] \qquad (22).$$

To simplify notations, only exciton states are shown in eq. 22.

The matrix elements represent the exciton-photon ($H_{ER}$) and exciton-phonon ($H_{EP}$) interaction that depend on the intermediate exciton states ($\alpha,\beta,\alpha'$) as well as on $\vec{q}$, $\vec{k}_i$ and $\vec{k}_s$. The sum has to be performed over the intermediate exciton states and the final state, i.e. the phonon density of states as represented by the phonon wavevector $\vec{q}$. The 23 other terms stem from permuting the order in which $H_{ER}$ and $H_{EP}$ occur. Examination of all 24 terms shows that only the one explicitly written in eq. 22 results in a resonance enhancement of the cross-section R at an indirect exciton state due to the pole in the denominator at $E_L = E_\beta + \hbar\omega_1(\vec{q})$. Thereby $\alpha$, $\alpha'$ and $\beta$ have to be identified as direct and indirect exciton states with energies $E_\alpha$, $E_{\alpha'}$ and $E_\beta$, respectively, while $\hbar\omega_1$ and $\hbar\omega_2$ are the momentum-conserving phonons necessary to accomplish resonance.

At this point a remark concerning resonant in relation to non-resonant RS in silver halides may be appropriate. Like in non-resonant RS, in the resonant case two phonons of wavevector $\vec{q}_1 = -\vec{q}_2$ are involved in the scattering process. Due to resonance with the intermediate state, however, only certain pairs of these are selectively enhanced. At the indirect exciton these phonons are the same as those required for momentum-conservation in exciton absorption and emission discussed in section 3.

### 4.2 Evaluation of the 2TO(L) RRS Cross-section

RRS takes advantage of the resonances occurring in R if e.g. $E_L$ is tuned near and across some intermediate state. As described above, in AgBr at 1.8K strong resonance enhancement is observed for the 2TO(L) process at the indirect exciton edge, so that $\hbar\omega_1$ and also $\hbar\omega_2$ in eq. 22 may be identified with the TO(L) phonon energy. For low temperature implying phonon emission only, energy and wavevector conservation for this process follows from the schematic exciton dispersion in Fig. 14 (process a). With regard to the exciting laser line, the scattered line is shifted by $E_L - E_S = 2\hbar\omega_{TO(L)}$. To evaluate the cross-section, for this case the resonant term in eq. 22 further may be simplified due to the following reasonable assumptions (Weber and von der Osten, 1976):

1. Since the direct exciton states in silver halides are higher in energy by about 1.5 eV compared to the indirect edge (Fig. 1), i.e. $E_\alpha, E_{\alpha'} \gg E_\beta \simeq E_L, E_S$, the terms $(E_{\alpha'} - E_S)$ and $(E_\alpha - E_L)$ may be regarded constant.

2. In agreement with experimental observation only the lowest indirect exciton state $\beta$ has to be considered while higher exciton states ($n > 1$) may be neglected.

3. Since $\vec{k}_i \simeq \vec{k}_s \simeq 0$ and the exciton-phonon matrix element of the allowed TO(L) process (section 3.1) is constant, all matrix elements in the nominator may be regarded constant.

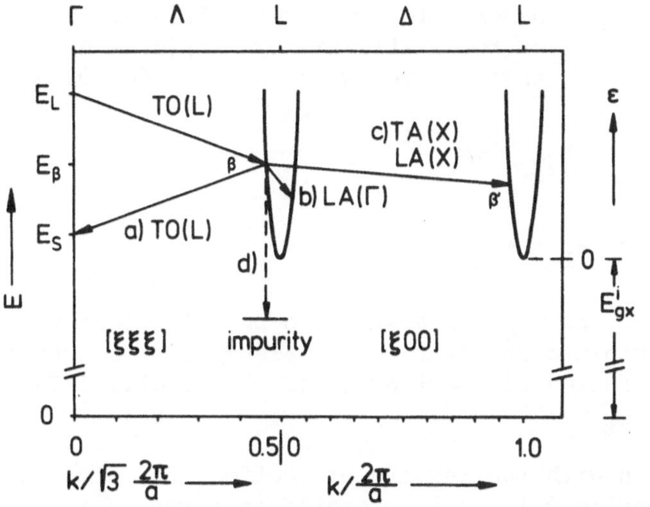


Fig. 14. Schematic representation of exciton dispersion and relevant scattering processes as described in the text. L and L' denote non-equivalent points in the Brillouin zone.

With these assumptions the 2TO(L) Raman cross-section may be written as

$$R^{2TO(L)} \sim |M_{abs}|^2 \, |M_{em}|^2 \sum_{\vec{q}} \left| \frac{1}{E_\beta + \hbar\omega(\vec{q})_{TO(L)} - E_L} \right|^2 \qquad (23),$$

where the exciton-phonon matrix elements for the indirect absorption and emission processes in eq. 22 are explicitly shown. The sum over the phonon wavevectors $\vec{q}$ may be transformed into an integral over the exciton band by noting that the phonon is dispersionless in the relevant region of the Brillouin zone (Fig. 2) and replacing $E_\beta = E^{1s}(\vec{k})$ with $\vec{k} = \vec{k}_L - \vec{q}$ by the kinetic energy ε according to eq.9

$$\sum_{\vec{q}} \rightarrow \int_0^{\varepsilon_o} \frac{\sqrt{\varepsilon}\; d\varepsilon}{|E^i_{gx} + \varepsilon + \hbar\omega_{TO(L)} - E_L|^2} \qquad (24),$$

where the integration has to be performed over the whole exciton band. Since resonance always occurs for $E_L \geq E^i_{gx} + \hbar\omega_{TO(L)}$ (Fig.14), the integral diverges due to the fact that any broadening of the exciton state is neglected. Replacing phenomenologically ε by

$\varepsilon + i\gamma_{tot}(\varepsilon)$ in the denominator, where the lifetime broadening is related to the exciton total lifetime by $\gamma_{tot}(\varepsilon) = \hbar/\tau_{tot}(\varepsilon)$, eq.24 may be computed to finally give (Weber and von der Osten, 1976)

$$R^{2TO(L)} \sim |M_{abs}|^2 |M_{em}|^2 \frac{\varepsilon^{1/2}}{\gamma_{tot}(\varepsilon)}, \quad \varepsilon > 0 \tag{25}$$

with $$\varepsilon = E_L - E^i_{gx} - \hbar\omega_{TO(L)} \tag{26}.$$

In deriving eq. 25 $\gamma_{tot}$ is assumed to be sufficiently small compared to the kinetic energy $\varepsilon$. This assumption is justified from the exciton lifetime of about 60 ns (Stolz et al., 1976) corresponding to $\gamma_{tot} \simeq 10^{-5}$ meV.

Comparison with the results of sections 3.3 and 3.5 suggests that eq. 25 may readily be interpreted in terms of RRS being a process of "absorption followed by emission" (Klein, 1973). In this situation the scattering probability is expected to be the product of the absorption probability times the quantum efficiency for a radiative transition of the exciton to the ground state.From eq. 11 the absorption probability for the indirect allowed exciton is given by the absorption coefficient $\alpha \sim |M_{abs}|^2 \cdot \varepsilon^{1/2}$. The quantum efficiency is $\tau_{tot}/\tau_{rad} = \gamma_{rad}/\gamma_{tot}$ with $\tau_{rad}$ being the radiative lifetime. Since the matrix element $|M_{em}|^2$ that determines $\gamma_{rad}$ was assumed to be energy-independent and constant, eq. 25 is obtained. RRS in context with an intermediate indirect exciton state thus may either well be interpreted as "hot luminescence" probing the non-thermal population of the exciton state. Although there are controversies found in the literature (for references see e.g. Solin and Merkelo, 1976) as to whether RRS and "hot luminescence" in general are identical processes, this definitely is the case in the considered example.

Eq. 25 is crucial in studying the exciton by means of RRS. It suggests that $\gamma_{tot}(\varepsilon)$ can be investigated by varying the kinetic energy $\varepsilon$ and by detecting the scattered intensity which is proportional to $R^{2TO(L)}$. Its energy dependence reveals the various relaxation processes by which the exciton is affected. The variation of $\varepsilon$ acc. to eq. 26 is accomplished by tuning the incident photon energy $E_L$ of a tunable laser across the exciton edge into the absorption.

### 4.3 Results for AgBr and AgCl

The most important component in performing a RRS experiment is a dye laser tunable in the energy range of interest, the rest of the equipment being standard in normal RS spectroscopy. In regions of small absorption the scattered light is observed under $90^o$ with respect to the incoming laser beam, while a $180^o$ geometry has to be used in case of strong absorption. To obtain signals proportional to the scattering cross-section, the intensity is usually normalized to the incoming laser intensity. Also the measured intensity in most cases has to be corrected for absorption of the incident and scattered light taking into account the scattering geometry (Richter, 1976).

As an example in Fig. 15 a series of RRS spectra is shown taken at slightly different $E_L$ or $\varepsilon$. In each spectrum the scattered intensity is plotted vs. the Stokes shift $E_L - E_S$. The spectra are arranged with their baselines according to the left-hand scale for the incoming photon energy $E_L$. This plot allows to discriminate Raman lines occurring at $E_L - E_S = \Delta E$ = const (lines a, c1, c2) or, for scattering involving dispersive phonons (line b) at $E_L - E_S = \Delta E(E_L)$, against luminescence lines that acc. to $E_S$ = const shift linearily with $E_L$ (line $S_1$). By tuning $E_L$ merely a few meV into the indirect exciton absorption, remarkable changes in the appearance of the spectra come about. Concerning the 2TO(L) scattering (line a), the model developed in section 4.2 is completely confirmed, the variation in the scattered intensity reflecting strong exciton relaxation. The complete identification of the various relaxation processes is put forward as they manifest themselves in well-defined additional scattering lines (b, c1, c2) occurring in the spectra. Based on the Raman energy shifts, in comparison with the known phonon dispersion curves, the intensities and resonance thresholds, various relaxation processes are identified and schematically shown in Fig. 14 (Windscheif and von der Osten, 1980):

1. Intravalley scattering by dispersive acoustic phonons near the zone center (process b) and similarly by non-dispersive phonons (TO(Γ), LO(Γ)),

2. intervalley scattering by acoustic phonons of negligible dispersion with wavevector near the X-point (process c), and

3. radiative recombination of excitons and trapping at impurities and defects (process d).

These processes determine the relaxation of the exciton, thereby affecting the exciton lifetime. By considering them in detail, their relative contributions to the total lifetime broadening $\gamma_{tot}(\varepsilon)$ may be deduced and used to describe the 2TO(L) intensity in a consistent picture. As presumed in Fig. 14, due to the low temperatures of the experiments, only phonon emission processes have to be considered in the analysis.

<u>Intravalley scattering.</u> This type of scattering is of particular interest among the various relaxation processes since it involves dispersive phonons near the Brillouin zone center. The phonons participating are LA(Γ) phonons. As seen from Fig.15

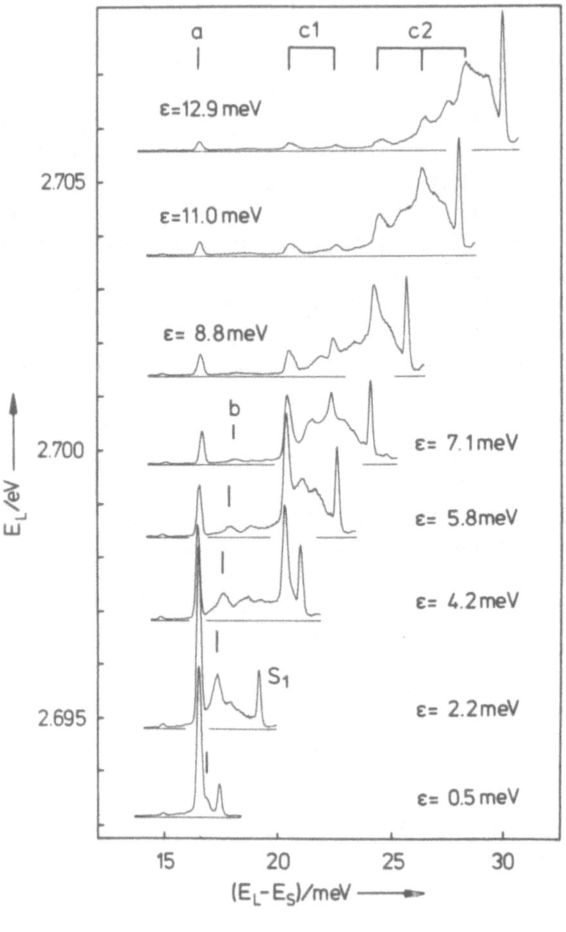


Fig. 15. RRS spectra of AgBr at 1.8K for different kinetic exciton energies ε. Notation: a) 2TO(L) scattering; b) intravalley scattering by dispersive LA(Γ) phonons; c1, c2) intervalley scattering by TA(X), LA(X) and 2TA(X), TA(X)+LA(X), 2LA(X) phonons, respectively. Forbidden TO(L)+TA(L) scattering at 15 meV. $S_1$: luminescence line.

with increasing $E_L$ a scattering line (b) splits off the 2TO(L) peak resulting in an energy-dependent Raman shift $E_L - E_S = 2\hbar\omega_{TO(L)} + \hbar\omega_{LA(\Gamma)}(\varepsilon)$.

The contribution $\gamma_1(\varepsilon)$ to the total lifetime broadening $\gamma_{tot}(\varepsilon)$ for intravalley scattering may be calculated by Fermi's "golden rule" giving

$$\gamma_1(\varepsilon) \sim \sum_{\vec{q}} | < M > |^2 \, \delta\left( \frac{\hbar^2 \vec{k}_i^2}{2M^*} - \frac{\hbar^2 \vec{k}_f^2}{2M^*} - \hbar v|\vec{q}|\right) \qquad (27).$$

Here <M> is the matrix element for exciton-phonon interaction via the deformation potential which acc. to Toyozawa (1958) is given by

$$< M > \equiv <1s(\vec{k}_f),\, \vec{q}|H_{EP}|1s(\vec{k}_i),0> \sim |\vec{q}|^{1/2} \qquad (28).$$

It describes transitions from the initial (wavevector $\vec{k}_i$) into the final exciton state ($\vec{k}_f$) by emission of a phonon of wavevector $\vec{q}$ (Fig. 16). Here $\vec{q}$ is determined by the conservation of

$$\text{wavevector} \qquad \vec{k}_i = \vec{k}_f + \vec{q} \qquad (29)$$

$$\text{and energy} \qquad \frac{\hbar^2 \vec{k}_i^2}{2M^*} = \frac{\hbar^2 \vec{k}_f^2}{2M^*} + \hbar v|\vec{q}| \qquad (30),$$

where a linear phonon dispersion (sound velocity v) is assumed. Carrying out the summation in eq. 27 that runs over all possible phonon wavevectors and noting that $M^* v \ll \hbar \vec{k}_i$ in the wavevector region of interest eq. 27 can be shown to give

$$\gamma_1(\varepsilon) = c_1 \cdot \varepsilon \qquad (31)$$

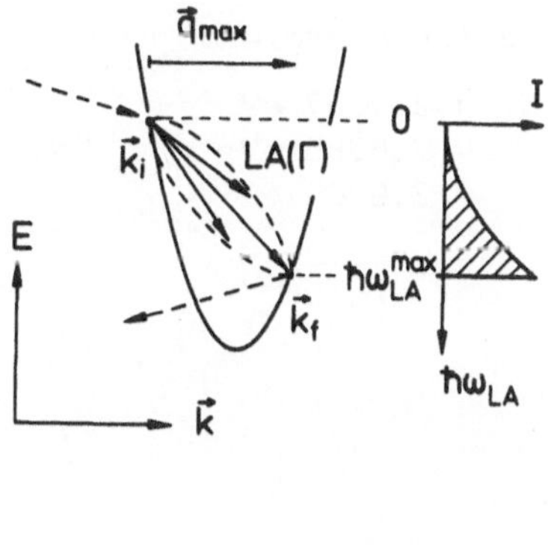


Fig. 16.

Schematic representation of exciton dispersion and intravalley scattering by long wavelength LA phonons. The final states f are along the conic section (dashed). The lineshape of the scattering spectrum (see text) is also shown.

($c_1$ = proportionality constant), implying the increasing importance of exciton intravalley relaxation at higher exciton energies.

Note that scattering by TA(Γ) phonons is not observed. This is probably due to the small deformation potential produced by the shear strain that is induced by a transverse acoustic phonon, in comparison to that produced by the lattice dilatation originating from LA(Γ) phonons.

Intervalley scattering. This process turns out as a surprisingly efficient exciton relaxation mechanism (Windscheif et al., 1978). As illustrated in Fig. 14 (process c), the scattering of the exciton takes place between non-equivalent valleys that correspond to different L-points (L, L') in the Brillouin zone. Considering low temperature, a wavevector conserving phonon from the vicinity of the X-point is created as readily seen from inspection of the Brillouin zone. Accordingly the total Raman shift is $E_L - E_S = 2\hbar\omega_{TO(L)} + \sum_i \hbar\omega_{X,i}$ where $\hbar\omega_{X,i}$ stands for the energy of the participating phonons we observe, i.e. TA(X) and LA(X). Higher order processes involving overtones and combinations are also observed (Fig. 17). From Fig. 14 it is evident that these processes are activated at energies of the initial exciton state

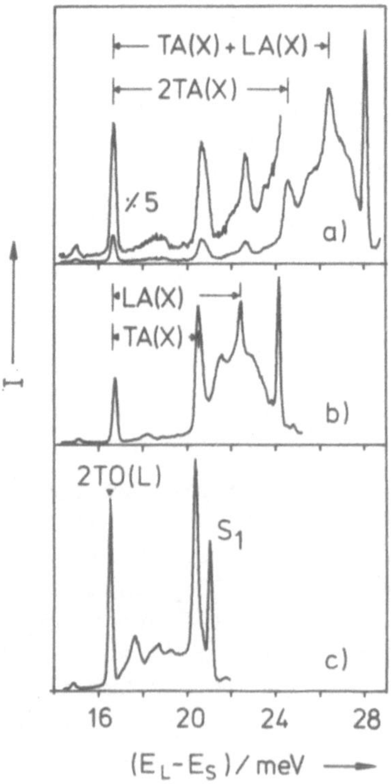


Fig. 17.

Intervalley scattering involving X-point phonons in AgBr at 1.8K. Dispersive intravalley scattering due to long wavelength LA phonons between 17 and 20 meV.

Excitation energies:

a) $E_L$ = 2.7037 eV
b) = 2.6998 eV
c) = 2.6969 eV

$\varepsilon = \sum_i \hbar\omega_{X,i}$. This criterion becomes clear also from Fig. 18, where the total experimental Raman shifts of the peaks are plotted vs. $E_L \sim \varepsilon$. The diagonal line corresponds to the position of the $S_1$ fluorescence line in Fig. 15 that is due to a weakly (0.3 meV) localized exciton state. All Raman lines that correspond to processes ending with emission of a TO(L) phonon appear on the right-hand side of this line clearly demonstrating the existence of the threshold, the arrow at the abscissa marking the TO(L) assisted 1s exciton absorption.

To derive the contribution $\gamma_2(\varepsilon)$ of the intervalley process to the total lifetime broadening $\gamma_{tot}(\varepsilon)$ of the exciton state it is crucial to realize that scattering of the exciton by odd-parity phonons is forbidden between two L-points (Weber and von der Osten, 1976) becoming gradually allowed for tuning into the Brillouin zone. We therefore assume the matrix element responsible for intervalley scattering to be proportional to the difference between the wavevector $\vec{q}$ of the scattering phonon and the wavevector $\vec{k}_X$ at the X-point of the zone so that

$$\langle M\rangle \equiv \langle 1s(k_f),\ \vec{q}\ |H_{EP}|\ 1s(\vec{k}_i),0\rangle \sim |\vec{q} - \vec{k}_X| \qquad (32).$$

The probability for intervalley scattering of an exciton from the initial (energy $\varepsilon$, wavevector $\vec{k}_i$) into the final exciton state ($\varepsilon'$, $\vec{k}_f$) will then be given by

$$\gamma_2(\varepsilon) \sim \sum_{\vec{k}_f} |\langle M\rangle|^2\ \delta(\varepsilon - \varepsilon' - \hbar\omega_X) \qquad (33).$$

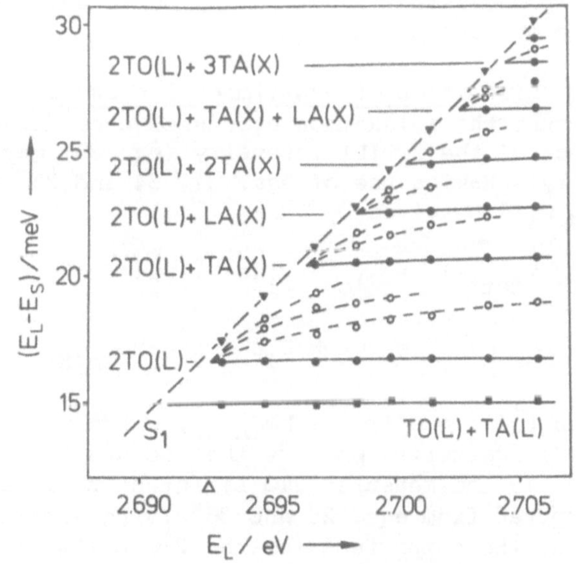


Fig. 18.

Raman shift of various scattering processes in AgBr vs. excitation energy. Dispersive phonon scattering by dashed lines. The diagonal marks position of $S_1$ luminescence line in Fig. 15.

Neglecting the dispersion of the participating X-phonon, the summation can be carried out (Windscheif et al., 1978) giving

$$\gamma_2(\varepsilon) = c_2\ (\varepsilon - \hbar\omega_X)^{1/2}\ (2\varepsilon - \hbar\omega_X) \qquad (34),$$

where $\gamma_2(\varepsilon) = 0$ for $\varepsilon \leq \hbar\omega_X$.

Radiative recombination and trapping. The probability for radiative recombination of the exciton can be estimated from the absorption constant (Dexter, 1958) giving $\tau_{rad} \simeq 10^{-4}$ s for AgBr. This value compares with the measured exciton lifetime $\tau_{tot} \simeq 60$ ns (Stolz et al., 1976) indicating that the effect of radiative recombination onto the lifetime broadening is negligibly small compared to other recombination processes.

A more relevant process, however, is trapping of excitons at impurities or lattice imperfections (process d in Fig. 14). The trapped excitons do no longer contribute to the RRS spectra. Their recombination determines the fluorescence observed in AgBr on the long wavelength side of the FE emission, examples being the line $S_1$ in Fig. 15 and the luminescence lines shown in Fig. 25. Some of the impurity states are considerably lowered in energy with regard to the FE state. In first approximation therefore the trapping probability is assumed as independent on $E_L$ and $\varepsilon$, respectively, giving a contribution to $\gamma_{tot}(\varepsilon)$

$$\gamma_0 = c_0 = \text{const} \qquad (35).$$

It is to be pointed out, however, that we have also observed strongly resonant trapping at weakly localized defect states (section 4.6) suggesting that energy-dependent trapping probabilities should also be considered.

Discussion of experimental cross-section. Taking into account the relaxation processes considered above, the energy dependence of the 2TO(L) intensity (eq. 25) may now be evaluated explicitly. Making use of eqs. 31, 34 and 35, the total lifetime broadening will be given by

$$\begin{aligned}\gamma_{tot}(\varepsilon) &= \gamma_0 + \gamma_1(\varepsilon) + \gamma_2(\varepsilon)\\ &= c_0 + c_1\varepsilon + c_2(\varepsilon - \hbar\omega_{TA(X)})^{1/2}(2\varepsilon - \hbar\omega_{TA(X)}) \qquad (36)\end{aligned}$$

with $\gamma_2(\varepsilon) = 0$ if $\varepsilon \leq \hbar\omega_{TA(X)}$, where we have neglected the weaker LA(X) scattering process that occurs at still higher energies. In Fig. 19 the measured 2TO(L) intensity is plotted together with that computed from eqs. 25 and 36 giving excellent agreement. In addition, the comparison for the 2TO(L)+TA(X) line is shown in which

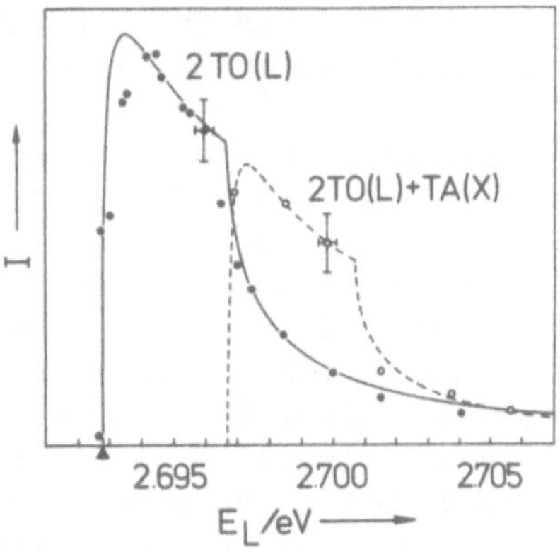


Fig. 19.

RRS intensity of the 2TO(L) and 2TO(L)+TA(X) process vs. $E_L$. Experimental points shown together with fit acc. to eqs. 25 and 36. Arrow at abscissa marks absorption threshold (Windscheif et al., 1978).

case a similar expression like eq. 25 holds. The fitting parameters used, normalized to $c_0$ are listed in Table 3. Depending on the exciton energy they imply different relative strengths of the various relaxation channels. The interpretation of Fig. 19 now is obviously clear: the 2TO(L) process is activated at the exciton absorption threshold increasing in its intensity with $\varepsilon^{1/2}$. Scattering by LA(Γ) phonons gradually reduces the intensity leading to the occurrence of the maximum in cross-section. At still higher energies, the onset of intervalley scattering at $\varepsilon = \hbar\omega_{TA(X)}$ then

Table 3. Normalized fitting parameters $c_0$ to $c_2$ describing trapping, LA(Γ) intravalley and TA(X) intervalley scattering of excitons, and relative matrix elements for allowed scattering derived from RRS (see text). Notation for matrix elements: 1 = TA(L) for AgCl and TO(L) for AgBr, 2 = LA(L) for both AgCl and AgBr.

| | $c_0$ | $c_1/(\text{meV})^{-1}$ | $c_2/(\text{meV})^{-3/2}$ | $\left\lvert\frac{M_{em,2}}{M_{em,1}}\right\rvert^2$ |
|---|---|---|---|---|
| AgCl | 1 | 0.05 | 0.0065 | 0.35 |
| AgBr | 1 | 1.3 | 1.25 | 0.19 |

drastically further reduces the intensity that at the same time emerges in the 2TO(L)+TA(X) line as shown.

Very recently, RRS at low temperatures was successfully extended to the indirect exciton edge in AgCl (Nakamura et al., 1981). Experimentally these measurements are difficult since the exciton gap is hardly accessible to tunable cw dye-lasers necessary for excitation. On the other hand, in comparison to AgBr scattering in AgCl is of interest since self-trapping of excitons (section 3.5) is expected to affect RRS. As seen from Fig. 20, the scattering observed in AgCl involves pairs of momentum-conserving TA(L) and LA(L) phonons, but is considerably weaker than in AgBr. In fact, it turns out to be of the same strength as second-order non-resonant RS seen to be broadly superimposed in the spectra. The parameters $c_0$ to $c_2$ obtained from the analysis of the AgCl data along the lines described above are also listed in Table 3. While for AgBr they all are found to be of the same order of magnitude, in AgCl one finds the energy independent trapping process to be predominant. Interesting enough, the relative probabilities for intra- and intervalley

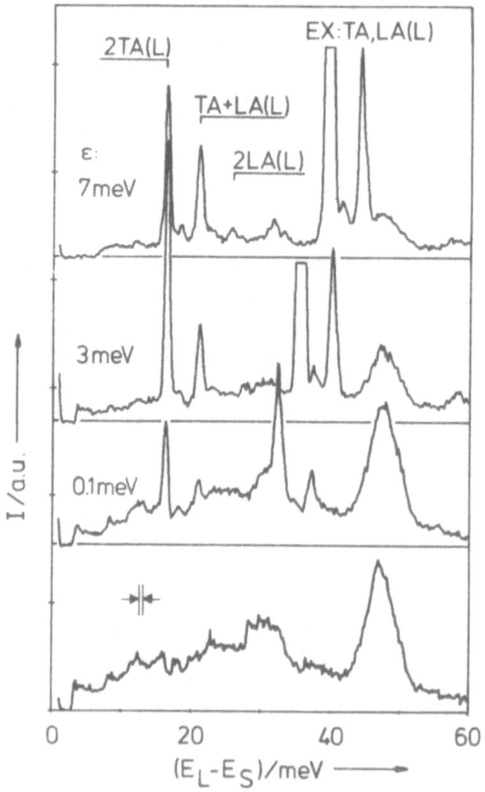


Fig. 20.

RRS spectra of AgCl at 1.8K excited near the indirect exciton absorption edge involving TA(L) and LA(L) phonons.

ε: kinetic energy of exciton

EX: phonon-assisted bound exciton luminescence

The spectrum at the bottom excited slightly below exciton absorption is due to non-resonant RS (Nakamura et al., 1981).

scattering is smaller by orders of magnitude indicating that self-trapping obviously is an efficient relaxation mechanism for the FE in AgCl. Besides RRS the broad recombination luminescence shown in Fig. 9 (bottom) is also observed, proving the coexistence of the FE and the STE in AgCl, the FE being in the unrelaxed state. Applying eq. 25, from the ratio of scattering intensities of the TA(L)+LA(L) and 2TA(L) lines below the LA(L) absorption edge the relative matrix elements for symmetry-allowed scattering are obtained for AgCl and similarly for AgBr. They are included in Table 3, the value for AgBr showing good agreement with the corresponding quantity from fitting the absorption edge (Table 2).

To unravel the effect of higher (n>1) exciton states and exciton-LO phonon interaction, the RRS cross-section for the 2TO(L) line in AgBr was investigated up to higher exciton energies ($\varepsilon \approx 40$ meV). The predominant process in this regime is intravalley scattering by zone center LO phonons that strongly comes into

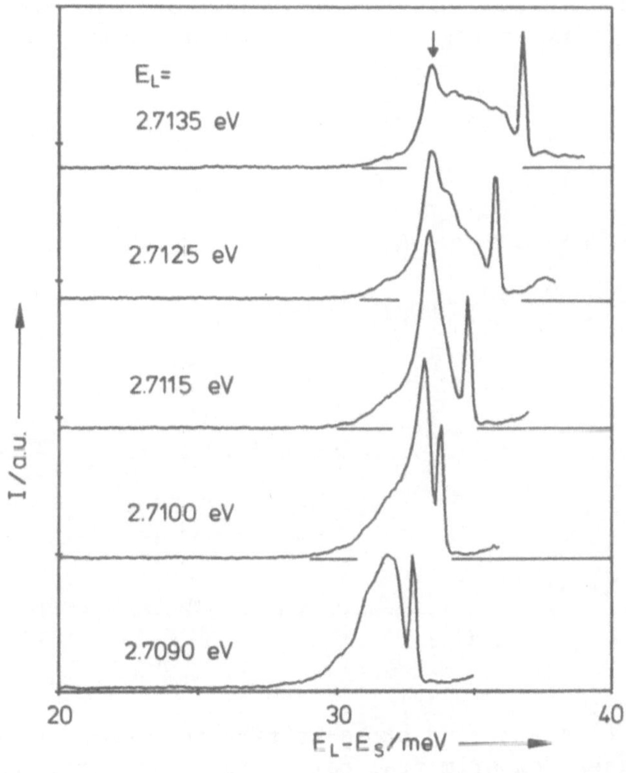


Fig. 21. RRS in AgBr at 1.8K for high excitation energies $E_L$. Arrow marks 2TO(L)+LO(Γ) intravalley scattering process. The narrow line at the right-hand side corresponds to $S_1$ in Fig. 15 (Sliwczuk, 1981).

resonance at multiples of $\hbar\omega_{LO(\Gamma)}$ = 17 meV above the 1s exciton absorption threshold. Accordingly, the scattering lines (2TO(L), 2TO(L)+LA(Γ) etc.) of Fig. 15 practically repeat at these energies with Raman shifts larger by $n\hbar\omega_{LO(\Gamma)}$ (n = 1,2 ... ) but increasingly broadened due to multiple scattering. As an example, in Fig. 21 the 2TO(L)+LO(Γ) process at $E_L - E_S \simeq 33$ meV is illustrated and seen to run through resonance at $\varepsilon = \hbar\omega_{LO(\Gamma)}$ corresponding to $E_L \simeq 2.7100$ eV.

The 2TO(L) intensity for this extended range of energies is represented in Fig. 22. Although some effect due to strong LO(Γ) scattering is observed at one and two LO phonon energies above threshold (arrows), the line intensity starts already to reduce at about $\varepsilon \geq 11$ meV (see deviation of fit from experimental points in Fig. 22) becoming smaller by orders of magnitude at higher energies. This implies that another relaxation channel for the 1s-exciton opens up, the origin of which is not understood. Basically scattering into higher exciton states may have to be considered, but does not explain the observation quantitatively. Since we don't observe a corresponding additional absorption edge in the energy range of interest, exciton scattering into an exciton band at a different point in the Brillouin zone as suggested for AgCl

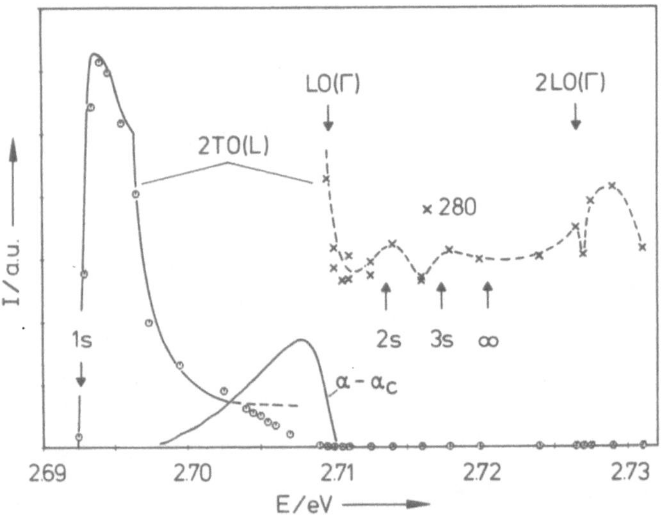


Fig. 22. 2TO(L) scattering cross-section for AgBr. Circles, crosses: experimental data. Full line through experimental points: fit from Fig. 19. Arrows mark thresholds due to 1s, 2s ... exciton states obtained from the hydrogen exciton model assuming $E^{i}_{bx}$ = 28 meV. Onset of LO(Γ) and 2LO(Γ) intravalley scattering is also indicated. For $\alpha - \alpha_c$ see text.

(Stulen and Ascarelli, 1976) may also be ruled out as possible explanation. It is interesting to realize, however, that the reduction in scattering intensity energetically coincides with the part of the exciton absorption edge that remained unexplained by the fit performed in section 3.3. For comparison, we have included into Fig. 22 the difference between experimental and computed absorption coefficient, $\alpha - \alpha_C$, from Fig. 7, which suggests the common origin of the two observations.

In concluding this section it is to be pointed out that with increasing $E_L$ or $\varepsilon$, respectively, any structure in the RRS spectra is smeared out due to the superposition of a multitude of intra- and intervalley exciton scattering processes. This situation e.g. is approximately realized for $E_L$ = 2.709 eV in Fig. 21, before the onset of strong LO scattering then again results in pronounced structure, becoming broadened again at still higher $E_L$ etc. At sufficiently high energies the spectrum finally develops into the Maxwell-Boltzmann lineshape discussed in 3.5. Due to the low temperatures only phonon emission occurs in exciton relaxation so that in spite of exciton-phonon collisions no thermal equilibrium will be established between excitons and phonons. The lineshape of the emitted light thus completely is determined by the exciton scattering processes.

## 4.4 Effective Exciton Mass

Scattering by dispersive intravalley phonons introduced in 4.3 in principle allows to determine the effective exciton mass $M^*$ (Windscheif et al., 1977). Looking back to the situation sketched in Fig. 16, it becomes obvious that for the involved phonon energies $\hbar\omega_{LA(\Gamma)}$ there exists a continuous range from zero to a maximum value $\hbar\omega_{LA(\Gamma)}^{max}$ representing the peak of the scattering line. The lineshape, i.e. the spectral intensity distribution $I(E_S)$ follows from eq. 27 if we take the sum over the direction of $\vec{q}$ only, $|\vec{q}|$ being determined by $\hbar\omega_{LA(\Gamma)} = \hbar v|\vec{q}|$. The evaluation of the sum for fixed $E_L$ then gives

$$I(E_S) \sim \frac{q^2}{k_i} \sim \hbar\omega_{LA(\Gamma)}^2 \qquad (37)$$

with $E_S = E_L - 2\hbar\omega_{TO(L)} - \hbar\omega_{LA(\Gamma)}$, as schematically indicated in Fig. 16. Using the wavevector and energy requirements in eqs. 29 and 30, it is straightforward to derive the maximum phonon energy as

$$\hbar\omega_{LA(\Gamma)}^{max} = 2M^* v^2 \left( \sqrt{\frac{2\varepsilon}{M^* v^2}} - 1 \right) \qquad (38)$$

resulting in the expected energy dependence of the Raman shift. Since v is known from independent data ($v = 3.06 \cdot 10^3$ m/s for AgBr at low temperatures) and $\varepsilon$ is determined by the incident photon energy $E_L$, eq. 38 may be used to fit the experimental Raman shift of this process in AgBr in Fig. 23. The full line represents the fit to the experimental data using $M^* = (1.5 \pm 0.4)\ m_o$ ($m_o$ = electron mass). As seen from the RRS spectra in Figs. 15 and 17, higher-order processes involving two and more LA(Γ) phonons are also observed. They may be treated in similar fashion. In Fig. 23 the full line through the experimental points for the 2TO(L)+2LA(Γ) process is a fit my means of the same effective mass value giving good accordance.

From the fitting procedure an average value for $M^*$ is obtained only, since any anisotropy brought about by the hole mass is neglected. Provided only the longitudinal (heavy) hole and the electron mass will be important, our result would be in good accordance using the band masses and neglecting polaron effects ($M^* = 1.25\ m_o + 0.215\ m_o$, see Table 1). This proceeding would be justified (Behnke and Büttner, 1978) if the polaron radius is comparable to the exciton radius which, however, is uncertain in AgBr (see section 3.4). On the other hand, the actual experimental lineshape observed for LA(Γ) intravalley scattering (Figs. 15 and 17) is only approximately described by eq. 37. Thus there is reason to believe the lineshape is essentially determined by the mass anisotropy with the polaron masses being the proper choice. From corresponding RRS data in AgCl a preliminary value for the

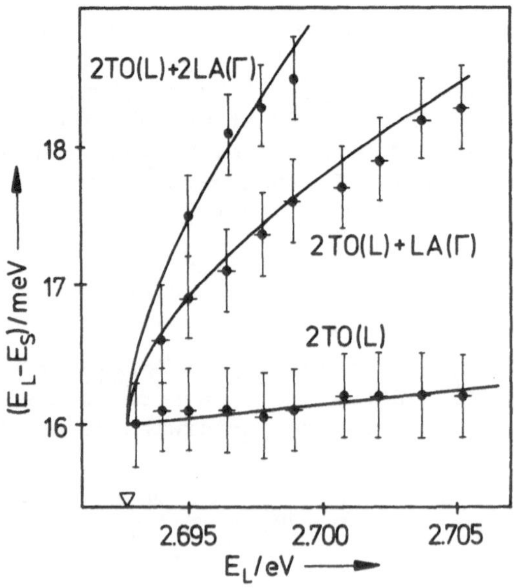


Fig. 23.

Raman shift vs. $E_L$ for long wavelength LA intravalley scattering in AgBr at 1.8K. The full lines are fits through the experimental data as discussed in the text (Windscheif et al., 1977).

effective exciton mass of $M^* \simeq 2.5\ m_0$ may be estimated in reasonable agreement with expectation (Nakamura, 1981).

### 4.5 Phonon-Symmetry Forbidden Scattering

The detection of a weak RRS line in AgBr involving the TA(L) phonon (Fig. 15) offers the possibility to study symmetry-forbidden processes and to derive the relative strength of the corresponding matrix element compared to allowed processes. The advantage of RRS in comparison to conventional absorption is evident. While in the absorption spectrum the extremely weak forbidden component for most part of the exciton energies is superimposed by strong allowed absorption and therefore totally obscured, in RRS the two components give rise to two distinct scattering lines (e.g. TA(L)+TO(L) and 2TO(L)) that may be studied separately. The information about the desired matrix element as well as the relaxation processes is contained in the scattered intensity which in turn is accessible without principle difficulty.

Considering RRS as equivalent to a process of "absorption followed by emission" (section 4.2) and starting out from eq. 25 it is straightforward to derive an expression for the energy-dependent scattering cross-section. In case of a symmetry-forbidden transition, the matrix element, being zero in lowest order, acc. to eq. 6 has to be expanded in powers of $\vec{K}$. In case of the TA(L)+TO(L) scattering process the expansion has to be performed for either $M_{abs}$ or $M_{em}$ depending on if absorption or emission is accomplished by the TA(L) phonon. As already discussed in section 3.3 (eq. 15), this then leads to a $\varepsilon^{3/2}$ dependence for the scattering cross-section. Taking into account the different threshold energies for the TA(L)+TO(L) and TO(L)+TA(L) processes, one obtains

$$R^{TA(L)+TO(L)} + R^{TO(L)+TA(L)} = K^{TA} \left( \frac{\varepsilon_1^{3/2}}{\gamma_{tot}(\varepsilon_1)} + \frac{\varepsilon_2^{3/2}}{\gamma_{tot}(\varepsilon_2)} \right) \qquad (39)$$

where $K^{TA}$, besides other constants, contains the first-order matrix element $|M^1_{TA(L)}|^2$ of the TA(L) assisted absorption and emission process, respectively. This energy dependence has to be compared to that for allowed scattering which from eq. 25 may be accordingly written as

$$R^{2TO(L)} = K^{TO} \cdot \frac{\varepsilon_2^{1/2}}{\gamma_{tot}(\varepsilon_2)} \qquad (40)$$

with $K^{TO}$ containing the zeroth-order matrix element $|M^0_{TO(L)}|^2$ of the TO(L) process.

The experimental data for the two processes in AgBr (Fig. 24)

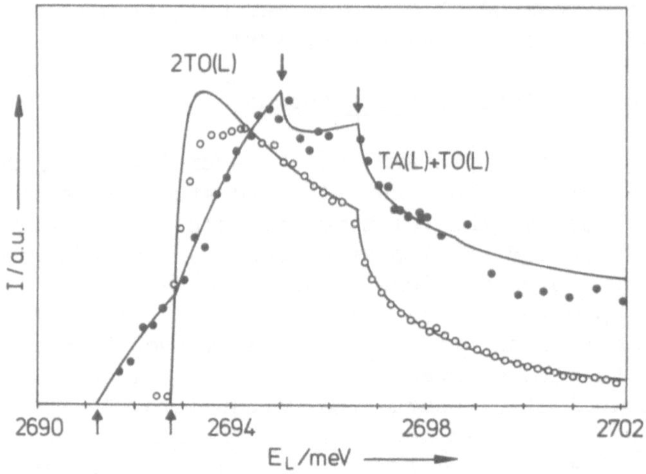


Fig. 24. RRS cross-section for phonon-symmetry allowed (2TO(L)) and forbidden (TA(L)+TO(L)) processes in AgBr at 1.8K. The experimental data (points, circles) are shown together with fits acc. to eqs. 39 and 40. A multiplication factor of 50 was used for the intensity of the TA(L)+TO(L) component. Upward and downward arrows mark absorption thresholds and onset of intervalley exciton relaxation, respectively, for the various components (Sliwczuk et al., 1982).

support the predictions above, the full lines being fits according to eqs. 39 and 40. In considering the various exciton relaxation processes that contribute to the total lifetime broadening, for both cases the same set of parameters $c_0$ to $c_2$ were used, slightly different from those in Table 3. From the fit the ratio of matrix elements is found to be $|M^1_{TA(L)}|^2/|M^0_{TO(L)}|^2 \sim K^{TA}/K^{TO} = 5.5\cdot10^{-3}$ meV$^{-1}$ (Sliwczuk et al., 1982).

## 4.6 Resonant Exciton Trapping

Impurities are known to introduce localized exciton states into the spectrum of the host crystal that for excitation at high energies give rise to bound exciton recombination luminescence. Even in nominally pure AgBr and AgCl several transitions from weakly localized excitons are observed (Fig. 25). Due to the weak localization (localization energies < 10 meV) their wavefunctions differ only slightly from those of the FE (see e.g. Thomas, 1968) so that the bound exciton recombination is accomplished by the same momentum-conserving L-point phonons as for the FE. Although the exact origin of these transitions is still uncertain (Weber, 1976),

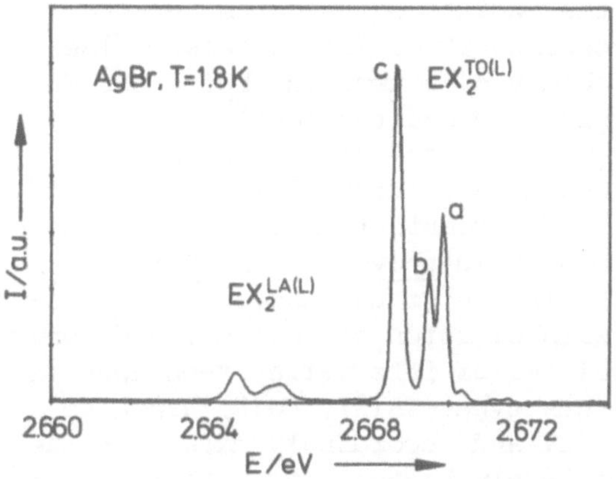


Fig. 25.

Recombination luminescence of weakly bound excitons accomplished by momentum-conserving TO(L) and LA(L) phonons. Also a weak TA(L) assisted component observed at the high energy side. Lines a, b, c suggested to be due to different extrinsic defects (Sliwczuk, 1981).

preliminary studies of doped AgBr suggest that $Cd^{2+}$ probably acts as one of several recombination centers.

One principal difficulty in studying bound excitons by conventional luminescence is that it will only probe the ground state of the system, while information about excited states in general has to come from far-infrared measurements. Also a multitude of electronic processes resulting from above-band-gap excitation are superimposed obscuring details of relaxation. To reveal ground and excited states, excitation spectroscopy may be used allowing for selective excitation. In this technique the luminescence that occurs at fixed photon energy is monitored as function of excitation photon energy, the form of the excitation spectrum in general resembling the absorption spectrum.

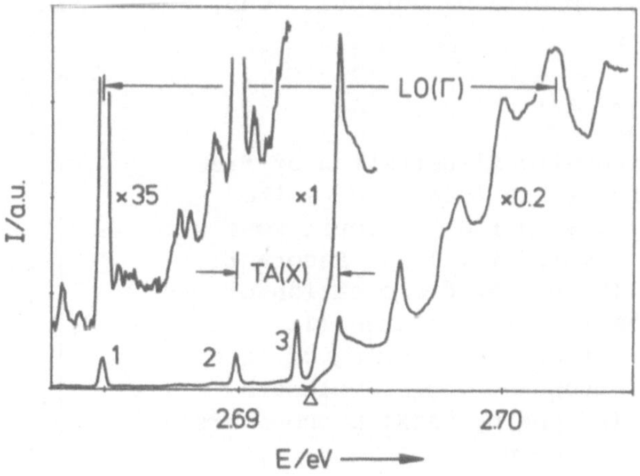


Fig. 26.

Excitation spectrum of c-component of the TO(L) assisted bound exciton luminescence in Fig. 25. Arrow at abscissa marks exciton absorption threshold. For details see text (Sliwczuk, 1981).

As one example, in Fig. 26 the excitation spectrum for the c-component of the $EX_2^{TO(L)}$ luminescence (Fig. 25) is shown. The origin of the complex structure observed is twofold: below the FE absorption threshold the ground plus two excited localized states are detected (line 1 and lines 2, 3, respectively). The peaks correspond to TO(L) assisted resonant excitation of these states followed by immediate TO(L) assisted recombination via the ground state (Fig. 27). The structure monitored above threshold is explained by resonant trapping of FE in the ground and excited localized states. It is associated with emission of well-defined zone center optical and acoustic as well as of intervalley X-phonons. Two representative processes of this type (TA(X), LO(Γ) are also schematically illustrated in Fig. 27 and accordingly marked in the excitation spectrum, that may completely be analyzed with regard to even the weak structure and lineshapes. The excitation spectra for the other components a and b in Fig. 25 are found to differ slightly, supporting the different origin of these lines (Sliwczuk, 1981).

These results are intimately connected to RRS of excitons (section 4.3). They suggest that by using excitation spectroscopy, in silver halides the exciton trapping mechanisms may be studied in detail leading to a profound understanding of bound excitons in these materials. At the same time, they demonstrate that strongly energy-dependent exciton trapping occurs that might have to be considered in the analysis of RRS.

### 4.7 RRS in Magnetic Fields

In the study of spin properties of excitons, magneto-absorption and -luminescence are known to be powerful techniques. This is especially true for the case of direct excitons that exhibit discrete and narrow line spectra. The magnetic field in these cases often results in line splittings and intensity changes that may readily be detected (Cho, 1979). In contrast, for indirect

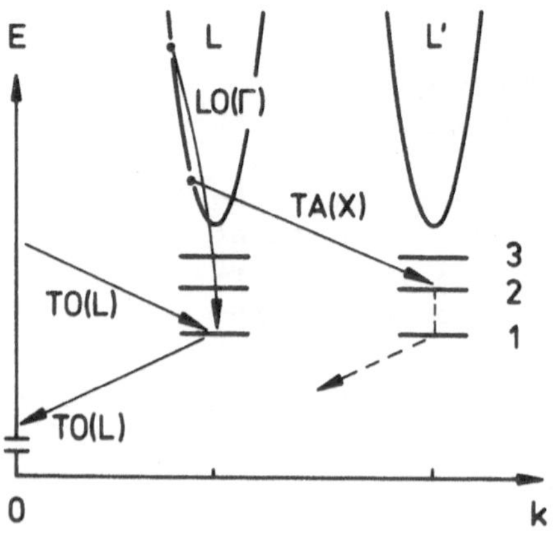


Fig. 27.

Schematic illustration of free and weakly localized exciton states in the Brillouin zone (compare Fig. 14). Resonant TO(L) assisted excitation of the localized states (1 ... 3) and resonant trapping of free excitons by long wavelength LO(Γ) and by TA(X) phonons are also shown.

excitons magnetic field effects are far more difficult to analyze mainly due to the continuous absorption and hardly observable weak additional structure introduced by the field (Kurita and Kobayashi, 1978; Baba and Masumi, 1978). As explained in context with phonon-symmetry forbidden scattering processes (section 4.5), here again RRS is superior to conventional magneto-optical methods. It is the occurrence of a resonantly enhanced narrow scattering line that allows to detect the threshold energies of the split exciton states. Also, contributions from various exciton components are revealed with high sensitivity by following up the scattered intensity, the variation as function of field strength reflecting the degree of mixing between states due to the magnetic field. By studying RRS in AgBr in fields up to 20 Tesla most recently the magnetic properties of the indirect exciton were successfully investigated (Stolz et al., 1982). Experimentally, for several fixed excitation energies $E_L$, the scattering cross-section is measured as function of the magnetic field, the results providing consistent values of the effective electron and (longitudinal) hole g-factors ($g_c$, $g_v^{\parallel}$) and the exchange splitting energy $\Delta$.

If the indirect exciton is taken to consist of a hole and an electron of $L_{4,5}^-$ and $\Gamma_6^+$ symmetry, respectively (section 2), at zero magnetic field four exciton states exist (compare Fig. 28). Characterizing the exciton wavefunctions in the missing electron scheme by the effective hole and electron spin $\sigma_z = \pm 1/2$ ($L_{4,5}^-$) and $S_z = \pm 1/2$ ($\Gamma_6^+$), these are the doubly degenerate mixed singlet-triplet states $\phi_1^o = |1/2, 1/2\rangle$ and $\phi_4^o = |-1/2, -1/2\rangle$ and the degenerate pair of pure triplet states $\phi_2^o = |1/2, -1/2\rangle$ and $\phi_3^o = |-1/2, 1/2\rangle$. $\phi_1^o$, $\phi_4^o$ at zero field give rise to the indirect absorption edge, while $\phi_2^o$, $\phi_3^o$, split off by exchange interaction, are optically forbidden. The energies $E_j$ including a diamagnetic shift $\sim \vec{B}^2$ (Cho et al., 1975) and the wavefunctions $\phi_j$ ($j = 1 \ldots 4$) in the magnetic field are given by

$$E_{1,2} = \frac{1}{2}\left\{-(\Delta+g_v^{\parallel}\mu_B B_z) \pm \sqrt{(\Delta+g_c\mu_B B_z)^2 + g_c^2\mu_B^2(B^2-B_z^2)}\right\} + \frac{m_o^2}{2E_{bx}^i\mu^2}\mu_B^2\vec{B}^2 \tag{41a}$$

$$E_{3,4} = \frac{1}{2}\left\{(-\Delta+g_v^{\parallel}\mu_B B_z) \mp \sqrt{(\Delta-g_c\mu_B B_z)^2 + g_c^2\mu_B^2(B^2-B_z^2)}\right\} + \frac{m_o^2}{2E_{bx}^i\mu^2}\mu_B^2\vec{B}^2 \tag{41b}$$

and

$$\phi_{1,2} = \alpha^S_{1,2}\ \phi^o_1 + \alpha^T_{1,2}\ \phi^o_2 \tag{42a}$$

$$\phi_{3,4} = \alpha^T_{3,4}\ \phi^o_3 + \alpha^S_{3,4}\ \phi^o_4 \tag{42b}$$

Here $\mu$, $E^i_{bx}$ and $\mu_B$ are the reduced effective exciton mass, indirect exciton binding energy and Bohr magneton. $E_{1,4}$ in eqs. 41a,b is taken to be zero for B = 0 at the TO(L) assisted absorption threshold. $\vec{B} = (B_x, B_y, B_z)$ is the magnetic field referred to <111> as z-axis. $\alpha^S_j$ and $\alpha^T_j$ denote the singlet and triplet amplitudes that can be shown to have the form

$$\frac{1}{|\alpha^S_1|^2} = 1 + 4\ [E_1 - \tfrac{1}{2}(g_c - g^{\parallel}_v)\mu_B\ B_z]^2 / g^2_c \mu^2_B (B^2 - B^2_z) \tag{43}$$

and similarly for the other coefficients. Considering simple parabolic bands with the same effective exciton mass for all states and realizing that each exciton contributes to the overall cross-section only through its spin singlet component $\alpha^S_j$ the 2TO(L) scattering cross-section acc. to eq. 25 may be written as

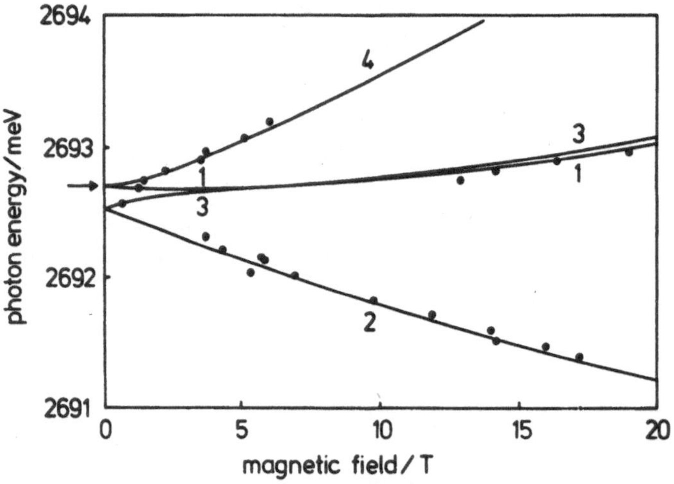


Fig. 28. Energies of singlet-triplet (1,4) and pure triplet (2,3) exciton states obtained from RRS in AgBr at 1.8K vs. magnetic field strength. Experimental data (points) are shown together with a fit (full lines) acc. to eq. 41 a,b. Arrow marks TO(L) assisted indirect absorption threshold at zero field determined to be at 2.69271 eV (Stolz et al., 1982).

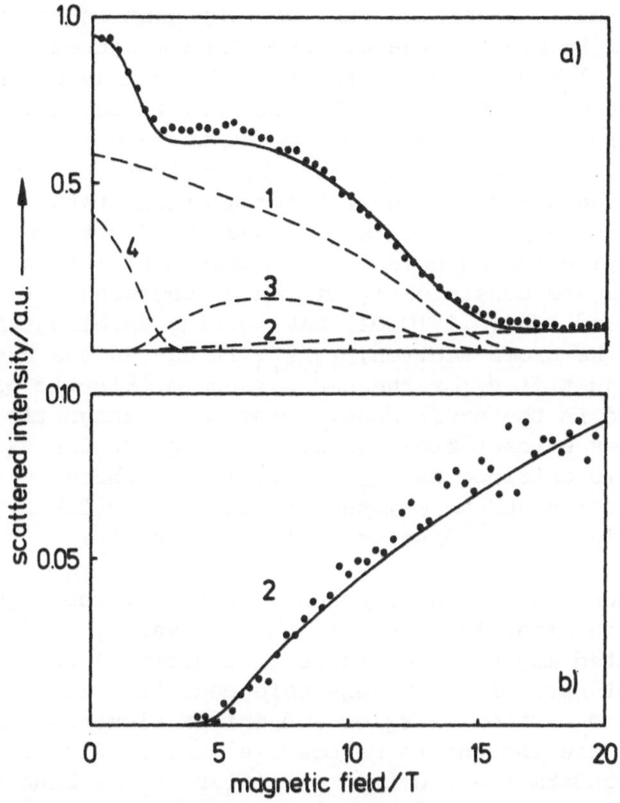


Fig. 29. 2TO(L) RRS intensity vs. magnetic field strength for excitation slightly above (a: $E_L$ = 2.69283 eV) and below (b: $E_L$ = 2.69214 eV) zero-field exciton absorption threshold. Points: experimental data. Full lines: computed from eq. 44 as described in the text. Dashed lines correspond to individual contributions from the various exciton states in Fig. 28 (Stolz et al., 1982).

$$R^{2TO(L)} \sim \sum_{j=1}^{4} |\alpha_j^S|^4 \cdot \frac{\varepsilon_j^{1/2}}{\gamma_{tot,j}(\varepsilon_j)} \qquad (44)$$

where, acc. to eq. 26, now

$$\varepsilon_j = E_L - E_{gx}^i - E_j - h\omega_{TO(L)} \qquad (45).$$

Fig. 28 shows a comparison between the experimentally observed exciton energies as function of the magnetic field with $E_1$ to $E_4$ computed from eqs. 41 a,b. The experimental data are obtained from the threshold energies at which, for fixed excitation energy

$E_L$, scattering due to a certain exciton component is activated. As typical examples in Fig. 29 the cross-section measured for $E_L$ slightly above (a) and below (b) the zero-field absorption threshold is represented together with the fit acc. to eq. 44. The contributions from various exciton components (j = 1 ... 4) are separately illustrated to demonstrate the field-induced exchange of intensity between singlet and triplet exciton states. A polarization factor for the states $\phi_3$ and $\phi_4$ had thereby to be included to achieve quantitative agreement with experiment (Stolz et al., 1982). All data are consistently fitted by choosing $g_c = 1.46 \pm 0.05$, $g_v^{\parallel} = 2.61 \pm 0.05$ and $\Delta = (0.17 \pm 0.01)$ meV, giving excellent agreement. In calculating the cross-section, $\gamma_{tot,j}$ in eq. 44 was assumed to be constant as justified for the small range of kinetic exciton energies covered in the experiment. From the diamagnetic shift of the energy states evident from considering Fig. 28 the exciton binding energy is obtained as $E_{bx}^{i} = (32 \pm 5)$ meV, where we have used the polaron electron and hole masses to give $\mu = 0.229\ m_o$, a result reasonable in view of the discussion in section 3.4.

These values have to be compared to results from magneto-absorption (Matsushita, 1973; Kurita and Kobayashi, 1978) and optically detected magnetic resonance (Marchetti, 1981), from which fairly different sets of parameters were deduced. Since the RRS results are based both on energies and matrix elements, the data presented above are thought to be most reliable. At the same time they directly confirm the ordering of states in the band structure calculation in Fig. 1 with $L_{4,5}^-$ higher in energy than $L_6^-$ and also $\Sigma$ (Matsushita, 1973), since different order would lead to quite different RRS results.

## 5. CONCLUDING REMARKS

The methods described in this lecture and applied to the investigation of excitons in AgCl and AgBr were also successfully used to study mixed silver halides of $AgBr_{1-x}Cl_x$ (Koumvakalis and von der Osten, 1979; Fujii et al., 1982). Like in the examples presented above, here they provide new and detailed information on both the unrelaxed and relaxed exciton states as well as on phonon properties, in particular regarding their dependence on composition. Altogether the results obtained in silver halides demonstrate that resonant light scattering in context with conventional spectroscopic techniques may be used as an extremely powerful tool in the investigation of excitons and exciton relaxation. It still will be desirable to employ, as a step forward, more direct techniques like time-resolved spectroscopy, as now available for the time regime of interest, to gain deeper insight into the dynamics of excitons.

REFERENCE

Ascarelli, G., and Baxter, J.E., 1972, Solid State Commun., 10:315.
Baba, T., and Masumi, T., 1978, Solid State Commun., 27:1113.
Bassani, F., Knox, R.S., and Fowler, W.B., 1965, Phys. Rev. 137:A1217.
Behnke, G., and Büttner, H., 1978, phys. stat. sol., (b) 90:53.
Brown, F.C., 1973, in: "Treaties on Solid State Chemistry", Vol. 4, B. Hannay, ed., Plenum Press, New York, p.333.
Cardona, M., 1975, and other authors in: Topics in Applied Physics, "Light Scattering in Solids", ed. M. Cardona, Springer-Verlag, Berlin.
Carrera, N.J., and Brown, F.C., 1971, Phys. Rev., B 4:3651.
Cho, K., 1979, in: Topics in Current Physics, "Excitons", K. Cho, ed., Springer-Verlag, Berlin, p.15.
Cho, K., Suga, S., Dreybrodt, W., and Willmann, F., 1975, Phys. Rev., B 11:1512.
Dexter, D.L., 1958, in: Solid State Physics, Vol. 6, ed. F. Seitz and D. Turnbull, Academic Press, New York, p.353.
Dorner, B., von der Osten, W., and Bührer, W., 1976, J. Phys. C: Solid State Phys., 9:723.
Elliott, R.J., 1957, Phys. Rev., 108:1384.
Fesefeldt, H., 1930, Z. Physik, 64:328.
Fischer, K., Bilz, H., Haberkorn, R., and Weber, W., 1972, phys. stat. sol. (b), 54:285.
Fujii, A., Stolz, H., and von der Osten, W., 1982, to be published.
Ganguly, A.K., and Birman, J.L., 1967, Phys. Rev., 162:806.
Haken, H., 1958, Fortschr. Physik, 6:271.
Hayes, W., Owen, I.B., and Walker, P.J., 1977, J. Phys. C: Solid State Phys., 10:1751.
Hodby, J.W., 1971, J. Phys. C: Solid State Phys., 4:L8.
Höhne, M., and Stasiw, M., 1968, phys. stat. sol., 28:247.
Joesten, B.L., and Brown, F.C., 1966, Phys. Rev., 148:919.
Kanzaki, H., 1980, Photogr. Sci. Eng., 24:219.
Kanzaki, H., and Sakuragi, S., 1969, J. Phys. Soc. Japan, 27:109.
Kanzaki, H., Sakuragi, S., and Sakamoto, K., 1971, Solid State Commun., 9:999.
Klein, M.V., 1973, Phys. Rev., B8:919.
Klein, P.B., Masui, H., Song, J.J., and Chang, R.K., 1974, Solid State Commun., 14:1163.
Knox, R.S., 1963, "Theory of Excitons", Academic Press, New York.
Komatsu, T., Karasawa, T., Iida, T., Miyata, K., and Kaifu, Y., 1981, J. Luminescence, 24/25:679.
Koumvakalis, N., and von der Osten, W., 1979, phys. stat. sol., (b), 92:441.
Kurita, S., and Kobayashi, K., 1978, J. Phys. Soc. Japan, 44:1583.
Loudon, R., 1963, Proc. Roy. Soc., A275:218.
Loudon, R., 1964, Adv. Phys., 13:423.
Marchetti, A.P., 1981, J. Phys. C: Solid State Phys., 14:961.
Matsushita, M., 1973, J. Phys. Soc. Japan, 35:1688.

Nakahara, J., Kobayashi, K., Fujii, A., 1974, J. Phys. Soc. Japan, 37:1312.
Nakamura, K., 1981, unpublished.
Nakamura, K., Windscheif, J., and von der Osten, W., 1981, J. Luminescence, 24/25:425.
von der Osten, W., 1974, Phys. Rev., B9:789.
von der Osten, W., 1982, in: Landolt-Börnstein New Series III,17b, "Semiconductors", O. Madelung, ed., Springer-Verlag, Berlin.
von der Osten, W., and Dorner, B., 1975, Solid State Commun.,16:431.
von der Osten, W., and Weber, J., 1974, Solid State Commun., 14:1133.
von der Osten, W., Weber, J., and Schaack, G., 1974, Solid State Commun., 15:1561.
Overhof, H., 1981, private communication.
Phillips, J.C., 1970, Rev. Mod. Phys., 42:317.
Richter, W., 1976, in: Solid State Physics, ed. G. Höhler, Springer-Verlag, Berlin, p.121.
Sliwczuk, U., 1981, Diplomarbeit, Universität-GH Paderborn, unpublished.
Sliwczuk, U., Stolz, H., and von der Osten, W., 1982, Verh. DPG, 5:751.
Solin, J.R., and Merkelo, H., 1976, Phys. Rev., B14:1775.
Stolz, H., 1976, Diplomarbeit, TH Darmstadt, unpublished.
Stolz, H., and von der Osten, W., 1980, J. Phys. Soc. Japan, 49 Suppl. A:543.
Stolz, H., von der Osten, W., and Weber, J., 1976, in: Proc. 13th Int. Conf. Phys. of Semiconductors, Rome, p.865.
Stolz, H., Waßmuth, W., von der Osten, W., and Uihlein, Ch., 1982, (to be published).
Stulen, R.H., and Ascarelli, G., 1976, Phys. Rev., B13:5501.
Tamura, H., and Masumi, T., 1973, Solid State Commun., 12:1183.
Thomas, D.G., 1968, in: Localized Excitations in Solids", R.F. Wallis, ed., Plenum Press, New York, p.239.
Toyozawa, Y., 1958, Progr. Theor. Phys., 20:53.
Toyozawa, Y., 1961, Progr. Theor. Phys., 26:29.
Toyozawa, Y., 1974, in: "Proc. 4th Int. Conf. Vacuum Ultraviolet Radiation Physics", E.E. Koch, R. Haensel, C. Kunz, ed., Pergamon, Vieweg, New York, Braunschweig, p.317.
Toyozawa, Y., 1980, Techn. Rep. ISSP, Ser.A, No.1036:1.
Vijayaraghavan, P.R., Nicklow, R.M., Smith, H.G., and Wilkinson, M.K., 1970, Phys. Rev., B1:4819.
Weber, J., 1975, Thesis, TH Darmstadt, unpublished.
Weber, J., 1976, phys. stat. sol., (b) 78:699.
Weber, J., and von der Osten, W., 1976, Z. Physik, B24:343.
Windscheif, J., and von der Osten, W., 1980, J. Phys. C: Solid State Phys., 13:6299.
Windscheif, J., Stolz, H., and von der Osten, W., 1977, Solid State Commun., 24:607.
Windscheif, J., Stolz, H., and von der Osten, W., 1978, Solid State Commun., 28:911.

# ELECTRON-HOLE LIQUID CONDENSATION IN SEMICONDUCTORS

T. L. Reinecke

Naval Research Laboratory

Washington, D. C. 20375 USA

A description is given of the basic energetics and thermodynamics of the condensation of excitons into electron-hole liquid in semiconductors. The ground state energetics is reviewed, and recent work on the thermodynamics of the condensation, especially those features which can be related simply to the basic energetics, is described. In particular, theoretical approaches to the condensation phase diagram and to its critical point are given, and simple scaling relations between the ground state properties and the critical parameters are developed.

## INTRODUCTION

At low temperatures excitons and free carriers in indirect gap semiconductors are observed to condense into a unique state of matter characterized by micron sized droplets of dense, metallic electron-hole liquid (EHL).[1,2,3] The electron hole system generally is formed by optical excitation of the semiconductor. It involves electrons and holes at the conduction and valence band edges of the fundamental gap of the semiconductor. This condensation has now been observed in Ge and in Si, in systems formed by straining Ge and Si along several uniaxial directions, and also in a number of compound semiconductors including GaP, AℓAs, and SiC.

Physically as the carrier density increases the excitons become screened out, and the liquid state is stabilized by the many-body correlations in the interacting metallic electron-hole system. At finite temperature the excitons and electron-hole liquid are seen to coexist suggesting a first order phase transition. Electron-hole recombination in these indirect gap semiconductors is phonon assisted, and therefore the recombination time is long $\gtrsim\mu$sec. The thermalization time for carriers in the bands, on the other hand, is much shorter ~nsec, and therefore the system is effectively in thermodynamic equilibrium on the time scale of the electron-hole recombination lifetime. Thus for most purposes the system can be thought of as composed of electrons and holes with effectively infinite lifetimes in equilibrium with one another.

The phenomenon of electron-hole liquid condensation from excitons in semiconductors has been of great interest both theoretically[1] and experimentally[2] during the past ten years. Among the most fundamental and sustaining reasons for this interest are: (i) it provides the best system known for detailed comparisons between theory and experiment in advancing the understanding of interacting electronic systems, and (ii) the condensation of excitons into the EHL involves the most quantum fluid known. In addition to these attributes, the electron-hole luminescence spectrum provides a direct, detailed, and quantitative picture of the energetics of the constituents of the system.

The properties of interacting electronic systems both at zero temperature and at finite temperature have been one of the central concerns of many body physics for decades.[4] The correlation energy which arises from the long-ranged Coulomb interactions in these fermi systems has been the particular focus of attention. No exact solution for the correlation energy has been obtained, but approximate theoretical approaches appropriate to fairly high densities have been developed. The validity of these methods of calculation cannot be established rigorously, and previously it has not been possible to make quantitative comparisons with experiment. The EHL system furnishes a unique opportunity for such comparisons between theory and experiment primarily because of the accuracy of the effective mass for carriers in semiconductors. By comparison, in the case of simple metals it is not straightforward to make a quantitative separation of the effects of the interactions between the electrons and the host ions from the effects of the electron-electron interactions. In semiconductors, on the other hand, the effective mass accounts for the electron-ion interaction to a high degree of accuracy. The carriers then become effectively free particles interacting via a screened Coulomb interaction. In addition, the varying band structures of semiconductors provide EHL systems which have varying densities, a feature which is important in comparing to theory. Further, these densities span a region of particular interest to theory ($r_s \cong \frac{1}{2}$-4) where $r_s$ is the average interparticle spacing measured in terms of the excitonic Bohr radius.

The electron-hole liquid system is unique in that it is composed of the lightest particles of any fluid known, and therefore it is expected to show strong quantum effects. A measure of the quantum character is deBoer's quantum parameter $\Lambda$, which is essentially the ratio of the deBroglie wavelength to the interparticle spacing.[6] For EHL $\Lambda \sim 80$, whereas for $^3He$, the most quantum of other fluids, $\Lambda \sim 3$.[2] The quantum corrections for most fluids other than $^3He$ which undergo phase transitions are very small.[6] Therefore the condensation of excitons into EHL provides a uniquely important system in which to study quantum effects in phase transitions.

The ground state energetics of electrons and holes in semiconductors are now quite well understood in the low and high density regimes.[3] At low densities electrons and holes form excitons. As the density increases the excitons become screened out. At sufficiently high density the electrons and holes exist as extended quasiparticles in the metallic EHL system. This state is stabilized by the many-body exchange-correlation energy, and it generally has a lower energy per particle than the exciton.

The finite temperature thermodynamic properties of the excitons and EHL of this system have not come to be understood until more recently, and they are not yet established in such rigorous detail as the ground state properties. Nevertheless substantial advances have been made recently in understanding the finite temperature properties of this system in terms of its ground state energetics. The principal intent of the present article is the description of some of these developments. In particular theoretical approaches to the coexistence phase diagram for exciton-EHL condensation will be described. The critical point for this condensation will be discussed. The concept of simple scaling relations between the ground state properties and the critical parameters will be developed.

The topics discussed here were chosen because of their timeliness, because they relate to basic properties of this interacting electronic system, and as a result of the author's interests. The unifying theme is the relationship of the thermodynamic properties of the condensation with the basic ground state energetics of the system. Descriptions of other phenomena and work can be found in the references, particularly in the reviews in Refs. 1 and 2.

In the present discussion the electrons and holes will be treated as if they have effectively infinite lifetimes on the scale of the thermalization times in the system, the electrons and holes will be taken to have relaxed into their lowest respective bands, and the system will be taken to be in spatial equilibrium. For the properties discussed here this picture provides a very good approximation. The effects of deviations from these conditions will be mentioned briefly in the last section.

The second section deals with the basic energetics of the system, the third section with its thermodynamics, phase diagram, and critical point, and the fourth section with scaling relations between the ground state and critical parameters. In the fifth section some concluding remarks are made.

## GROUND STATE ENERGETICS

The ground state energetics of the electron-hole system in the indirect gap semiconductors discussed here is sketched in Fig. 1. Here the energy is shown as a function of an increasing uniform density $\rho$ of electron-hole pairs. At low density the electron and hole form a hydrogenic bound state called an (Wannier) exciton which is bound by energy $E_x$.[1] (In fact the lowest energy state generally is a bi-excitonic molecule bound further by an energy typically ~0.1 $E_x$.) As the density increases the energy per pair is believed to increase due to interactions between excitons or bi-excitons.[1] The energetics in this intermediate density regime (shown by the dashed line in Fig. 1) is not yet well understood, but the form of the curve shown in Fig. 1 is consistent with theoretical considerations[1] and with the experimental data. Finally at sufficiently high densities the excitons screen one another out, and the electrons and holes become extended quasi-particles in a dense metallic electron-hole liquid. Typical densities for EHL are ~$10^{17}$-$10^{18}$cm$^{-3}$ or $r_s$ ~O(1) in terms of the excitonic Bohr radius.

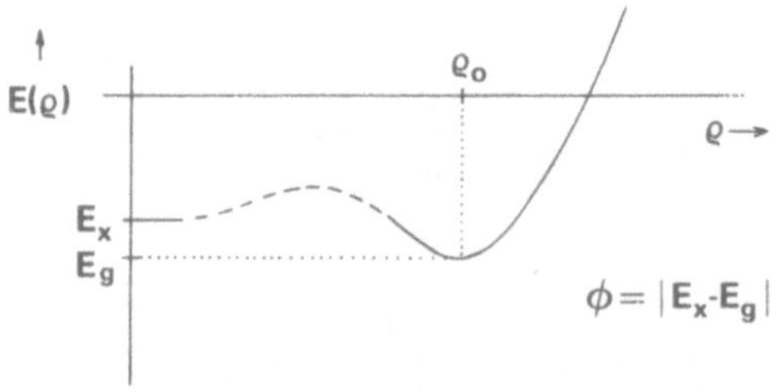


Fig. 1. Schematic representation of the energy $\varepsilon$ of a system of a uniform density $\rho$ of electron-hole pairs. $E_g=\varepsilon_0$ is the ground state energy of the electron-hole liquid of density $\rho_0$ which is bound by a condensation energy $\phi$ with respect to the exciton of energy $E_x$.

The energetics in this range of high densities are understood quite well on the basis of the theory of dense interacting electronic systems. Here the effects of the Pauli exclusion principle and of the Coulomb interactions give rise to large negative exchange-correlation energy thus producing a deep local energy minimum $\varepsilon_o$ in the energy with a corresponding ground state density $\rho_o$. The difference between the EHL ground state energy $\varepsilon_o=\varepsilon(\rho_o)$ and $E_x$ is the condensation energy $\phi$. It is important to remember in the following that the basic energetics of the system between the limits of low and high density is not yet well understood. This includes the region where exciton-exciton interactions become significant and also that in which the Mott metal-insulator transition of excitons to free carriers occurs. The absence of a detailed knowledge of the energetics in this region of densities places restrictions on approaches currently available to treat the thermodynamics of the system.

The ground state energetics of carriers in semiconductors are governed by two basic effects: (i) semiconductor band structure, and (ii) carrier interactions. Semiconductors exhibit a wide variety of band structures, and in general the interactions between carriers are both direct Coulombic interactions and also those mediated by (LO) phonons. For the most part the Ge and Si systems will be discussed in detail here. They exhibit all of the essential features of interest here, and the most detailed experimental results are available for them. The cases of compound semiconductors and of LO-phonon coupling will be commented on briefly at the end of this section.

## Band Structure

The electronic states of covalent semiconductors are intermediate between the extended states of a simple metal and the highly localized states of an ionic solid. The strong carrier-ion interaction gives rise to considerable distortion of their $\varepsilon(\underset{\sim}{k})$ carrier dispersion relations from that for the free carriers. The structure of the bands near the fundamental gap of Ge for example is illustrated in Fig. 2. For Ge the bottom of the conduction band consists of four equivalent minima at the L point of the Brillouin zone (the zone edges in the <111> directions), and the valence band maximum occurs at the $\Gamma$ point (the zone center). Even at the carrier densities in EHL ($\sim10^{17}cm^{-3}$) the carriers occupy only very small regions of $\underset{\sim}{k}$ space near these extrema points. In these small regions of $\underset{\sim}{k}$ space the carrier dispersion relations can be taken to be approximately parabolic. These effectively parabolic bands for Ge are shown by the dashed bands in Fig. 3. The electrons then occupy four degenerate highly ellipsoidal valleys with dispersion ($m_\ell/m_t\sim20$ in Ge)

$$\varepsilon_e(\underset{\sim}{k}) = \frac{\hbar^2 k_x{}^2}{2m_{e\ell}} + \frac{\hbar^2 k_y^2 + \hbar^2 k_z^2}{2m_{et}} \tag{1}$$

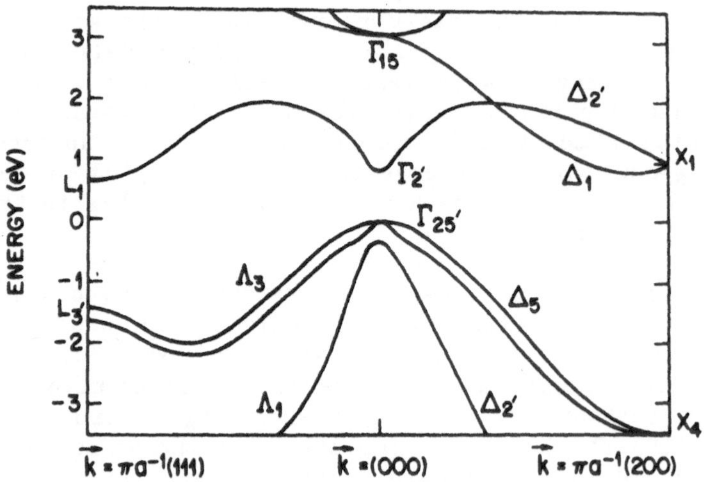


Fig. 2. The electronic band structure of Ge in the vicinity of the fundamental energy gap.

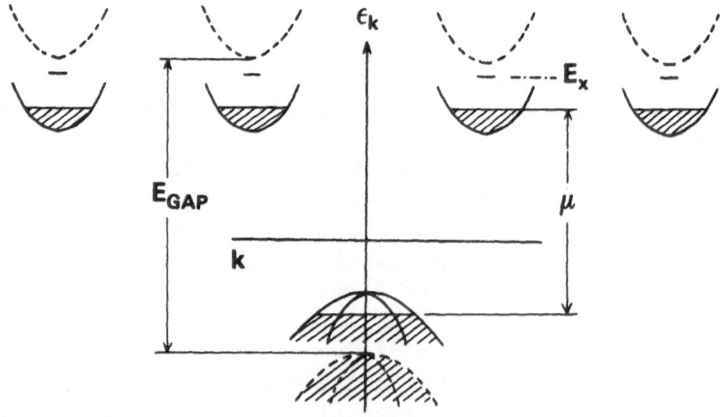


Fig. 3. Effective parabolic bands of Ge near the conduction and valence band edges (dashed lines). The solid curves show the bands for high carrier density in EHL shifted by the exchange-correlation energy and filled to respective fermi levels.

The holes occupy coupled heavy (H) and light (L) hole bands of the form

$$\varepsilon_{\pm}(\underset{\sim}{k}) = Ak^2 \pm [B^2k^4 + C^2(k_x^2k_y^2 + k_y^2k_z^2 + k_z^2k_x^2)]^{\frac{1}{2}} \tag{2}$$

which are degenerate at k=0. A frequently used approximation which gives isotropic H and L hole bands replaces B, C by B',C' where $B' = (B^2 + 1/6C^2)^{\frac{1}{2}}$ and C'=0 giving $m_{hL,H}^{-1} = A \pm B'$. It should be noted however that the heavy and light hole bands remain coupled by matrix elements of the coulomb interaction.

The band structure of Si is very similar except that the six degenerate conduction valleys are less ellipsoidal, and they lie near but not at the zone boundary in the <100> directions.[7] The band structure of many semiconductors near the edges of the fundamental gaps are known to high accuracy from cyclotron resonance and related experimental studies. This is especially true of Ge and Si. The band parameters for Ge and Si are listed in Table I. The result of the above effective mass approximation in semiconductors is to produce a system of essentially "free" electrons and holes moving with modified masses and interacting with one another via a coulomb interaction screened by a static background dielectric constant. The complexities of the carrier-ion interactions are thus removed.

A set of systems based on Ge and Si with uniaxial strains is often useful as a set of model systems. They have widely varying band structures, and their effective masses are known well.[8] These effects in Ge are illustrated in Fig. 4. In this case a uniaxial strain along <111> decreases the energy of one of the conduction ellipsoids with respect to the others, and for modest strain the valence maximum remains approximately degenerate. For large uniaxial strain the hole band degeneracy is split, and the maximum of the valence band is a single ellipsoidal hole band. These systems are often denoted by Ge(1;2) and Ge(1;1) respectively with Ge(4;2) denoting unstrained Ge. The first index gives the conduction (band) degeneracy and the second gives the valence band degeneracy. In Si the systems for zero, intermediate, and large strain along <100> are Si(6;2), Si(2;2), and Si(2;1). Here the conduction band degeneracy is not completely raised by uniaxial strain because the conduction band minima lie away from the zone edges. From symmetry it is seen that large uniaxial strain along <110> and <111> in Si gives the systems Si(4;1) and Si(6;1) respectively.

The way in which detailed information concerning the properties of the exciton-EHL system can be obtained from the luminescence spectra can be seen best by describing the energetics in terms of a band picture as shown in Fig. 3. The effectively parabolic bands appropriate for a low density of carriers are shown by

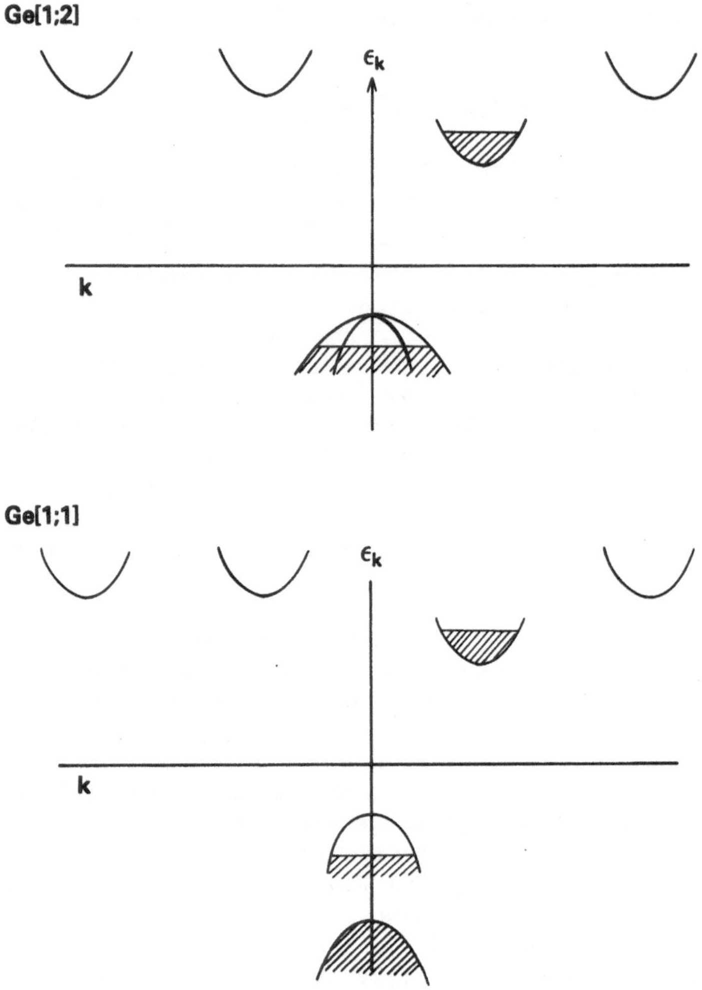


Fig. 4. The effective band structure of Ge with strain along the <111> direction. Ge(1;2) for intermediate strain and Ge(1;1) for large strain.

dashed lines. An electron-hole pair excited across the band gap will form an exciton the energy of which is bound by $E_x$. At the high densities (high excitation) appropriate to the EHL state the electrons and holes are once again in extended quasi-particle states, and the band picture is appropriate. At these densities the bands are filled up to some effective Fermi energies $E_{Fe}$ and $E_{Fh}$. The effect of carrier interactions which give the exchange and correlation energies is to lower the states in each band nearly rigidly[1] as shown. The energy to add an electron-hole pair to the

Table I. Values of the band structure parameters and excitonic Rydberg units for Ge and Si

| Parameter | Ge | Si |
|---|---|---|
| $m_{e\ell}$ | 1.58 | 0.9163 |
| $m_{et}$ | 0.082 | 0.1905 |
| $m_{hH}$ | 0.347 | 0.523 |
| $m_{hL}$ | 0.042 | 0.154 |
| $\kappa$ | 15.36 | 11.4 |
| $R_x$(meV) | 2.65 | 12.85 |
| $a_x$(Å) | 177 | 49 |

system is then given by a chemical potential $\mu$ which at T=0 is identical to the ground state energy pair

$$\mu = \partial/\partial\rho\ (\rho\varepsilon(\rho)) = \varepsilon(\rho_o) = \varepsilon_o$$

($\varepsilon'(\rho_o) = 0$ at T=0). In the EHL each electron can recombine with each hole. For a transition which is phonon allowed, the lineshape and position of the EHL luminescence is given by

$$I(h\nu)=M^2\int\int d\varepsilon_e d\varepsilon_h D_e(\varepsilon_e)D_h(\varepsilon_h)f_e(\varepsilon_e)f_h(\varepsilon_h)\delta(\varepsilon_e+\varepsilon_h+\varepsilon_g'-h\nu)$$

where the matrix element for the transition has been taken to be approximately constant, $D_i$ and $f_i$, i=e,h, are the densities of states and (temperature dependent) fermi distributions for electrons and holes respectively, $\varepsilon_g'$ is the renormalized gap energy, and carrier energies are measured from the respective band edges. By means of detailed fits to the EHL lines and exciton lines in recombination spectra as shown for example in Fig. 5, accurate values of the Fermi energies $E_{Fe}$, $E_{Fh}$, the EHL pair density $\rho$, the condensation energy $\phi$, the EHL chemical potential $\mu$, and the carrier temperature T are obtained. Crudely speaking the Fermi energies and pair density are obtained from the width of the EHL line, the condensation energy $\phi$ from the distance between the top of the EHL line and the exciton line, and the temperature from the shape of the high energy side of the EHL line. In short, the luminescence spectra of excited semiconductors give a remarkably detailed and complete description of the coexisting exciton-EHL system.

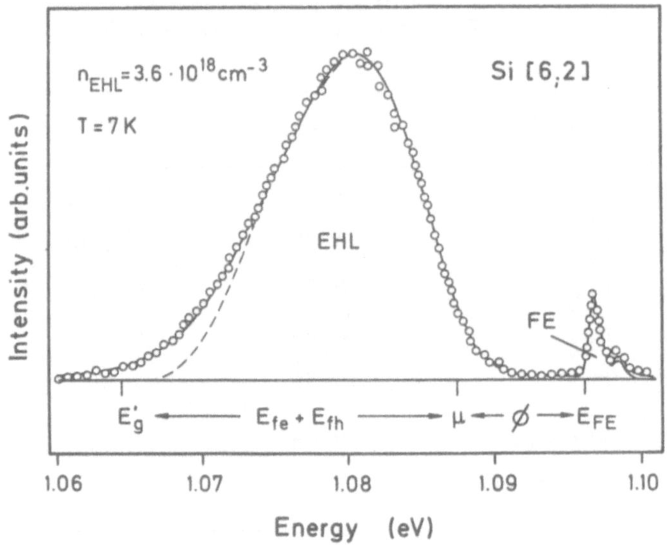


Fig. 5. Luminescence spectrum of Si from Ref. 20 showing the electron-hole liquid (EHL) line and free exciton (FE) line. Solid lines give detailed lineshape fits from which the electron and hole Fermi energies $E_{Fe}$, $E_{Fh}$, EHL density, the EHL chemical potential μ, and condensation energy ϕ are obtained.

The effects of interactions between the carriers will now be considered.

<u>Excitons</u>

The basic properties of excitons in Ge and Si systems will be described only briefly here. Excitons in semiconductors are discussed in greater detail in other articles in this volume. The electron and hole produced by band gap excitation interact with one another by means of the Coulomb interaction screened by the background dielectric constant κ of the material. In the effective mass approximation the Hamiltonian is given by

$$H = - \frac{\hbar^2 \nabla_e^{\ 2}}{2\underline{m}_e} - \frac{\hbar^2 \nabla_h^2}{2\underline{m}_h} - \frac{e^2}{\kappa |\underline{r}^e - \underline{r}^h|}$$

where in general the masses may have tensor character as indicated above. The simplest approximation to the exciton in these systems is given by a spherically symmetric wavefunction. The ground state then is the 1s state, and the exciton binding energy $E_x$ is given by the simple Rydberg formula

$$R_x = \mu_o e^4/2\kappa^2\hbar^2 \tag{3}$$

where $\mu_o$ is the reduced optical mass

$$\mu_o^{-1} = \frac{1}{3}(2m_{et}^{-1} + m_{e\ell}^{-1}) + m_{hH}^{-1} + m_{hL}^{-1} \tag{4}$$

and the corresponding Bohr radius is

$$a_x = \kappa\hbar^2/\mu_o e^2 \tag{5}$$

Values for $R_x$ and $a_x$ are given in Table I for Ge and Si.

From more detailed treatments of excitons in these systems it is known that the multivalley conduction bands and coupled hole bands complicate their descriptions.[1] Their effect is to increase somewhat the values of $E_x$ in these multiband systems. The spherical wave approximation nevertheless gives a good first order description for excitons in Ge and Si, and is quite good for semiconductors with relatively simple band structures such as Ge and Si with large uniaxial strains.

It is seen from the fact that $E_x/E_g \ll 1$, where $E_g$ is the energy of the fundamental gap, that the effective mass approximation is justified well for excitons in these systems. The same is true of carriers in the EHL state for which the pair binding energy is of the order of $R_x$. Physically this means that the wavefunctions of these states vary slowly on the scale of the lattice constant.

A set of atomic units for which $a_x$ is the measure of length and $2R_x$ is the measure of energy gives a natural set of units for these interacting systems, and it will be used here. In particular the dimensionless unit of density is given by an $r_s$ for which

$$\rho = (\frac{4\pi}{3} r_s^3 a_x^3)^{-1}$$

The correspondence between the low physical densities ($\sim 10^{17}$cm$^{-3}$) and high effective densities ($r_s \sim O(1)$) in EHL is seen to result from the facts that the Coulomb interaction is screened by a large static dielectric constant $\kappa$ and that the effective masses of carriers in semiconductors are small.

Electron-Hole Liquid

The electrons and holes in the dense EHL are in extended quasi-particle states moving with effective masses and interacting via the Coulomb interaction screened by a static background dielectric constant κ according to the many particle Hamiltonian

$$H = -\sum_i \hbar^2\nabla_i^2/2m_e - \sum_j \hbar^2\nabla_j^2/2m_h$$

$$- \sum_{ij} (e^2/\kappa)/|r_i^e - r_j^h| + \tfrac{1}{2}\sum_{i\neq j} (e^2/\kappa)/|r_i^e - r_j^e|$$

$$+ \tfrac{1}{2}\sum_{i\neq j} (e^2/\kappa)/|r_i^h - r_j^h| \qquad (6)$$

The carriers fill the electron and hole bands up to effective Fermi energies $E_{Fe}$, $E_{Fh}$ according to the Pauli excursion principle. The kinetic energy is the energy of the electrons and holes in the absence of interactions and gives a positive contribution to the total energy which corresponds to the band filling (Fig. 2). The kinetic energy per pair in Ge and Si is

$$\varepsilon_K(\rho) = \frac{3}{10}(3\pi^2)^{2/3}\,\mu_o\,(\nu^{-2/3}m_{de}^{-1} + m_{dh}^{-1})\,\rho^{2/3} \qquad (7)$$

where $m_{de}$, $m_{dh}$ are the effective density of states masses

$$m_{de} = (m_{et}^2\, m_{e\ell})^{1/3} \qquad (8a)$$

$$m_{dh} = m_{hH}\,(1+\gamma^{3/2})^{2/3} \qquad (8b)$$

$\nu$ is the conduction band degeneracy, and $\gamma = m_{hL}/m_{hH}$. The electron and hole densities are assumed to be equal giving charge neutrality in the EHL. Expressions in this section are written explicitly for the Ge and Si systems.

The Coulomb interaction term in the Hamiltonian eq. (6) have a factor of $r_s$ when written in atomic units. Therefore the effects of the Coulomb interactions can be written as a perturbation expansion in $r_s$, and they become more important for densities decreasing from very high values. The exchange interaction is the first order contribution of the Coulomb interaction. In terms of the usual many-body diagrams it is given by Fig. 6(a)[9,10]

$$E_x(\rho) = -\frac{3}{4}(3/\pi)^{1/3}\,[(\Phi(m_{et}/m_{e\ell})/\nu^{1/3} + \Psi(\gamma)]\rho^{1/3} \qquad (9)$$

where Φ arises from the ellipsoidal shape of the electron valleys and Ψ from the coupled H and L hole bands. Φ and Ψ are weak functions of their arguments. Physically the exchange energy is the

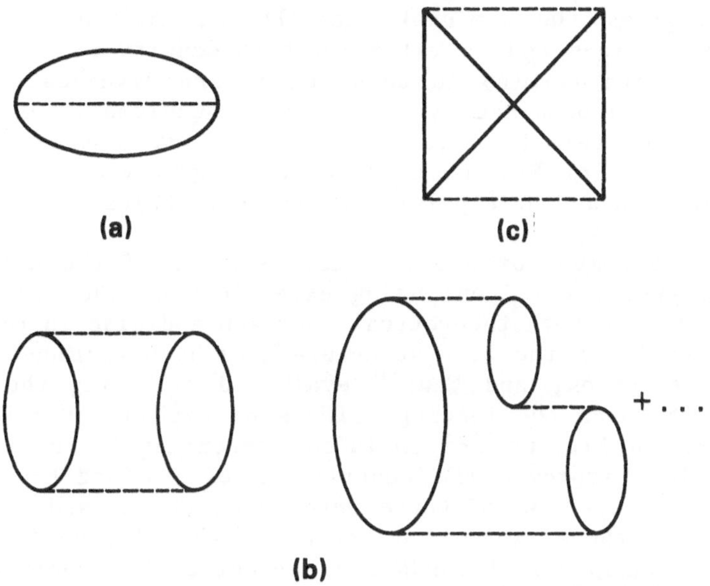


Fig. 6. Graphs contributing to the exchange-correlation energy of EHL. (a) gives the exchange energy, (b) the correlation energy in the R.P.A., and (c) the leading exchange correction to the R.P.A. correlation energy.

gain in Coulomb energy corresponding to the effect of the Pauli exclusion requirement that the wavefunctions of like particles of parallel spin avoid one another. The kinetic and exchange energy together constitute the Hartree-Fock energy.

The correlation energy $\varepsilon_c$ for interacting electronic systems is defined as the difference between the exact ground state energy and the Hartree-Fock energy. In EHL $\varepsilon_c$ includes the Coulomb energy gained by the spatial correlation of electrons with holes and by correlations between the wavefunctions of like particles of opposite spin. The effects of correlation in EHL are especially large because both electrons and holes are fully dynamic and quantum. The existence of a large correlation energy in EHL is essential in order to understand why EHL is strongly bound with respect to the exciton.

The correlation energy of electronic systems in general cannot be calculated exactly, but approximate techniques have been developed[4] which give good results in the density regime of $r_s \lesssim 6$. All of these approaches are based on the Random Phase Approximation

(R.P.A.) which is shown diagrammatically by the series in Fig. 6b. The R.P.A. gives the dominant contribution of the long ranged Coulomb correlations and is valid for high density $r_s \lesssim 1$. Physically the leading term(s) in this series are divergent, and the convergent sum is obtained by replacing one Coulomb interaction in the lowest order term by a screened Coulomb interaction. For EHL the R.P.A. series of Fig. 6b consists of graphs with all combinations of electron and hole polarizabilities (bubbles).

For EHL a number of detailed calculations of the correlation energy have been carried out using extensions of the R.P.A. They differ somewhat in both theoretical approach and also in the extent to which details of the band structure have been included. Brinkman, Rice, Anderson, and Chui[10] evaluated $\varepsilon_c$ using the Hubbard approach[11] in which the leading effects of exchange for like particles shown in Fig. 6c are included approximately in the R.P.A. approach. This improves the description of e-e and h-h correlations but does not change those between electrons and holes. In calculating $\varepsilon_c$ these workers approximated the ellipsoidal conduction bands with spherical bands and neglected the coupling of the hole bands. These band structure features have been included fully in the Hubbard approximation in later calculations.[12,13] Combescot and Nozières[9] applied the Pines-Nozières approach[14] to $\varepsilon_c$ of EHL and included the band structure in detail. This approach is physically very similar to the Hubbard approach (Fig. 6c) and uses an interpolation between the small and large wavevector contributions to $\varepsilon_c$. In a later series of calculations Vashishta and co-workers[13,15] have applied the Singwi-Tosi-Land-Sjölander[16] approach to $\varepsilon_c$ in EHL. Here a set of coupled equations for the polarizabilities of the carrier components are solved numerically. Effectvely sperical carrier bands were used explicitly, and additional details of the band structure were treated approximately.[13] These calculations include higher order electron-hole correlations in an approximate way, and they are the most sophisticated calculations to date. These higher order multiple scattering effects do not make an especially large contribution to the ground state energy and density, but they make a larger contribution to e.g. the electron-hole correlation function.

The results of these calculations for the ground state energy and density in Ge and Si are shown in Table II along with the corresponding experimental values obtained from luminescence studies. It is seen that these detailed calculations give values in good agreement with one another and with experiment. It is thus seen that the current understanding of the basic energetics of EHL is good. The consistency between calculations and the agreement with experiment is particularly good for the later calculations which take the full band structure into account in the greatest detail. It has been found that different treatments of the band structure

Table II. Comparison of theory and experiment for ground state properties of EHL.

| Reference | $-\varepsilon_o$ (meV) | $\rho_o$ ($10^{17}cm^{-3}$) |
|---|---|---|
| Ge | | |
| Experiment | | |
| TRPH (Ref. 18) | 6.0±0.2 | 2.38 |
| LO (Ref. 19) | 6.2±0.2 | 2.38 |
| Theory | | |
| BRAC (Ref. 10) | 5.3 | 1.8 |
| CN (Ref. 9) | 6.1 | 2.0 |
| RS (Ref. 12) | 6.57 | - |
| BR (Ref. 17) | 6.1 | - |
| BMSV (Ref. 15) | 5.6 | 2.2 |
| VDS (Ref. 13) | 5.9 | 2.2 |
| Si | | |
| Experiment | | |
| FLMSR (Ref. 20) | 22.15±0.2 | 35.0±1.0 |
| HMM (Ref. 21) | 22.9 | 33.0 |
| VL (Ref. 22) | 23 | 30-35 |
| Theory | | |
| BRAC (Ref. 10) | 20.4 | 34 |
| CN (Ref. 9) | 21 | 34 |
| BR (Ref. 17) | 21.7 | - |
| BMSV (Ref. 15) | 21.8 | 29 |
| VDS (Ref. 13) | 22 | 32 |

tend to have larger effects on $\varepsilon_c$ than differences between the basic theoretical approaches. The values of $r_s$ for Ge and Si are ≅0.6 and 0.95 respectively, which means that they are effectively even more dense ("metallic") than are simple metals.

An understanding of the importance of band structure degeneracy and anisotropy in giving a stable EHL state bound with respect to the exciton at T=0 has emerged from these studies. Their effects occur primarily through the kinetic energy, eq. (7). Increasing degeneracy dramatically lowers the cost in kinetic energy per pair in the EHL state, and increasing anisotropy has a weaker effect in decreasing it. Changes in degeneracy and anisotropy, on the other hand, have a relatively small effect on $E_x$. The effect of band degeneracy on EHL properties can be seen dramatically in going from Ge and Si to Ge and Si under large uniaxial strain. For Ge, for example, the ground state energy $\varepsilon_o$ decreases from 5.3 meV to 3.0 meV and $\rho_o$ decreases from $2.38\times10^{17}cm^{-3}$ to $1.1\times10^{16}cm^{-3}$.[9,10,13,15] Even more dramatically, EHL in a model band structure with a single parabolic band of equal mass for each electron and hole is believed to be unbound with respect to the exciton.[1]

It has recently come to be appreciated that the effects of band structure on the combined exchange and correlation energy $\varepsilon^{xc}(\rho)$ for EHL is remarkably very weak.[23,24,25] Band degeneracy and anisotropy have significant effects on the exchange and correlation energies separately, but they nearly cancel in the sum $\varepsilon^{xc}(\rho)$. This behavior has been established largely on the basis of numerical calculations for differing band structures. Physically it is believed to result from the fact that in EHL the major contribution to the combined $\varepsilon^{xc}(\rho)$ arises from virtual plasmons (for which $E_{Fe}$, $E_{Fh}$ << $\hbar\omega_p$, the plasmon energy). The plasmon energy scales with the reduced optical mass $\mu_o$ in the same way as does the atomic unit of energy $2R_x$. Thus when expressed in appropriate atomic units $\varepsilon^{xc}(\rho)$ is approximately independent of system. This universal character of $\varepsilon^{xc}(\rho)$ has not been established rigorously, and it is expected to break down for densities well outside of the range of EHL densities. Nevertheless this approximate universality of $\varepsilon^{xc}(\rho)$ can be useful in considering the physics of different systems without the necessity of doing extensive numerical calculations.

Electron-hole liquid condensation has been observed in several indirect gap compound semiconductors including GaP, AℓAs, and SiC, and the possibility of its occurrence has been discussed for a wide variety of compound systems including those with direct gaps. The electron (hole)-LO phonon coupling is significant in such systems, and its effects on the ground state energetics have been the subject of several theoretical studies.[26,27,28] These studies treat the coupling between carriers and LO phonons in the Frölich form. Physically the leading effects of this coupling are found to result in the replacing of the free carrier masses with the polaron masses and the high frequency dielectric constant $\kappa_\infty$ with the appropriate low frequency value $\kappa_o$. This is sometimes called the "$\kappa_o$ approximation," and it is strictly valid in the regime where $E_{Fi}, \hbar\omega_p \ll \hbar\omega_\ell$ where $\omega_p$ is the plasma frequency and $\omega_\ell$ the LO phonon frequency. Keldysh and Silin[26] considered the exchange energy of EHL in the high density and low density limits ($E_{Fi} \gg \hbar\omega_\ell$ and $E_{Fi} \ll \hbar\omega_\ell$, i=e,h), and they find that the next terms beyond the $\kappa_0$ approximation decrease further the ground state energy. On this basis they argued that LO phonon coupling in compound systems increases the stability of the EHL phase with respect to the exciton due to an additional phonon mediated coupling between the carriers.

The most detailed calculations to date have been carried out by Beni and Rice[28] who have studied a variety of semiconductors using a treatment like that of the Hubbard approach for the correlation energy. Here a set of graphs which are similar to those in Fig. 6b,c in which the Coulomb interaction is replaced by the sum of a Coulomb interaction plus a Frölich phonon coupling. These authors also find additional binding of the EHL compared to the exciton due to phonon coupling. They point out however that in some cases (particularly where $\omega_p \sim \omega_\ell$) their method breaks down due to the absence of vertex corrections, and the $\kappa_0$ approximation is often better. In fact in many cases the exchange-correlation energy is found to be given quite well (to within ~5%) by the $\kappa_0$ approximation; and in addition when expressed in units of polaron masses and $\kappa_0$, it is given reasonably well by the universal form discussed above.[24] Thus considerable progress has been made in the theoretical understanding of the energetics of EHL in polar systems, but some points remain to be put on a more rigorous basis. A further uncertainty in results for compound systems is that the input effective masses for them are often not yet known in detail experimentally.

In summary, a rather good understanding has been achieved of the ground state energetics of the interacting electron-hole system in semiconductors, particularly in the elemental Ge and Si systems, both in the low density regime corresponding to free excitons and in the high density regime corresponding to electron-hole liquid.

## THERMODYNAMICS, CRITICAL POINT, AND PHASE DIAGRAM

The properties of the system of interacting excitons, electrons, holes and electron-hole liquid at finite temperatures is now considered. In particular, the condensation of carriers into EHL and the coexistence of these phases is discussed. Once again the carriers in the semiconductors are treated as electrons and holes of effectively infinite lifetimes in quasiequilibrium with one another.

There is a wealth of experimental data which indicates that the excitonic system should be pictured as undergoing a condensation into electron-hole liquid through a first order-like phase transition:[2] In particular at low densities and low temperatures the system is composed primarily of free excitons. As the density (optical pumping) is increased at a fixed temperature there is a sharp onset of features in the luminescence spectra corresponding

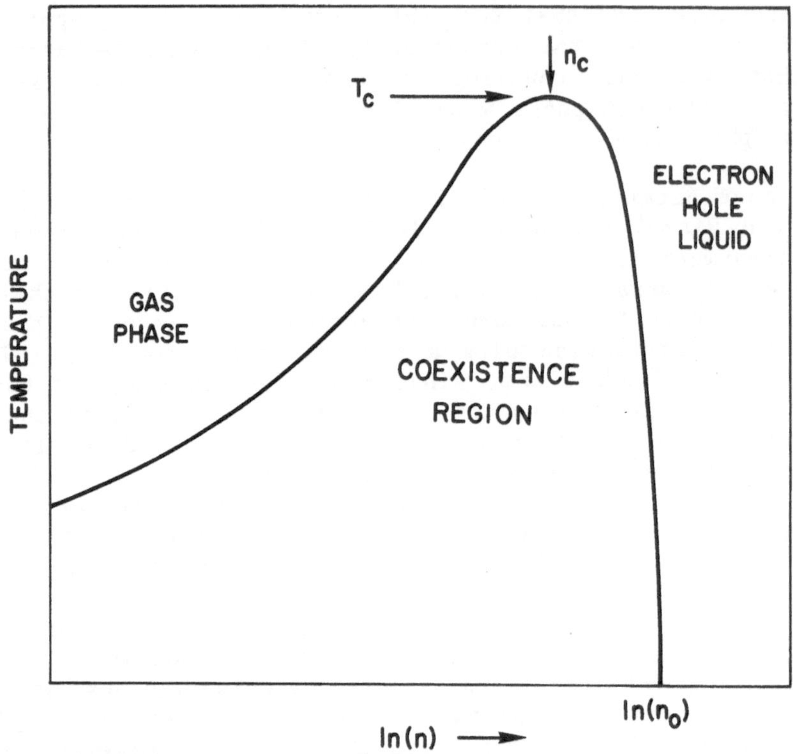


Fig. 7. Schematic representation of a condensation phase diagram for electron-hole liquid in units of the ℓn of the density n=ρ vs temperature. $n_o=\rho_o$ is the ground state density and $T_c$ the critical temperature.

to the EHL. For a wide range of densities above this onset, the gas phase and EHL are seen to coexist in these spectra. For temperatures above a critical temperature $T_c$ such a phase separation does not occur.

The most complete picture of EHL condensation is given by its phase diagram shown schematically in Fig. 7. The phase boundary separates a low density gas phase composed of excitons, biexcitons, electrons, and holes from a uniform high density phase of EHL, and from a region in which these two phases coexist. Phase separation does not occur for $T>T_c$. Experimentally the gas side can be determined from the onset of EHL for increasing density (optical pumping power) at fixed temperature or vice versa, and it has been studied experimentally for densities $\gtrsim 10^{12}cm^{-3}$ in Ge.[2] The liquid side is determined experimentally by fitting the EHL line in luminescence

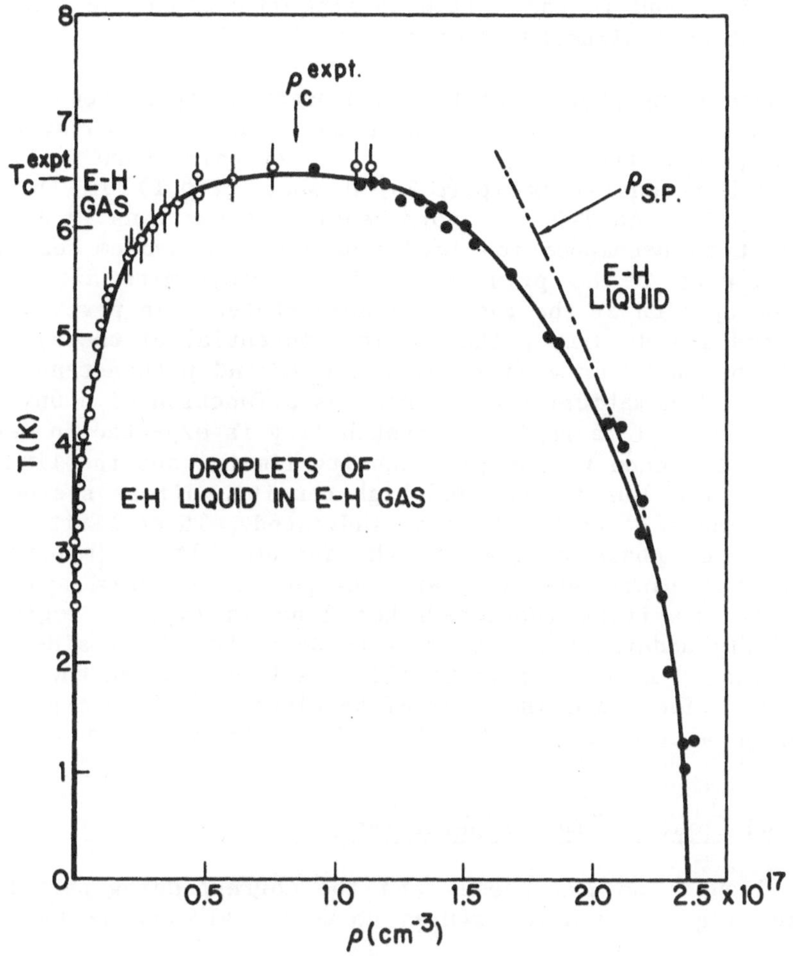


Fig. 8. Comparison of results of the droplet fluctuation approach (solid line) and experimental results for the phase diagram in Ge. Experimental data from Ref. 29, 33.

spectra as described above. Data obtained in this way for EHL condensation in Ge are shown in Fig. 8. In thermal equilibrium at finite temperature the gas phase contains free electrons and holes as well as excitons. For increasing gas density below the EHL condensation point a many-body Mott transition due to carrier screening is also expected. The basic physics of this transition remains a major unsolved problem, and the relevant experimental data remain unclear. Most data to date suggest that this transition does not affect the EHL condensation phase diagram in an important way,[20,29,30,31] and in the following its effects on the condensation will not be considered in detail.

The underlying physics of the condensation can be seen by considering the energy per electron-hole pair in a uniform system as shown in Fig. 1 or the chemical potential per pair shown in Fig. 9. The chemical potential $\mu=\partial(\rho f(\rho,T)/\partial\rho$ where $f(\rho,T)$ is the free energy per pair. At T=0 a uniform density of excitons is unstable with respect to decomposition into a non-uniform system containing EHL. At finite T the principles of thermodynamics dictate the relative occupation of the gas and liquid states. In particular at finite T and low density $\rho$ the chemical potential of the system is given well by that for a free exciton gas, and $\mu$ then tends to $-\infty$ for small $\rho$. Schematically $\mu$ is shown as a function of a uniform $\rho$ in Fig. 9. For $T<T_c$ a region of instability is expected to develop between the low density and high density limits, and the limits of stability of the low density and high density uniform systems are given by a Maxwell construction as indicated. These limits define the coexistence phase diagram of the system. It is important to note that the basic ground state energetics of this system at intermediate densities (dotted-dashed line in Fig. 9) region are not yet fully understood. In order to describe either side of the phase diagram, $\mu$ must be known at all densities between the gas and liquid side. Therefore the lack of knowledge of the energetics in this region ultimately limits the ability to describe the phase diagram.

The Gas and Liquid at Low Temperatures

The liquid and gas phases and the corresponding portions of the phase diagram for temperatures up to $T \lesssim \frac{1}{2}T_c$ can be understood on the basis of simple arguments. For low temperatures the gas in equilibrium with the EHL exerts little pressure on it and therefore does not affect significantly the free energy of the EHL. Then for temperatures low with respect to $E_{Fe}$, $E_{Fh}$ (in Ge $E_{Fh} \cong 45K$, $E_{Fe} \cong 29K$) the free energy per pair can be expanded in powers of $(k_BT/E_{Fi})^2$

$$f(\rho,T) \cong \varepsilon(\rho) - \tfrac{1}{2}\gamma(\rho)(k_BT)^2 \tag{10}$$

where

$$\gamma(\rho) = \hbar^{-2}(\pi/3\rho)(\nu^{2/3}m_{de} + m_{dh}) \tag{11}$$

The leading temperature dependence of the EHL density is given by minimizing $f(\rho,T)$ with respect to n

$$\rho(T) \cong \rho_0 - \tfrac{1}{2}[\gamma'(\rho_0)/\varepsilon''(\rho_0)](k_BT)^2 \tag{12}$$

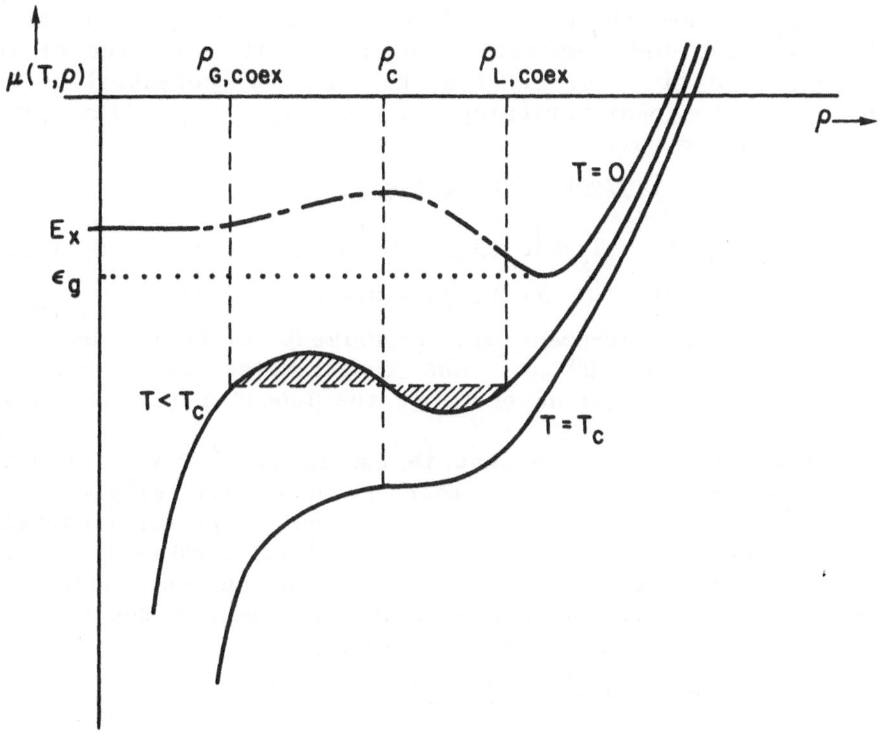


Fig. 9. Sketch of the uniform plasma approach to EHL condensation. The chemical potential μ is shown as a function of uniform density ρ for several temperatures T. The critical point is given by an inflection point in μ vs ρ.

Therefore the temperature dependence of the EHL density at low temperatures is given entirely in terms of the ground state properties of the system. Physically this $T^2$ variation of ρ arises from the single particle thermal excitations over the electron and hole Fermi levels which occur because of the Pauli exclusion principle. This $T^2$ variation has been observed in detail experimentally for $T \lesssim \frac{1}{2}T_c$ [2,31] and it describes the low temperature liquid side of the phase diagram as in Fig. 8.

For sufficiently low temperatures and low densities the gas phase is composed primarily of excitons, and to a good approximation its chemical potential is given by that for noninteracting excitons

$$\mu_x = E_x + k_B T \, \ln[(\rho_x h^3/g_x)(2\pi/M_x k_B T)^{3/2}]$$

where $\rho_x$, $g_x$, $M_x$ are the exciton density, degeneracy, and translational mass. The low temperature portion of the gas side of the phase diagram is given by equating the chemical potential of the EHL to that of the noninteracting excitons, $\mu_x = \mu_{EHL}$. This gives for the coexistence curve

$$\rho_{g,coex} \cong g_x(M_x k_B T/2\pi h^2)^{3/2} \exp(-\phi/k_B T) \tag{13}$$

where $\phi(T) = |\mu_{EHL}(T) - E_x| \cong |\varepsilon_0(\rho_0) - E_x|$ dropping the temperature dependence of $\phi$ at low T. The temperature dependence of $\rho_{g,coex}$ in eq. (13) is in good agreement with experiment for Ge for densities between $10^{12}cm^{-3}$ and $10^{15}cm^{-3}$ and temperatures up to ~4K.[2,32] (Finite lifetime corrections occur at the lowest T (see Ref.43).)

It is clear from Fig. 8 that in the region $T \gtrsim \frac{1}{2}T_c$ the phase diagram deviates strongly from that given by the simple low T descriptions. Physically in this region, the gas pressure modifies the free energy of the electron-hole liquid and conversely for the gas. Thus in order to understand either phase one must understand both phases and their interactions at intermediate densities as indicated by the Maxwell construction in Fig. 9.

Uniform Plasma Model for the Critical Point

A simple method to obtain an estimate of the critical point based on microscopic calculations was first suggested by M. Combescot.[34] This method is attractive because of its calculational simplicity and physical clarity and has been used[35,36] and elaborated[37] by others. It is based on consideration of limits of stability of a uniform electron-hole system as shown in Fig. 9. The critical point in this picture is given by an inflection point in the chemical potential as a function of density for various temperatures. The critical point is found to lie at sufficiently high densities and low temperatures that the system remains approximately degenerate there.

In this approach the free energy per pair in the uniform plasma in the vicinity of the critical point is described by a simple approximate free energy expression. It is given by the form in eq. (10) in which several approximations are made in the temperature dependence of $f(\rho,T)$: (i) For these temperatures only the leading term in $(k_B T/E_{Fi})^2$ is retained. Inclusion of higher order terms has been found to have only a small effect on the resulting critical point.[34,36] (ii) The temperature dependence of the combined exchange-correlation energy is neglected, and (iii) many-body effects in the masses appearing in $\gamma(\rho)$ are neglected. These latter two effects have been shown to be small by direct calculation[38] and from experimental data.[2] Physically the reason is that $E_{Fi} \ll \hbar\omega_p$, and thus the exchange-correlation lowers the bands rigidly with little distortion. Then the leading temperature dependence due to single particle excitations at the Fermi energies is relatively insensitive to this band lowering. The chemical potential is formed using this $f(\rho,T)$ in eq. (10), and the critical point is obtained from the inflection point

$$\partial\mu/\partial\rho|_c = \partial^2\mu/\partial\rho^2|_c = 0 \tag{14}$$

The most extensive studies of the critical point using this method were made by Reinecke et al.[37] Their results for Ge and Si and for several strained Si systems are shown in Table III along with the most recent experimental values. The detailed correlation energies of Vashishta et al.[13] were used.

Table III. Critical points from the uniform plasma approach to EHL condensation and from experiment.

| THEORY+ | Ge(4;2) | Si(6;2) | Si(6;1) | Si(4;1) | Si(2;1) |
|---|---|---|---|---|---|
| $T_c$(K) | 7.91 | 29.0 | 21.4 | 20.8 | 17.4 |
| $\rho_c$($10^{17}cm^{-3}$) | 0.44 | 4.1 | 1.3 | 1.1 | 0.75 |
| **EXPERIMENT** | **Ge(4;2)≠** | **Si(6;2)*** | **Si(6;1)*** | **Si(4;1)*** | **Si(2;1)*** |
| $T_c$(K) | 6.7±0.2 | 23.0±1.0 | 16.9±0.5 | 16.4±0.5 | 14.0±0.5 |
| $\rho_c$($10^{17}cm^{-3}$) | 0.6±.1 | 12 ± 2 | 3.6±0.8 | 2.9±0.6 | 1.8±0.3 |

+ Ref. 37
≠ Ref. 29, 33
* Ref. 20

The approximations used in $f(\rho,T)$ can be examined a posteriori using the results of this model. For Ge the value of $r_s$ at the critical point is $r_s \cong 1.0$ and for Si $r_s \cong 1.8$. Thus the system remains dense there, and a plasma description is appropriate. At the critical point the values of $(k_BT/E_{Fi})$ for electrons and holes for Ge are ≅ 0.8 and 0.5 respectively, and for Si they are ≅ 1.3 and ≅ 1.9 respectively. Therefore the assumption of degeneracy is only crudely valid. Explicit inclusion of higher order terms in $(k_BT/E_{Fi})$ have been found to have only a small effect,[34,36] and it is believed that the low temperature expansion is useful in an asymptotic sense.

These recent studies (Table III) have been used to make several points which tend to clarify this model approach for the critical point: (1) The theoretical results for $T_c$ are in reasonably good agreement with but consistently are somewhat higher than the experimental values. The latter feature can be understood by noting that this approach is based on a model free energy for a uniform electron-hole system which neglects statistical fluctuations that become important near the critical point. Therefore, in principle, this mean field treatment of the statistical mechanics gives an upper bound for the critical temperature. (2) Early work with this model suggested that it gave unreliable values of $T_c$ because of uncertainties in the correlation energy the third derivative of which is required to locate the critical point.[1,35,36] Reinecke et al.[37] have shown that previous results were in fact consistent with one another and have shown that $T_c$ is not unduly sensitive to uncertainties in the correlation energy. (3) Consider the values of critical densities $\rho_c$ obtained from this method. First, they are smaller than the experimental values; this results from the overestimate of $T_c$ in the determination of the critical point in eq. (14). Secondly, the chemical potential $\mu$ is a very flat function near the inflection point, and therefore $\rho_c$ is sensitive to minor uncertainties in $f(\rho,T)$. As a result the values of $\rho_c$ are less reliable than are those for $T_c$.

In summary, the uniform plasma model gives reliable estimates of the upper bounds for $T_c$ and represents well the variations of the critical point between systems.

## Droplet Fluctuation Approach to EHL Condensation

A simple, physically clear approach to describe the EHL phase diagram was introduced by Reinecke and Ying.[39,40] In it the entire phase diagram is expressed in terms of properties of the dense, degenerate EHL system at low temperature which can be calculated microscopically. This approach based on droplet-like fluctuations of EHL is an extension of a method used earlier to treat classical gases[41] and used more recently to discuss aspects of critical phenomena.[42]

The gas near the condensation point is pictured as composed of a distribution of noninteracting fluctuations of size $\ell$ at constant chemical potential $\mu$ and temperature T

$$\rho_g = \sum_{\ell} \ell \exp[-\Omega(\ell)/k_B T] \tag{15}$$

where $\Omega(\ell) = F(\ell) - \mu\ell$ for the droplet-like fluctuations. The exponential factors give the thermodynamic probability of the fluctuations. The free energy of the fluctuation is written approximately as

$$F(\ell) \cong F_B \ell + \sigma(T)\, a\ell^{2/3} + k_B T\tau\, \ell n(\ell) + \dots \tag{16}$$

where $F_B$ is the bulk free energy, $\sigma\ell^{2/3}$ the surface area, and the term in $\ell n(\ell)$ is a higher order entropy term.

For $F_B < \mu$ the probability of fluctuations of very large size diverges, and for $F_B > \mu$ it converges. Thus $\mu = F_B(T)$ is taken to indicate the onset of a stable liquid phase and the condensation point in this approach. In addition, as T approaches $T_c$ the surface tension $\sigma(T)$ decreases, and fluctuations increase. Finally at a temperature such that $\sigma(T)$ vanishes, stable droplet formation is no longer possible, and

$$\sigma(T = T_c) = 0 \tag{17}$$

determines $T_c$.

The liquid near the critical point is pictured symmetrically as a dense EHL containing bubble-like fluctuations of the gas phase. From Eqs. (15) and (16), the gas side of the coexistence curve is given by

$$\rho_{g,coex} = q_o \sum_{\ell} \ell^{1-\tau} \exp[-\sigma(T)a\ell^{2/3}/k_BT] \tag{18a}$$

and the liquid side by

$$\rho_{L,coex} = \rho_o(T) - q_o \sum_{\ell} \ell^{1-\tau} \exp[-\sigma(T)a\ell^{2/3}/k_BT] \tag{18b}$$

Here $\rho_o(T)$ gives the temperature variation of the liquid density due to single particle quantum excitations across the electron and hole Fermi surfaces. The critical density is given by the simple expression

$$\rho_c = \tfrac{1}{2}\rho_o(T = T_c) \tag{19}$$

The complete phase diagram given by eqs. (18a,b) is expressed in terms of only two quantities,[43] the temperature dependent surface tension $\sigma(T)$ and the low temperature density $\rho_o(T)$ which are properties of the dense low temperature EHL. This approach has the merit that in it model averages are made over configurations consisting of dense EHL and empty background, and therefore the free energy as a function of uniform density in the intermediate density regime is not required. In addition, critical fluctuations are included here in a simple, intuitively clear way. In practice the properties of the EHL appearing in these configurations are evaluated as if the system were in equilibrium, which is a good approximation for the large long-lived fluctuations occurring as T approaches $T_c$.

For temperatures up to $T_c$, $\rho_o(T)$ and $\sigma(T)$ can be expanded to lowest order in $k_BT/E_{Fi}$ as before giving

$$\rho_0(T) \cong \rho_0(1-\delta_\rho T^2) \qquad (20a)$$

$$\sigma(T) \cong \sigma_0(1-\delta_\sigma T^2) \qquad (20b)$$

Then the phase diagram is given in terms of only the ground state density $\rho_0$ and $\delta_\rho$ and the surface energy $\sigma_0$ and $T_c = \delta_\sigma^{-\frac{1}{2}}$.

This approach can be used in two rather different ways. First, experimental data for the phase diagram can be fitted to extract values of these parameters, and secondly these parameters can be calculated microscopically in order to give theoretical phase diagrams. In Fig. 8 the experimental data for the phase diagram in Ge have been fitted using $\rho_0$ and $\delta_\rho$ from independent low T measurements and $T_c = 6.7K$ is taken from these measurements. In this way an experimental estimate of the surface energy $\sigma_o \sim 1x10^{-4}$ erg/cm$^2$ is obtained.[39,40,44] These data are seen to be fit well by the droplet fluctuation approach, which gives confidence in this approach as a method to describe the phase diagram.

It should be noted that for temperatures away from the immediate vicinity of $T_c$ the sums in eqs. (18a,b) are dominated by droplets of relatively small size. In Ge for $T \lesssim 6K$, $\ell \gtrsim 20$ contribute $\lesssim 15\%$ of the sums. Thus, except very near $T_c$, that separation of the free energy into a bulk and a surface term is somewhat arbitrary, and the interpretation of $\sigma_0$ as the surface energy appropriate for a large droplet should be treated with some caution. The same is true with respect to microscopic results for $\sigma_\rho$ calculated for large drops which are used in forming theoretical phase diagrams.

In order to obtain theoretical phase diagrams including their critical points, microscopic calculations of the temperature dependent surface tension are required. The method used[39,40,45,46,47] is based on a generalization of the Hohenberg-Kohn approach for the inhomogeneous electron gas. In the present case the droplet free energy is expressed as a functional of the electron and hole densities $\rho_e(\underset{\sim}{r})$ and $\rho_h(\underset{\sim}{r})$ which vary through the droplet surface region. This gives

$$\begin{aligned} F[\rho_e,\rho_h] &= \tfrac{1}{2}\iint[\rho_h(\underset{\sim}{r})-\rho_e(\underset{\sim}{r})][\rho_h(\underset{\sim}{r}')-\rho_e(\underset{\sim}{r}')]/|\underset{\sim}{r}-\underset{\sim}{r}'|d^3r d^3r' \\ &+ \int g(\rho_e,\rho_h)d^3r + \int[g_e(\underset{\sim}{r})|\underset{\sim}{\nabla}\rho_e|^2+g_h(\underset{\sim}{r})|\underset{\sim}{\nabla}\rho_h|^2 \\ &+ g_{e,h}(\underset{\sim}{r})\underset{\sim}{\nabla}\rho_e\cdot\underset{\sim}{\nabla}\rho_h]d^3r + \ldots \end{aligned} \qquad (21)$$

where the free energy function has been expanded in gradients of the densities for the slowly varying densities in the surface region. For the bulk free energy density

$$g(\rho_e(\underset{\sim}{r}),\rho_h(\underset{\sim}{r}))=(3/10)(3\pi^2)^{2/3}\mu_o\{\rho_e^{5/2}m_{de}^{-1}[1-(5\pi^2/12)(k_BT/E_{Fe})^2]$$

$$+\ \dots\ +\rho_h^{5/2}m_{dh}^{-1}[1-(5\pi^2/12)(k_BT/E_{Fh})^2\ +\dots]\}+\varepsilon^{xc}(\rho_e,\rho_h) \quad (22)$$

where $\rho_i = \rho_i(\underset{\sim}{r})$. Here the leading order temperature dependence is retained in only the kinetic energy density, the exchange part of the exchange-correlation energy density $\varepsilon^{xc}$ is calculated exactly as in eq. (9), and the correlation energy is given by a Wigner-like fit to the numerical results of Vashishta et al.[13] Here

$$E_{Fi}(\underset{\sim}{r}) = [3\pi^2\rho_i(\underset{\sim}{r})]^{2/3}(\mu_o/2m_{di}) \ ,i = e,\ h \quad (23)$$

The leading contributions to the coefficients of the gradient terms are obtained from the kinetic energy functional. This is expected to be a good approximation for the high effective densities $r_s \lesssim 1.5$ in EHL. These coefficients are calculated by comparing the response of the system to several weak perturbations of long wavelength calculated in the free energy-density formulation with those obtained from the R.P.A. linear response. For low temperatures

$$g_i = g_i(0)[1 + (\frac{\pi^2}{3})(k_BT/E_{Fi})^2 + \dots] \qquad ,i = e,h \quad (24)$$

where the expressions for $g_i(0)$ proportional to $\rho_i^{-1}$. The coefficients are found to be functions of the details of the band structure including the ellipsoidal shape of the conduction bands and the coupling between the light and heavy hole bands.[46]

In performing the calculations the equilibrium carrier density $\rho_o(T)$ in eq. (18b) is obtained by minimizing the bulk free energy density in eq. (22) with respect to density. The surface tension is obtained here using a variational approach. The carrier densities are parameterized by

$$\rho_i(x) = \begin{cases} \tfrac{1}{2}\,\rho_o(T)\,\exp(-\beta_i x) \ , & x>0 \\ \tfrac{1}{2}\,\rho_o(T) - \tfrac{1}{2}\,\rho_o(T)\,\exp(\beta_i x)\ , & x<0 \end{cases} \quad (25)$$

where x is perpendicular to the surface located at x=0. The surface tension is obtained by minimizing

$$\sigma(\beta_i,T) = A^{-1}(F[\rho_e,\rho_h] - F[\rho_o,\rho_o]) \quad (26)$$

with respect to $\beta_i$, i=e,h.

The results of these calculations for the surface energy and for $T_c$ obtained from $\sigma(T)=\sigma_o[1-(T/T_c)^2]$ and for the corresponding critical densities are shown in Table IV along with the most recent experimental results obtained from EHL phase diagrams in several systems. The agreement between theory and experiment for the values of $T_c$ is seen to be remarkably good. The results for $\rho_c$ and $\sigma_o$ are also consistent with experiment (for them there are greater experimental uncertainties). The theoretical phase diagrams calculated for these systems are essentially indistinguishable from those obtained by fitting the data directly. This good agreement between theory and experiment for the values of $T_c$ and for the complete phase diagrams provides the most convincing justification for usefulness of the droplet fluctuation approach for describing

Table IV. Parameters of the phase diagram from droplet fluctuation approach to EHL condensation and from experiment.

| THEORY$^+$ | Ge(4;1) | Si(6;2) | Si(6;1) | Si(4;1) | Si(2;1) |
|---|---|---|---|---|---|
| $T_c$(K) | 6.73 | 23.3 | 16.5 | 15.8 | 14.1 |
| $\rho_c$($10^{17}$cm$^{-3}$) | 0.066 | 0.95 | 0.21 | 0.20 | 0.14 |
| $\sigma_o$ ($10^{-4}$erg/cm$^2$) | 1.84 | 35.0 | 7.8 | 6.4 | 3.7 |
| **EXPERIMENT** | **Ge(4;1)$^{\neq}$** | **Si(6;2)$^*$** | **Si(6;1)$^*$** | **Si(4;1)$^*$** | **Si(2;1)$^*$** |
| $T_c$(K) | 6.7±0.2 | 23.0±2.0 | 16.9±0.5 | 16.4±0.5 | 14.0±0.5 |
| $\rho_c$($10^{17}$cm$^{-3}$) | .06±.01 | 1.2±0.2 | .36±.08 | .29±.06 | .18±.03 |
| $\sigma_o$($10^{-4}$erg/cm$^2$) | ~1.0 | 28 ± 5 | 10 ± 3 | 8 ± 3 | 4 ± 2 |

\+ Ref. 40
≠ Ref. 29, 33
* Ref. 20

EHL condensation. By comparing these results with those from the uniform plasma approach in Table III the importance of including statistical fluctuations in calculating values of $T_c$ can be seen.

Brief remarks on several details of these calculations are now made. In order to treat the low density tail of the density profiles at finite T an interpolation between eqs. (22) and (24) and their corresponding classical limits were made.[40] Variations in the treatment of this tail region have only a small effect on $T_c$. The differences in $\beta_i$ for electrons and holes have only a small effect on the magnitude of $\sigma_o$ but give rise to a dipole layer and a charge on the droplet which are discussed elsewhere.[46,47] The leading gradient contribution to the kinetic energy function has been shown to be a good approximation to its exact treatment (at

T=0) by direct calculation.[12] Inclusion of the leading contributions from the exchange and correlation in the gradient coefficients remains somewhat controversial.[48] Doing so in these calculations gives somewhat larger values of $\sigma_o$.[40,49] More detailed experimental values of $\sigma_o$ than those in Table IV are available from recent nucleation experiments.[50] Estimates of the effects of the uncertainties in the calculation of σ(T) gives an estimated uncertainty in the values of $T_c \lesssim 8\%$.[40]

In summary, the droplet fluctuation approach has proven to be very useful in treating the entire phase diagram for EHL condensation. It reduces the description of the phase diagram to properties of the dense EHL. The results of microscopic calculations based on EHL energetics are in excellent agreement with experiment, especially for $T_c$.

## SCALING RELATIONS

The systematic variations of the ground state and critical properties of EHL with varying band structure and in particular the possibility of the existence of some simple relations between the ground state and critical properties has been of interest since the early studies of EHL.[2] This interest derives in part from the desire to understand better the physics of this unique system by means of varying is basic energetics (with varying band structure), and in part because as EHL came to be observed and sought in an increasingly wide variety of systems, such relationships would be a valuable guide to experiment.

Several simple relationships between the ground state and critical properties were proposed[2,52] and used in analyzing experimental results.[2,33] Based on the Principle of Corresponding States for classical gas condensation it was argued that all EHL condensation phase diagrams when densities and temperatures are scaled with the critical parameters $\rho_c$ and $T_c$ should have the same shape.[2,33] From this universal shape the simple relation

$$\phi/k_B T \cong \text{const} \tag{27}$$

was obtained. Two additional relations were proposed based on early experimental results[2,52]

$$\rho_c/\rho_o \cong \text{const} \tag{28}$$

$$T_c/\rho_o^{1/2} \cong \text{const} \tag{29}$$

A relation between $T_c$ and $\rho_o$ would be particularly useful to the experimentalist in that $\rho_o$ is one of the easiest parameters to measure, but the critical point is among the most difficult.

Recently however general arguments have been advanced[53] that these relations do not have a foundation in the physics of EHL. The Principle of Corresponding States is appropriate for gases which obey classical statistics and which have simple pairwise interactions of the form $v(r) = \alpha\Psi(\beta r)$ where $\alpha$ and $\beta$ are scales of the strength and length of the interaction which vary between systems, and $\psi$ is a general function for all systems.[6] This description is not expected to apply to EHL. It was pointed out[53] that in general there were two kinds of shortcomings in establishing the existence and forms of scaling relations between the ground state properties and the critical parameters: (1) Reliable, accurate results for the ground states and critical points of systems with widely vaying band structure had not been determined. In particular the experimental results were not sufficiently accurate or extensive to establish particular forms such as those in eqs. (28) and (29) for such relations. (2) The underlying physics of such relations was not known.

Theoretical Studies

Reinecke and Ying[53] approached this issue along two lines. First by making detailed calculations of the ground state and critical properties for a set of model systems with widely varying band structures and second by seeking to trace the origins of the relations to the basic energetics of EHL.

The phase diagrams for EHL condensation were calculated[40,53] for Ge(4;2), Ge(1;2), Ge(1;1), Si(6;2), Si(2;2), and Si(2;1) using the droplet fluctuation approach for the finite temperature region. Results for the Si systems are shown in Fig. 10 where the systematic changes in both the ground state and the critical point with decreasing degeneracy are seen. This set of model systems has widely varying band structures, and the band structure parameters are known well. In order to obtain results which are as nearly independent of modelistic theoretical approach for the critical point as possible, calculations of the critical points were also made using the uniform plasma approach.[37,40] Both approaches represent well the variations of parameters between systems.

The results of these theoretical studies are shown in Table V where the upper entry in each case was obtained using the droplet fluctuation approach, and the lower entry was obtained using the uniform plasma approach. The results from the two approaches show the same trends with band structure. The ratios $\phi/k_BT_c$ and $T_c/\rho_o^{\frac{1}{2}}$ are seen to have systematic variations with changing band structure (principally band degeneracy), and therefore they do not provide good scaling relations for EHL. Three scaling relations were proposed[53]

Table V. Theoretical values of ratios for EHL scaling relations.

| | Ge(4;2) | Ge(1;2) | Ge(1;1) | Si(6;2) | Si(2;2) | Si(2;1) |
|---|---|---|---|---|---|---|
| $\phi/k_B T_c$ | 3.07 (2.61) | 1.04 (0.835) | 1.75 (1.35) | 3.66 (2.96) | 2.67 (1.90) | 1.71 (1.37) |
| $T_c/\rho_0^{1/2}$ | 14.2 (16.7) | 18.3 (22.8) | 27.6 (35.7) | 13.0 (16.1) | 16.0 (22.5) | 21.2 (26.3) |
| $\rho_c/\rho_0$ | .290 (.193) | .284 (.163) | .288 (.222) | .295 (.150) | .321 (.196) | .316 (.172) |
| $\lvert\varepsilon_0\rvert/k_B T_c$ | 10.3 (8.76) | 10.9 (8.75) | 12.3 (9.52) | 11.0 (8.88) | 11.8 (8.40) | 12.0 (9.68) |
| $(\kappa T_c/\rho_0^{1/4})(\kappa/\mu_0)^{1/4}$ | 2.03 (2.38) | 1.97 (2.44) | 1.87 (2.41) | 1.95 (2.42) | 1.94 (2.71) | 1.94 (2.40) |
| $\sigma_0/\rho_0^{2/3} k_B T_c$ | .529 | .387 | .337 | .449 | .362 | .306 |

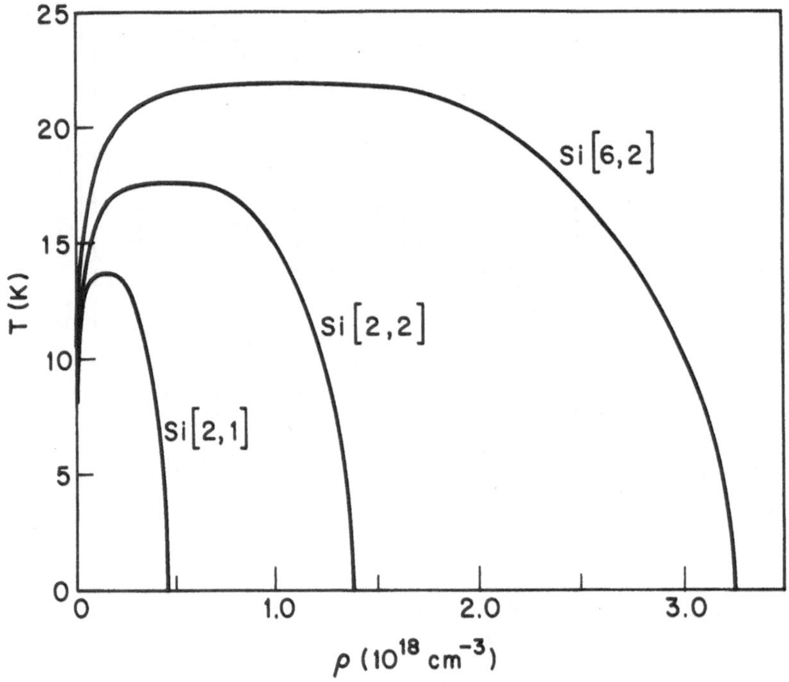


Fig. 10. Condensation phase diagrams calculated from the droplet fluctuation approach for unstrained Si (Si(6,2)) and for Si with intermediate (Si(2;2)) and large Si(2;1) strain along the (100) direction.

$$\rho_c/\rho_o \cong \text{const} \tag{30a}$$

$$|\varepsilon_o|/k_BT_c \cong \text{const} \tag{30b}$$

$$(\kappa T_c/\rho_o^{\frac{1}{4}})(\kappa/\mu_o)^{\frac{1}{4}} \cong \text{const} \tag{30c}$$

These ratios are seen to be approximately constant for these systems, and therefore they form useful scaling relations.

Finally the dimensionless ratio $\sigma_o/\rho_o^{2/3}\, k_BT_c$ is listed in Table V. From the droplet fluctuation approach in eqs. (18a,b) this ratio is seen to provide a simple measure of the shape of the phase diagram scaled to $\rho_c, T_c$ in the region $T \gtrsim \frac{1}{2}T_c$ with larger values corresponding to a flatter top for the diagram. The systematic variation of this parameter seen with band structure indicates that the shapes of EHL phase diagrams are not universal.

In order to trace the existence of these scaling relations to the physics of EHL and to understand the reasons for their forms, a simple calculation can be done.[53] The uniform plasma approach is followed. The free energy per pair is written

$$f(\rho,T) \cong A\rho^{2/3} + \varepsilon^{xc}(\rho) - D\rho^{-2/3}\tau^2 \tag{31}$$

where $A = (3/10)(3\pi^2)^{2/3}\ \mu_0(\nu^{2/3}m_{de}^{-1} + m_{dh}^{-1})(2R_x)$, $D=\frac{1}{2}(\pi/3)^{2/3}\mu_0^{-1} \times(\nu^{2/3}m_{de}+m_{dh})(2R_x)$ and $\tau=k_BT/2R_x$. In order to obtain analytical results, an analytical form for $\varepsilon^{xc}(\rho)$ is required. By direct comparison with detailed numerical results for $\varepsilon^{xc}(\rho)$ at T=0 it is found that the simple universal form

$$\varepsilon^{xc}(\rho) \cong -b\rho^p \tag{32}$$

with $p \cong \frac{1}{4}$ and $b \cong 3.5(2R_x)$ fits the numerical results and their first three derivatives very well.[53] The ground state is obtained from the minimum of $f(\rho,0)$, and the critical point is obtained from eq. (14). The scaling ratios are found to be

$$\rho_c/\rho_0 = f_1(p) \tag{33a}$$

$$|\varepsilon_0|/k_BT_c = f_2(p)[AD/(2R_x)^2]^{\frac{1}{2}} \tag{33b}$$

$$(\kappa T_c/\rho_0{}^p)(\kappa/\mu_0)^{1-3p} = bf_3(p)[AD/(2R_x)^2]^{-\frac{1}{2}}\mathrm{Kcm}^{3/4} \tag{33c}$$

where $f_i(p)$ are smooth functions.

Therefore the scaling ratios are approximate constants if two conditions are met: (1) The exchange-correlation energy is given to a good approximation by the simple form eq. (32) with $p=\frac{1}{4}$. This has been shown explicitly to be the case. (2) The factor $[AD/(2R_x)^2]^{\frac{1}{2}}$ is approximately independent of systems. $[AD/(2R_x)^2]^{\frac{1}{2}} \propto (m_{de}m_{dh})^{\frac{1}{2}}/(m_{de}+m_{dh})$, and it depends only weakly on the ratio $m_{de}/m_{dh}$. This factor varies only weakly for all common semiconductor band structures; for example for the systems in Table V, it varies by ≦5%.

Consider the forms of the scaling ratios. In eq. (30b) the physically appropriate energy scale with which to measure $T_c$ is the EHL ground state energy $\varepsilon_0$ rather than $\phi$ because the condensation at the critical point occurs from a dense plasma rather than from excitons. In the important relation between $T_c$ and $\rho_0$ in eq. (30c), the power of $\rho_0$ entering is $\rho_0^{\frac{1}{4}}$ because of the density dependence of $\varepsilon^{xc}(\rho)$ which is appropriate to the dense plasma near $T_c$. The factors of $\kappa$ and $\mu_0$ in this relation arise from scaling the atomic units of a universal $\varepsilon^{xc}(\rho)$ between systems.

Recent calculations of $\varepsilon^{xc}(\rho)$ in a number of polar systems can be fitted to a reasonable approximation by the simple universal form in eq. (32) provided that the low frequency dielectric constant and polaron masses are used.[53] This should particularly be the case for those systems for which $E_{Fi}$, $\hbar\omega_p << \hbar\omega_\ell$.[26,28] For at least those systems for which such a universal form obtains, the scaling relations eqs. (33a-c) are expected to hold by the arguments above. Recently calculations similar to the analytical result given above have been made for systems with arbitrary band non-parabolicity.[54] Substantial non-parabolicity occurs in such systems as the valence bands of Si and Ge with small and intermediate strain and in the camel's back-like conduction bands of GaP and AℓAs. It has been shown that for physically realizable degrees of non-parabolicity the scaling ratios also obtain.

It should be noted that these scaling relations in general are only approximately valid (to within $\lesssim 15\%$). Furthermore, if systems are found for which the conditions on band structure and on the exchange-correlation energy do not obtain, then the scaling relations are not expected to hold for them.

Experimental Verification

Recently Forchel et al.[20] have performed the most extensive experimental studies of the systematic variations of the properties of EHL to date. Their results provide the first definitive experimental test of the scaling relations. They studied the set of systems Si, Si(6;1), Si(4;1), and Si(2;1). The latter three systems are obtained by applying large stresses to Si along the <111>, <110>, and <100> directions. These systems show substantial variations of band structure within a single material.

The complete phase diagrams including both the liquid and gas sides were measured based on detailed pits of luminescence spectra for these four systems. Their data are shown in Fig. 11. The ground state properties $\varepsilon_o$ and $\rho_o$ and critical parameters $T_c$ and $\rho_c$ were obtained by fitting these data with the droplet fluctuation approach in which the parameters were variables. As shown in Tables II and IV these experimental results for the ground state properties and critical parameters are in good agreement with the results of microscopic calculations.

The scaling ratios formed from these ground state and critical parameters are listed in Table VI along with the experimental results for Ge. The corresponding theoretical values of the scaling ratios using the droplet fluctuation approach for the critical point are also shown. Once again it is seen that the ratios $\phi/k_BT_c$ and $T_c/\rho_o^{\frac{1}{2}}$ show systematic variations between the systems. These variations are well outside of the experimental uncertainties. On the other hand, the ratios given in eqs. (33a-c) are

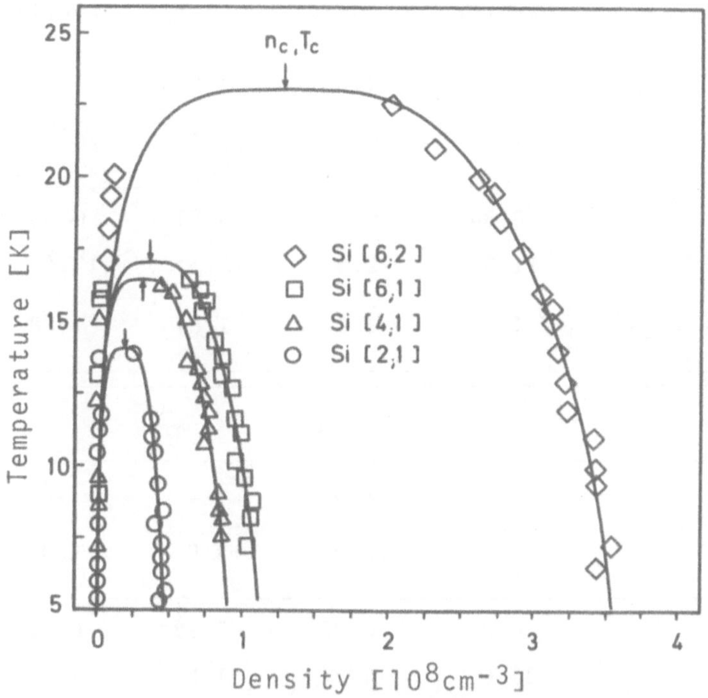


Fig. 11. Experimental results for the condensation phase diagrams in Si and in Si with large uniaxial strain in the (111), (110), and (100) directions. Solid lines are fits using the droplet fluctuation approach. Data are from Ref. 20.

seen to be approximately constant. Therefore the approximate scaling relations proposed and developed theoretically now have also been tested and established experimentally.

Very recently[55] the extent of the experimental verification of these scaling rules has been enlarged in two ways. (i) The range of variations in the band structure (underlying energetics) has been extended. On the basis of measurements for Ge(1;1) and GaP in addition to those the Si systems, the range of EHL densities has been increased to a factor of ~$10^3$ from $2x10^{16}cm^{-3}$ for Ge(1;1) to $1.5x10^{19}cm^{-3}$ for GaP, and the range of $T_c$ to a factor of ~20 from 3.0K for Ge(1;1) to 57K for GaP. (ii) Experimental verification of the scaling relations has been obtained for systems having significantly non-parabolic bands. These results are based on studies of

Table VI. Experimental results for ratios in EHL scaling relations.

| EXPERIMENT | Si(6;2)[+] | Si(6;1)[+] | Si(4;1)[+] | Si(2;1)[+] | Ge(4;2)[≠] |
|---|---|---|---|---|---|
| $\phi/k_BT_c$ | 4.7 | 3.6 | 3.4 | 1.7 | 4.2 |
| $T_c/\rho_o^{1/2}$ | 12.3 | 15.9 | 17.0 | 20.3 | 13.7 |
| $\rho_c/\rho_o$ | 0.34 | 0.32 | 0.31 | 0.35 | 0.31 |
| $\lvert\varepsilon_o\rvert/k_BT_c$ | 11.9 | 12.5 | 12.5 | 12.4 | 10.8 |
| $(\kappa T_c/\rho_o^{1/4})(\kappa/\mu_o)^{1/4}$ | 1.88 | 1.84 | 1.87 | 1.88 | 2.01 |
| THEORY[*] | | | | | |
| $\phi/k_BT_c$ | 4.6 | 2.6 | 2.2 | 1.6 | 7.07 |
| $T_c/\rho_o^{1/2}$ | 12.9 | 16.9 | 18.0 | 21.0 | 14.2 |
| $\rho_c/\rho_o$ | 0.29 | 0.25 | 0.29 | 0.32 | 0.29 |
| $\lvert\varepsilon_o\rvert/k_BT_c$ | 11.0 | 11.7 | 11.6 | 12.1 | 10.3 |
| $(\kappa T_c/\rho_o^{1/4})(\kappa/\mu_o)^{1/4}$ | 1.94 | 1.93 | 1.94 | 1.93 | 2.03 |

+ Ref. 20
≠ Ref. 29, 33
* Ref. 20

the Si systems with varying magnitude of uniaxial strain for which the coupled valence bands are non-parabolic and on studies of GaP for which the camel's back conduction band is non-parabolic.

In summary, simple approximate scaling relations between the ground state properties and the critical point have been obtained from detailed calculations on widely varying systems, their origin has been traced to the basic energetics of EHL, and they have verified in detail experimentally.

## CONCLUDING REMARKS

The ground state energetics of electron-hole liquid in indirect gap semiconductors has been discussed. Approaches for treating its finite temperature thermodynamic effects including the critical point and the phase diagram have been described.[56] Simple scaling relations between the ground state properties and the critical point have been developed.

The emphasis here has been on basic energetic and thermodynamic properties and their relationships. In particular, the thermodynamic properties have been described in terms of the basic energetics of the system. It has been seen that experimental results are in good agreement with the results of microscopic calculations for the ground state properties, for the condensation phase diagram including its critical point, and for simple scaling relations between the ground state and the critical parameters. The agreement between microscopic theoretical results and experiment for these basic properties is one of the most impressive achievements from the study of EHL condensation, and it makes a significant contribution to an improved understanding of interacting electronic systems and of quantum phase transitions.

In the treatment given here several time dependent features of electron-hole liquid condensation have been neglected. The electrons and holes have been taken to have effectively infinite lifetimes, to occupy their lowest conduction and valence bands, and to be in spatial equilibrium. For the present purposes these are good approximations. Nevertheless there are interesting effects in the EHL condensation which arise from these time dependent effects. Phenomenological statistical rate equations have been used to treat nucleation in the presence of finite carrier lifetimes.[51,57] Such treatments describe the existence and the properties of a distribution of finite sized electron hole droplets. They also describe hysteresis in the onset of condensation at low temperatures and the effects of finite lifetimes on the low temperature gas side of the phase diagram. The effects of impurities give rise to bound multiexciton complexes which may be associated with the initial onset of nucleation in some cases. In uniaxially strained Ge and Si the relatively long relaxation time of electrons from the higher bands to the lower band can give rise to EHL condensation in a quasi-equilibrium state in which there are "hot" electrons (in the upper bands) and "cold" electrons (in the lower bands);[58] the case of strained Ge Kirczenow and Singwi[59] have predicted that under certain circumstances a separation into a EHL phase involving "cold" electrons and a phase involving both "hot" and "cold" electrons can occur.

The excitation of semiconductors leading to EHL condensation is most often done using surface photo-excitation. The spatial and temporal evolution of the system to the equilibrium state is in general very complex and is now only beginning to be understood.[60] In highly excited direct gap semiconductors the carrier lifetimes generally are much shorter (~nsec) than those in the indirect gap systems discussed here. The behavior of such highly excited direct gap systems is as yet poorly understood.

The study of the condensation of excitons into electron-hole liquid in semiconductors has been of great interest for the past decade, and a wide variety of phenomena continue to challenge workers. It is hoped that our knowledge of the energetics and thermodynamics of this system will be expanded and deepened and that it will continue to form a foundation for the understanding of these phenomena.

## ACKNOWLEDGEMENTS

The author is grateful for the insights and encouragement particularly of S. C. Ying and also of A. Forchel, W. Schmid and M. H. Pilkuhn during work in this area.

## REFERENCES

The list of references is intended to be helpful to the reader rather than to be complete.

1. For a review of the theoretical literature see T. M. Rice, Solid State Phys. 32, ed. F. Seitz, D. Turnbull, H. Ehrenreich (Academic, New York, 1977), p. 1.
2. For a review of the experimental literature see J. C. Hensel, T. G. Phillips, and G. A. Thomas, Solid State Phys. 32, ed. F. Seitz, D. Turnbull, and H. Ehrenreich (Academic, New York, 1977), p. 88.
3. The possibility of a Bose-Einstein condensation of excitons or of a condensation of molecular excitons in semiconductors has occasionally been raised. To date there is no convincing evidence for them (see Refs. 1,2, R.J. Elliott, this volume.)
4. See for example David Pines, The Many Body Problem (W. A. Benjamin Inc., New York, 1962).
5. Ref. 2, p. 298.
6. D. ter Haar, Elements of Statistical Mechanics (Rinehart and Co., New York, 1954).
7. In addition, in Si the so-called split-off spin-orbit band is nearer the valence band maximum, but it is nonetheless generally neglected in EHL studies.
8. J. C. Hensel and K. Suzuki, Phys. Rev. B 9, 4219 (1974); J. C. Hensel and G. Feher, Phys. Rev. 129, 1041 (1963).
9. M. Combescot and P. Nozières, J. Phys. C 5, 2369 (1972).
10. W. F. Brinkman, T. M. Rice, P. W. Anderson, and S. T. Chiu, Phys. Rev. Lett. 28, 961 (1972); W. F. Brinkman and T. M. Rice, Phys. Rev. B 7, 1508 (1973).
11. J. Hubbard, Proc. Roy. Soc. (London) A243, 336 (1957).

12. J. H. Rose and H. B. Shore, Phys. Rev. B 18, 1884 (1978).
13. P. Vashishta, S. G. Das and K. S. Singwi, Phys. Rev. Lett. 33, 911 (1974).
14. P. Nozières and D. Pines, Phys. Rev. 111, 442 (1958).
15. P. Bhattacharyya, V. Massida, K. S. Singwi and P. Vashishta, Phys. Rev. B 10, 5127 (1974).
16. K. S. Singwi, M. P. Tosi, R. H. Land, and A. Sjölander, Phys. Rev. 176, 589 (1968).
17. G. Beni and T. M. Rice, Phys. Rev. B 18, 768 (1978).
18. G. A. Thomas, T. G. Phillips, T. M. Rice, and J. C. Hensel, Phys. Rev. Lett. 31, 386 (1973).
19. T. K. Lo, Solid State Commun. 15, 1231 (1974).
20. A. Forchel, B. Laurich, G. Moersch, W. Schmid, and T. L. Reinecke, Phys. Rev. Lett. 46, 678 (1981); A. Forchel, B. Laurich, J. Wagner, W. Schmid, and T. L. Reinecke, Phys. Rev. B 25, 2730 (1982).
21. R. B. Hammond, T. C. McGill, and J. W. Mayer, Phys. Rev. B 13, 3566 (1976).
22. M. A. Vouk and E. C. Lightowlers, J. Phys. C 8, 3695 (1975).
23. G. Kirczenow and K. S. Singwi, Phys. Rev. Lett. 41, 326 (1978).
24. T. L. Reinecke and S. C. Ying, Phys. Rev. Lett. 43, 1054 (1979).
25. P. Vashishta, and R. K. Kalia, Phys. Rev. B (to be published).
26. L. V. Keldysh and A. P. Silin, Zh. Eksp. Teor. Fiz. 69, 1053 (1975) [Sov. Phys. JETP 42, 535 (1976)].
27. M. Rösler and R. Zimmerman, Phys. Stat. Solidi b 83, 85 (1977).
28. G. Beni and T. M. Rice, Phys. Rev. B 18, 768 (1978).
29. G. A. Thomas, J. B. Mock, and M. Capizzi, Phys. Rev. B 18, 4250 (1978).
30. W. Miniscalco, C.-C. Huang, and M. B. Salamon, Phys. Rev. Lett. 39, 1356 (1977).
31. J. Shah, M. Combescot and A. H. Dayem, Phys. Rev. Lett. 38, 1497 (1977).
32. J. C. Hensel, T. G. Phillips, and T. M. Rice, Phys. Rev. Lett. 30, 227 (1973).
33. G. A. Thomas, T. M. Rice, and J. C. Hensel, Phys. Rev. Lett. 33, 219 (1974).
34. M. Combescot, Phys. Rev. Lett. 32, 15 (1974).
35. P. Vashishta, S. G. Das, and K. S. Singwi, Phys. Rev. Lett. 33, 911 (1974).
36. T. M. Rice, Proceedings of the Twelfth International Conference on the Physics of Semiconductors, Stuttgart, 1974, ed. M. H. Pilkuhn (B. G. Teubner, Stuttgart, 1974), p. 23.
37. T. L. Reinecke, M. C. Lega, and S. C. Ying, Phys. Rev. B 20, 5404 (1979).
38. T. M. Rice, Nuovo Cimento B 23, 226 (1974).
39. T. L. Reinecke and S. C. Ying, Phys. Rev. Lett. 35, 311 (1975).
40. T. L. Reinecke, M. C. Lega, and S. C. Ying, Phys. Rev. B 20, 1562 (1979).
41. J. Frenkel, Kinetic Theory of Liquids (Oxford University, Oxford, England, 1946), Chap. VII.

42. M. E. Fisher, Physics (N.Y.) 3, 255 (1967).
43. The parameter $q_o=\rho_o/\xi(\tau)$ where $\xi(\ )$ is the Riemann zeta function, and $\tau\cong2.2$ from results on the lattice gas (see Ref. 41). The effects of finite carrier lifetimes give variations from the present results on the low temperature gas side of the phase diagram (see Refs. 40, 51). They occur for $T\lesssim2K$ for Ge and $T\lesssim8K$ for Si, and their effects are not included in the present treatment.
44. T. L. Reinecke, Phys. Rev. B 18, 2947 (1978).
45. T. L. Reinecke and S. C. Ying, Phys. Rev. B 13, 1850 (1976).
46. T. L. Reinecke and S. C. Ying, Solid State Commun. 14, 381 (1974).
47. T. M. Rice, Phys. Rev. B 9, 1540 (1974); L. M. Sander, H. B. Shore, and L. J. Sham, Phys. Rev. Lett. 31, 533 (1973).
48. B. Y. Tong and L. J. Sham, Phys. Rev. 144, I (1966); J. Vannimenus and H. F. Budd, Solid State Commun. 15, 1739 (1974).
49. P. Vashishta, R. K. Kalia, and K. S. Singwi, Solid State Commun. 19, 935 (1976); R. K. Kalia and P. Vashishta, Solid State Commun. 24, 171 (1977); M. Rasolt and D. J. W. Geldart, Phys. Rev. 15, 979 (1977).
50. R. M. Westervelt, Phys. Status Solidi B 74, 727 (1976); 76, 31 (1976); B. Etienne, C. Benoît à la Guillaume, and M. Voos, Phys. Rev. B 14, 712 (1976); J. L. Staehli, Phys. Status Solidi B 75I, 451 (1976).
51. R. N. Silver, Phys. Rev. B 11, 1569 (1975); Phys. Rev. B 12, 5689 (1975).
52. D. Bimberg, M. S. Skolnick, and L. M. Sander, Phys. Rev. B 14, 2231 (1979).
53. T. L. Reinecke and S. C. Ying, Phys. Rev. Lett. 43, 1054 (1979).
54. T. L. Reinecke, to be published.
55. T. L. Reinecke, A. Forchel, H. Schweizer, B. Laurich, R. Hangleiter, J. Wagner, and W. Schmid, Proceedings of the XVI Int'l Conference on the Physics of Semiconductors, Montpellier, 1982 (to be published).
56. Recently Singwi and Tosi (K. S. Singwi and M. P. Tosi, Phys. Rev. B 23, 1640 (1981)) have developed another method by which to estimate the critical point for EHL condensation from its microscopic energetics. They express the temperature dependent surface tension in terms of the bulk compressibility.
57. Ya. E. Pokrovskii, Physics Status Solidi (a) 11, 385 (1972).
58. H.-h. Chou, G. K. Wong and B. J. Feldman, Phys. Rev. Lett. 39, 959 (1977).
59. G. Kirczenow and K. S. Singwi, Phys. Rev. Lett. 41, 326 (1978); Phys. Rev. Lett. 42, 1007 (1979).
60. G. Mahler, G. Maier, A. Forchel, B. Laurich, H. Sanwald, and W. Schmid, Phys. Rev. Lett. 47, 565 (1981).

# SPECIAL TOPICS

# TRANSIENT LUMINESCENCE, TRANSPORT AND PHOTOCONDUCTIVITY IN CHALCOGENIDE GLASSES

K. L. Ngai
Naval Research Laboratory
Washington, D.C. 20375 USA

Abstract

The chalcogenide glasses are perhaps the most thoroughly studied amorphous semiconductor systems. In the past two decades their measured properties have been repeatedly compared with model predictions and served as guides in the search for fundamental concepts and principles necessary for amorphous semiconductors. An example is the synthesis of the Mott-CFO model which is a low carrier density and high carrier mobility picture that involves the concepts of band tail states and mobility edge. On the other hand, Emin proposed that the charged carriers in chalcogenide glasses form small polarons and necessarily implies low carrier mobility because of small band width caused by atomic displacements associated with self-trapping. The small-polaron model can explain, in addition, the Hall effect sign anomaly and the difference in activation between the Peltier heat and conductivity. In spite of the fact that the Mott-CFO model and the small-polaron model are orthogonal to each other, the experimental data prior to 1978 have not been able to discriminate which is the correct model. In this work, transient optical and transport data which have rapidly accumulated since 1979 are considered. These recent data enable us to narrow down the possible states that can exist in chalcogenide glasses. As a result of this analysis, a minimal set of states has been proposed that can explain the totality of optical and transport data. It confirms that transport occurs by small polaron hopping. The proposed minimum set of states is consistent with Anderson's bipolaronic ground state. The transient transport data, transient optical data, the dynamical dielectric relaxation data, and the volume and enthalpy recovery data of chalcogenide glasses are shown to conform to a universal pattern predicted by a recent unified model of relaxation at low frequencies/long times of condensed matter in general.

## 1. Introduction

In the last two decades we have witnessed the emergence and the development of the field of amorphous semiconductors. It has been a multi-national effort with the involvement of many laboratories and research institutions. Naturally, as more and more experimental data become available and phenomena unfamiliar in the context of the physics of the crystalline state are observed, there is a need for a physical model for organization and explanation of the data. To arrive at a model one needs to isolate the physical principles, mechanisms and concepts that are judged to be necessary basic ingredients for the construction of the model. The credibility of the model then depends on the success it has in understanding existing data and consistency of its predictions with data that are taken after the model has been advanced.

The chalcogenide glasses are perhaps the most widely studied amorphous semiconductors. Their transport and optical data are reproducible from laboratory to laboratory and are one but many of the reasons that chalcogenide glasses play a central role in the development of models for amorphous semiconductors. One model commonly referred to as the Mott-CFO model[1] invokes the concept of mobility edges and band tail states. They envisaged conductivity by variable-range hopping in the band tail states at low temperatures and by transport of carriers in extended states excited beyond the mobility edge. The latter implies it is a high mobility picture. An alternative model is due to Emin[2-9] who suggested that in chalcogenide glasses charged carriers form small polarons. A small polaron is an extra electron or a hole severely localized within a potential well that it creates by displacements of atoms that surround it (self-trapping).[10] Predictions of the small polaron model has been worked out extensively by Emin. Not only is it more elegant and simpler in explaining the data that can be explained by the Mott-CFO model, but also it stands out alone in being able to explain the Hall effect sign anomaly[4] which has been observed in various amorphous semiconductors, including amorphous silicon, germanium, arsenic and many chalcogenide glasses. It predicts that the Peltier heat has activation energy $E_\pi$ smaller than that $E_\sigma$, of the conductivity, while the Mott-CFO model would predict $E_\pi \cong E_\sigma$. In chalcogenide glasses, it has been observed that $E_\pi \neq E_\sigma$ in agreement with the small-polaron model. The small polaron model for conduction is certainly consistent with the bipolaronic ground state picture of Anderson.[11] The low-temperature optically induced properties of a defect free system in which both electrons and holes from small polarons are in total agreement with observations including two ESR signals, induced absorption characteristics and luminescence.[12] On the other hand, the Mott-CFO model has to invoke further defects such as the dangling bonds[13] or the valence-alternation-pairs[14] in order to account for the optically induced properties. For example, Mott,

Davies and Street[1] assume localized states in the gap of chalcogenide glasses exist at dangling bonds which can take zero, one, or two electrons (referred to as $D^+$, $D^0$ and $D^-$ into respectively) to be able to account for the photo-induced properties.

In spite of the fact that these two models are diametrically opposite each other in physical contents, until recently there are no critical experiments that can tell us which one is closer to the truth. Earlier, insights into the electronic structure of chalcogenide glasses are gained from CW photoluminescence (PL)[15,16] and photo-induced absorption and ESR studies.[12] Very recently time-resolved optical techniques[17-21] and polarization memory[22,23] have been introduced to the studies of chalcogenide glasses and their crystalline counterparts.[24] These transient optical data together with the transient transport data obtained by Pfister and coworkers[25-28] earlier are considered in this work. It will be shown that these data lead us to favor the small polaron hopping conduction mechanism and to a set of states that include: (1) localized excitons with strong lattice relaxation first suggested by Murayama and coworkers[17,22,23] and which may not be distinguishable from self-trapped excitons suggested by Bösch et al.;[8] (2) separated electron and hole small polarons;[4,9] and (3) states associated with transition metal impurities.[29,30]

We shall demonstrate that the time dependences of the transient transport and transient optical data conform remarkably well to a universal pattern. This is true also for the dynamical (i.e. frequency dependent) dielectric relaxation[31] and for the enthalpy or volume recovery[32] of chalcogenide glasses. Although all these are low frequency or long time relaxation processes yet the nature of one is very different from the other. For example, transient luminescence involves optical recombination relaxation, transient transport involves hole hopping relaxation and dielectric relaxation involves some dipolar unit relaxation, etc. Considering these drastic differences in nature of the relaxation and the primary species that are involved in the relaxation processes, such universality (to be discussed in the following sections) is spectacular. Such universal behavior at low frequencies/long times is not unique to the chalcogenide glasses.[33-42] We shall give other examples to support the empirical fact that condensed matter at low frequencies/long times behave in a surprisingly predictable and universal manner. This global universality is of great interest of course. However, it is outside the scope of this summer school and will not be discussed in detail here. Returning to the chalcogenide glasses, it is reasonable to require any model of electronic structure, in order to be complete, to account for the transient characteristics of the optical and transport data and for their universal pattern. Prior to 1979 this additional requirement is not a pressing issue because at least transient optical data were not yet available. Thus a field like electronic properties of chalcogenide glasses or amorphous semiconductors in general now has the need of understanding low frequency/long time relaxation as well.

## 2. Nature of States in Chalcogenide Glasses

In this section we consider the recent advances made in optical and transport studies of chalcogenide glasses and crystalline counterparts. The purpose is to give, via conclusions drawn from these experimental facts, the entirety of the set of states and the physical mechanism that are necessary and sufficient for what seemingly is a complete understanding of the totality of the experimental data available so far.

Earlier studies of the electronic structure of chalcogenide glasses are by CW photoluminescence (PL) measurements.[15,16,43] These are complemented later on by the optically induced ESR and induced absorption measurements.[12] The CW luminescence spectra have large Stokes shifts and half widths and, as we have discussed in the Introduction, they can be interpreted as well by both the MDS model[1,13] (or equivalently the KAF model)[14] and the small polaron model. New insights came from transient luminescence spectra. The first work of this kind came from Murayama et al.[17] Although it is not genuinely time-resolved spectroscopy, it reports for the first time luminescence with decay time of about 10 nsec or shorter in a-$As_2S_3$ at 4.2K on the high energy side of the CW luminescence observed by Kolomiets et al.[15] The CW luminescence $A_1$ observed by Kolomiets et al. is characterized by a non-exponential decay[17] with the decay times of 10-1000 μsec and has peak position of 1.18 eV. The intensity of $A_2$ is about 5% that of $A_1$. This fast photoluminescence $A_2$ has a PL spectrum which depends on the excitation energy $E_x$. $A_2$ has peak energy increasing from 1.4 eV to 1.7 eV as the excitation energy $E_x$ is increased from 2.2 eV to 2.6 eV, and remaining constant for $E_x > 2.6$ eV. Hence the Stokes shift of $A_2$ is 0.84 eV and is almost independent of $E_x$ for $E_x <$ 2.6 eV. The intensity of $A_2$ is strongest when $E_x$ is at the Urbach tail of the optical absorption and is proportional to the intensity of excitation. Murayama, Suzuki and Ninomiya[22,23] subsequently introduced linear polarized light for excitation and measured the polarization degree of the fast luminescence $A_2$ as a function of $E_x$. They observed the intensity of the emitted light component with polarization parallel to that of the exciting light is stronger than the component with perpendicular polarization. Thus the luminescence remembers the polarization of the exciting light. The memory decreases with $E_x$, and it is extinguished at $5x10^{-6}$ sec in a-$As_2S_3$ indicating that the luminescence with decay time longer than $5x10^{-6}$ sec has no memory for the polarization of the exciting light. Similar to a-$As_2S_3$, a fast luminescence with large Stokes shift is observed in a-$As_2Se_3$, a-$GeS_2$ and a-$GeSe_2$ and all have memory for the polarization of the exciting light. These properties imply that $A_2$ is not due to the recombination of electron-hole pairs in the process of motion of the electrons and holes to defects in the neighborhood of the pairs as suggested by Street,[21] because the luminescence due to the recombination of pairs in motion would not retain any memory for the polarization of the exciting light. The fast luminescence has been interpreted as being due to the recombination of localized excitons[17] with various excitation energies because of internal potential fluctuations due to disorder. Here a localized exciton is defined as an electron-hole pair with the electron and the hole located at the same site, i.e. the electron and hole do not separate after excitation. In order to explain the large Stokes shift, the exciton must have associated with it strong lattice relaxation. In this respect, the localized exciton proposed by Murayama et al. may not be distinguishable from the self-trapped exciton proposed by Bösch, Epworth and Emin.[8] The latter authors envisage those carriers which do not separate after excitation from excitons which ultimately become self-trapped. It is natural to associate the Urbach tail with these localized excitonic states. Abe and Toyozawa[44] have in fact discussed the Urbach tail in amorphous semiconductors to arise from Gaussian distributed site energies for the band tail states. It has also been argued that the center of the localized exciton may not be due to defects such as $D^+$, $D^-$ and $D^0$ of MDS because the luminescence depending on the excitation was obtained up to the absorption coefficient of about $10^4 cm^{-1}$ and radiative quantum efficiency is large in comparison with the supposed density of these defects.[17]

This interpretation of the fast luminescence $A_2$ as localized excitons with strong lattice relaxation or self-trapped is further supported by luminescence data taken in high quality yellow orpiment (Siberia) e.g. crystalline $As_2S_3$. Two different luminescence[24] processes have been observed. The higher energy luminescence $C_2$ with peak energy at 1.53 eV has its photoluminescence excitation spectra peaks at 2.58 eV with half width of 0.2 eV which corresponds closely to the absorption band at 2.5 eV in the optical absorption tail region of c-$As_2S_3$. Therefore it may originate from excitation between the band edge and a shallow impurity band. Exciton with strong lattice relaxation and localized at an impurity is formed. Its recombination gives rise to the high energy luminescence in c-$As_2S_3$. The similarities between $C_2$ and $A_2$ implies that the localized exciton interpretation is probably correct for both.

True time-resolved photoluminescence spectra from a-$As_2S_3$ at 2K with 10 nsec time resolution for delay times up to 200 nsec was first obtained by Bösch and Shah.[18] These authors used 2.54 eV exciting light and saw two luminescence processes. The high energy process has peak position at 1.75 eV and a very short lifetime and disappears very rapidly with increasing temperature. This is the fast luminescence $A_2$ observed by Murayama et al. and shall not be discussed any further in this section. The low energy luminescence peak which we call $A_1$ here appears at an energy of ~1.45 eV on a 10 nsec time scale is also seen by Street.[21] It shifts to considerably lower energy ~1.15 eV with increasing time delay until 500 nsec. Between $10^{-6}$ sec and $10^{-4}$ sec, the peak position of $A_1$ remains virtually stationary. According to Street and Shah[20] there is an additional shift of the luminescence peak to lower energies at times greater than $10^{-4}$ sec. The peak position of $A_1$ at 1.15 eV correspond closely to that of the luminescence peak excited by continuous wave (CW) excitation observed by Kolomiets et al., and can be identified with each other. As we shall see in the following section where luminescence decay data are presented, this identification is even more compelling.

Several suggestions have been made as to the origin of the $A_1$ luminescence. Street[21] suggests that $A_1$ is due to the recombination between a tail state electron and the $D^0$ formed by the capture of a hole after photoexcitation by a dangling bond in the $D^-$ state. This suggestion appears unlikely if we compare $A_1$ with that of the luminescence peak $C_1$ at about 1.2 eV in crystalline $As_2S_3$. As pointed out by Murayama et al.,[24] there is a striking resemblance in shape and position between $A_1$ and $C_1$. This strong resemblance suggests that they must have the same origin and rules out the involvement of dangling bond in both $A_1$ and $C_1$ luminescence. Another model explanation due to Higashi and Kastner[19] (HK) attributes $A_1$ to recombination of $D^+D^-$ pair (also called valence-alternation-pair) of varying separation. Street[21] has commented that this $D^+D^-$ pair model is inconsistent with the time resolved shift of the peak position of $A_1$ due to a large Coulomb term. Higashi and Kastner[45] maintained however that their model can be consistent with the spectral evolution of $A_1$. One should recall that HK was led to the $D^+D^-$ pair model by their total light decay data[19] of luminescence in a-$As_2S_3$. These earlier measurements obtained with $E_x$ = 2.57 eV has shown that the luminescence decays approximately as $t^{-1}$ (more exactly as $t^{-0.89}$, see Fig. 2 of Ref. 19) from $10^{-7}$ to $10^{-3}$ sec and then decays more rapidly with roughly an exponential dependence. Unfortunately this decay has not been confirmed by Murayama and Ninomiya (MN).[46] These latter authors reported that the decay of the intensity of the luminescence in a-$As_2S_3$ with the energies of 1.16 and 1.57 eV, excited by 2.39 eV light and measured from $10^{-8}$ sec to $2x10^{-3}$ sec at 4.2K, deviates greatly from the $t^{-1}$ decay. Instead the data reveals three types of nonexponential luminescence decay with "average" decay times of $2x10^{-8}$, $2x10^{-6}$ and $2x10^{-4}$ sec. Subsequently, HK repeated their total light decay measurements with $E_x$ at lower energies of 2.4 to 1.9 eV. Their results[37] are in essential agreement with that of MN. These data sets will be discussed in detail in the next section. For the present purpose it suffices to say that the approximate $t^{-1}$ decay is observed at high $E_x$ only. Thus the $D^+D^-$ model cannot explain the decay for 1.9 eV < $E_x$ < 2.4 eV.

To explain the origin of $A_1$, we resort to the small polaron model. It has already been proposed by Bösch et al.[8,9] (see also work by A. Sumi as quoted in Ref. 10) that, in chalcogenide glasses, after photoexcitation, the electron and hole which separate rapidly will individually relax to form small polarons. They tend to hop toward their ground state before recombining. Those that radiatively recombine before reaching the ground state will luminesce at a higher energy than those which reach the ground state.[9] Furthermore, since the overlap and the matrix elements for recombination decrease with increasing separation, the centroid of the luminescence spectrum will shift to lower energy with time as the pairs with smaller separation recombine first. Thus the Stokes shift increases in time as more luminescence carriers reach their minimum energy separation. If we assume that most of the separated small polaron (SSP) pairs have reached their ground state for $t > 10^{-6}$ sec., then at times between $10^{-6}$ sec and $10^{-4}$ sec which, as we shall see in the next section, is the effective decay time of $A_1$, the $A_1$ peak position should remain stationary. The time dependence of the spectral shift expected from the SSP pair recombination are in agreement with time resolved data.[20,21] It is obvious that radiation emitted from separated small polaron pair recombinations should have no memory of the polarization of the exciting light. From the data of Murayama et al.[22,23] we know that extinction of polarization memory occurs at times greater than $10^{-5}$ sec. Since $A_1$ has an effective decay time of $2x10^{-4}$ sec (see next section), it should dominate the decay at times $>10^{-5}$ sec. Hence identifying $A_1$ with SSP pair recombinations will lead to consistency with the polarization memory data for $t > 10^{-5}$ sec. Other evidence in support for SSP comes from c-$As_2S_3$ photoluminescence studies by Murayama and Bösch.[24] We have already mentioned that they have found a striking similarity between $A_1$ with peak position at 1.15 eV and an emission band $C_1$ with peak energy at 1.25 eV observed by them in c-$As_2S_3$. The excitation spectra of $C_1$ are found to increase with $E_x$ until beyond 2.8 eV where it levels off and eventually decreases. This behavior leads to the conclusion that $C_1$ originates from band to band transition.[24] Band to band transition naturally will introduce separated electron-hole pairs each of which subsequently forms a SSP pair in c-$As_2S_3$. The resemblance of $A_1$ to $C_1$ suggests that they originate from the same process, and we are thus led to conclude that $A_1$ comes from SSP pair recombination now that $A_1$ has been identified with the CW luminescence observed by Kolomiets et al. In this connection we should recall that Emin[4] has already proposed in 1977 that an optically generated electron-hole pair which thermalizes to form metastable pairs of separated small polarons can explain all the low temperature optically induced properties of chalcogenide glasses.

For example, recombination of the SSP pair give the CW luminescence, the small polaron pair give rise to two ESR signals and absorption bands, etc., as first observed by Bishop, Strom and Taylor.[12]

The effects of doped impurities such as Cu, Tℓ, I, Ag, In, Ga, and K in chalcogenide glasses upon photoluminescence and optically induced paramagnetic states are inconsistent with the MDS or the KAF model.[48,49] Mott suggests that the introduction of impurities will eliminate charged dangling bond centers of one sign ($D^+$ or $D^-$ which are hypothesized to present with comparable concentrations in the undoped glass) or form charged centers which are compensated by D centers of the opposite sign. If large changes in the density and relative fractions of $D^+$ and $D^-$ caused by doping are going to occur and if the $D^+$ and/or the $D^-$ defects are the radiative recombination centers, then the photoluminescence spectrum and efficiency would also be altered dramatically. On the contrary, Bishop et al.[29] have found that photoluminescence efficiency or intensity of optically induced ESR undergo no significant change for a variety of dopants until dopant concentrations exceed ~1 at.%. On the other hand these experimental results are consistent with the SSP model for CW luminescence and induced ESR. This is so because small polaron formation is intrinsic, and occurs in the presence or absence of impurities except perhaps when the impurity concentration becomes too high as to modify the host lattice.

Bishop and Taylor[29] discovered also that iron is the only dopant yet found to cause drastic change in the luminescence in a-$As_2S_3$ and a-$As_2Se_3$ by quenching the photoluminescence efficiency. There is an overall reduction in the CW $A_1$ luminescence peak strength upon doping a "pure" $As_2S_3$ or $As_2Se_3$ glass by 120 ppm of Fe. We shall return to further discuss this effect of Fe on the $A_1$ luminescence in the next section. Even in nominally pure chalcogenides, it has been found invariably in low-temperature ESR spectra the presence of the g=4.2 resonance characteristic of $Fe^{3+}$. This led them to conclude that iron is a persistent, inadvertent impurity in most chalcogenide glasses prepared under a broad range of conditions. This conclusion is consistent with the results of previous studies of optical absorption edge in Fe-doped $As_2S_3$ glass by Tauc et al.[30] who demonstrated that the addition of iron produces optical absorption tails extending from the Urbach edge toward lower energies. All these results point to the importance of the roles played by states associated with transition metal impurities such as iron in chalcogenide glasses. We shall see in the next section that Murayama and Ninomiya[46] have actually found a new luminescence, $A_3$, with decay time of $\sim 2\times10^{-6}$ sec which has polarization memory and is due to recombination of electron-hole pairs localized at transition metal impurities such as $Fe^{2+}$.

Briefly summarized, the discussions in this section of recent optical data in chalcogenide glasses enable us to ascertain the existence of a minimal set of states that seem at this time to be necessary and sufficient for a complete understanding of the totality of experimental data accumulated to this date. The minimal set includes: (1) localized excitons with strong lattice relaxation or self-trapped, (2) separated small polarons and (3) states associated with transition metal impurities which are inadvertently present.

## 3. Transient Photoluminescence Transport and Photoconductivity

In the previous section we have narrowed down the possible set of states that we need to consider in chalcogenide glasses. The present section is devoted to the subject of time-dependence of optical and transport properties. The times we are interested in are typically longer and much longer than a nanosecond. Hence we are in the relaxation regime for description of the phenomena. As we shall see not only these transient data give valuable insight into the electronic structure of chalcogenide glasses, but also, when viewed as relaxation phenomena of their own, the fact that they conform to a universal pattern is surprising as well as interesting.

### Luminescence Decay

We start with luminescence decay. The decay of the intensity of the luminescence was first investigated by Higashi and Kastner[19] in 1979 after Murayama, Ninomiya, Suzuki and Morigaki[17] reported the discovery that a part of the luminescence (i.e. $A_2$) decays in time as short as $10^{-8}$ sec. HK found the luminescence decays approximately as $t^{-0.89}$ for time delays from $10^{-8}$ to $10^{-3}$ sec. This result has led HK to the model in which the photoluminescence comes from valence alternation $D^+D^-$ pairs of variable separation. The variable separation between the electron and the hole wave function would then leads to a broad distribution of radiative recombination rates or time constants and, according to HK, this can account for the observed approximate $t^{-1}$ decay from $10^{-8}$ to $10^{-3}$ sec. In 1981 Murayama and Ninomiya (MN)[46] reinvestigate the luminescence decay in a-$As_2S_3$ from $10^{-8}$ sec to $2x10^{-3}$ sec at 4.2K. MN studied only the decay of the intensity of the luminescence with the energies of 1.16 and 1.57 eV only, and not the decay of the integrated total light intensity as done by HK. MN's result is shown in Fig. 1 and it is evident by inspection that it departs drastically from the approximate $t^{-1}$ behavior of HK. The shape of

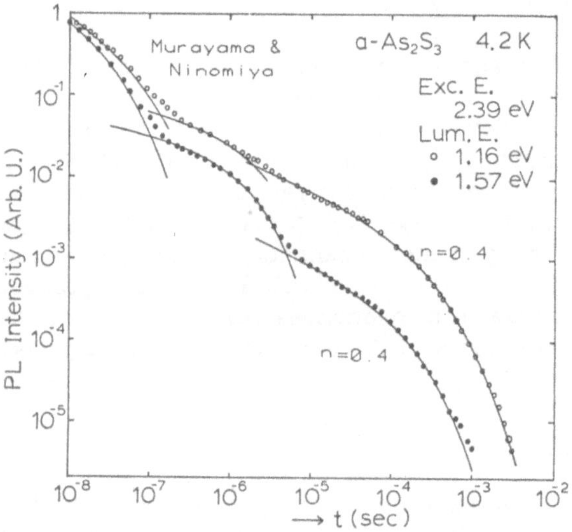
Murayama &
Ninomiya
a-As2S3 4.2 K
Exc. E.
2.39 eV
Lum. E.
1.16 eV
1.57 eV
n=0.4
n=0.4
PL Intensity (Arb. U.)
→ t (sec)

Figure 1

the decay reveals three luminescence processes. Each of the three processes has intensity decaying nonexponentially. Murayama mentioned this data to me when we met at the 1981 Amorphous Semiconductors Meeting in Grenoble. I suggested to him that since the decay is nonexponential it would be worthwhile for him to take a look into my recent model for low frequency/long time relaxation for condensed matter in general and fit his data with the empirical decay function

$$\psi(t) = \frac{d}{dt} \exp[-(t/\tau_p)^{1-n}] \tag{1}$$

where $0 < n < 1$ and $\tau_p$ can be considered for the present purpose as an effective decay time. This he did, and the results can be seen in Fig. 1. The three kinds of luminescence decays according to the empirical law of Eq. (1) with effective decay times $\tau_p$ of $2x10^{-8}$, $2x10^{-6}$ and $2x10^{-4}$ sec. In the last section we have discussed the fast decay luminescence $A_2$ which has memory for polarization of the exciting light which decays in times as short as $10^{-8}$ sec. In Fig. 1 this fast luminescence $A_2$ corresponds to the one with $\tau_p = 2x10^{-8}$ sec. It is nice to see the $A_2$ process first observed by Murayama et al. as early as 1977 and interpreted as recombination of localized excitons with strong lattice relaxation or self-trapped to be time resolved in 1981.

The second luminescence process which we called $A_3$ has effective decay time of $2x10^{-6}$ sec which has polarization memory. $A_3$ is strongly excited by photons with energy in the weak absorption tail[47] due to $Fe^{2+}$. From this property it may be concluded that $A_3$ is due to the recombination of electron-hole pairs localized at the transition metal impurities such as $Fe^{2+}$. Again, it is nice to see an iron impurity related luminescence process being time resolved.

The third luminescence process has $\tau_p = 2x10^{-4}$ sec. and $n = 0.4$ as labelled in Fig. 1. It has no memory for the polarization of the exciting light. These facts together with the properties of $A_1$ discussed in Section 2 implies that this third luminescence process has to be identified with $A_1$. Let us recall that for $t > 10^{-6}$ sec, recombination of separated small polaron pairs each separated by a distance corresponding to the minimum energy or ground state of the pair. Thus we should expect a priori that each pair should have the same radiative recombination time $\tau$ since the distance of separation for each pair (~5A) is practically the same for each pair in the glass because of short range order are known to be preserved, not to mention the possibility of even medium-range order suggested by Phillips.[50] If all the SSP pairs have the same recombination time $\tau$, then we expect the decay to be exponential which correspond to the special case of n=0 in Eq. (1). The observed time dependence is clearly nonexponential. It is well described by the derivative of a fractional exponential as in Eq. (1) with n=0.4. The mechanism for the modification from the exponential to a fractional exponential can be found in the author's model[33] for low frequency/long time relaxation of condensed matter. Detail of this model has been described elsewhere and it is inappropriate for me to duplicate it here. Its principal result will be stated here however as that: all decay or relaxation processes in glasses with relaxation time $\tau$ will be modified by a fundamental mechanism from the decay law of exp $(-t/\tau)$ to

$$\exp\{-\exp(-n\gamma)(1-n)^{-1}\,\omega_c^{-n}\,t^{1-n}/\tau\} \tag{2}$$

where $0<n<1$, $\gamma = 0.577$ a constant, and $\omega_c$ a cut-off frequency. The exponent n in Eqs. (1) and (2) can be viewed as a combined measure for departure from crystallinity and the strength of coupling between the relaxation process (e.g. radiative recombination between a SSP pair in $A_1$ process) and the low energy excitations of the glassy state. We expect n to approach zero value in crystals. Fractional exponential relaxation function can be derived from the relaxation rate, instead of being of the time independent form $W_o=1/\tau_o$, now acquires a time dependence of the form $W_o(\omega_c t)^{-n}$. We are familiar with the concept of time independent relaxation rate which gives the usual exponential decay. The reason why in general, and especially in glasses, polymers and amorphous materials, the relaxation rate becomes time dependent can be seen from the following argument. In glasses the relaxing species like a SSP pair is not an isolated noninteracting element. In fact it can be strongly interacting with the glassy matrix. For the SSP the coupling occurs via the strong lattice relaxation associated with the formation of the small polarons. Radiative recombination relaxation after time $\tau_o$ means the disappearance of the small polaron pair which then causes a sudden (i.e. we are considering times much longer than $\tau_o$, and the radiative recombination which occurs in a time duration of $\tau_o$ can be considered sudden) perturbation on the glassy structure. The sudden perturbation excites and deexcites the energy level structures of the glass, and it is these accompanying processes with small excitation and deexcitation energies that makes the transition rate time dependent. That particular form $W(t) = W_o(\omega_c t)^{-n}$ for the relaxation rate comes about because the low energy excitations in condensed matter following Wigner's random matrix arguments have very general properties that guarantee the condition for "infrared divergent" excitation and deexcitation from one energy level to another. For the technical details a forthcoming article by A. K. Rajagopal and the author will be more illuminating than that given in Ref. 33. Note that we have treated the relaxation process classically. The relaxation process can be radiative recombination of SSP pair, dipolar relaxation or small polaron hopping relaxation, etc.

Turning back to the present subject of chalcogenide glasses, we have thus seen that for the $A_1$ luminescence the decay follows the law of Eq. (1) predicted by the relaxation model of the author. It contradicts with the approximate $t^{-1}$ form that follows from the $D^+D^-$ model with variable separation between the valence alternation pair of HK. The model of Street[21] which attributes $A_1$ to recombination between a tail state electron and $D^o$ has not made any prediction on the decay. So far only the shift of the luminescence peak has been calculated from this model.[21] The approximate $t^{-1}$ decay data, now known to be incorrect, was used as input for the calculation.[21] It is fair to say that these other models are not able to explain the decay of $A_1$ of Fig. 1. On the other hand the SSP recombination together with the author's model for nonexponential relaxation gives a satisfactory account for various characteristics of $A_1$. Higashi and Kastner[47] later also found the same result in a-$As_2S_3$ as that of MN in their total light decay for excitation photon energies in the range of 1.9-2.4 eV. Their data are shown in Fig. 2 for a-$As_2S_3$ and in Fig. 3 for a-$As_2Se_3$. The $A_1$ luminescence with time constant of approximately $2x10^{-4}$ sec is time resolved in Fig. 2. Its decay also follows the law of Eq. (1) with n=0.4 in agreement with MN's result. In Fig. 2 we observe by inspection that as $E_x$ is lowered from 2.2 eV down to 1.9 eV a luminescence process with effective decay time of $10^{-6}$ sec becomes more prominent. Thus this luminescence process is strongly excited by photons with energy in the weak absorption tail due to $Fe^{2+}$. This is one reason why it is believed to be due to recombination of electron-hole pairs localized at the transition metal impurity. In the case of a-$As_2Se_3$ (Fig. 3) the excitation photon energies used are 1.78 and 2.15 eV. Since the gap energy of a-$As_2Se_3$ is smaller than that of a-$As_2S_3$, these excitation energies are relatively high. Only one luminescence process with $\tau_p \cong 10^{-4}$ sec is evident and its decay also follows Eq. (1) with n=0.4. The high values of $E_x$ explain the absence of $A_3$ caused by recombination of electron-hole pair localized at the iron impurity because even the lowest $E_x$ of 1.78 eV is still in the Urbach tail region and at significant energy above the iron related weak absorption tail.

In the brief description given in the last paragraph of the origin of n, we mention that n goes to zero if the glass is replaced by its crystalline counterpart. Thus, we expect that the $C_1$ luminescence in c-$As_2S_3$, the correspondent of the $A_1$ luminescence in a-$As_2S_3$, would have an n close to zero. In Fig. 4 the total light decay data on c-$As_2S_3$ and c-$As_2Se_3$ obtained by Higashi and Kastner are reproduced. For c-$As_2S_3$, the exciting photon energy is 2.71 eV which we know already from the work of Murayama and Bösch that the $C_1$ luminescence is strongly excited. The $C_1$ decays with an effective time constant of $\cong 4x10^{-5}$ sec and conforms well to the law given by Eq. (1) with n≦0.1, much smaller than the value of n=0.4 for $A_1$. The case of c-$As_2Se_3$ is similar.

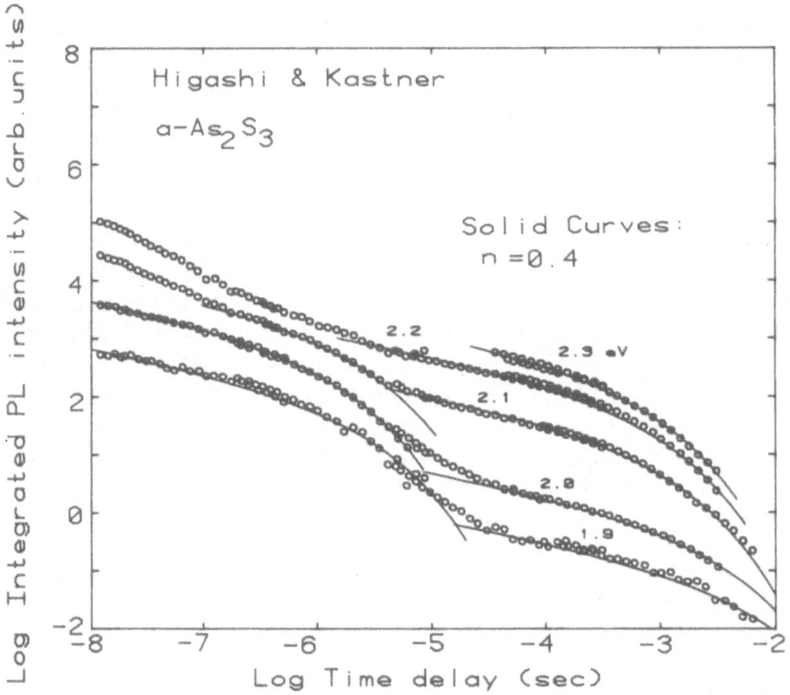
Higashi & Kastner
a-As2S3
Solid Curves:
n =0.4
2.2
2.3 eV
2.1
2.0
1.9
Log Integrated PL intensity (arb.units)
Log Time delay (sec)
8
6
4
2
0
-2
-8
-7
-6
-5
-4
-3
-2

Figure 2

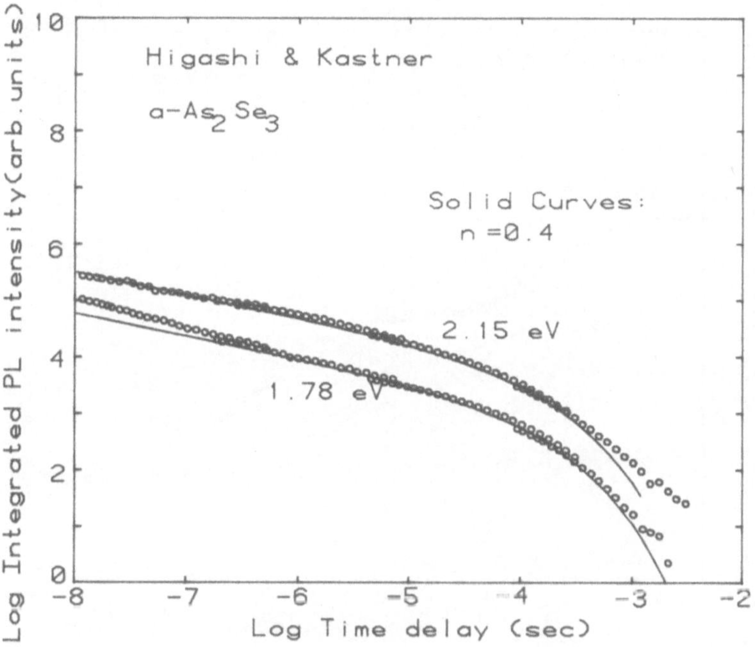
Higashi & Kastner
a-As2 Se3
Solid Curves:
n =0.4
2.15 eV
1.78 eV
Log Integrated PL intensity(arb.units)
Log Time delay (sec)
10
8
6
4
2
0
-8
-7
-6
-5
-4
-3
-2

Figure 3

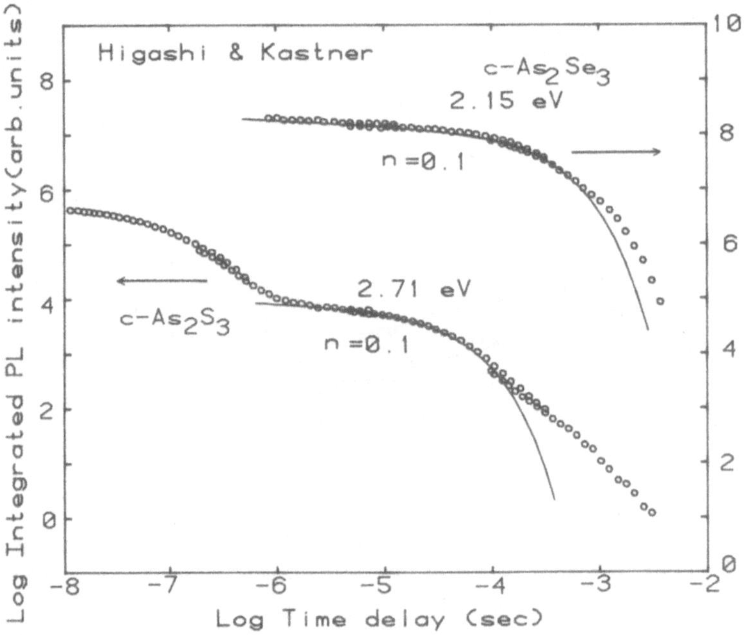
Higashi & Kastner
c-As2Se3
2.15 eV
n =0.1
2.71 eV
c-As2S3
n =0.1
Log Integrated PL intensity(arb.units)
Log Time delay (sec)
-8
-7
-6
-5
-4
-3
-2
0
2
4
6
8
10

Figure 4

Transient Transport

Hole drift mobilities in semiconducting chalcogenide glasses have been studied for many years. The earliest experiments on well-characterized semiconducting glasses such as selenium[51,52] and arsenic triselenide[53,54] discovered that, at least at low temperatures, the drift mobility measured by transient transport techniques depended on the electric field. In a series of elegant papers Pfister and Scher and their collaborators[55,56] explained this anomalous result in terms of a wide (algebraic) distribution of statistical event times which introduced a dispersive character to the transport processes. Because the dispersion did not fit the temperature dependence predicted by the multiple trapping formalism,[57,58] Pfister and Scher suggested a mechanism called trap-controlled hopping in which deep traps ($E \cong \frac{1}{2}E_g$ where $E_g$ is the gap energy) were the predominant factors in limiting the transport processes. Doping experiments[59] on samples of $As_2Se_3$ have raised some significant questions concerning the role of defects in doped samples and hence the general applicability of the trap controlled hopping process to doped $As_2Se_3$. From this, we can appreicate that transient transport or time-of-flight (ToF) experimental data will put to stringent tests of any model on transport and electronic structure of chalcogenide glasses, as well as on the origin of the dispersive (i.e. non-Gaussian) character of the transient transport.

Before considering the chalcogenide glasses where the identity of the correct model for transport and electronic structure is still not totally settled, we consider first dispersive hole transport data of a-$SiO_2$ where there is general agreement to the hole transport mechanism, and the hole in the non-bonding 2p orbital of an oxygen forms a small polaron, and hole transport is via small polaron hopping. The dispersive hole transport in a-$SiO_2$ was first observed by Boesch, et al.[60] at times longer than $10^{-4}$ sec. Later Hughes performed high precision time resolution hole-transport.[61,62] He has shown that after an initial very short time interval following the X-ray pulse, the hole relaxes to form a small polaron (self-trapped) in the non-bonding 2p orbital of an oxygen. Transport at times shorter than $10^{-6}$ sec takes place by the hopping of the small hole polaron from one oxygen to another. Its mobility has magnitude, temperature dependence and electric field dependence all in agreement with the prediction of small polaron theory. In particular, the mobility is low, thermally activated (with activation energy $E_\mu \cong 0.14$ eV, see Fig. 5) at temperatures above one third the Debye temperature and breaking to a non-activated behavior at lower temperatures.[2] For $t \gtrsim 10^{-5}$ sec, dispersive transient current is observed with

$$I(t) \sim t^{-(1-\alpha)} \quad \text{for } t < t_T$$

and

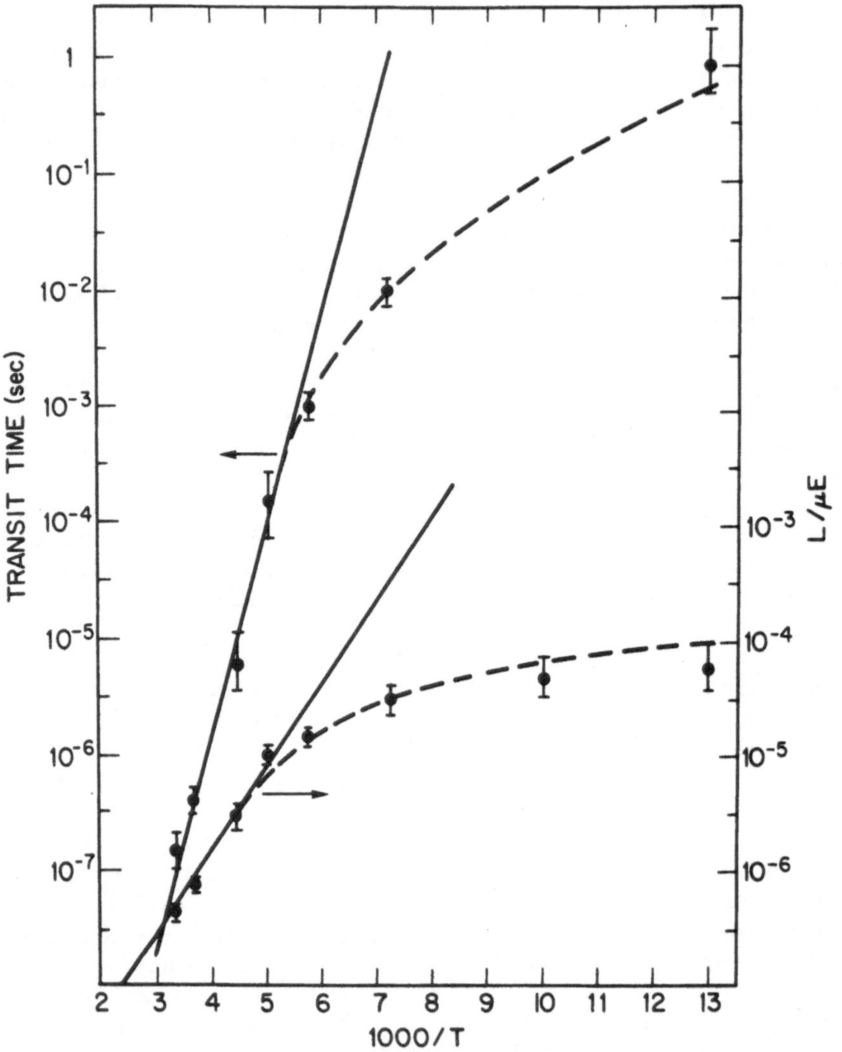


Fig. 5. Transit time $t_T$ and L/μE data of Hughes replotted. The solid lines indicate Arrhenius behavior for T > 200K with activation energies of 0.37 eV and 0.14 eV. The value of 0.14 eV is slightly different from the value of 0.16 eV obtained by Hughes. At every T the ratio of the local slope of the lower dashed curve to that of the upper dashed curve is $\alpha \equiv 0.3$ (see text). The two dashed curves have at every T their local slopes differing by a factor of $\alpha \cong 0.3$ (see text).

$$I(t) \sim t^{-(1+\alpha)} \text{ for } t > t_T$$

where $0 < \alpha < 1$ and $t_T$ is the "transit time." The dependence of $t_T$ on electric field E and sample thickness L through $t_T \propto (E/L)^{-1/\alpha}$ is also observed. These are in accord with the general result of the continuous time random walk (CTRW) model of Scher and Montroll[63] obtained for an algebraic waiting time distribution function $\psi(t) \propto t^{-(1+\alpha)}$. It has been hypothesized[25] that the CTRW occurs in a-$SiO_2$ because the hole, after initially diffusing by small polaron hopping, becomes trapped at some structural defect. Further transport is by tunneling from defect to defect randomly located in the lattice, leading to a non-Gaussian waiting time distribution function that is assumed to have the form of $t^{-(1+\alpha)}$. The temperature dependence of $t_T$ (Fig. 1) is thermally activated above 200K and can be fit to a simple Arrhenius plot with activation energy $E_A^* = 0.37$ eV. The large difference between $E_A^*$ and $E_\mu$ would seem to indicate that the transient transport mechanism is different from the earlier transport mechanism of small polaron hopping; and is the primary reason for invoking traps in transient transport.[54] The hole transport dispersion in a-$SiO_2$ is remarkably stable with temperature.[50,51] The dispersion parameter $\alpha$ as appears in $I(t) \sim t^{1-\alpha}$ for $t < t_T$ is shown by McLean, et al.[60] for their own oxide films to have the constant value of 0.22 for the temperature range of 87-377K. The same is true for data of Hughes where he found, for his own oxide samples, $\alpha = 0.3$ for $140K<T<298K$. This property rules out the mechanism of a distribution of trap depths for dispersive transport.

There are some remarkable correlations between the small polaron transport (SPT) and the dispersive transient transport (DTT). Both SPT and DTT are markedly non-Arrhenius below 200K (Fig. 5) with a decrease in both activation energies (AEs) as T decreases. The activation energies above 200K are related through the relations: $E_A^* \cong E_\mu/(1-n)$, where $n \equiv (1-\alpha) = 0.7$. Most remarkably this relation continues to hold for $T < 200K$ which now takes the form $\partial(\log t_T)/\partial(1/T) = [\partial(1/\mu)/\ \partial(1/T)]/(1-n)$. That is, in an Arrhenius plot, the local slopes of DTT and SPT differ by a factor of $(1-n)$, and n characterizes the dispersive current, i.e. $I(t) \sim t^{-n}$ for $t < t_T$. This is quantitatively illustrated in Fig. 5 by the two non-Arrhenius dashed curves which interpolate the $(L/\mu E)$ and the $t_T$ data respectively for $T < 200K$, and at any T the ratio of the local slope of the former to the local slope of the latter has the constant value of $(1-n) = 0.3$. Such correlations are not expected from the tunneling/ hopping among traps model for DTT. The $T < 200K$ data also contradicts the "trap-controlled" hopping model designed for a-$As_2Se_3$ which predicts the activation energy for $t_T$ to be $E_\mu + E_t$ with $E_t$ the trap depth. The value of $\alpha$ depends on the preparation of the oxide films. McLean, et al. have $\alpha=0.22$ with a measured $E_A^*=0.6$ eV, while Hughes has $\alpha=0.3$ and $E_A^*=0.37$ eV. The large discrepancy between the two activation energies of $t_T$ imposes some difficulties in the trap models. Now what Hughes has measured for $E_\mu$ is for hole hopping in the non-bonding 2p oxygen orbitals, an intrinsic quantity which should be the same for both samples. Then, if we use the relation $E_A^*=E_\mu/\alpha$ for McLean et al.'s sample we find $E_\mu/\alpha = (0.14/0.22)$ eV = 0.64 eV which is almost identical to the measured value of $E_A^*$.

Our approach to dispersive transient transport follows from the mechanism for time dependent relaxation rate discussed earlier for radiative recombination of SSP pairs. For dispersive transient transport the relaxation process is the hopping of the small polaron from one site to another, which gives rise to a sudden perturbation acting on the level structure of the glass. The constant small polaron hopping relaxation rate $\tau_0^{-1} = W_0 = \tau_\infty^{-1}\exp(-E_A/kT)$ is modified by "infrared divergent" excitations and deexcitations of level structures to be

$$W(t) = W_o \exp(-n\gamma)(\omega_c t)^{-n} \quad . \tag{3}$$

Then the decay of the probability Q(t) that the hole remains on a particular site as described by

$$\frac{dQ}{dt} = -W_o \exp(-n\gamma)(\omega_c t)^{-n} Q \tag{4}$$

and $\psi(t) = -\frac{dQ}{dt}$ can be written as

$$\psi(t) = \tau_o^{-1}\, e^{-n\gamma}(\omega_c t)^{-n} \exp\{-\tau_o^{-1}\, e^{-n\gamma}\, t^{1-n}/(1-n)\, \omega_c^n\} \tag{5}$$

To derive the transient current I(t) with $\psi(t)$ given by Eq. (5), we shall use the time-scale transformation technique for relaxation proposed by Teitler, Rajagopal and the author.[41] We can rewrite (3) as

$$W(t) = W_o \frac{d\theta}{dt} \tag{6}$$

with

$$\theta = A_n t^{1-n} \tag{7}$$

and

$$A_n = (1-n)^{-1} e^{-n\gamma} \omega_c^{-n} \tag{8}$$

Equation (6) represents a genuine time-scale transformation with $d\theta/dt>0$ at all times t greater than zero. In the θ-time frame, the variable Q relaxes according to

$$\frac{dQ}{d\theta} = -W_o Q \tag{9}$$

In the θ-frame, hopping relaxation at any site will have the same time independent transition rate $W_o$. The waiting time distribution has the form $\psi(\theta) = W_o \exp(-W_o\theta)$. An observer with the θ-clock will therefore witness conventional Gaussian statistics and transport. That is, the dispersion σ of the carrier sheet initially created at the sample surface and the mean displacement ℓ from the same surface obey the well-known relations $\sigma \propto \theta^{\frac{1}{2}}$ and $\ell \propto \theta$. The transient current caused by the drifting charge sheet remains constant even though there is continued spreading about the mean ℓ. The current I(θ) begins to drop as soon as the leading edge of the carrier packet reaches the other surface where carriers are collected by the electrode. The time when the peak of the charge packet reaches the collector electrode can be identified as the transit time $\theta_T$. I(θ) remains constant for $\theta \ll \theta_T$, and decays rapidly for $\theta \gg \theta_T$.

The transit time $\theta_T$ is given by the conventional relation:

$$\theta_T^{-1} = \mu_D L/E \tag{10}$$

where L is the sample thickness, E is the applied electric field and $\mu_D$, the drift mobility, is related to the diffusion constant $D_o$ via the Nernst-Einstein relation $\mu_D = (e/kT)D_o$ and $D_o$ in turn is given by $D_o = W_o C_o^2/2$ for one-dimensional diffusion with jump length $C_o$.

With I(θ) known, the transient current in the laboratory t-frame, I(t), will be given by

$$I(t) = I(\theta)\,\frac{d\theta}{dt} \quad . \tag{11}$$

For the fractional time scale transformation of Eq. (7),

$$I(t) = (1-n)A_n t^{-n} I(A_n t^{1-n}) \quad . \tag{12}$$

Whereupon, from the dependence of I(θ) as discussed, we have $I(t) \propto t^{-n}$ for $A_n t^{1-n} \ll \theta_T$ and a more rapid drop-off for $A_n t^{1-n} \gg \theta_T$. From the well-known results of Gaussian transport in the θ-frame, we have $\theta_T^{-1} = (e/kT)(W_o C_o^2/2)L/E$. Hence the corresponding transit time in the laboratory time frame defined by

$$A_n t_T^{1-n} = \theta_T \tag{13}$$

is given by the expression

$$\begin{aligned} t_T &= \{(kT/e)((1-n)e^{n\gamma}\omega_c^n)(L/EW_o)\}^{1/1-n} \\ &\equiv \tilde{W}_o^{-1}(L/E)^{1/(1-n)} \exp(E_A/(1-n)kT) \end{aligned} \tag{14}$$

The transit time $t_T$ exhibits the unusual superlinear thickness and field dependences. If the hopping transition rate $W_o$ is thermally activated with a microscopic activation energy $E_A$, it follows from inspection of Eq. 14 that $t_T$ will also be thermally activated with an apparent activation energy $E_A^*=E_A/(1-n)$. Identifying $E_A$ with $E_\mu$ and $\alpha$ with $(1-n)$, we can see that our model predicts the observed empirical relation $E_A^*=E_\mu/\alpha$ (Fig. 5), and implies that the DTT occurs from small polaron hopping also. No traps are involved. It can reconcile the large difference of $E_A^*$ between measurements on two samples since it is only an apparent activation energy, while the product $(\alpha E_A^*)$ is nearly equal for both samples as it should be, because it is the intrinsic small polaron mobility activation energy, $E_\mu$.

P. Craig Taylor and the author[34,65] have considered the transient transport in chalcogenide glasses. In undoped or doped chalcogenide glasses, the transient current exhibits the asymptotic forms $I(t) \propto t^{-n}$ and $I(t) \propto t^{-(2-n)}$ for $t \ll t_T$ and $t > t_T$ respectively. As far as the experimental results are concerned, the two most important features of Eq. (14) are the superlinear dependence of $t_T$ on (L/E) and the appearance of an "effective" activation energy $E_A^*=E_A/(1-n)$ which depends on both the microscopic activation energy $E_A$ and the dispersion parameter n. In the CTRW model of Scher and Montroll the dispersive character of the transport process is contained in the parameter $\alpha$ where $\alpha$ corresponds roughly to $(1-n)$ in the present model. Of course, in the CTRW approach $\alpha$ enters only in the superlinear dependence of (L/E) and does not affect the activation energy.

The application of Eq. (14) to transient hole transport experiments[55] in $As_2Se_3$ yields an "apparent" activation energy $E_A^* \cong 0.6$ eV. The measured dispersion varies considerably from sample to sample,[56,26-28] but the average value on well characterized samples is $n=1-\alpha=0.45$. To determine n one can either use the asymptotic behavior of the current as a function of time or the superlinear dependence of $t_T$ on (L/E). For several reasons,[56] the latter approach is by far the more accurate and has been employed exclusively in this work.[65]

The first striking conclusion which one draws from $E_A^*$ and n for pure $As_2Se_3$ is that the true microscopic activation energy $E_A$ is ~0.33 eV in this amorphous solid. But 0.3 eV is just the activation energy estimated for small polaron hopping[2,66] in $As_2Se_3$. One is thus led to the possibility that, in spite of the apparently large activation energy $E_A^*$ for the drift mobility, the transport mechanism need not involve the presence of deep traps or defects.

This possibility is particularly intriguing in light of the

fact that combined measurements of drift mobility and various spectroscopic techniques,[59] such as photo-induced ESR and photoluminescence (PL), on doped films of $As_2Se_3$ are not consistent with simple interpretations based on defects. In particular, if a defect model is invoked, then the doping results require that the defect centers associated with PL and photo-induced ESR must be different from the trapping centers which limit transient hole transport.[11] Also there is no good explanation for the lack of observation in ESR of the traps observed in transient hole transport experiments. Both of these difficulties, as well as several others, are removed in a very natural way if the explanation for the transport measurements does not invoke the presence of traps.

In Fig. 6 we show the dependence of the measured activation energy $E_A^*$ on doping for several impurities in glassy $As_2Se_3$ as measured by the transient transport[26-28,67] technique (solid symbols). For pure $As_2Se_3$ we have used the exhaustive measurements of Pfister which yielded an average activation energy of $E_A^*=0.6$ eV and an average dispersion of $\alpha=(1-n)=0.55$. The experimentally observed spread in $E_A^*$ in samples of pure $As_2Se_3$ is indicated by the cross-hatched region of Fig. 6.

Also plotted in Fig. 6 (open symbols) are the activation energies predicted by the relation $E_A^*=E_A/(1-n)$ where $E_A=0.33$ eV is taken as a constant activation energy for (polaronic) conduction in all samples. The inset of Fig. 6 emphasizes the correlation of $E_A^*$ with n predicted by the present model (solid line) and the agreement obtained with this prediction on a number of doped samples. Because these predictions involve the dispersion parameter n which is notoriously difficult to determine and varies from sample to sample ($0.45 \leq n \leq 0.55$ for pure $As_2Se_3$), an estimate of the scatter expected from this single source of error is indicated by the dashed lines in the inset to Fig. 6. The actual scatter, which in addition involves both variations from sample to sample and uncertainties in the measured activation energies, is expected to be somewhat greater than the bounds as indicated.

The agreement between the experimental points and the predictions of the model is by no means perfect, but the trends are well established. When the dispersion decreases (n decreases or $\alpha$ increases), as it does upon doping with Ga, In and Tℓ, then the measured activation energy decreases. Conversely, when n increases, as it does for doping with Cu, then the activation energy increases. There are some quantitative difficulties in fitting the data for long transit times ($t_T > 1$ sec) such as occur on doping with $>10^{18}$ $cm^{-3}$ of Ni or Mn, but the trends are still preserved. Because experimental difficulties involved in measuring long transit times significantly increase the uncertainty of the results, we have omitted these data from Fig. 6.

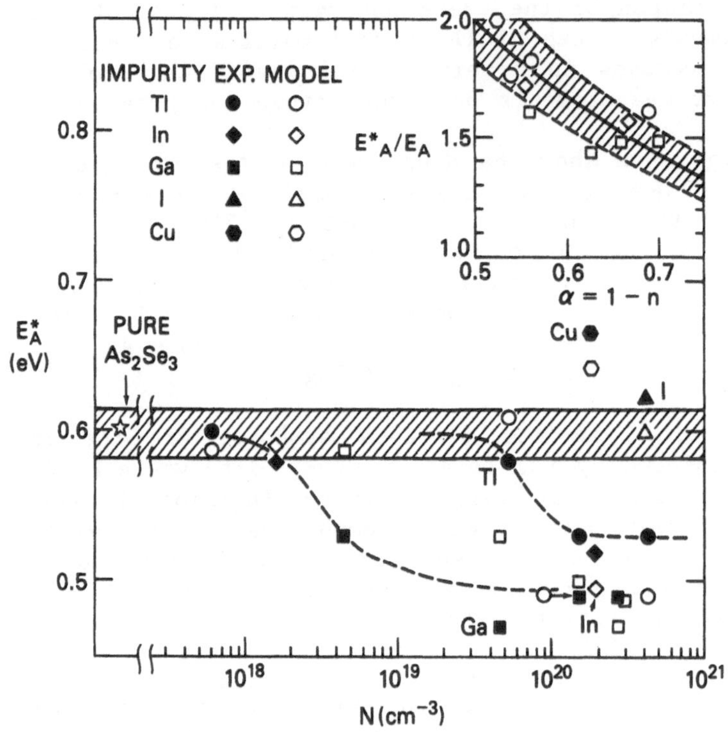


Fig. 6. Dependence of the experimentally measured activation energy for transient hole transport at 300K (solid data points) and calculated activation energy (open data points) as functions of impurity content. Data for Tℓ, In, Ga, Cu are from ref. 28; data for I and $3x10^{20}$ Ga are from refs. 26 and 27, respectively. Pure $As_2Se_3$ data are from ref. 25. The inset shows the model calculations of the dependence of $E_A^*$ on dispersion for several dopants.

The present model does not address the microscopic significance of the disorder parameter n, but the sensitivity of n to small additions of impurities can be understood qualitatively. The very low energy excitations, for which n serves as an empirical "coupling constant," must involve a great many atoms so that the addition of a small number of impurities which become bonded to the network could have a significant effect on n. The impurities do not form defects, in the usual sense of the word, because they go into the network so as to satisfy all of their normal bonding requirements. Consequently, the impurities do not necessarily act as traps for electrons or holes.

The doping experments of Pfister et al.[27,28] on Na, Tℓ, In and Ga in $As_2Se_3$, also showed that the dc mobility $\mu_{dc}$ paralleled the values of $\mu_h$ over several orders of magnitude in both $\mu_{dc}$ and concentration. This startling fact, which is very difficult to understand if defects are invoked to interpret the drift mobility measurements, is expected within the framework of the present model because the variable $\tilde{W}_o^{-1}$ of Eq. (14) is effectively proportional to the dc mobility (more accurately, log $\tilde{W}_o \propto$ log $\mu_{dc}$). Thus, provided that the activation energy is essentially constant, as it is for Tℓ below $10^{19}$ $cm^{-3}$ and for In and Ga below $10^{18}$ $cm^{-3}$, the change in $t_T$ is due to a change in the dc mobility. When the activation energy changes this term will dominate, and the correlation with $\mu_{dc}$ may be masked.

It is not our intention to imply that defects or traps are never important in determining the transport properties of $As_2Se_3$ or doped $As_2Se_3$; after all, traps are often the determining factors in transport processes in crystalline semiconductors. Nonetheless, the appearance in the $As_2Se_3$ system of thermally activated transport processes which can be described by a constant microscopic activation energy ($E_A \cong 0.33$ eV) equal to the small polaron hopping energy estimated from low frequency thermoelectric power and conductivity measurements,[2,66] strongly suggests that deep traps do not play a dominant role in most instances.

In the transient transport measurements[25-28] of the $As_2Se_3$ system, there is no evidence for a change in the dispersion parameter with temperature. One would thus expect Arrhenius behavior for the hole drift mobility, $\mu_h \propto t_T^{-1}$, as indicated in Eq. (14).

It is interesting to reconsider the difference in n value for the two a-$SiO_2$ samples in the light of the effect of doping in chalcogenide glasses. The sample of McLean is grown under different conditions from that of Hughes. One is a dry oxide and the other is a wet oxide. The wet oxide can be viewed as doped dry oxide with doping achieved by growth in the presence of water.

Transient Photoinduced Absorption and Photoconductivity

The first time-resolved photoinduced-absorption (PA) spectra was obtained by Orenstein and Kastner (OK)[68] in a-$As_2Se_3$ from $10^{-5}$ sec to $10^{-2}$ sec in the temperature range of 20K to 395K. In this latest work used on photoconductivity, they have reduced the excitation intensity and low excitation energies in order to avoid difficulties that they ran into before. The induced optical absorption is by carriers which have become localized after photoexcitation. OK observed the PA spectra shifts to higher energy with time indicating the possibility of a distribution of state energies for the optically induced carriers. The energy shift is roughly equal for exponentially increasing intervals of time. The model they suggested is initially following pulsed photoexcitation the states are uniformly popoulated by carriers. However, as time goes on, carriers in shallower states will be released preferentially via thermal excitation to some as yet unspecified mobile states, and will be retrapped elsewhere. It is argued that since the retrapping process is random, the mean occupation number for the deeper states will therefore increase as time evolves and hence the time dependence of the PA spectra can be explained. OK masured also the transient photocurrent which decays with time as the power law of $t^{-n(T)}$. In the temperature range of 395-200K, n is a function of temperature and has the form of $n(T)=1-T/T_0$ with $kT_0 \cong 50$ meV. OK noticed that if they assume that the density of states of the localized levels decreases exponentially below a certain energy $E_0$,[69,70] i.e. $g(E)=g_0 \exp(-(E_0-E)/kT_0)$, then the multiple trapping process will give a photocurrent which has a power law decay with $n(T)= 1-T/T_0$ as observed. This is an important assumption. It needs to be justified. Otherwise the model of OK cannot be considered complete for it has merely replaced one difficult problem (i.e. the origin of the $t^{-n(T)}$ transient) by another (i.e., why an exponential density of localized levels). To avoid any possible confusion for the reader the induced absorption has nothing to do with the Urbach tail which in addition to other explanations for its cause, can be explained also by an exponential tail of density of states of the conduction and valence bands. Abe and Toyozawa have an elegant model of obtaining exponential band tail from Gaussian site diagonal disorder through interplay with the transfer energy. This model may not be applicable to the discussion of the localized photoinduced states in the gap that are responsible for the transient PA. Even we start with the reasonable assumption of a Gaussian distribution of the localized levels, these localized levels in OK's model apparently communicate with each other only by release into high mobility state and subsequently retrapped. This is unlike the Abe and Toyozawa's model described by a tight-binding Hamiltonian with site-diagonal randomness such that the joint distribution of the upper and lower band site energies is Gaussian. Thus at this time it is not clear how the pivotal assumption of exponential density of states used by OK can be justified.

Nevertheless OK has suggested in view of their transient photoconductivity (PC) data that the time-of-flight measurements of Pfister have to be questioned. Their reason is that the photoconductivity has the dispersion parameter n being a function of T, while Pfister's data has T-independent n. The direct comparison between PC and ToF data is inappropriate. The energy of the exciting light for OK's PC has never been explicitly stated but from information given we gather it is about 1.6-1.7 eV. On the other hand, Pfister had to use highly absorbed light in order to produce carriers at one surface of the sample. The energy of the exciting light used by him must be superbandgap and hence is much higher than that used by OK. From Sections 2 and 3 we can recall the presence of several luminescence processes and the dependences of their intensities on $E_x$ are all different. It is obvious that OK's PC and Pfster ToF most probably involve different sets of carrier states. Thus direct comparison between the PC and ToF experimental data as done by OK is not possible. They are different things.

Further, recent experiments[71] below 200K show deviations from the high temperature behavior observed by OK. Instead, below 200K, $n(T)>1$ which contradicts the assumption of an exponential $g(E)$.

One possible explanation is that, with $E_x \cong 1.6$-$1.7$ eV in a-$As_2Se_3$, electron-hole pair that are responsible for the $A_2$ or $A_3$ luminescence are photo-induced. The hole may have a Gaussian distribution of energies, consistent for example with the fact that $A_2$ shifts rigidly with $E_x$ as seen in a-$As_2S_3$. When it is converted to the hole small polaron state by thermal activation, it will conduct via small polaron hopping. As in the ToF experiment where SSP are produced, the hole small polaron will, according to the fundamental mechanism of the author, have a time-dependent hopping relaxation rate as given by Eq. (3). This leads to a time-dependent mobility $t^{-n}$. Multiple trapping of the hole small polaron by the Gaussian density of states will then introduce a temperature dependence to n as well as $n>1$ at lower temperatures.

In concluding this section we mention also the caution voiced by Bösch[72] on the danger of sweeping interpretations and conclusions made by OK based on their experiment alone that can probe only one among many sets of states in chalcogenide glasses.

## 4. Structural Relaxations

The dielectric relaxation data[31] in a chalcogenide glass of 0.33% As-Se are available at temperatures above $T_g$ (Fig. 7). The solid line, the dielectric loss $\varepsilon''$ predicted from the Fourier transform of the function $Ct^{-n}\exp(-t/\tau_n)^{1-n}$ for n=0.65, is in good agreement with data. The same class of function explains also the volume[40] or enthalpy[32] recovery of a-$As_2Se_3$. It was deduced from

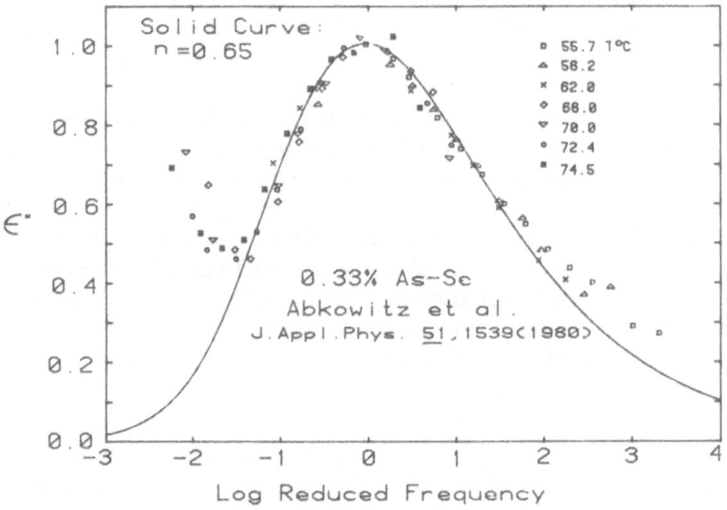
Solid Curve:
n =0.65
55.7 T°C
58.2
62.0
68.0
70.0
72.4
74.5
0.33% As-Se
Abkowitz et al.
J. Appl. Phys. 51, 1539(1980)
ε"
Log Reduced Frequency

**Figure 7**

experimental data by Moynihan[73] that n=0.33 for enthalpy relaxation and n=0.2 for volume relaxation. I may add also that temperature variation of n, especially near $T_g$, accounts also at least partly for the WLF behavior for relaxation and the Vogel behavior for viscosity.[40,41] A comprehensive model for glass transition that can account quantitatively all aspects of this phenomenon will appear in the near future.

## 5. Universality of Low Frequency Relaxations: Examples Other than Chalcogenide Glasses

We have seen from discussions in the earlier paragraphs that, in the chalcogenide glasses system, all relaxations at low frequencies/long times proceed according to the same laws of

$$Q = Q_0 \exp[-(t/\tau_p)^{1-n}] \quad , \; 0<n<1 \tag{15}$$

and

$$\tau_p = [(1-n)\omega_c^n \tau_0]^{1/(1-n)} \tag{16}$$

irrespective of the nature of the relaxation and the relaxation species that are involved. Thus it can be the luminescence decay which is optical relaxation, or the transient transport which is carrier hopping relaxation, or dielectric, mechanical, enthalpy and volume relaxations. The pair of laws (15) and (16) are a direct consequence of one essential piece of physics. Namely that the familiar relaxation rate $W_0 = \tau_0^{-1}$ of the relaxation species is always modified to have the time dependence of

$$W(t) = W_0(\omega_c t)^{-n} \quad , \; 0<n\leq 1$$

for $\omega_c t \gg 1$. The cause of this time dependence is due to coupling of the relaxation species to low-lying excitations which is preponderant in glasses.

It is important to realize that these unexpected universalities for chalcogenide glasses are only special examples of a much more general and larger class of universal laws that have been found to hold for various relaxation properties of many forms of condensed matter. A review article[33] written by the author in 1979 has presented some evidence for this universality. Since 1979, I have taken the development further. The relaxation phenomena that have been thoroughly investigated for universality include dielectric relaxation, ac conductivity, creep, stress relaxation, viscoelastic relaxations, internal friction, relaxations observed by photon correlation spectroscopy, nuclear magnetic resonance relaxations, spin-echo measurements, transient electrical transport, transient optical luminescence, volume and enthalpy relaxations and recoveries, differential scanning calorimetry (DSC), steady flow viscosity, stress-strain relationship and its dependence on strain-rate, ultrasonic attenuation, noise, diffuse, diffusion controlled chemical reactions and electronic recombinations, etc. Materials involved include liquids, supercooled liquids, liquid crystals, inorganic glasses, electrolytes, ionic conductors, insulators, dielectrics, gate insulators of electronic devices, electrets, semiconductors, amorphous semiconductors, xerographic materials, polymer melts and solutions, amorphous polymers, rubbers, plastics, epoxies, ceramics, piezoelectrics, pyroelectrics, ferroelectrics, metals, amorphous metal, metallic glasses, lubricants, biopolymers, coal, oil shales, etc. The wealh of physics, the number of interesting physical phenomena and unsolved problems related to relaxation phenomena are staggering. I shall enumerate some as examples: (1) the glass transition, the Vogel-Tamann-Fulcher behavior for viscosity and the Williams-Landel-Ferry relation for viscoelastic quantities; (2) physical aging of glasses and polymers; (3) enthalpy and volume recovery of glasses and polymers following a temperature or pressure jump, the Kovac's $\tau$ paradox, the asymmetry, nonlinearity and memory effects in the recovery process; (4) endotherms in DSC experiment, $T_g$ overshoot, dependence of structure on the endotherms on thermal history, annealing time, heating and cooling rates; (5) stress-strain relation, its dependence on strain rate, memory effects, yield stress dependence on strain rate; (6) ductile-brittle transition of polymers and glasses; (7) the characteristic dependence of steady flow viscosity, viscoelastic relaxations, plateau modulus and recoverable compliance on molecular weight (or degree of polymerization) in polymer melts and concentrated solutions, linear and branched polymeric systems; (8) the origin of non-Debye nature of relaxations of various kinds; (9) dispersive electron transient transport or time of flight measurements in amorphous semiconductors, insulators and in polymers, departure from Gaussian transport, the nature of the transport mechanisms; (10) transient luminescence (i.e. radiative electron-hole recombination) in amorphous semiconductors, departure from a single radiative relaxation time, the identity of the states in the optical gap in amorphous semiconductors; (11) asymmetric spin-lattice relaxation time $T_1$ minimum when log $T_1$ is plotted versus inverse temperature, departure from classical BPP (Bloembergen, Purcell and Pound) behavior; (12) flicker 1/f noise; (13) noise that arises from diffusion of carriers, departure from that expected of classical diffusion; (14) time-dependences of creep compliance and stress relaxation departure from behavior of a single Maxwell element or Voight element; (15) relaxations in spin glasses; (16) flash photolysis and transient recombination in hemoglobins, myoglobin and related systems; (17) frquency dependences of ac conductivity and electric modulus (e.g. inverse of the complex dielectric susceptibility) and dispersive transient transport due to electronic hopping in amorphous semiconductors, to ion hopping in ionic conductors, silicate glasses, Na-$\beta A\ell_2O_3$, and to soliton hopping in polymeric conductors such as polyacetylene.

A more up-to-date review of these new developments can be found in a forthcoming paper by the author.[74] The preponderance of the evidence for universality in relaxation properties in various forms of condensed matter including the chalcogenide glasses and a-$SiO_2$ that have been discussed in these lectures should not be ignored in any attempt to arrive at a satisfactory microscopic understanding of the electronic structure, state and transport mechanism and optical properties of the chalcogenide glasses.

## 6. Conclusions

In this work we have reviewed a lot of experimental data gathered on the chalcogenide glasses and emphasized especially on the transient optical data which have rapidly accumulated since 1979. These recent data enable us to narrow down the possible kinds of states that can exist in the chalcogenide glasses. As a result of these analyses, a minimal set of states has been proposed that seem necessary at this time to explain the totality of optical and transport data. We are convinced of the existence of localized excitons with strong lattice relaxation or even self-trapped, states associated with the persistent and inadvertent transition metal impurities such as iron, and separated electron and hole small polarons. Strong evidence from transient transport studies on both doped and undoped chalcogenide glasses as well as on a-$SiO_2$ supports Emin's model of transport by small polaron hopping. This proposed minimum set of states is certainly consistent with Anderson's bipolaronic graound state. However, we do not need to invoke the existence of a mobility edge or the defects VAPs and IVAPs. At least they are not necessary to explain the data of chalcogenide glasses. Recently, the concept of minimum metallic conductivity, $\sigma_{min}$, which is intimately related[75] to the concept of conduction at a mobility edge, i.e. $\sigma=\sigma_{min} \exp(-E/kT)$, have been found to be in disagreement with experiment.[76] Also recent calculations[77] on the effective Hubbard U for the bonding coordination defects in glassy Se do not confirm that VAPs and IVAPs are negative U systems, rather they have a positive U.

We have discussed also the various low frequency/long time relaxation properties of the chalcogenide glasses and demonstrated that they fall into a universal pattern which is shared by many other condensed matter. Two universal laws are obeyed by not only the relaxations that are electronic in character, but also by other relaxations which are non-electronic in character. A satisfactory understanding of this universality is the key to many problems in glasses and polymers.

## 7. Acknowledgements

The work would not be possible without the discussions and collaborations with a number of workers in the field including

M. Bösch, S. G. Bishop, U. Strom and P. C. Taylor and especially Kazuro Murayama. Part of this work was carried out during the visit of the author to the Aspen Center for Physics in July, 1982. Discussions with D. Emin and P. W. Anderson on the subject of chalcogenide glasses at the Aspen Center for Physics make the writing of this article easier. The valuable assistance given to the author by Pauline T. Iaconangelo and Mr. G. Fong in preparation of the final manuscript is also gratefully acknowledged. This work is supported in part by ONR under Task NR 319-059.

## References

1. N. F. Mott and E. A. Davis, Electronic Processes in Non-crystalline Materials, Second Edition (Clarendon, Press, Oxford, England, 1979).

2. D. Emin, Adv. Phys. 24, 305 (1975).

3. D. Emin, J. Non-cryst. Solids 35 & 36, 969 (1980).

4. D. Emin, Amorphous and Liquid Semiconductors, ed. W. E. Spear (Edinburgh), p. 249 and p. 261 (1977).

5. D. Emin, Phys. Rev. Lett. 32, 303 (1974).

6. D. Emin, C. H. Seager and R. K. Quinn, Phys. Rev. Lett. 28, 813 (1972).

7. D. Emin, Physics Today, June issue, p. 34 (1982).

8. M. A. Bösch, R. W. Epworth and D. Emin, J. Non-cryst. Solids 40, 587 (1980).

9. D. Emin, J. de Physique, Colloque C4, 535 (1981).

10. E. I. Rashba, Optika i Spectrosk. 2, 75 and 88 (1957); Y. Toyozawa, Progr. Theor. Phys. 26, 29 (1961); A. Sumi, J. Phys. Soc. Jap. 43, 1286 (1977).

11. P. W. Anderson, Phys. Rev. Lett. 34, 953 (1975); J. de Physique, Colloque (1976).

12. S. G. Bishop, U. Strom and P. C. Taylor, Phys. Rev. Lett. 34, 346 (1975).

13. R. A. Street and N. F. Mott, Phys. Rev. Lett. 35, 1293 (1975).

14. M. Kastner, D. Adler and H. Fritzsche, Phys. Rev. Lett. 37, 1504 (1976).

15. B. T. Kolomiets, T. N. Mamontova and A. A. Babaev, J. Non-cryst. Solids 4, 289 (1970).

16. R. A. Street, T. M. Searle and I. G. Austin, Amorphous and Liquid Semiconductors, ed. J. Stuke and W. Brenig (Taylor & Francis ltd., London, 1974), p. 953.

17. K. Murayama, T. Ninomiya, H. Suzuki and K. Morigaki, Solid State Commun. 24, 197 (1977).

18. M. Bösch and J. Shah, Phys. Rev. Lett. 42, 118 (1979).

19. G. S. Higashi and M. Kastner, J. Non-Cryst. Solids 35-36, 921 (1980).

20. J. Shah, Phys. Rev. B 21, 4751 (1980).

21. R. A. Street, Solid State Commun. 34, 157 (1980).

22. K. Murayama, H. Suzuki and T. Ninomiya, J. Non-cryst. Solids 35-36, 915 (1980).

23. K. Murayama, K. Kimura and T. Ninomiya, Solid State Commun. 36, 349 (1980).

24. K. Murayama and M. A. Bösch, J. de Physique C4, 343 (1981).

25. G. Pfister and H. Scher, Adv. Phys. 27, 747 (1978).

26. G. Pfister, A. R. Melnyk and M. E. Scharfe, Solid State Commun. 21, 907 (1977).

27. G. Pfister, M. Morgan and K. S. Liang, Solid STate Commun. 30i, 227 (1979).

28. G. Pfister and M. Morgan, Phil. Mag. B 41, 209 (1980).

29. S. G. Bishop and P. C. Taylor, Phil. Mag. B 40, 483 (1979).

30. J. Tauc, F. J. DiSalvo, G. E. Peterson, and D. L. Wood, Amorphous Magnetism, ed. H. O. Hooper and A. M. de Graaf (Plenum Press, NY, 1973), p. 119.

31. M. Abkowitz, J. Appl. Phys. 51, 1539 (1980).

32. C. T. Moynihan et al., Annals N.Y. Acad. Sci. 279, 15 (1976).

33. K. L. Ngai, Comments Solid State Phys. 9, 127 (1979); 9, 141 (1980).

34. K. L. Ngai, Recent Developments in Condensed Matter Physics, Vol. I, Invited Papers, ed. J. T. Devreese (Plenum, NY, 1981), p. 527.

35. K. L. Ngai and F. S. Liu, Phys. Rev. B 24, 1049 (1981).

36. K. L. Ngai, X. Huang and F. S. Liu, in Physics of MOS Insulators, ed. G. Lucovsky (Pergamon, NY, 1980), p. 44.

37. K. L. Ngai, in Tetrahedrally Bonded Amorphous Semiconductors, AIP Conf. Proceedings No. 73, (1981), p. 293.

38. K. L. Ngai, Solid State Ionics 5, 27 (1981).

39. K. L. Ngai, Polymer Preprints 22, 287 (1981).

40. J. T. Bendler and K. L. Ngai, Polymer Preprints 22 (No. 2), 287 (1981).

41. S. Teitler, A. K. Rajagopal and K. L. Ngai, NRL Memo Report No. 4757 (1982), and Phys. Rev. A, November issue (1982).

42. K. L. Ngai and R. W. Rendell, Polymer Preprints 23, Sept 1982.

43. S. G. Bishop and D. L. Mitchell, Phys. Rev. B 8, 5696 (1973).

44. S. Abe and Y. Toyozawa, J. Phys. Soc. Japan 50, 2185 (1981).

45. G. Higashi and M. Kastner, Phys. Rev. B 24, 2275 (1981).

46. K. Murayama and T. Ninomiya, Jap. J. Appl. Physics (to be published). The results of these authors can be seen from Fig. 1 in these lecture notes.

47. G. Higashi and M. Kastner, Preprint (1982). Figs. 2-4 reproduce their data.

48. N. F. Mott, Phil. Mag. B 34, 1101 (1976).

49. M. Kastner, Phil. Mag. B 37, 127 (1978).

50. J. C. Phillips, J. Non-cryst. Solids 43, 37 (1981).

51. W. E. Spear, Proc. Phys. Soc. (London) B70, 669 (1957); B76, 826 (1960); J. L. Hartke, Phys. Rev. 125, 1177 (1962); H. P. Grunwald and R. M. Blakney, Phys. Rev. 165, 1006 (1968).

52. M. D. Tabak, Phys. Rev. B2, 2104 (1970).

53. B. T. Kolomiets and E. A. Lebedev, Sov. Phys. Semicond. 1, 244 (1967); J. M. Marshall and A. E. Owen, Phil. Mag. 24, 1281 (1971).

54. M. E. Scharfe, Phys. Rev. B2, 5015 (1970); D. M. Pai and M. E. Scharfe, J. Non-Cryst. Solids 8-10, 752 (1972).

55. H. Scher and E. W. Montroll, Phys. Rev. B12, 2455 (1975).

56. G. Pfister and H. Scher, Phys. Rev. B15, 2063 (1977).

57. F. W. Schmidlin, Phys. Rev. B16, 2362 (1977).

58. J. Noolandi, Phys. Rev. B16, 4466 and 4474 (1977).

59. G. Pfister, K. Liang, M. Morgan, P. C. Taylor, E. J. Friebele and S. G. Bishop, Phys. Rev. Lett. 41, 1318 (1978).

60. F. B. McLean, H. E. Boesch and J. M. McGarrity, IEEE Trans. Nuc. Sci. NS-23, 1506 (1976); and private communication from F. B. McLean.

61. R. C. Hughes, Phys. Rev. B 15, 2012 (1977).

62. R. C. Hughes and D. Emin, Proc. Int'l Conf. Physics of $SiO_2$ and its Interfaces, ed. S. Pantelides (Pergamon, NY, 1978), p. 14.

63. H. Scher and E. W. Montroll, Phys. Rev. B 12, 2455 (1975).

64. G. Lucovsky, Phil. Mag. B 39, 531 (1979).

65. P. C. Taylor and K. L. Ngai, Solid State Commun. 40, 525 (1981).

66. C. H. Seager and R. K. Quinn, J. Non-cryst. Solids 17, 386 (1975).

67. M. Abkowitz, Phys. Rev. 22, 3843 (1980).

68. J. Orenstein and M. Kastner, Phys. Rev. Lett. 46, 1421 (1981); for earlier work on the same subject by the same authors see Phys. Rev. Lett. 43, 161 (1979); J. Noncryst. Solids 35/36, 951 (1980).

69. F. W. Schmidlin, Phys. Rev. B 16, 2362 (1977).

70. J. Noolandi, Phys. Rev. B 16, 4466 (1977).

71. D. Monroe, M. Kastner and J. Orenstein, J. de Physique, C4, 559 (1981).

72. M. Bösch, Phys. Rev. Lett. 48, 1228 (1982); 48, 649 (1982).

73. C. T. Moynihan, Annals N.Y. Acad. Sci. 279, 15 (1976).

74. K. L. Ngai, "Evidences for Universal Behavior of Condensed Matter at Low Frequencies/Long Times," Proceedings of "Discussion Meeting on Non-Debye Relaxation in Condensed Matter," held at Indian Institute of Science, Bangalore, India, Sept. 1982, ed. T. V. Ramakrishnan (Special Publication of the Indian Academy of Sciences).

75. N. F. Mott, J. Non-Cryst. Solids 35/36, 1321 (1980).

76. M. A. Paalanen, T. F. Rosenbaum, G. A. Thomas and R. N. Bhatt, Phys. Rev. Lett. 48, 1284 (1982).

77. D. Vanderbilt and J. D. Joannopoulos, Phys. Rev. Lett. 49, 823 (1982).

# SURFACE STATE ELECTRONS ABOVE A LIQUID HELIUM FILM AND THE SURFACE POLARON PROBLEM

S. A. Jackson

Bell Laboratories, Murray Hill, NJ 07974

## A. Surface State Electrons on Liquid Helium

### Bulk Helium

Electrons on a liquid helium surface form a classical two-dimensional (2D) electron gas because an electron above the surface is bound in a hydrogenic-like well formed by the image potential the electron feels from the helium and repulsive barrier (due to the Pauli exclusion principle) felt by the electron a few Angstroms from the surface. With respect to motion perpendicular to the helium surface this potential gives rise to a series of S-like hydrogenic electron bound states but whose binding energies are four orders of magnitude less than the Rydberg energy (=13.6 eV).

This situation is depicted in figure (1), where z measures the electron's height above the liquid helium surface. For z>b (a few Angstroms), the potential is essentially given by

$$V_{image} = \frac{-Qe^2}{z} \tag{1}$$

where

$$Q = \frac{1}{4}\frac{(\varepsilon_2-\varepsilon_1)}{(\varepsilon_2+\varepsilon_1)} = \frac{1}{4}\frac{(\varepsilon_2-1)}{(\varepsilon_2+1)} \tag{2},$$

and $\varepsilon_2$, $\varepsilon_1$ are respectively the helium and vacuum dielectric constant. In hydrogenic limit ($V_o \to \infty$, $b \to o$), the bound state energies become

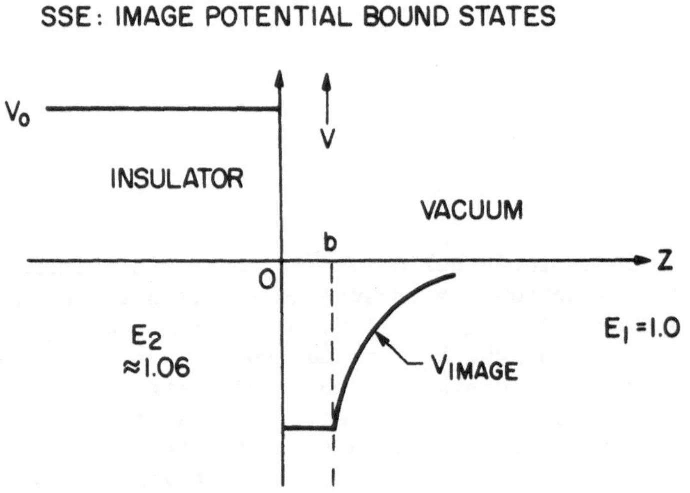


Figure 1. Effective potential seen by an electron above the surface of liquid helium.

$$E_z = \frac{-Q^2}{2n^2} E_H \qquad (3),$$

where $E_H$ = Rydberg = 13.6 eV. The ground state binding energy is then $\sim$ .7 meV. The binding energy is very small compared to the repulsive potential, $V_o$(= 1.0 eV), because Q $\tilde{\sim}$ $10^{-2}$ for electrons above a liquid helium film ($\varepsilon_2 \tilde{\sim}$ 1.06). Thus, there is little penetration of the electron wavefunction into the helium. This result holds even for the more correct form of the binding energies

$$E_z = \frac{-Q^2}{2K^2} E_H \qquad (4)$$

where[2] the eigenvalue K is obtained from continuity of the wave functions at z=b, and for $b \to 0$, $V_o \to \infty$ becomes an integer.

The hydrogenic approximation is thus sufficient.

The form of the wave function is

$$\psi = N e^{i\underset{\sim}{k}\cdot\underset{\sim}{r}} \phi(z) \qquad (5)$$

where (for the ground state),

$$\phi(z) = \text{constant} \quad z e^{-\gamma_o z} \qquad (6),$$

$$\gamma_o = \frac{mQe^2}{\hbar^2} \qquad (7),$$

and $\underset{\sim}{r}$ is the position vector of the electron in the plane of the liquid helium surface. $N_o$ is a normalization constant. For the lowest subband $\langle z \rangle$ $\sim$ 100 Å; for the second subband $\langle z \rangle$ $\sim$ 450 Å.

The energy level spacing for motion perpendicular to the surface is large compared to the temperature (at liquid helium temperatures) so that to a first approximation there is no motion in this direction, and attention can be focused on the in-plane dynamics of the carriers. In general the electrons are not localized in the plane of the surface and remain free to move parallel to the liquid helium surface except for coupling to the thermally excited ripples of the surface.

The existence of these image potential-bound, surface state electrons (SSE) was independently predicted by Sommer[1], Cole and Cohen[2], and by Shikin[3]. A number of experiments have confirmed their existence. These include mobility[4] and trapping lifetime[5] measurements; also cyclotron resonance[6,7], bound state spectroscopy[8] and plasmon dispersion and damping[9].

For low temperatures ($T < 0.7K$), electron-ripplon scattering limits the surface mobility, which has a limiting value of about $10^7$ $cm^2/V$-sec, and decreases as $1/T$ as the temperature increases. Above this temperature, scattering by helium gas atoms dominates and the mobility decreases exponentially with temperature[10].

The thermodynamic behavior of the 2D electron gas as a function of a real density of electrons, $n_s$, has been extensively discussed by Grimes[10]. At low $n_s$ and high temperature, the SSE behave as a 2D classical electron gas. For higher densities and lower temperatures, more liquid-like behavior is expected. At yet lower temperatures and higher densities, a 2D electron crystal has been observed to form[11].

## Thin Film Modifications

If, instead of bulk liquid helium, the electrons are placed atop a thin (∿100 Å) liquid helium film (see figure 2), the image potential is then modified and becomes[3].

$$V = \frac{-Qe^2}{z} - Q_1 e^2 \sum_{\ell=1}^{\infty} \frac{(-a)^{\ell-1}}{z+\ell d} \quad (8),$$

where the second term is an effective image potential (due to an infinite number of image charges) from the substrate. Here

$$Q_1 = \frac{\varepsilon_2(\varepsilon_3-\varepsilon_2)}{(\varepsilon_2+\varepsilon_1)^2(\varepsilon_2+\varepsilon_3)} \quad (9)$$

and

$$a = \frac{(\varepsilon_2-\varepsilon_1)(\varepsilon_3-\varepsilon_2)}{(\varepsilon_2+\varepsilon_1)(\varepsilon_3+\varepsilon_2)} \quad (10).$$

Here $\varepsilon_3$, $\varepsilon_2$, $\varepsilon_1$ are respectively the substrate, helium and free space(vacuum) dielectric constants. In general the series in equation (8) converges rapidly; hence, we may use

$$V = \frac{-Qe^2}{z} - \frac{Q_1 e^2}{z+d} \quad (11).$$

Then the ground state binding energy is approximately

$$E^{(0)} = \frac{-Q^2}{2} E_H - \frac{Q_1 e^2}{d} \quad (12).$$

The ground state wave function becomes

$$\phi(z) = \text{constant} \quad ze^{-\gamma z} \qquad (13),$$

where

$$\gamma = \gamma_o + \frac{3mQ_1 e^2}{2\hbar^2 (\gamma_o d)^2} \qquad (14).$$

Therefore the average distance, at low temperatures, at which the electron resides above the surface is now much smaller, $\langle z \rangle \sim 20$ Å; and the electron can approach the helium surface closely enough to perhaps form a dimple under itself (due again to the Pauli exclusion principle).

Variability of Electron-Surface Coupling

Even in the electron liquid regime, cyclotron resonance and mobility measurements give information about the one-electron properties of this system. For relatively thin (100 Å - 500 Å) liquid helium films, the one-electron properties of this ideal system are of interest because of the variability of the coupling of the electron to the surface ripplons. This coupling comes about because of the change in energy of the electron as it rides on surface waves in the presence of a perpendicular electric field, $\mathcal{E}$, which is made up of an image field and an applied field. The use of different substrates for the liquid helium film, of different film thicknesses, or different strengths of applied field allows a change in the effective coupling over several orders of magnitude.

For bulk helium (in zero external field), $E \simeq 10^2$V/cm, and the electron-ripplon coupling is weak. As the helium thickness is reduced to $\sim$ 100 Å, large image potential fields contribute so that the total field increases to $E \simeq 10^5$V/cm. This means that the electron-ripplon coupling is strong and new now-perturbative polaronic effects may occur, as hinted at in the previous section.

B. Electron Dimples on the Surface of Liquid Helium

Because of the Pauli exclusion principle, if an electron is close to the liquid helium surface, it will tend to form a dimple under itself. Any deformation of the surface may be analyzed in terms of ripplon modes. Ripplons are quantized capillary-gravity waves on the liquid helium surface.

The Liquid Hamiltonian

We treat the liquid helium as an ideal, incompressible, non-viscous semi-infinite liquid bounded by the plane $z = 0$. Then the

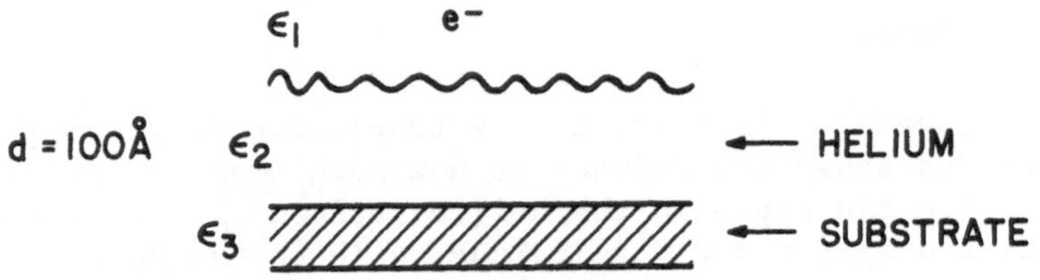


Figure 2. Electron above a thin film of helium which is on a substrate of dielectric constant, $\varepsilon_3$.

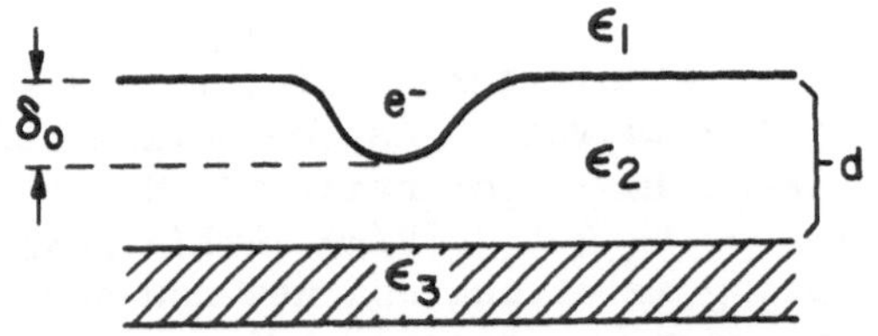


Figure 3. Surface deformation (and effective potential well) caused by an electron above the liquid helium surface (due to the Pauli Exclusion Principle).

liquid kinetic energy may be written (since liquid velocity, $\vec{v} = \vec{\nabla}\phi$),

$$T = \frac{1}{2}\rho\int(\vec{\nabla}\phi)^2 d\tau = \frac{1}{2}\rho\int d\vec{s} \cdot \phi\vec{\nabla}\phi \qquad (15)$$

where $\rho$ is the mass density of the liquid helium.
If $u(\underset{\sim}{r})$ represents the normal displacement of the surface (where $\underset{\sim}{r}$ = 2D vector in the plane of the flat surface), then the velocity associated with it is

$$\dot{u}(r) = [\vec{\nabla}\phi(\underset{\sim}{r}) \cdot \hat{n}]_{\vec{r} \text{ on } \vec{s}} \qquad (16).$$

The potential energy has contributions from body forces (gravity) and surface tension and may be written,

$$V = \int d^2\underset{\sim}{r} \int_{-\infty}^{u(\underset{\sim}{r})} dz\, \rho g z + \tau \int d^2\underset{\sim}{r}\, [1 + (\vec{\nabla}u)^2]^{\frac{1}{2}}$$

$$\approx \int d^2\underset{\sim}{r}\, [\frac{\rho g u^2}{2} + \frac{\sigma}{2}(\nabla u)^2] \qquad (17)$$

for long wavelength displacements.

We can expand $u(\underset{\sim}{r})$ in normal modes, $\underset{\sim}{k}$, which are quantized due to periodic boundary conditions. Then we have

$$u(r) = A^{-\frac{1}{2}} \sum_k Q_k\, e^{i\underset{\sim}{k}\cdot\underset{\sim}{r}} \qquad (18),$$

$$\dot{u}(r) = A^{-\frac{1}{2}} \sum_k \dot{Q}_k\, e^{i\underset{\sim}{k}\cdot\underset{\sim}{r}} \qquad (19).$$

If we define

$$\pi_{\underset{\sim}{k}} \equiv \frac{\rho}{k}\dot{Q}_{-\underset{\sim}{k}} \qquad (20),$$

then (with $|\underset{\sim}{k}| = k$),

$$H = T + V$$

$$= \sum_{\underset{\sim}{k}} \left\{ \pi_{\underset{\sim}{k}}\pi_{-\underset{\sim}{k}} \left(\frac{k}{2\rho}\right) + Q_{\underset{\sim}{k}}Q_{-\underset{\sim}{k}} \left[\frac{1}{2}(\rho g + \sigma k^2)\right] \right\} \qquad (21).$$

Quantizing these modes means,

$$[Q_{\underset{\sim}{k}}, \pi_{\underset{\sim}{k}'}] = i\hbar\delta_{\underset{\sim}{k},\underset{\sim}{k}'} \tag{22}$$

where

$$Q_k = \left(\frac{\hbar k}{2\pi\omega_k}\right)^{\frac{1}{2}} \left(a^+_{-\underset{\sim}{k}} + a_{\underset{\sim}{k}}\right), \tag{23a}$$

$$\pi_k = i\left(\frac{\hbar\rho\omega_k}{2k}\right)^{\frac{1}{2}} \left(a^+_{\underset{\sim}{k}} - a_{-\underset{\sim}{k}}\right) \tag{23b}.$$

The liquid Hamiltonian then has the usual form

$$H = \sum_{\underset{\sim}{k}} \hbar\omega_{\underset{\sim}{k}} (a^+_{\underset{\sim}{k}} a_{\underset{\sim}{k}} + \tfrac{1}{2}) \tag{24}$$

where the ripplon dispersion relation is

$$\omega^2_{\underset{\sim}{k}} = gk + \frac{\sigma}{\rho} k^3 \tag{25}$$

<u>The Effect of Surface Deformation on the Image Potential</u>

In the presence of surface deformation we cannot treat as simply as before the image potential from the helium. Surface deformation gives rise to an effective polarization potential

$$V(\underset{\sim}{r},z) = \frac{-Qe^2}{\pi} \int d^2\underset{\sim}{r}' \int_{-\infty}^{u(r')} \frac{dz'}{[(\underset{\sim}{r}'-\underset{\sim}{r})^2+(z'-z)^2]^2} \tag{26},$$

and Q is defined in equation (2). If $z \rightarrow \zeta + u(\underset{\sim}{r})$, where $\zeta$ measures the distance of the electron above the "flat" surface (see figure (3)), the polarization potential is

$$V \underset{\sim}{\sim} \frac{-Qe^2}{\zeta} - U(r,\zeta) \tag{27}$$

If, as is true here, the capillary wavelength $\lambda_c \left(= 2\pi\sqrt{\frac{\sigma}{\rho g}}\right) >> \gamma^{-1}$ (effective Bohr radius), then $U(r,\zeta) \rightarrow 0$, and we are left with the image potential as discussed earlier. We also have a contribution from the image potential from the substrate. Recalling equation (11),

$$\frac{-Q_1 e^2}{z+d} \simeq \frac{-Q_1 e^2}{(\zeta+d)} + \frac{Qe^2u}{(\zeta+d)^2} \qquad (28).$$

Remembering that $\langle\zeta\rangle \sim 20$ Å, the effective potential due to the perpendicular pressing fields is

$$V = eE_{\perp}\, u(\underset{\sim}{r}) \qquad (29)$$

where

$$E_{\perp} = E_{\perp\text{external}} + \frac{eQ_1}{(\langle\zeta\rangle+d)^2}$$

$$\simeq E_{\perp\text{external}} + \frac{eQ_1}{d^2} \qquad (30)$$

if $\langle\zeta\rangle \ll d$.

Collecting the results of equations 27, 28, and 29,

$$V = \frac{-Qe^2}{\rho} - \frac{Q_1 e^2}{(\rho+d)} + eE_{\perp}u(\underset{\sim}{r}) \qquad (31).$$

The z-motion problem goes through as before (with corrections to the level structure due to the pressing field), but we now have an explicit contribution to the motion parallel to the plane of the liquid helium due to the deformation of the liquid surface (recall that u(r) depends on the 2D positron vector, $\underset{\sim}{r}$, in the plane of the liquid helium surface.

The liquid Hamiltonian maintains the form given in equation (24), but now the ripplon dispersion relation is

$$\omega_k{}^2 = (g'k^2 + \frac{\sigma}{\rho}\, k^3)\ \tanh kd \qquad (32),$$

where $\rho$, $\tau$, and g' are respectively the density, surface tension, and acceleration of the liquid due to its Van der Waals coupling to the substrate, and where d = thickness of liquid helium film. The displacement normal mode amplitude becomes

$$Q_k = \left(\frac{\hbar k\ \tanh kd}{2\rho\omega_k}\right)^{\frac{1}{2}} \left(a^{+}_{-\underset{\sim}{k}} + a_{\underset{\sim}{k}}\right) \qquad (33).$$

C. The Surface State Electron Problem as a Polaron Problem

The interaction of electrons of mass, m, with the surface (or rather with ripplons of frequency $\omega_{\underset{\sim}{k}}$) can be described by the Hamiltonian

$$H = \frac{p^2}{2m} + \sum_{\underset{\sim}{k}} a^+_{\underset{\sim}{k}} a_{\underset{\sim}{k}} \hbar\omega_{\underset{\sim}{k}} + U \qquad (34).$$

In the presence of a strong pressing field $U = e\underset{\sim}{E}\cdot\underset{\sim}{u}$ with $\underset{\sim}{u}$ the displacement of the surface. When we expand in ripplon modes,

$$U = \frac{1}{A^{\frac{1}{2}}} \sum_{k} (a_{\underset{\sim}{k}} + a_{\underset{\sim}{k}}^{+}) e^{i\underset{\sim}{k}\cdot\underset{\sim}{r}} Q(k) \qquad (35)$$

$$\text{with } Q(k) = \left[\frac{\hbar k \tanh kd}{2\rho\omega_{\underset{\sim}{k}}}\right]^{\frac{1}{2}} e\underset{\sim}{E} \qquad (36).$$

Equation (35) is valid whenever the distortion of the helium surface is small compared to d and the forces from the image charge in the substrate dominate the forces arising from polarization of the helium. These two conditions certainly are well satisfied for 10 Å < d < $10^3$ Å. In this case

$$\underset{\sim}{E} = \frac{e^2}{4d^2}\left(\frac{\varepsilon-1}{\varepsilon+1}\right) + E_{\text{external}} \qquad (37),$$

for a substrate of dielectric constant ε. The ripplon frequency for such films (d ≃ 100 Å) is given by equation (32). For real substrate materials and for d ≃ 100 Å, $g'/g = 3.8\times10^8$ (g being the acceleration due to gravity). The capillary constant $k_c = (\rho g'/\sigma)^{\frac{1}{2}} = 4.24 \times 10^5\ \text{cm}^{-1}$.

The Hamiltonian in equation (34) is clearly recognizable as a polaron Hamiltonian. In this case however, our problem is a two-dimensional polaron problem. We will find that the dimensionality of the system coupled with the form of the ripplon dispersion relation will eliminate the possibility of divergences at long and short wavelengths.

The effective lattice "polarization" in this problem is due to the surface dimple formation. As such, this problem bears close resemblance to deformation potential-type polarons which have been discussed by a number of authors[12,13]. The electron-"phonon" coupling is embodied in Q(k), defined in equation (36). In fact, the natural coupling constant which emerges is[15]

$\alpha \equiv (eE)^2/8\pi\sigma\left(\frac{\hbar^2 k_c^{\ 2}}{2m}\right)$ . The greater the pressing field energy is relative to surface tension the stronger the coupling of the electron to the surface. This makes sense because α is essentially a measure of dimple depth, hence trapping potential depth.

For weak coupling we would expect the electron to move about freely on the surface with only a slightly enhanced mass. For sufficiently strong coupling however, the electron causes the medium to deform enough that the dimple forms an attractive potential for the electron, which then becomes trapped and thus sustains the lattice deformation. It is then self-trapped. Since the dimple potential is of short range and we know that there are only a finite number of bound states in such a potential, then effectively, at a value of the coupling constant α where the potential can have at least one bound state we expect the electron to become self-trapped. This then is the polaron state, which we proceed to investigate in more detail using the Feynman path integral techniques, introduced by a Schroedinger equation treatment of the problem.

## II. CALCULATION OF THE EFFECTIVE MASS AND GROUND STATE ENERGY USING PATH INTEGRAL TECHNIQUES

### A. Path Integral Techniques in Polaron Problems

Once we have the polaron Hamiltonian given in equations (34) and (35) we can proceed in one of two ways. We can go back to a first quantized form for the Hamiltonian written in terms of electron and liquid wavefunctions,

$$H = \int d^2\underset{\sim}{r}\left[(\underset{\sim}{\nabla}X(\underset{\sim}{r}))^2 + e\varepsilon_\perp\, u(\underset{\sim}{r})\,|X(\underset{\sim}{r})|^2\right] + H_L \qquad (38),$$

where

$$H_L = \int d^2\hat{r}\left[\frac{1}{2}\rho g'\, u^2(\underset{\sim}{r}) + \frac{1}{2}\sigma(\underset{\sim}{\nabla}u)^2\right], \qquad (39)$$

with all coefficients as defined previously, and $X(\underset{\sim}{r})$ the electron's wave function. We can then write down a set of coupled Schroedinger equations for the electron and liquid wavefunctions, which if solved, would tell us how a surface polaron or electron dimple forms, as a function of the parameters in the problem. Sander[16] has already published what amounts to a strong coupling calculation of the ground-state energy of this system, by assuming the existence of a helium dimple and using a trial variational form for it, where the dimple depth and shape are variational parameters, i.e. ($\delta_0$ and α are variational parameters).

$$u(\underset{\sim}{r}) = u(r) = \frac{-\delta_0}{\cosh^2 \alpha r} \tag{40}$$

If the ground state wavefunction is written as

$$X(r) = G(r)/\sqrt{r} \tag{41},$$

the Schroedinger equation becomes

$$\frac{d^2G}{dr^2} + \left(\frac{1}{4r^2} - \frac{2meE_\perp u}{\hbar^2}\right) G + \frac{2m\varepsilon_e}{\hbar} G = 0. \tag{42}$$

The ground state energy and wavefunction are

$$\varepsilon_e = \frac{-\hbar^2\alpha^2}{2m} \left[\left(\frac{1}{4} + \frac{2meE_\perp}{\hbar^2\alpha^2} \delta_o\right)^{\frac{1}{2}} - 1\right]^2$$

$$= \frac{-\hbar^2\alpha^2}{2m} (a-1)^2 , \tag{43}$$

$$G = D \tanh^{\frac{1}{2}} (\alpha r) \left[1 - \tanh^2(\alpha r)\right]^{(a-1)/2} \tag{44}$$

with a defined by equation (43). For a sufficiently thin film Sander finds a ground state consisting of an electron trapped in a dimple whose size is roughly the capillary length ($\lambda_o = k_c^{-1}$; for 100 Å films, $\lambda_o \simeq 250$ Å), and the binding energy is about 8K, with $\delta_o \approx 10$ Å.

However approach gives no information about the effective mass or about where in coupling constant the self-trapping transition occurs. So, if we wish to see what happens to the effective mass of the electron-helium system as we go from weak to strong electron-ripplon coupling, and if we wish to treat the finite temperature case, we must use the Feynman Path integral approach to the polaron problem.

The Feynman approach to the polaron problem consists of eliminating the phonon (ripplon) degrees of freedom from the problem in favor of a non-linear retarded interaction of the electron with itself. The transformation function (or for finite temperature - the partition function) can then be expressed as a path integral over the electron coordinate. This procedure has been extensively discussed in a number of places[17]. Therefore we will only outline how the calculations are done. The path integral arises in the calculation of the transformation function

$$K(r''t''; r't') = \langle r''t''|r't'\rangle \tag{45}$$

which, for t">t', is the probability amplitude for the system to be at the point r" at time t", if it was at r' at time t'. The function K satisfies the Schroedinger equation; thus,

$$K = \langle r'' \mid e^{-iH(t''-t')/\hbar} \mid r' \rangle \tag{46}$$

$$= \Sigma e^{-i\frac{En}{\hbar}(t''-t')} \phi_n(r'')\, \phi_n(r') \tag{47}$$,

where the $\phi_n$ are eigenstates of the system. If we know K, we, in principle, know the eigenvalues and eigenfunctions of H. If we let $\tau = i(t''-t') \rightarrow \infty$,

$$K \sim e^{-E_g \tau/\hbar} \tag{48}$$,

where $E_g$ is the ground state energy.

If, in terms of the electron's coordinates, we have $H = \frac{p^2}{2m} + V(\underset{\sim}{r})$, then it can be shown that

$$K(r'', t''; r't'') = \int \mathcal{D}\underset{\sim}{r}(t)\; e^{\frac{i}{\hbar}S} \tag{49}$$

where with the Lagrangian L derived from the Hamiltonian H,

$$S = \int_{t'}^{t''} dt\; L(x, \dot{x}, t) \tag{50}$$

Equation (49) is shorthand for

$$K(r'',t'';r't'') = \lim_{n\to\infty} \frac{1}{A} \int \cdots \int \frac{dx_1}{A} \cdots \frac{dx_n}{A} \exp \left\{ \frac{i}{\hbar}\epsilon \sum_{0}^{n} L(x_j, \dot{x}_j, t_j) \right\} \tag{51}$$.

Equation (49) is the path integral representation of the transformation function with boundary conditions r(t") = r" and r(t') = r'. A series of inequalities allows us to obtain a variational upper bound on the ground state energy, $E_g$. For large imaginary times, $\tau$, we have

$$\int e^{S}\, \mathcal{D}r(\tau) \underset{\tau\to\infty}{\sim} e^{-E_g\tau}$$

$$\geq e^{\langle S-S_o \rangle} \int e^{S_0}\, \mathcal{D}\underset{\sim}{r}(\tau) \tag{52}$$;

with

$$<F> \equiv \frac{\int e^{S_0} F \, \mathcal{D}\underset{\sim}{r}(\tau)}{\int e^{S_0} \, \mathcal{D}\underset{\sim}{r}(\tau)} \tag{53}$$

Therefore

$$E_g \leq E = E_0 - s \quad , \tag{54}$$

where

$$\int e^{S_0} \, \mathcal{D}\underset{\sim}{r}(\tau) \sim e^{-E_0 \tau} \tag{55}$$

and

$$<S - S_o> \sim s\tau \tag{56}$$

The Feynman variational approach then consists of calculating $E_0$, from some trial action, $S_0$, which is also used to calculate $<S-S_o>$. The variational energy, E, is then minimized with respect to the parameters which specify the trial action.

The ripplon coordinates are eliminated by calculating the transformation function in which the helium is in the uncoupled ripplon vacuum at $\tau'$ and again at $\tau''$[17]. The transformation function is given in equation (49), where now the action takes the form[18],

$$S = \frac{-m}{2}\int_0^\infty \left\{\frac{d\underset{\sim}{r}(\tau)}{dt}\right\}^2 + \frac{1}{2}\sum_{\underset{\sim}{k}} |Q(k)|^2 \iint_o^{\tau_o \to \infty} d\eta \, d\rho \, e^{i\underset{\sim}{k}\cdot[\underset{\sim}{r}(\eta)-\underset{\sim}{r}(s)]}$$

$$x e^{-\hbar\omega_k \tau} \tag{57}$$

If, as is usual, we employ a quadratic trial action, $S_0$, based on a model of the electron interacting with a single particle of mass M via a spring with spring contact $\kappa$, we have two variational parameters (v, w) defined by $\kappa/m = (v^2-w^2)$ and $\frac{M}{m} = \frac{(v^2-w^2)}{w^2}$.

The Lagrangian is

$$L = \frac{1}{2}m \, \frac{d\vec{x}}{dt}^2 + \frac{1}{2}M \, \frac{d\vec{y}}{dt}^2 - \frac{1}{2}\kappa \, (\underset{\sim}{x}-\underset{\sim}{y})^2 \, , \tag{58}$$

where $\vec{x}$ is the 2D coordinate of the electron, and $\vec{y}$ is the 2D coordinate of the fictitious particle.

B. Ground State Energy and Effective Mass

The trial energy of the system in this two parameter model is given by

$$E = \frac{1}{2}\frac{(v-w)^2}{v} - A \ , \tag{59}$$

where (in the units $m = \hbar = 1$)

$$A = \int d\tau \int \frac{d^2k}{(2\pi)^2} \, |Q(k)|^2 e^{-\omega_k \tau} \, e^{-\frac{k^2}{2}F(\tau)} \quad . \tag{60}$$

In equation (60)

$$F(\tau) = \frac{v^2-w^2}{v^3}(1-e^{-v\tau}) + \frac{w^2}{v^2}\tau \tag{61}$$

is the time dependent response function of the two oscillator system described by equation (58), and Q(k) is the coupling function defined in equation (36).

Because of the form of the phonon dispersion and the complexity of the coupling Q(k) the evaluation of the integrals except at strong coupling must be done numerically. However, we found by a careful examination of the analytic form and numerical results for ($d \simeq 100$ Å) that the problem is well approximated by assuming[15] $\tanh kd \simeq kd$ and,

$$\omega_k = sk, \ s = (g'd)^{\frac{1}{2}}, \ k < k_c \ . \tag{62}$$

In this cutoff approximation the integrals over k and $\tau$ may be performed analytically in the strong-coupling and weak-coupling limits.

We define a coupling constant $\alpha \equiv (eE)^2/[8\pi\sigma(\hbar^2 k_c^2/2m)]$ - its significance will be elaborated upon later; energy in units of $k_c^2/2$. In the strong-coupling limit ($\alpha \to \infty$) $v \sim \alpha^{\frac{1}{2}}$, $w/v \to 0$, and $F(\tau) \simeq 1/v$. Thus

$$E = \frac{1}{2}v - \alpha v(1 - e^{-1/v}) \tag{63}$$

minimizing with respect to v yields $v = \alpha^{\frac{1}{2}}$ and

$$E = -\alpha + \sqrt{\alpha} + \ldots \ ; \tag{64}$$

i.e., the strong-coupling limit [$F(\tau) = 1/v$] gives a power series in $\sqrt{\alpha}$. It is important to point out that there is no minimum for

$\alpha < \frac{1}{2}$. This is true because the range of the strong-coupling effective potential is short and a critical value of the coupling constant is needed to bind an electron in its well. The corrections due to the time dependence of $F(\tau) \simeq 1/v + w^2/v^2\tau + (1/v)e^{-v\tau}$ arise from two physically distinct effects: recoil $(w^2\tau/v^2)$ and internal excited states $[(1/v)e^{-v\tau}]$. A minimization of the energy including these two terms to lowest order shows that another physically relevant parameter $\eta \equiv \omega_c/(k_c^2/2)$ comes in. (For 100 Å films $\eta \approx 5\times10^{-3}$.) The energy

$$E_{sc} = (-\alpha + \alpha^{\frac{1}{2}} + \ldots) - \frac{19}{40}\eta - \frac{9}{16}\frac{\eta^2}{\alpha^{\frac{1}{2}}} + \ldots \tag{65}$$

The term in parentheses is identical to Eq. (64) and comes from setting $F(\tau)=1/v$. The term linear in $\eta$ comes from corrections due to the internal excitation energy, i.e., $e^{-v\tau}$, and the $\eta^2$ term comes from the recoil of the entire object. The parameter $\eta$ obviously measures the degree of abiaticity of the system, i.e., the maximum frequency with which the ripplons can respond to the motion of the electron.

In the weak-coupling regime $v/w \to 1$; i.e., $v/w \equiv 1+\varepsilon$. To order $\varepsilon$

$$F(\tau) = \frac{w^2}{v^2}\tau + \frac{2\varepsilon}{v}(1-e^{-v\tau}) \tag{66}$$

and

$$E_{wc} = -(\alpha\eta) - \frac{\pi^2}{2}(\alpha\eta)^2 + \ldots \tag{67}$$

In the weak coupling limit the effective coupling constant is reduced by $\eta$. Thus the weak-coupling expansion $(\eta \simeq 10^{-3})$ appears to be valid for $\alpha >> 1$. However, we know from our strong-coupling results that the system will essentially switch from a quasifree object to a self-trapped object at $\alpha \sim 1$.

The numerical results for the energy are displayed along with the approximate strong- and weak-coupling results in figure (4). The sharp change at $\alpha \simeq 1/2$ is evident. When the polaron is weakly coupled it is delocalized, and we expect that its mass will be of order 1. When it is strongly coupled, it is localized, and its mass will be of the order of several helium-atom masses. Since the transition in coupling constant is extremely rapid, we would expect an even more dramatic variation of the effective mass. In figure (5) we plot the model mass $m_o = v^2/w^2$ and the so-called Feynman mass (which comes from calculating the energy as a function of velocity)[17], or the response of the system to an applied field[18,19],

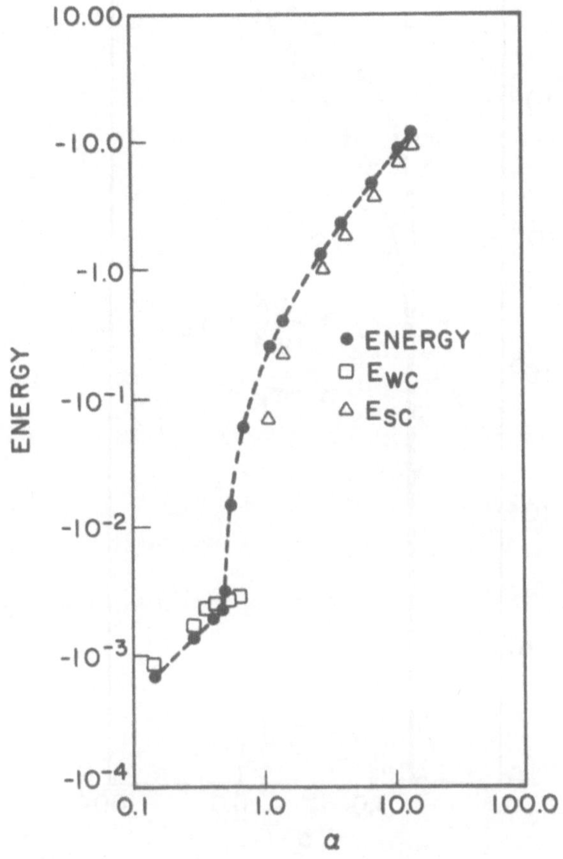


Figure 4. Energy in units of $\hbar^2 k_c^2/2m$ vs. coupling constant α. Points are numerical results; lines are guides to the eye.

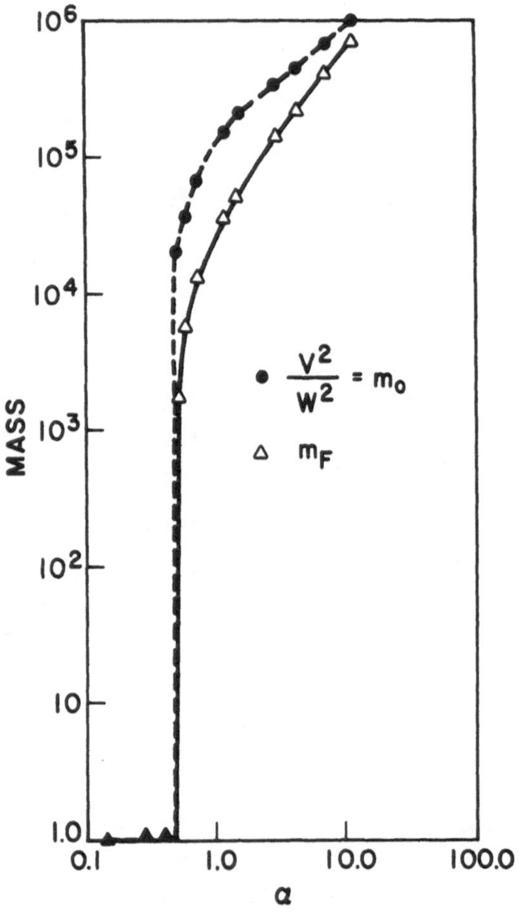


Figure 5. Model mass ($m_o$) vs. coupling constant α. Feynman mass ($m_F$) vs. coupling constant α. Mass in units of free electron mass. Points are numerical results; lines are guides to the eye.

$$m_F = 1 + \frac{(e\delta)^2 d}{8\pi\rho s} \int dq q^4 \int d\tau \tau^2 e^{-w_q \tau} e^{-q^2/2F(\tau)} \tag{68}$$

The strikingly rapid four-orders-of-magnitude change in this quantity for a <10% change in $\alpha$ is evident.

We have shown that the existence of a "localization" transition in an ideal system which is, nonetheless, physically realizable. Such a rapid transition from a quasi-free to a quasi-localized state has been discussed for deformation potential systems by Toyozawa[12], Toyozawa and Shinozaka[13] and others[14, 23]. (See discussion below). We refer to the transition as "localization" because of the rapidity and size of the change in the effective mass over a very narrow range of coupling constant (<10%). This implies that in a mobility measurement, the diffusion constant would decrease very rapidly. In fact, in the strong-coupling limit, the model mass is proportional to $\alpha/\eta^2$, and the width in coupling constant over which is transition occurs is $\Delta\alpha \sim e^{-1/\eta}$. As $\eta \to 0$, this mass $\to\infty$ and the magnitude of the jump in mass at the transition becomes larger and the transition occurs over a narrower range of $\alpha$. Therefore, the transition looks more and more like a real localization transition.

The beauty of the present system is threefold. First, the system corresponds very closely to a continuum electron-phonon model. Secondly, the value of the coupling constant where the transition occurs is in a physically accessible region. Finally, the coupling constant is variable by changing the substrate or by changing the film thickness in the presence of an applied field.

These calculations have assumed a single electron at zero temperature. The validity of these assumptions will depend on the experimental conditions. The scale of the energy $\hbar^2 k_c^2/2m$ for $d \simeq 100$ Å is for the strongly coupled state 8 K. Therefore, we would expect our model to be valid for $n < 10^8$; i.e., the inter-particle Coulomb energy $V_c \equiv (e^2/\varepsilon^2) n^{\frac{1}{2}} \simeq 1$ K ($n = 10^8$), and for $T < 1$ K.

Our results agree with a two-dimensional deformation potential [12,14] treatment of the problem in which the total energy of the system (minimized with respect to the dilation of the medium, $\Delta$) is

$$E = \frac{h^2}{mR^2} - \frac{D^2}{4CR^2} \tag{69}$$

where R is the radius of the deformation which traps the electron, C is an elastic constant and D is an effective electron-phonon coupling. This suggests that there is a minimum electron-phonon coupling (D) required for the electron to become self-trapped.

Our results also agree with Toyozawa's[13] stability index ($\sigma$) discussion of an electron in a deformable lattice. If $\sigma>0$, there is stiff stability of the quasi-free state with a discontinuous transition to the self-trapped state. The stability index is defined as

$$\sigma \equiv d - 2\lambda - 2 \tag{70}$$

where d is the dimensionality of the system and $\lambda$ reflects the range of the electron-phonon interaction. If the interaction Hamiltonian

$$H_L \sim \sum_k \gamma_k \, e^{ik\cdot r} \tag{71}$$

and

$$\gamma_k \sim \gamma_o \left(\frac{k_o}{k}\right)^{\lambda} \quad \text{as } k \to 0 \tag{72}$$,

the $\lambda<0$ means a short-ranged interaction while $\lambda>0$ means a long ranged interaction. For our helium surface polaron, $\lambda=-\frac{1}{2}$, d=2 and $\tau=1$. There should be (and is) a discontinuous transition to the self-trapped state. Toyozawa's treatment emphasizes the importance of the long wavelength behavior of the electron-phonon interaction. The longer the range of the interaction relative to the dimensionality of the system, the more continuous will be the transition to the self-trapped state.

C. Temperature Dependent Effective Mass

It is of obvious interest to investigate what happens at finite temperatures. Does the dimple-like state "melt" at some temperature. If so, how rapidly does this occur? We now proceed to answer these questions[20].

As in the T=0 problem the Hamiltonian is

$$H = \frac{p^2}{2m} + \sum_{\underset{\sim}{k}} a_{\underset{\sim}{k}} + a_{\underset{\sim}{k}} \, h\omega_{\underset{\sim}{k}} + U \tag{34}$$

where U is defined by equations (35) and (36). In path integral form, the free energy is defined by

$$e^{-\beta F} = Z_r \int \mathcal{D}r(\tau) \, e^{S} \tag{73}$$

where $z_r$ is the free ripplon partition function and S is the action arising from the Hamiltonian in equation (34). If $S_o$ is a trial action, then the variational principle for the free energy is

$$F \leq F_r + F_0 = kT \langle S-S_o \rangle \quad (74),$$

where the average < > is defined by equation (53). We again pick $S_0$ as arising from the harmonic oscillator Lagrangran in equation (58), with variational parameter defined as before. The trial free energy is then

$$F = F_0 - B - A \quad (75)$$

where

$$F_0 = \frac{2}{\beta} \ln \left[ \frac{\sinh\left(\frac{\beta v}{2}\right)}{\sinh\left(\frac{\beta w}{2}\right)} \right] - \frac{2}{\beta} \ln\left(\frac{v}{w}\right) \quad (76a)$$

$$B = \frac{(v^2-w^2)}{2v} \left[ \coth\left(\frac{\beta v}{2}\right) - \frac{2}{\beta v} \right] \quad (76b)$$

$$A = \int_0^\beta d\tau \int \frac{d^2k}{(2\pi)^2} |Q(k)|^2 e^{-k^2/2\, F(\tau)}$$

$$x \left[ (N+1)\, e^{-\omega_k \tau} + N e^{\omega_k \tau} \right] . \quad (76c)$$

and where

$$N = 1/(e^{\beta\omega_k}-1).$$

Here

$$F(\tau) = \frac{w^2}{v^2} \tau (1 - \tau/\beta)$$

$$+ \frac{v^2-w^2}{v^3} \left[ 1-e^{-v\tau} + \frac{2(1-\cosh v\tau)}{(e^{\beta v}-1)} \right] \quad (77)$$

is the temperature-and time-dependent response function of the system. It should be noted that we do not include in equation (76) the free particle free energy and free ripplon free energy ($F_r$) because they do not depend on the variational parameters (v,w) which determine the effective mass (see below). However, in order to make any detailed comparison of the total free energy with the ground state energy, they would have to be accounted for.

It is possible to define an effective electron mass for this system in the following way. It is well known[18,19] that if a free particle is put into a constant external field, f, then the free energy changes by an amount

$$F^{(m)} = -\frac{\hbar^2\beta^2 f^2}{24m^*} \tag{78}$$

If we then apply a constant external field to our system, we can define the effective electron mass as

$$\frac{1}{m^*} = \frac{-24}{\beta^2}\frac{1}{\hbar^2}\lim_{f^2\to 0}\frac{\partial F}{\partial f^2} \tag{79}$$

which becomes, when the path integrals are performed,

$$\frac{m^*}{m} = \left[\frac{w^2}{v^2} + \left(1 - \frac{w^2}{v^2}\right)\frac{6}{\beta v}\left(\coth\left(\frac{\beta v}{2}\right) - \frac{2}{\beta v}\right)\right]^{-1} \tag{80}$$

which reduces to the model mass, $\frac{v^2}{w^2}$, at T=0.

We define a reduced inverse temperature $x_0 = \beta v_c$. We minimize the free energy in equation (76) and use the values of v and w so obtained to calculate the effective mass from equation (80) for various values of $x_0$ and $\alpha$.

The results of our minimization are given in figures (6) and (7), where we have plotted the effective mass (from equation (80)) versus $\alpha$ for various temperatures, $x_0$; and versus $x_0$ for various values of $\alpha$. The value of $x_0$=100 corresponds to T=17 mK since $v_c$=1.7K for a 100 Å thick helium film. We also have, for comparison, our earlier results for the zero temperature mass in figure (5). The figures show that the mass decreases toward the free electron mass for all values of the coupling constant as the temperature increases (as expected). For high temperature ($x_0$=1) the mass remains free electron-like even as the coupling constant increases. Figure (6) indicates that the strongly-coupled dimple-like state has already "melted" for $\frac{kT}{E_B} < 1$ ($E_B$[21] is the zero temperature binding energy of the localized state and at $x_0 = 100$, $\frac{kT}{E_B} \sim 0.75$); and the sharp transition at $\alpha \approx \frac{1}{2}$ has been considerably broadened and reduced in magnitude (compare figures (6) and (5)).

We can obtain an approximate expression for the effective mass in the strongly coupled state, at very low temperature, by using in equation (80) the values of the variational parameters

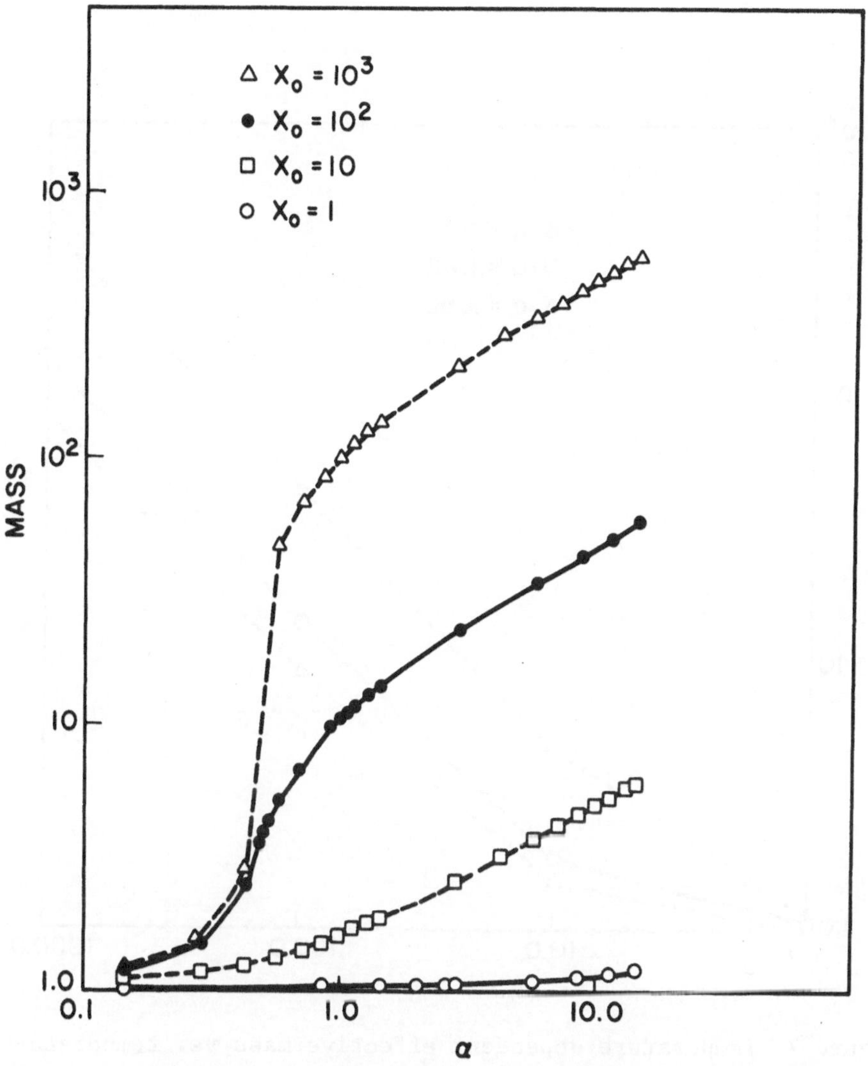


Figure 6. Temperature dependent effective mass vs. coupling constant α for various temperatures, $x_0 = k_c^2/2kT$. Points are numerical results; lines are guides to the eye.

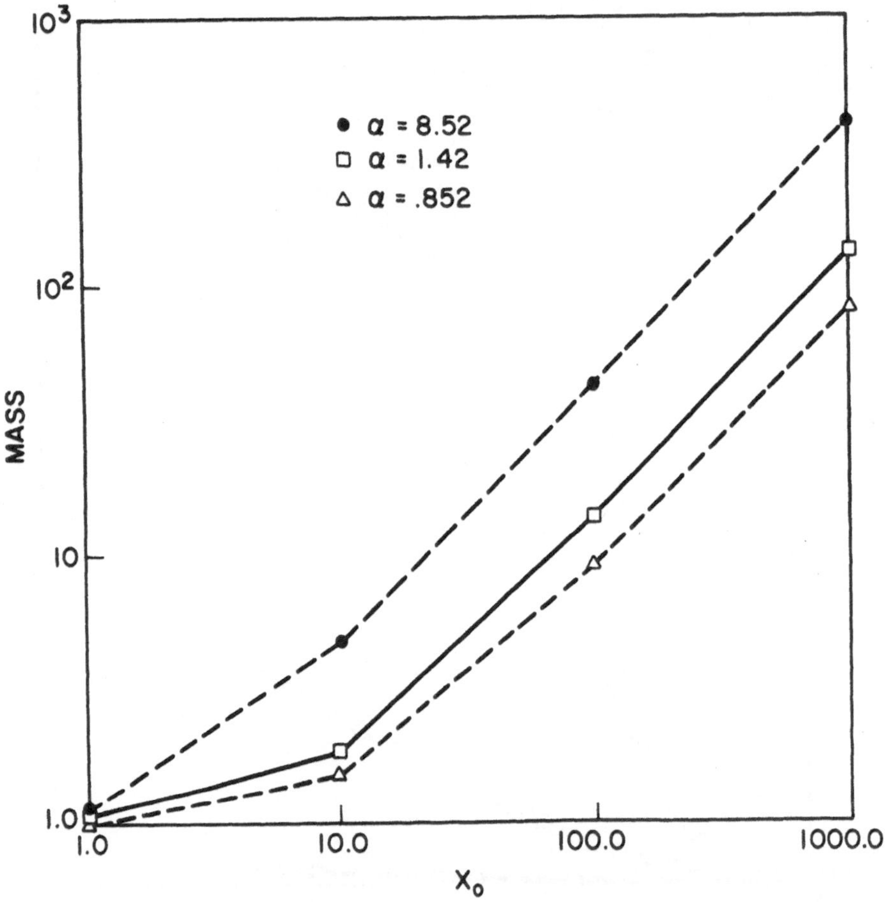


Figure 7. Temperature dependent effective mass vs. temperature for various coupling constants.

obtained from the T=0 results[15] with corrections from recoil and internal excited states included. We obtain

$$\frac{v}{v_c} = \sqrt{\alpha} \tag{81}$$

or

$$\beta v = x_o \sqrt{\alpha} \tag{82}$$

and

$$\lambda^2 = \frac{w^2}{v} = \frac{9}{4}\frac{\eta^2}{\alpha}\left(1-3\frac{\eta}{\sqrt{\alpha}}\right) \tag{83}$$

where $\eta = \omega_o/k_c{}^2/2$ ($\omega_c = sk_c$ for the linearized ripplon dispersion relation). Then

$$\frac{m^*}{m} \sim \frac{1}{\left(\lambda^2 + \frac{6}{x_o\sqrt{\alpha}}\right)} \tag{84}$$

Even though the range of validity of this expression is for a somewhat lower temperature than we have investigated numerically, it does show the following. The zero temperature model mass, $\lambda^{-2}$, (being proportional to $\alpha/\eta^2$) $\to \infty$ as $\eta \to 0$. However, for finite temperature, the effective mass as we have defined it here remains finite because of the temperature dependent correction term indicated in equation (84). Therefore, the jump in mass, at $\alpha \sim \frac{1}{2}$, remains finite as $\eta \to 0$ (i.e. as the ripplon velocity vanishes) and the "localization" features of the transition are smeared.

The fact that the dimple-like structure has begun to melt at T=17mK (for a 100 Å film), diminishing the jump in mass, indicates that the temperature has a big effect and that experiments must be done at quite low temperatures. However, this temperature is not unreasonable, suggesting that experiments to look for the "localization" transition described here and in our earlier work[15] are difficult, but possible.

## III. MOBILITY OF POLARONIC ELECTRONS ON LIQUID HELIUM FILMS

If a frequency dependent electric field is applied to our SSE on a liquid helium film, then a mobility measurement would tell us how the energy which the electric field imparts to an electron on the helium surface is dissipated. In our almost ideal system (no defects, etc.) the mobility is limited by ripplon scattering at low temperatures. The ripplons play two roles. At low frequencies of the applied field, the ripplons provide a thermal reservoir, which interacts with the particle (the polaron in the strong coupling regime), moving it from place to place. Energy is dissipated by diffusion, and the carriers are brought into thermal equilibrium at the original temperature of the medium. At higher frequencies different phenomena may occur. For strong electron-ripplon coupling the electron is in the self-trapped state; it is bound in a potential which it has created by distorting the medium around it. If the distortion were fixed, the electron would truly be in a potential well and would then have various excited states in that potential. In this case, however, the medium moves, giving a finite lifetime to the states. The lifetime of the states depend on the ratio of the phonon frequencies to the excitation frequencies ($\sim\omega_c/\omega_{ex}$). If this ratio is small i.e., for a deep, adiabatic well, the excited states would still be fairly well defined. Therefore we would, in general, expect the lifetimes also to depend on the size of the electron-phonon coupling. We also would expect there to be resonances in the dissipative part of the impedance at frequencies corresponding to excitation energies in the well.

If we want the response of the system for arbitrary electron-ripplon coupling strength, for arbitrary temperature and for all frequencies of the applied fields; this necessitates a density matrix approach, again handled by path integral techniques. Since our acoustic-type ripplon modes are easily excited, each of the above-mentioned conditions may mean that many phonons are interacting with the electron most of the time, with possible interferences between them[22].

### A. Path Integral Treatment

Our path integral treatment follows that of Feynman, Hellwarth, Iddings and Platzman[22], which we briefly outline. If a frequency dependent electric field, $E=E_o e^{i\nu t}$, is applied to the 2D electron system in the plane of the helium surface, the object of interest is the impedance function, $z(\nu)$, defined from the current induced by the motion of the electron

$$j(\nu) = \frac{1}{[z(\nu)]} E_o e^{i\nu t} \quad . \tag{85}$$

Since $j = \langle \dot{x} \rangle$, then,

$$\langle x \rangle = \frac{E}{i\nu z(\nu)} \tag{86}$$

(where < > means expectation value) which gives for the average position of the electron at time $\tau$,

$$\langle x(\tau) \rangle = -\int_{-\infty}^{\infty} i\; G(\tau-\sigma) E(\sigma) d\sigma \tag{87}$$

The fourier transform of $G(\tau)$ is directly related to the impedance

$$G(\nu) = \frac{1}{[\nu z(\nu)]} \;. \tag{88}$$

Since

$$\langle x(\tau) \rangle = \mathrm{Tr}\;[x\; U(\tau,a)\; \rho_a\; U'^{-1}(\tau,a)] \tag{89}$$

where $\rho_a$ is the density matrix, and

$$U(\tau,a) = \exp\{\; -i\int_a^\tau [H_s \sim \underset{\sim}{X}_s \cdot \underset{\sim}{E}(s)]\}ds \tag{90}$$

is the unitary time development operator for the complete Hamiltonian $H - \underset{\sim}{X} \cdot \underset{\sim}{E}$. A time ordered operator notion is used, with all unprimed operators to the left, latest times being farthest to the left, then the matrix $\rho$, then all primed operators to the right of $\rho$, with the last time being farthest to the right. Primed operators are ordered oppositely to the unprimed.

Then using the fact that $-iG(\tau-\sigma)$ is the response to a delta function field, $E$,

$$E(s) = E\delta(s-\sigma) + \eta\delta(s-\tau) \tag{91}$$

$$E'(s) = E\delta(s-\sigma) - \eta\delta(s-\tau) \tag{92}$$

and picking for the density matrix

$$\rho_a \sim e^{-\beta \sum_{\underset{\sim}{k}} \hbar\omega_{\underset{\sim}{k}}\, a^{+}_{\underset{\sim}{k}} a_{\underset{\sim}{k}}} \tag{93}$$

(due to the thermal reservoir assumptions about the ripplons discussed earlier), we may derive

$$G = \frac{1}{2} \frac{\partial^2 g}{\partial\eta\partial\varepsilon} \bigg|_{\eta = \varepsilon = 0} \tag{94}$$

where g is a response function

$$g = T_r \left[U(b,a) \ \rho_a \ U'^{-1}(b,a)\right] \tag{95}$$

(for $b \to \infty$, $a \to -\infty$), which is derivable from a double path integral

$$g = \int\int e^{i\phi} \mathcal{D}\underset{\sim}{X}(\tau) \ \mathcal{D}\underset{\sim}{X}'(\tau) \tag{96}$$

If we use a trial action, $\phi_o$, based on a harmonic oscillator Lagrangian (see equation (58)) and expand the path integral as before, we obtain for the impedance (in terms of a memory function)

$$-i\nu z(\nu) = \nu^2 - X(\nu) \tag{97}$$

where

$$X(\nu) = \int_0^\infty [1-e^{i\nu u}] \mathrm{Im} S(u) du \ , \tag{98}$$

$$S(u) = \int \frac{d^2\underset{\sim}{k}}{(2\pi)^2} \ |Q(k)|^2 k^2 e^{-\frac{k^2}{2} D(u)} [e^{i\omega_k u} + 2P(\beta\omega_k)\cos\omega_k u] \tag{99}$$

and

$$D(u) = \frac{v^2-w^2}{v^3} \left[1-e^{ivu}+4P(\beta v)\sin^2\left(\frac{vu}{2}\right)\right] + \frac{w^2}{v^2} \ [-iu+u^2/\beta] \tag{100}$$

$X(\nu)$ contains all the information about electron-ripplon coupling. Unlike the longitudinal optical phonon (in 3D) case, there exists in this case dissipation for any frequency and for T=0. The variational parameters v, w are as defined in the effective mass calculation and are chosen to be those which minimize the free energy (or ground state energy for T=0).

At very low frequencies (for T=0) we have

$$-i\nu z(\nu) = \nu^2 \left[1+ \frac{(eE_\perp)^2 d}{8\pi\rho s} \int_0^{k_c} dk k^4 \int_0^\infty d\tau\tau^2 e^{-\omega_k\tau} e^{-\frac{q^2}{2} F(\tau)}\right] \tag{101}$$

where $F(\tau)$ is given in equation (61). The expression in brackets is recognized to be the Feynman polaron mass $m_F$, mentioned in section II. Hence

$$-i\nu z(\nu) = \nu^2 m_F \tag{102}$$

and from equation (86) the electron's displacement is

$$\langle x\rangle \sim \frac{E}{\nu^2 m_F} \quad . \tag{103}$$

Therefore we expect a sharp drop in the value of the measured diffusion constant at the value of the electron-ripplon coupling where the transition to the self-trapped state occurs (see figure (5)).

At higher frequencies other effects of electron-ripplon coupling come into play. We are particularly interested in the internal structure of the polaron, i.e., the possible presence of polaron excited states and the effect of shape changing transitions on their widths. By shape changing transitions we mean changes in the structure of the polaron well itself. Since these are dissipative effects, we focus in the strong coupling regime on $\mathrm{Im}X(\nu)$, where $X(\nu)$ is defined in equation (98). Following our introductory discussion we seek answers to the following questions. Does $\mathrm{Im}X(\nu)$ exhibit a resonance structure in $\nu$ corresponding to polaron excited states? How do the widths of the resonances depend on the coupling constant $\alpha$ and the parameter $\eta$? How does this problem compare with a conventional bound state problem?

Using equation (98), we obtain

$$\mathrm{Im}X(\nu) = \int_0^\infty du \sin \nu u \; \mathrm{Im}S(u) \tag{104}$$

where $S(u)$ is defined in equation (99). We consider the case $T=0$ which should (at sufficiently low temperatures) give us the answer to the questions we posed. Then $\lim_{\beta\to\infty} S(u) = S_o(u)$, where

$$S_o(u) = \int \frac{d^2k}{(2\pi)^2} \, |Q(k)|^2 k^2 e^{-\frac{k^2}{2} D_o(u)} e^{i\omega_k u} \tag{105}$$

and

$$D_o(u) = \frac{v^2-w^2}{v^3}(1-\cos vu) - i\left[\frac{v^2-w^2}{v^3}\sin vu + u\,\frac{w^2}{v^2}\right] \tag{106a}$$

$$= D_0^R(u) + i\, D_0^I(u) \; . \tag{106b}$$

From equations (105) and (106), we obtain

$$\mathrm{Im}S_0(u) = \int \frac{d^2\underset{\sim}{k}}{(2\pi)^2} \, |Q(k)|^2 k^2 e^{-\frac{k^2}{2} D_0^R(u)} \; E \tag{107}$$

where

$$E = \sin\left\{\frac{k^2}{2}\left[\frac{v^2-w^2}{v^3}\,\sin vu + \frac{w^2}{v^2}\,u\right] + \omega_k\, u\right\} \tag{108}$$

Up to constant factors

$$\mathrm{Im}X(\nu_o) \simeq \frac{v_o^3}{\lambda^2(1-\lambda^2)^2} \sum_m \frac{1}{m!} \left[\frac{y_1^2}{v_0}(1-\lambda^2)\right]^{m+2} \exp\left\{-\left[\frac{y_1^2}{v_0}(1-\lambda^2)\right]\right\}$$

$$\times \frac{1}{2\sqrt{\left(\frac{\eta}{2\lambda^2}\right)^2 + \frac{(\nu_o - mv_0)}{\lambda^2}}} \; , \tag{109}$$

where $y_1 = \sqrt{\left(\frac{\eta}{2\lambda^2}\right)^2 + \frac{(\nu_o - mv_0)}{\lambda^2}} - \frac{\eta}{2\lambda^2}$ (110)

The results implied by equation (109) are shown in figure (8). Equation (109) clearly illustrates the following. The dissipative part of the impedance consists of a series of resonances with harmonic oscillator spacing (with $v_0$ being the fundamental frequency, in units of $k_c^2/2$). The widths of the resonances (reflecting the lifetime of the excited states) are proportional to $\eta$. As the coupling constant approaches the value where the mass transition (illustrated in figure (5)) occurs, there are more peaks in a given frequency range, suggesting that a continuous form for Im X is approached, probably reflecting the fact that the shape of the well is changed rapidly as the transition is approached. A continuum of phonon frequencies would then be expected to contribute. The peaks diminish in intensity as the frequency increases. This reflects the fact that the higher lying (excited) states are less localized than the ground state. The general results of 1) the existence of resonances, 2) their widths, 3) their intensity and 4) their spacing as the mass transition is approached in coupling constant follow what we would expect to happen in a system like this. The exact harmonic oscillator spacing of the peaks is an artifact of our having used a harmonic oscillator trial action. The harmonic oscillator

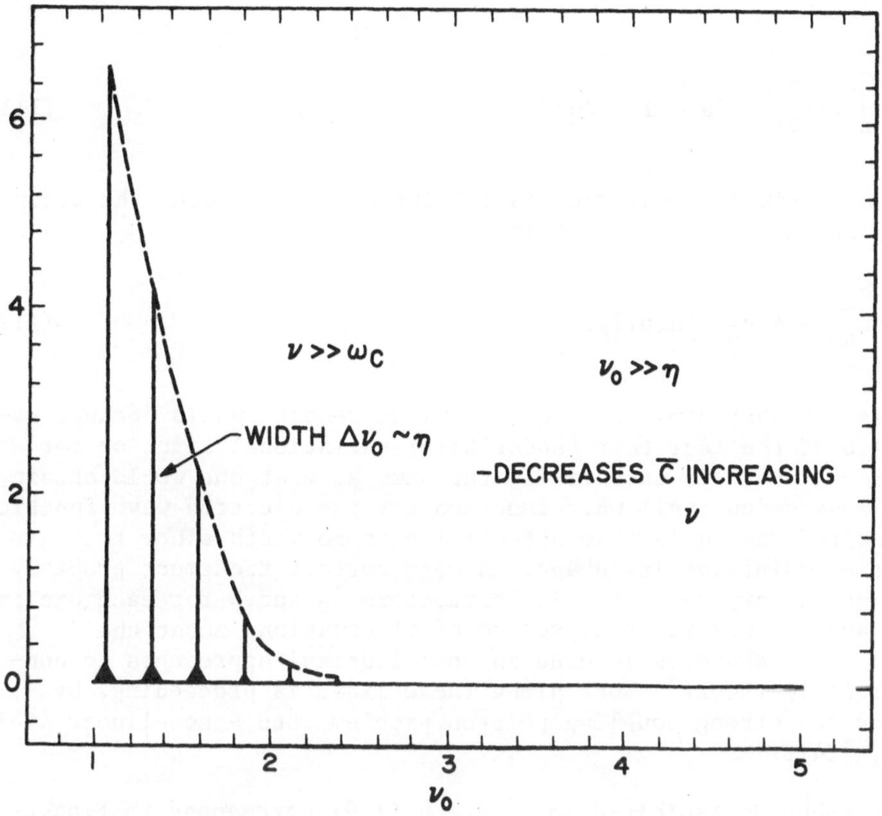


Figure 8. $ImX(\nu)/\nu$ vs. $\nu$ (in units of $\nu_c = \hbar^2 k_c{}^2/2m$) for $\alpha = 0.4$, $\eta = 10^{-3}$.

parameters can be varied to give a good fit to the ground state. We would expect it to be less good for the excited states which are more sensitive to the range of the potential (the harmonic oscillator potential is of infinite range).

One might expect that a variational potential such as that used in equation (40) would perhaps give a better treatment of the excited states since it is of finite range. If we solve for the excited states, the solutions are Gegenbauer polynomials in $[1+\sinh^2(\alpha r)]$. The energy levels are given by

$$\varepsilon_n = \frac{-h^2\alpha^2}{2m} (a - 1 - 2n)^2 \tag{111}$$

where a is given in equation (43). The spacing between the ground state and the excited states is

$$\Delta\varepsilon_{no} \approx 4nv_o - 4n(n+1). \tag{112}$$

This is somewhat similar to our previous result and is perhaps reflective of the fact that Sander's[16] variational solution for the ground state is essentially the same as what one would obtain using a Gaussian trail wave function for the electron wave function. The excited states in this potential have no width since the potential maintains its shape. A more correct treatment probably would be to vary the potential parameters $\delta_0$ and $\alpha$ for each excited state and/or look at the spectrum of fluctuations about the ground state shape as is done in semiclassical approaches to non-linear field theory. Work along these lines is proceeding, by mapping the strong coupling polaron problem onto a non-linear field theory [24].

The threshholds exhibited in equation (109) correspond to Franck-Condon excited states, which do not include lattice relaxation about the excited states. Devreese et al [25] have discussed the importance of calculating the optical absorption

$$\mathrm{Re}(1/z(\mu))$$

in order to get the correct structure corresponding to lattice relaxation about the excited states [26]. In this case we have

$$\mathrm{Re}(1/z(\mu)) = \frac{\mathrm{Im}X(\mu)}{(\mu^2 - \mathrm{Re}X(\mu))^2 + (\mathrm{Im}X(\mu))^2}$$

so there is a major contribution from the vanishing of $[\mu^2 - X(\mu)]$. The decrease in the number of threshhold peaks in $ImX(\mu)$ as the coupling becomes greater is also probably a reflection of the fact that one cannot excite the system to arbitrarily high lying states without including the effect of lattice relaxation, so at some point the excitation to successively higher lying Franck-Condon states must break down.

References

[1] W. T. Sommer, thesis, Stanford University (1964) (unpublished).
[2] M. W. Cole and M. H. Cohen, Phys. Rev. Lett. 23, 1238 (1969); M. W. Cole, Phys. Rev. B2, 4239 (1970).
[3] V. B. Shikin, Soviet Phys. - JETP 31, 936 (1970); V. B. Shikin and Yu P. Monarkhu, Soviet Phys. JETP, 38, 373 (1974); ibid. J. of Low Temp. Phys. 16, 193 (1974).
[4] W. T. Sommer and D. J. Tanner, Phys. Rev. Lett. 27, 1345 (1971).
[5] R. Williams, R. S. Crandall and A. H. Willis, Phys. Rev. Lett. 26, 7 (1971); R. Williams and R. S. Crandall, Phys. Rev. A4, 2024 (1971); R. S. Crandall and R. Williams, Phys. Rev. A5, 2183 (1972).
[6] T. R. Brown and C. C. Grimes, Phys. Rev. Lett. 29, 1233 (1972).
[7] V. S. Edelman, JETP Letters 24, 468 (1976); ibid. 25, 394 (1977).
[8] C. C. Grimes and T. R. Brown, Phys. Rev. Lett. 32, 280 (1974); C. C. Grimes, T. R. Brown, M. L. Burns and C. L. Zipfel, Phys. Rev. B13, 140 (1976).
[9] C. C. Grimes and G. Adams, Phys. Rev. Lett. 36, 145 (1976).
[10] C. C. Grimes, Surface Science 73, 379 (1978); ibid., J. de Physique LT 15, C6-1352 (1978).
[11] C. C. Grimes, Surface Science 98, 1 (1980); ibid., Physica 106A, 102 (1981).
[12] Y. Toyozawa, Prog. Theo. Phys. 26, 29 (1961); ibid., "Polarons and Excitons", (Proc. Scottish Univ. Summer School), p. 211 (1963) Plenum.
[13] Y. Toyozawa and Y. Shinozaka, ISSP Technical Report, Ser. A, No. 992 (1979) (unpublished).
[14] D. Emin, Adv. Phys. 22, 57 (1973); D. Emin, T. Holstein, Phys. Rev. Lett. 36, 323 (1976).
[15] S. A. Jackson and P. M. Platzman, Phys. Rev. B24, 499 (1981); ibid., Surface Science 113, 401 (1982).
[16] L. M. Sander, Phys. Rev. B11, 4350 (1975).
[17] R. P. Feynman, Phys. Rev. 97, 660 (1955); T. D. Schultz, "Polarons and Excitons", (Proc. Scottish Univ. Summer School), p. 71 (1963) Plenum.
[18] M. Saitoh, J. Phys. Soc. Japan 49, 878 (1980).

[19] R. P. Feynman and A. R. Hibbs, "Quantum Mechanics and Path Integrals", McGraw-Hill (1965).
[20] S. A. Jackson and P. M. Platzman, Phys. Rev. B25, 4886 (1982).
[21] $E_B$ is defined as the difference in the effective energy just above and just below the sharp transition in mass at T=0. Although this definition appears arbitrary, $E_B$ defined in this way is always ∿22 mK.
[22] R. P. Feynman, R. W. Hellwarth, C. K. Iddings and P. M. Platzman, Phys. Rev. 127, 1004 (1962); P. M. Platzman, "Polarons and Excitons", (Proc. Scottish Univ. Summer School), p. 123 (1963) Plenum.
[23] G. Whitfield and P. B. Shaw, Phys. Rev. B14, 3346 (1976).
[24] S.A. Jackson (to be published).
[25] J. Devreese, J. De Sitter and M. Goovaerts, Phys. Rev. B5, 2367 (1972).
[26] E. Kartheuser, R. Evrard and J. Devreese, Phys. Rev. Lett. 22, 94 (1969).

AUTHOR INDEX

MATERIAL INDEX

SUBJECT INDEX